3rd International Conference on Future of Engineering Systems and Technologies

AA007279

IOP Conference Series: Materials Science and Engineering
Volume 1228

Online
18 – 19 December 2021

Editors:

Vikas Kumar
Neeraj Gupta
Subhash Singh
Kavita Rani

ISBN: 978-1-7138-4686-4
ISSN: 1757-8981

Printed from e-media with permission by:

Curran Associates, Inc.
57 Morehouse Lane
Red Hook, NY 12571

Some format issues inherent in the e-media version may also appear in this print version.

This work is licensed under a Creative Commons Attribution 3.0 International Licence.
Licence details: http://creativecommons.org/licenses/by/3.0/.

No changes have been made to the content of these proceedings. There may be changes to pagination and minor adjustments for aesthetics.

Printed with permission by Curran Associates, Inc. (2022)

For permission requests, please contact the Institute of Physics
at the address below.

Institute of Physics
Dirac House, Temple Back
Bristol BS1 6BE UK

Phone: 44 1 17 929 7481
Fax: 44 1 17 920 0979

techtracking@iop.org

Additional copies of this publication are available from:

Curran Associates, Inc.
57 Morehouse Lane
Red Hook, NY 12571 USA
Phone: 845-758-0400
Fax: 845-758-2633
Email: curran@proceedings.com
Web: www.proceedings.com

TABLE OF CONTENTS

IOP Conf. Series: Materials Science and Engineering 1228 (2022) 011001 doi:10.1088/1757-899X/1228/1/011001

Preface

It is our immense pleasure to virtually organize (due to the Covid-19 situation) the conference on **Future of Engineering systems and Technologies (FEST)** during December 18-19, 2021 by '**Fusion and Integration of Science and Technology (FIST), New Delhi'** in association with '**Indira Gandhi Delhi Technical University for Women, Delhi, India**'. The conference was organized in association with the dedicated faculty members from Amity University India, I.K. Gujral Punjab Technical University Jalandhar; National Institute of Technology Srinagar; Manipal University India and many other national/International reputed Institutes/Universities.

The conference **FEST 2021** aimed to provide a platform for scientists and engineers to share their research results, exchange ideas on engineering systems, theories, technologies and applications; and dissemination of knowledge among engineers, researchers, scientists and entrepreneurs. **FEST 2021** also aimed to establish research relations and to find global partners for future collaboration. The conference had covered all the emerging research areas of engineering and technologies.

The proceedings present a selection from papers submitted to the conference from academia, research organisations and industries. The papers selected depended on their quality and their relevancy to the conference. The volume tends to present to the readers the recent advances in the field of engineering systems and technologies.

Our sincere gratitude goes to our invited speakers and all the authors for their participation and sharing their intellectual experiences. We would like to express our sincere appreciation to the members of the organizing committee, advisory committee, technical committee, reviewers and all the volunteers of **FEST 2021**. Our special thanks goes to IOP Publishing as our publication partner to publish the conference proceedings in MSE.

The organizing team believe that all the delegates had enjoyed an exciting, thought-provoking and unforgettable virtual meeting at **FEST 2021**.

Organising Team (FEST 2021)

Editors

Dr. Vikas Kumar Dr. Neeraj Gupta Dr. Subhash Singh Dr. Kavita Rani	Prof. (Dr.) Nathi Ram Chauhan Dr. Naveen Kumar Sharma Dr. Rohit Sharma Prof. (Dr.) Anish Gupta

Content from this work may be used under the terms of the Creative Commons Attribution 3.0 licence. Any further distribution of this work must maintain attribution to the author(s) and the title of the work, journal citation and DOI.
Published under licence by IOP Publishing Ltd

IOP Conf. Series: Materials Science and Engineering 1228 (2022) 011001 doi:10.1088/1757-899X/1228/1/011001

Organizing Committee

CHIEF PATRON	
Dr. (Mrs). Amita Dev	Vice Chancellor, IGDTUW Delhi
PATRON	
Prof. R.K. Singh	Registrar, IGDTUW Delhi
CONFERENCE CHAIR	
Prof. (Dr.) Nathi Ram Chauhan	Professor & Head, MAE Department, IGDTUW Delhi
ORGANIZING CHAIR(s)	
Dr. O.K. Singh	IGDTUW Delhi
Dr. Neeraj Gupta	NIT SRINAGAR
Dr. Ambrish Maurya	NIT Patna
ORGANIZING SECRETARY(s)	
Dr. V. K. Chawla	IGDTUW Delhi
Dr. Vikas Kumar	ETU, Addis Ababa
Dr. Om Hari Gupta	NIT Jamshedpur
Dr. Naveen Kumar Sharma	I K G Punjab Technical University, Jalandhar (Punjab)
JOINT ORGANIZING SECRETARY(s)	
Prof. Om Parkash	Amity University Haryana
Dr.(Mrs.). Tina Chaudhary	IGDTUW Delhi
Mr. Ravinder Kumar	IGDTUW Delhi
Dr. Kavita Rani	Lloyd institute of Engineering and technology, Greater Noida
CONVENER (S)	
Dr. Sonam Devgan Kaul	University of Toronto , Canada
Dr. Vinay Kumar Jadoun	MIT Manipal, Karnatka
Dr. Anish Gupta	ABES Engineering College, Delhi NCR

IOP Conf. Series: Materials Science and Engineering 1228 (2022) 011001 doi:10.1088/1757-899X/1228/1/011001

Mr. Urfi Khan	IGDTUW Delhi
TECHNICAL PROGRAM COMMITTEE (TPC)	
Dr. Anuradha Tomar (chair)	Eindhoven University of Technology, The Netherlands
Dr. Jaspreet Hira (chair)	IIT Delhi
Dr. Sanjay Yadav	CSIR – National Physical Laboratory, New Delhi, India
Er. Eswara Satyanarayana G	BHEL Hyderabad
Dr Ankit Chakravarti	Arba Mich University, Ethiopia, East Africa
Dr. Sandeep Sharma	National Institute of Technology Hamirpur, Himachal Pradesh
Dr. B Rajanarayan Prusty	VIT Vellore
Dr. Talari Ganesh	National Institute of Technology Hamirpur
Dr. Neeraj Gupta	Amity University Haryana
Mr. Rohit Sahu	G.L. Bajaj Institute of Technology & Managment , Greater Noida
Dr. Shipra Aggarwal	IGDTUW Delhi

IOP Conf. Series: Materials Science and Engineering 1228 (2022) 011001 doi:10.1088/1757-899X/1228/1/011001

Contact Details

Dr. Vikas Kumar Associate Professor Department of Automotive Technology Division of Mechanical Technology Ethiopian Technical University, Addis Ababa Mb: +919536345391 email: viksmied@gmail.com	Dr. Nathi Ram Chauhan Professor and Head Department of Mechanical and Automation Engineering, IGDTUW, New Delhi, India Mb: +919810106840 Email: hodmae@igdtuw.ac.in nrchauhan@igdtuw.ac.in
Dr. Neeraj Gupta Assistant Professor Department of Electrical Engineering NIT Srinagar, Hazratbal, Kashmir, India Mb: +918958054648 email: neeraj5822@gmail.com neerajgupta@nitsri.net	Dr. Naveen Kumar Sharma Assistant Professor Department of Electrical Engineering I K G Punjab Technical University, Jalandhar (Punjab), India Mb: +919882281525 email: naveen31.sharma@gmail.com dr.naveensharma@ptu.ac.in
Dr. Subhash Singh Assistant Professor Department of Production and Industrial Engineering NIT Jamshedpur, India Mb: +91 9045704384 Email: subhash.prod@nitjsr.ac.in	Dr. Rohit Sharma Assistant Professor Department of Mechanical Engineering Amity University Uttar Pradesh Noida, India Mb: +919996397602 email: r25sharma@gmail.com rsharma19@amity.edu
Dr. Kavita Rani Associate Professor Department of CSE Lloyd Institute of Engineering and Technology (LIET), Greater Noida Mb: +919536345391 Email: dhurankavitahce@gmail.com kavita.dhuran@lloydcollege.in	Dr Anish Gupta Professor and Additional Director Computer Science & Engineering Chandigarh University, Mohali, India Mb: +918510932323 email: gupta.anish1979@gmail.com anish.e11115@cumail.in

*For general queries: email: fest@fistinternational.com

IOP Conf. Series: Materials Science and Engineering 1228 (2022) 011002 doi:10.1088/1757-899X/1228/1/011002

Peer Review Statement

All papers published in this volume have been reviewed through processes administered by the Editors. Reviews were conducted by expert referees to the professional and scientific standards expected of a proceedings journal published by IOP Publishing.

- **Type of peer review:** Double Anonymous
- **Conference submission management system:** Easy Chair
- **Number of submissions received:** 91
- **Number of submissions sent for review:** 78
- **Number of submissions accepted:** 36
- **Acceptance Rate (Submissions Accepted / Submissions Received × 100):** 39.6
- **Average number of reviews per paper:** 2.5
- **Total number of reviewers involved:** 29
- **Contact person for queries:**
  Name: Dr. Vikas Kumar
  Email: fest@fistinternational.com
  Affiliation: FIST New Delhi, India

Content from this work may be used under the terms of the Creative Commons Attribution 3.0 licence. Any further distribution of this work must maintain attribution to the author(s) and the title of the work, journal citation and DOI.
Published under licence by IOP Publishing Ltd

Designing of Higher-Order Series-Parallel resonant converter with improved ZVS under variable source condition for constant DC-bus System

Shivam Sharma[1], Gagandeep Kaur[2], Naveen K. Sharma[3]

[1] Research Scholar, Department of Electrical Engineering, IKGPTU, Kapurthala, Punjab, India
[2] Associate professor, Department of Electrical Engineering, IKGPTU, Kapurthala, Punjab, India
[3] Assistant professor, Department of Electrical Engineering, IKGPTU, Kapurthala, Punjab, India

Abstract: Every Renewable based energy source requires an additional buffer unit to deliver power to either a standalone system or a grid system. Most of the time unit buffer either provides constant voltage DC-power or AC-power. Higher efficiency in the conversion process can only be achieved if ongoing losses of the buffer system are low. Square sine control approach for Higher-Order Series-Parallel resonant DC-DC Converter is presented in the paper which provides excellent results in both highly and lightly loaded conditions. This control also provides stable bandwidth for input voltage source acceptance which is mostly to be considered as a PV-based source. The strategy also enhances the reliability of the converter by providing a stable and most delightful way of soft switching triggering so as to maintain the output voltage by un-affecting the switching cycle of the Solid-State Semi-conducting Device (SSSD). Selection for the components used in converter design was also studied with a susceptive case study. Further to this control for the entire converter unit is maintained to single variable control so as to reduce the complexity of the system and enhance efficiency. The simulated results of the programmable single variable square sine-wave control is presented for the higher-order resonant converter operating under variable input condition.

Keyword: Solid-State Semi-conductor device (SSSD); Inductor-Capacitor-Inductor (LCL); Inductor-inductor-Capacitor (LLC); Higher-Order Series-Parallel Resonant Converter (HOSPRC); Renewable-based Energy Source (RES); Modified Series-Parallel Resonant Converter (MSPRC); Zero Voltage Switching (ZVS);

1. Introduction

Generation of electric power from renewable and sustainable energy sources are becoming necessity for changing environment. Both in financial as well climatic zone renewable based sources are taking thinking edge to ward their research. Exploring different renewable based energy sources is one of the developing factors, followed by developing most prominent design top extracts the renewable energy from raw form and convert it to electrical based energy. Following the present lead, development is taking place in designing most economical converter unit both in initial as we running cost. Any of renewable source requires both DC-DC converters and DC-AC converter. Both stability and cost in case of converter design move parallelly, for better stability more no. of conversion levels is required which finally increases the cost of entire conversion module. Maximum percentage of renewable sources in present energy market is solar Photo-voltaic. Connection of solar PV with main grid require stable conversion module which is either on the PV-side (DC-DC) or on grid side (DC-AC). Both the conversion units need to be stable and have low running losses. Enhancement of Power output from Solar PV system can only be increased either by providing better efficient PV module or by reducing running losses in the converter unit.

Content from this work may be used under the terms of the Creative Commons Attribution 3.0 licence. Any further distribution of this work must maintain attribution to the author(s) and the title of the work, journal citation and DOI.
Published under licence by IOP Publishing Ltd

One of the main components of any converter unit is their SSSD which itself work under non-ideal condition and decrease overall efficiency either by creating loss under on-state condition or in switching condition [1]. More than 5% of power delivered by SSSD is converted to which due to these two losses [2]. For better life span of SSSD, removal of heat generated by devices due to self-losses must be removed for further in-convenience, Improper cooling provided to SSSD create red hot spots inside the module periphery and reduces manufacturer specified rating [3]. With better cooling and high-grate fabrication of SSSD, losses remain the same. Due to non-linear structure of SSSD, on-state losses can be minimized to some extend but cannot be removed entirely [4]. For increasing efficiency of converter module, focus is shifted toward the designing appropriate converter unit and control strategy for minimizing switching losses to almost 95% [5]. High-frequency resonant converter design is most appropriate for reducing on-going switching loss [6-8]. Simplest method of resonant converter is series based either LCL or LCC converter with well documented configuration [9-10]. Series resonant based DC-DC converter provides better efficiency in full load configuration and only related with one drawback which is poor output voltage regulation at light loads [11-12] whereases' parallel resonant converter have poor performance efficiency at heavy loads [13]. Both series and parallel resonant converter designs have alternative merits as well demerits which can be counter by clubbing both techniques to increase performance efficiency of converter unit under any type of load configuration [14-15].

Converters with the capability of soft switching, high sweeping frequency working results in reduced component and overall converter size for desired voltage gain. Detailed comparison related to various techniques used in DC-DC resonant converter is discussed in [16]-[18]. Most of the control strategies are based on either duty cycle control or frequency control or both. With improved efficiency band in switching phase of the converter design, all the strategies somehow bypass the acknowledgment leads to wave-shape of working parameters and their influence over the entire efficiency portfolio of converter and its reliability.

In any type of distribution system, renewable based energy source (RES) can only provide power if power is well synchronized with grid parameters. For uni-directional power flow, per unit impedance of RES must be greater than per unit impedance of entire grid. For better power extraction from any RES, converter unit must have high stable region and low on-going losses. DC-DC converter unit is required to maintain DC-power shoved inverter section constant and stable in any type of disturbance [19]. To address this requirement, high-frequency converter topologies and techniques are designed among them resonant converter-based design is very favorable and economically viable as it do not need multi-level or matrix-based DC-DC conversion modules [20-22]. In electric vehicle as well grid tied RES modules, bidirectional isolated DC-DC converters are utilizing major portion of electronics market [23]. A dual active bridge-based (DAB) DC-DC converter is best suited for series/parallel resonant topology under bidirectional DC power flow which is best suited for Battery energy storage system [24-26]. Due to high no. of switching SSSD utilized in DAB initial investment on converter module is more in-comparison to traditional single level DC-DC converter [27]. In all application practices, more focus of more energy extraction from RES is shifted either toward the fabrication/ design of RES or improvising technique and design of DC-DC converter for better power transfer efficiency.

In section-1 introduction of DC-DC converter and its module design along with research outcome and market consumption is reviewed. Further in section-2 proposed work is discussed along with graphical representation of selection of reactive elements for DAB zero voltage switching design. Mathematical modelling of proposed design along with selected reactive elements for soft switching processor is discussed in section-3. Simulation results along with all possible outcomes are discussed in section-4 and finally conclusion is made for above research in section-5.

IOP Conf. Series: Materials Science and Engineering 1228 (2022) 012001 doi:10.1088/1757-899X/1228/1/012001

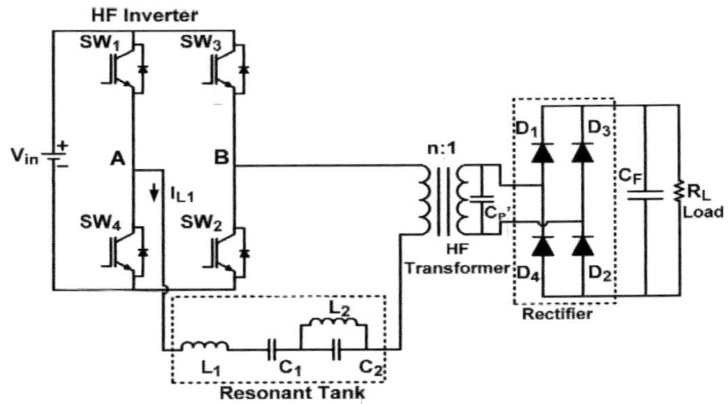

Fig.1: Dual Active-bridge higher order resonant converter with 1-D control

2. Proposed technique of DAB Resonant converter

The generalized control technique has been applied over every design aspect of HOSPRC. fig.1 shows the basic circuit configuration of HOSPRC discussed in [29] Proposed converter can operate at or around resonant frequency for the purpose of less switching losses and greater efficiencies. Although working with second order filter above or below resonant frequency, circuit parameters change which further leads to reduction in line current as overall impedance increases. While operating at or above resonant frequency, resonant converter can shape up the wave structure of current or voltage across the load to almost sinusoidal in nature [29]. Selection of series or parallel components for both resonant converters not only follow the resonating rule but also depends on Q-factor for wave-shaping of parameters [30].

In [28], researcher had assorted two different phase shift techniques and implemented them over common platform of DAB converter for better enhancement in the bandwidth for ZVS switching. In this paper our approach will be to utilize the design of MSPRC presented in [29] and perform parameter test over the design so as to get optimum values of RMS line current and making sure that wave-shape of current should be sinusoidal so as to maintain the level of harmonic power loss in the converter module to minimal value. As per the mathematical representation provided in [31], the relation between the normalized frequency (ratio of sweeping frequency to resonance frequency) and the system gain is shown in fig. 2 and generating different waves under different Quality factors (Q). Each value of 'Q' represents the percentage of load connected to system.

2.1. Factors of Q & Resonating frequency over second order filter system

As the definition implies, Q is the measure of peaking of any LC circuit and concept of Q is only valid for second or above order filter design. Quality factor can also describe about the wave-shaping of voltage and current across load. For better understanding related to quality factor, resonant frequency and wave-shaping, consider the second order filter as described in (1).

$$2^{nd}\ order\ filter = \frac{N_r}{-f^2 + \frac{jff_P}{Q} + f_P^2} \qquad (1)$$

Expression for numerator will depend upon the nature of filter either band-pass, Band-stop, etc. Fig. (3) Shows the magnitude of current against the sweeping frequency under ideal condition with only 2nd order design. Relation of finding the natural frequency and the Quality factor for different load situations is presented in (2) – (3).

IOP Conf. Series: Materials Science and Engineering 1228 (2022) 012001 doi:10.1088/1757-899X/1228/1/012001

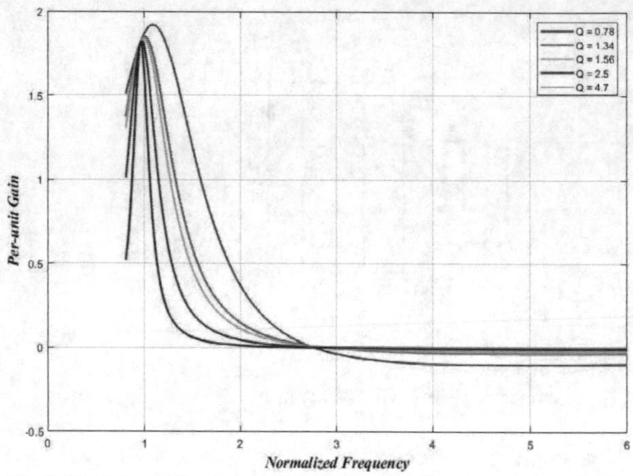

Fig.2: Variation of system gain in [31] under linear variations in Sweeping frequency.

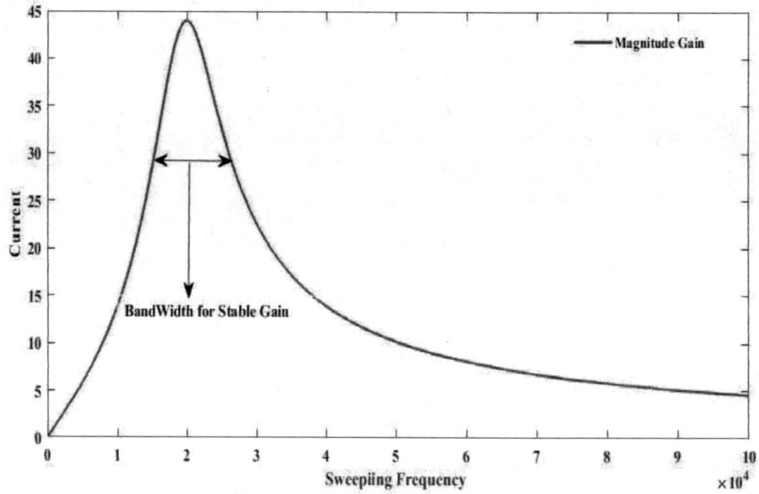

Fig.3: magnitude of current against the sweeping frequency under ideal condition with only 2nd order design.

$$f_P = \frac{1}{2\pi\sqrt{LC}} \tag{2}$$

$$Q = \frac{2\pi f_P L}{R_L} = \frac{1}{2\pi f_P C R_L} = \frac{\sqrt{\frac{L}{C}}}{R_L} \tag{3}$$

2.2. Selection of L & C for desired frequency and Load-current wave-shape

Under resonance condition, frequency of line current will be natural frequency but the wave-shape may not be obtained as desired. Wave-Shape of any resonating current solely depends on the relation of L&C as presented in (3). For proper understating related to the wave-shaping of load current, considering a test case in which resonance frequency is selected as unity and maintained to the same value throughout the changing value of L & C. Although natural frequency will remain the same which depends on the product of LC but the value of –factor will change as it depends on the ration of LC. Table-1 shows the different value of & C under constant resonant frequency of 1Hz. Fig.4 shows the wave-shape of load current under different values of Q-factor while resonant frequency remains same to unity. All the four cases used in fig.4, are taken from table.1.

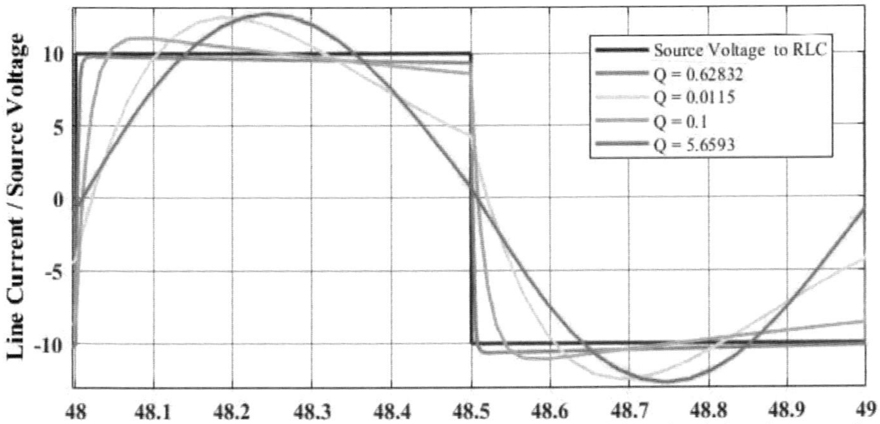

Fig.4: Wave-shape of resonant current under different value of Q maintaining the sweeping frequency to be same.

Table 1. Presumed value of Reactive along with Q

LC	L	C	Q
0.02533	1	0.02533	**0.62832**
0.02533	0.02533	1	**0.0115**
0.02533	0.15951	0.15915	**0.1**
0.02533	9	0.00281	**5.6593**

Analysing from the results of fig. 4, for proper sinusoidal wave-shape of load current ratio of L&C must follow the below condition.

$$\frac{L}{C} > 3^2$$

Therefore, in case of second order filter, ratio of L&C must follow the above condition irrespective of natural frequency to achieve almost sinusoidal wave-shape of load current. In general perspective let us consider a second order system shown in the fig.5 with natural frequency to be 100 Khz and load of 10 Ohms in resistance perspective.

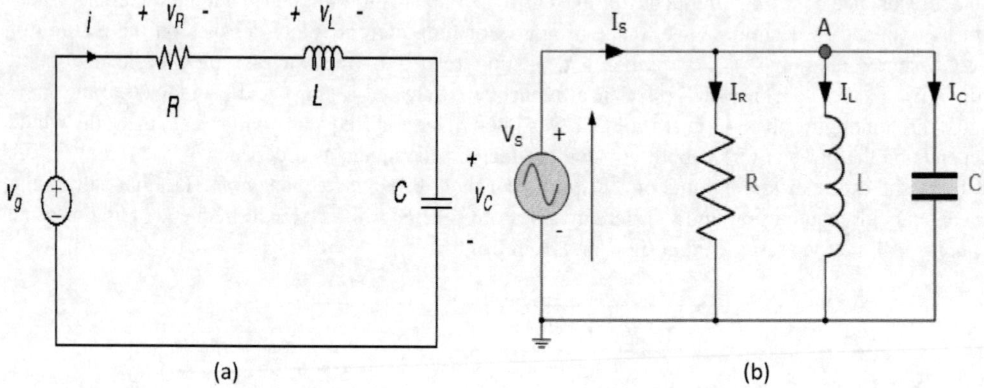

Fig. 5: (a) Second order RLC series; (b) Parallel circuit with sweeping frequency of V_g to be 100 Khz

3. Mathematical modelling of analyzed HOSPRC through conventional and proposed control technique

Sole purpose of this paper is to analyze design constraint of MSPRC and modifies it by varying designing aspects and control strategies. Fig. 6 shows the impedance or single line diagram of MSPRC as detailed in [31]. Visualizing the factor that magnetizing reactance of HF transformer is negligible which means only two components of HF is available i.e., primary leakage reactance and secondary leakage reactance. Both this reactance is considered to be the art of Ls.

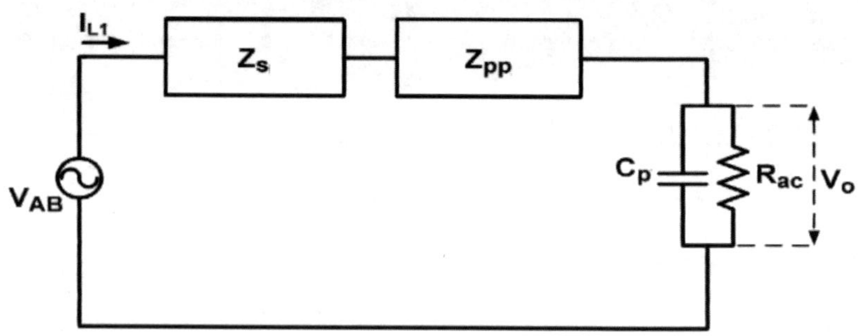

Fig. 6: Detailed single line diagram of MSPRC

3.1. Conventional switching frequency control

With variation in switching frequency, the variation of impedance of entire circuit changes which can further alter the output voltage. In the case of switching frequency control large bandwidth is required for voltage gain control, as change in frequency leads to change in line impedance which variate the magnitude and phase of line current.

If the above analysis of frequency control is done over MSPRC presented in fig.1, then with changing magnitude of current, flux generated in HF transformer changes which variates the secondary voltage and hence finally control the output voltage. As we look forward, if considering that rectifier unit is lossless then output voltage is completely proportional to the current at secondary terminal. Going further, primary current or line current magnitude controls the output voltage as evaluated in (4).

IOP Conf. Series: Materials Science and Engineering 1228 (2022) 012001 doi:10.1088/1757-899X/1228/1/012001

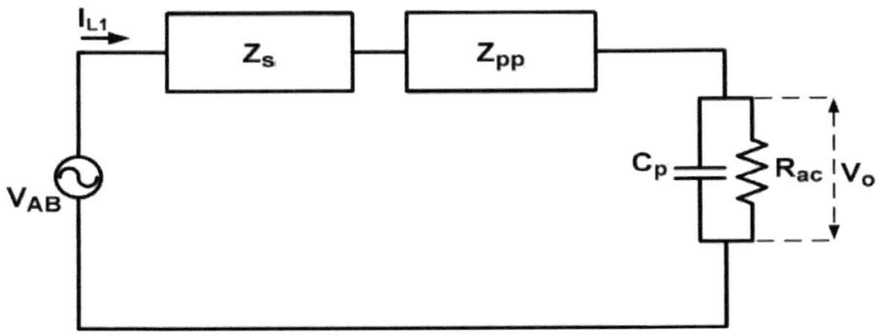

Fig.7: Single Line representation of higher order resonant converter including load section referred to the resonant section.

$$V_{in} = V_{Primary} \tag{4}$$

In switching frequency control, Voltage Gain (M) of converter is given by (5) where V_{INV} is input voltage provided to both resonating circuits and referred load. V_{OUT} is the voltage of the rectifier unit.

$$M = \frac{Z_{Primary}}{Z_{Inverter}} = \frac{V_{Output}}{V_{Inverter}} \tag{5}$$

Where Z_P is the parallel combination of R_{AC} and C_P, Z_{IN} is the cascaded combination of Z_P, Series resonant circuit and parallel resonant circuit shown in fig. 7. Although for actual realization with practical circuit, some of the assumptions are made to eliminate the ideality of selected circuit to some extend for better generalization. Expression for all parameters responsible for the selection of reactive elements and design of HOSPRC are represented from (6) – (11).

$$Q = \frac{w_o L_S}{R_{AC}} = \frac{1}{R_{AC}}\sqrt{\left(\left(\frac{L_S}{2C_S}\right)a\left(b + \sqrt{b^2 - 4}\right)\right)} \tag{6}$$

$$Z_s = jwL_s + \frac{1}{jwC_s} \tag{7}$$

$$Z_{PP} = \frac{-jwL_{PP}}{w^2 L_{PP} C_{PP} - 1} \tag{8}$$

$$Z_P = \frac{-jR_{AC}}{wR_{AC}C_P - j} \tag{9}$$

$$Z_{IN} = Z_s + Z_{PP} + Z_P \tag{10}$$

$$\varphi = f(w) = \tan^{-1}\left[\frac{wL_S - {}^{1}/_{wC_S} - {}^{wL_{PP}}/_{w^2 L_P C_{PP} - 1} - {}^{wC_P R_{AC}^2}/_{1 + wC_P^2 R_{AC}^2}}{{}^{R_{AC}}/_{1 + w^2 R_{AC}^2 C_p^2}}\right] \tag{11}$$

Angle φ is the angle by which the inverter voltage leads the line or resonant current. As voltage is in quasi-square wave in nature, consider the fact ‘Q’ is designed to make line current almost sinusoidal in wave-shape. As per Fourier analysis of wave-shaping, if current is other than sinusoidal in wave then all the odd components of fundamental frequency will be present in the current. This leads to distortion of current wave-shape and hence introduction to the new type of loss i.e., harmonic loss.

To achieve soft switching while utilizing conventional frequency control, switching frequency must be the multiple of resonating frequency so as for the current to be almost zero at the time of switching. In (12), changing switching frequency leads to change in the value of impedance and hence further output voltage. For the view in simplex form, conduction losses and resistance of each reactive element are neglected as considered to be constant and independent of frequency.

IOP Conf. Series: Materials Science and Engineering 1228 (2022) 012001 doi:10.1088/1757-899X/1228/1/012001

$$f\,(V_{OUT}) = f\,(V_{IN}) - \{f\,(V_{Parallel\ Resonant}) + f\,(V_{Series\ Resonant}) + f\,(S_{witching\ Loss})\} \quad (12)$$

Where $f\,(V_{OUT})$ = function of output voltage include nominal frequency; $f\,(V_{IN})$ = Function of Input voltage; $f\,(V_{Series\ Resonant})$ = function of series resonant tank circuit includes nominal frequency line current; $f\,(V_{Parallel\ Resonant})$ = function of parallel resonant tank circuit includes nominal frequency and line current; $f\,(V_{Switching\ Loss})$ = Function switching drop includes line current and factors of $\frac{dV_{Series\ Resonant}}{dI_{Line}}$ & $\frac{dV_{Parallel\ Resonant}}{dI_{Line}}$.

3.2. Switching frequency and duty-cycle control

This control works on both parameters of frequency and duty-cycle so as to variate the output voltage. The range of switching frequency and duty cycle is limited when selected individually, although clubbing both parameters result in control of voltage gain with increased range. As in [32], switching losses reduces with reduction in switching frequency and conduction losses get reduced by reduction in duty ratio which further increases the converter efficiency. For controlling the switching loss ZCS should utilized and it can be achieved with more reliability if worked on switching technique. Reliability of ZCS is questioned at the time of light loads when 'Q' varies and leads to increment in switching loss.

3.3. Square-Sine control

In comparison to above controlled techniques, square sine control is the perfect way to minimize the switching loss by maintain the switching frequency all the time same as resonant frequency at every load condition. As we understand from previous topics, for the controlling of wave-shape 'Q' plays most significant role. Making line current wave-shape almost sinusoidal results in reduced value of THD and hence further negligible high frequency components. In square sine technique, wave-shape of line current is created almost like sine square wave with less of narrowing as shown in fig. 8. and voltage gain is maintained to be only function of duty cycle, which results in reduction in voltage drops against series and parallel resonant circuits. Switching drops will also be neglected as working frequency of system to be resonant frequency which implies, output voltage function will be completely depend on input voltage function as discussed in (12).

4. Simulation & Results

The square sine control technique is utilized in MSPRC under different load conditions in simulation environment with all practical situations. I generalized way, power switches selected are provided with conduction drop and delay at the time of changing state of the power switch is also maintained so as make sure that switches are working under practical and non-ideal environment. As we can see in the fig. (9), at the time of full load condition duty ratio is shifted toward 48% which is the maximum peak of duty ratio for stable functioning line current is in phase with the input voltage which makes it clear that drop across the power switch at the time of switching is zero. All the power parameters used in the converter design with proper specifications are presented in Table (2). Efficiency provided by the technique at the time of full load is 98.32% which greater than the efficiency provided by same mode with different technique in [32] and [33]. Wave Shape of line current and the voltage across switching element referred to input voltage is shown in fig. 8 in which it can be clearly visualize about zero voltage switching.

Table 2. Specification of Improvised Converter design.

INPUT VOLTAGE RANGE	260 V - 180 V
RATED INPUT VOLTAGE	220 V
NOMINAL OUTPUT VOLTAGE	400 V
DELIVERED POWER	2800 W – 3200 W
RESONANT FREQUENCY	117 KHz

IOP Conf. Series: Materials Science and Engineering 1228 (2022) 012001 doi:10.1088/1757-899X/1228/1/012001

SERIES TANK PARAMETER	$L_S = 5.58*10^{-5}$ H; $C_S = 3.51*10^{-8}$ F
PARALLEL TANK PARAMETER	$L_P = 55.8*10^{-5}$ H; $C_P = 70.2*10^{-8}$ F
PARALLEL CAPACITOR	$C_{PP} = C_S/20$
FILTER CAPACITOR	10µF
SWITCHES (TYPE, DOCUMENT NO.)	IGBT (Without body diode)
ISOLATION TRANSFORMER	High Frequency, Ferrite core, PM 74, n = 0.40

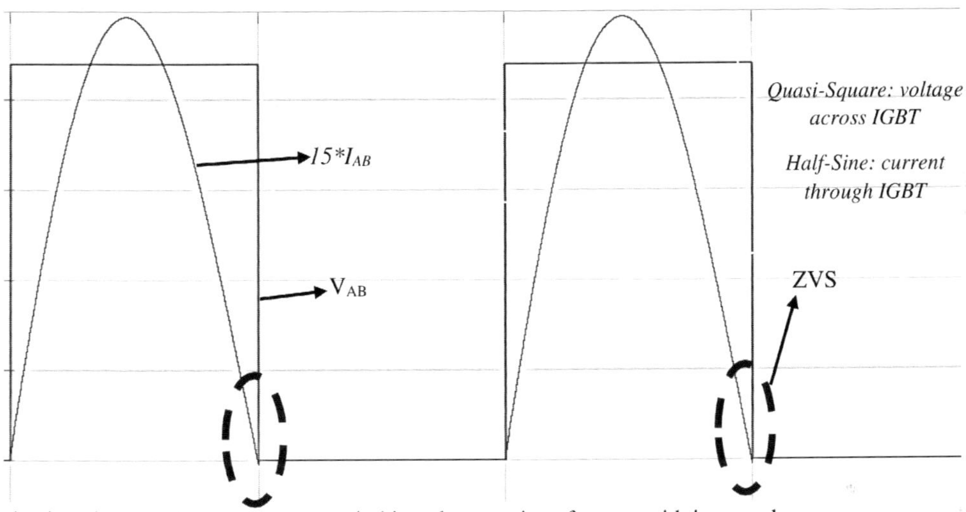

Fig. 8: voltage and current across switching elements in-reference with input voltage

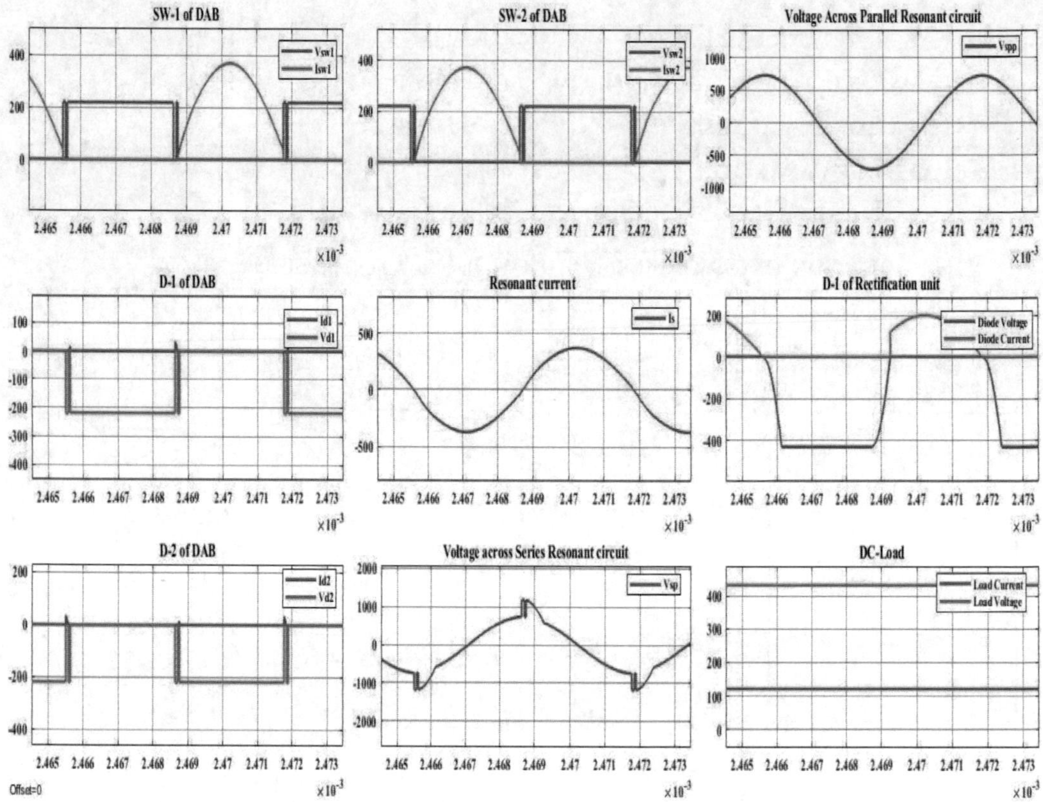

Fig. 9: Final Result of HOSPRC with Duty-Phase control technique

5. Conclusion

In DC-DC converter topology, overall converter efficiency can only be improved by reducing the loss created by SSSD and all reactive elements either used for filtering the output voltage surges or for reducing switching current. In ideal condition, power loss across entire converter unit is zero and efficiency is 100%. Although 100% efficiency is difficult step to achieve, but reducing the obvious losses inside the unit may leads to efficiency equivalent to AC power transformer. In this design of HOSPRC, value of line current is maintained to limited value so as to reduce the drop across reactive elements and creating wave-shape of current almost sinusoidal to reduce effect of harmonics on reactive elements and HFT. With suitable chosen design and switching topology, this model will not only provide less power loss across reactive circuits but also reduces value of current across every SSSD at the time of switching.

References

[1] J. O'Connell and L. J. Giacoletto, ""Semiconductor voltage controllable reactance"," in *IRE Transactions on Electron Devices*, vol. 3, no. 2, pp. 114-114, April 1956, doi: 10.1109/T-ED.1956.14160.

[2] A. Uhlir, "The Potential of Semiconductor Diodes in High-Frequency Communications," in *Proceedings of the IRE*, vol. 46, no. 6, pp. 1099-1115, June 1958, doi: 10.1109/JRPROC.1958.286892.

[3] T. Yamamoto, "New applications of thermoelements for cooling semiconductor devices," in *Proceedings of the IEEE*, vol. 56, no. 2, pp. 230-231, Feb. 1968, doi: 10.1109/PROC.1968.6250.

[4] L. Davis, L. G. Rubin and W. D. Straub, "Rapid determination of some electrical properties of semiconductors," in *Transactions of the IRE Professional Group on Electron Devices*, vol. 1, no. 2, pp. 34-42, April 1954, doi: 10.1109/T-ED.1954.14010.

[5] L. Jin, B. Liu and S. Duan, "ZVS Soft Switching Operation Range Analysis of Three-Level Dual-Active Bridge DC–DC Converter Under Phase Shift Control Strategy," in *IEEE Transactions on Industry Applications*, vol. 55, no. 2, pp. 1963-1972, March-April 2019, doi: 10.1109/TIA.2018.2872121.

[6] K. K. Sum, Recent developments in resonant power conversion, Intertech Communications Inc., 1988.

[7] A. K. S. Bhat and S. B. Dewan, "A generalized approach for the steady state analysis of resonant inverters", *IEEE Trans. Industry Appicat.*, vol. 25, pp. 326-338, Mar./Apr. 1989.

[8] A. K. S. Bhat, "A unified approach for the steady-state analysis of resonant converters", *Proc. Int. Conf. on Power Electron.*, pp. 191-197, 1990-Apr.

[9] R. Severns, "Topologies for three element resonant converters", *Conf. Rec. IEEE Applied Power Electron.*, pp. 712-722, 1990-Mar.

[10] A. K. S. Bhat and S. B. Dewan, "Analysis and design of a high-frequency link converter using LCC-type commutation", *IEEE Trans. Power Electron.*, vol. 2, pp. 291-301, Oct. 1987.

[11] Y. Kuwata, Y. Ohashi and K. Yotsumoto, "Characteristics of a new series-resonant converter with a parallel resonant circuit", *IEEE INTELEC*, pp. 204-210, 1987.

[12] S. Muroyama and K. Sakakibara, "Characteristics of a new series resonant converter", *IEEE INTELEC*, pp. 111-116, 1982.

[13] A. K. S. Bhat, "Analysis and design of a parallel resonant converter with the resonating capacitor on a tertiary winding," *Conference Record of the 1991 IEEE Industry Applications Society Annual Meeting*, 1991, pp. 990-995 vol.1, doi: 10.1109/IAS.1991.178357.

[14] R. Oruganti and F. C. Lee, "State-Plane analysis of parallel resonant converter", *IEEE Power Electronics Specialists Conference Record*, pp. 56-73, 1985.

[15] E. B. G. Nijhof and W. B. Rosink, "Control circuit design for the series resonant power supply with GTO switch", *Proceedings of Powercon 10*, 1983.

[16] Y. Shi, R. Li, Y. Xue and H. Li, "Optimized operation of current-fed dual active bridge DC-DC converter for PV applications", IEEE Trans. Ind. Electron., vol. 62, no. 11, pp. 6986-6995, Nov. 2015.

[17] R. W. De Donker, D. M. Divan and M. H. Kheraluwala, "A three-phase soft-switched high-power-density DC/DC converter for high-power applications", IEEE Trans. Ind. Appl., vol. 27, no. 1, pp. 63-73, Jan./Feb. 1991.

[18] A. K. Jain and R. Ayyanar, "PWM control of dual active bridge: Comprehensive analysis and experimental verification", IEEE Trans. Power Electron., vol. 26, no. 4, pp. 1215-1227, Apr. 2011.

[19] M. Maalandish, S. H. Hosseini, M. Sabahi and P. Asgharian, "Modified MPC based grid-connected five-level inverter for photovoltaic applications", Compel-Int. J. Comput. Math. Elect. Electron. Eng., vol. 37, no. 2, pp. 971-985, 2018.

[20] B. Yang, F. C. Lee, A. J. Zhang and G. Huang, "LLC resonant converter for front end DC/DC conversion", *Proc. IEEE Appl. Power Electron. Conf. Expo.*, pp. 1108-1112, 2002.

[21] Y. Gu, Z. Lu, L. Hang, Z. Qian and G. Huang, "Three-level LLC series resonant DC/DC converter", *IEEE Trans. Power Electron.*, vol. 20, no. 4, pp. 781-789, Jul. 2005.

[22] W.-Y. Choi, J.-M. Kwon and B.-H. Kwon, "High-performance front-end rectifier system for telecommunication power supplies", *Proc. Inst. Elect. Eng.*, vol. 153, no. 4, pp. 473-482, 2006.

[23] S. Inoue and H. Akagi, "A bidirectioanl isolated dc–dc converter as a core circuit of the next-generation medium-voltage power conversion system", *IEEE Trans. Power Electroni.*, vol. 22, no. 2, pp. 535-542, Mar. 2007.

[24] "Power Conversion Apparatus for dc/dc Conversion Using Dual Active Bridge", 2005.

[25] M. Marchesoni and C. Vacca, "New dc–dc converter for energy storage system interfacing in fuel cell hybrid electric vehicles", *IEEE Trans. Power Electron.*, vol. 22, no. 1, pp. 301-308, Jan. 2007.

[26] J. L. Duarte, M. Hendrix and M. G. Simoes, "Three-Port bidirectional converter for hybrid fuel cell systems", *IEEE Trans. Power Electron.*, vol. 22, no. 2, pp. 480-487, Mar. 2007.

[27] M. H. Kheraluwala, R. W. Gascoigne and D. M. Divan, "Performance characterization of a high-power dual active bridge dc-to-dc converter", *IEEE Trans. Ind. Appl.*, vol. 28, no. 6, pp. 1294-1031, Nov. 1992.

[28] X. Li and Y.-F. Li, "An optimized phase-shift modulation for fast transient response in a dual-active-bridge converter", IEEE Trans. Power Electron., vol. 29, no. 6, pp. 2661-2665, Jun. 2014.

[29] Suryawanshi, Hiralal & Tarnekar, S.G.. (1996). Modified LCLC-type series resonant converter with improved performance. Electric Power Applications, IEE Proceedings -. 143. 354 - 360. 10.1049/ip-epa:19960437.

[30] Forsyth, Andrew & Ho, Y.K.E.. (1996). Dynamic characteristics and closed-loop performance of the series-parallel resonant converter. Electric Power Applications, IEE Proceedings -. 143. 345 - 353. 10.1049/ip-epa:19960436.

[31] H. M. Suryawanshi et al., "Hybrid Control of High-Efficient Resonant Converter for Renewable Energy System," in IEEE Transactions on Industrial Informatics, vol. 14, no. 5, pp. 1835-1845, May 2018, doi: 10.1109/TII.2017.2756703.

[32] F. Blaabjerg, R. Teodorescu, M. Liserre and A. V. Timbus, "Overview of control and grid synchronization for distributed power generation systems", IEEE Trans. Ind. Electron., vol. 53, no. 5, pp. 1398-1409, Oct. 2006.

[33] Y.-P. Hsieh, J.-F. Chen, T.-J. P. Liang and L.-S. Yang, "Novel high step-up DC–DC converter with coupled-inductor and switched-capacitor techniques for a sustainable energy system", IEEE Trans. Power Electron., vol. 26, no. 12, pp. 3481-3490, Dec. 2011.

IOP Conf. Series: Materials Science and Engineering 1228 (2022) 012002 doi:10.1088/1757-899X/1228/1/012002

A Critical Review of Coating on AISI 1045 Steel Substrate by High Velocity Oxygen Fuel (HVOF)

Md Sarfaraz Alam [1, a], **Anil Kumar Das** [2, b]

[1,2] *Department of Mechanical Engineering, National Institute of Technology Patna, Bihar, India*

[a] *Corresponding author:* *mda.phd19.me@nitp.ac.in*, [b] *akdas@nitp.ac.in*

Abstract. Coating of surface is a cost-effective way to make materials, equipment, and machine components with desired surface characteristics including high hardness, resistance against erosion, wear and corrosion. To obtain the necessary characteristics, several coatings are employed. The surface characteristic of the AISI1045 steel is not optimum, despite its desired strength, resistance against wear, hardness or corrosion. Thermal spraying is currently a leading way in which materials are avoided for days to wear abrasively, corrosion at high temperatures, erosive wear and stress because of increased material life in use. Spraying using high-velocity oxygen fuel (HVOF) is a kind of thermal spraying noted for its resistance against wear, micro-hardness and thick micro structured coatings. The aim of this review paper is to provide a state-of-the-art about the research done by various scholars to improve physical, mechanical, wear, erosion and corrosion characteristics of AISI 1045 by the application of different or combination of coating powders utilising the HVOF Spray method.

***Keywords**: HVOF; AISI1045Steel; Corrosion Resistance; Wear Resistance; Coating Material; Hardness*

1. Introduction

Corrosion, wear and stress are all factors that may cause the most of engineering components to deteriorate or eventually fail in service. As a result, it depends on the design and features of their surface as well as their bulk material qualities. Coating is the process of improving the properties of engineering components and materials by using classic and novel surfacing methods. For many years, thermal spraying techniques have been widely utilised for component preservation and reclamation across all major engineering industries. Thermally sprayed coatings have increased in quality and applicability range as a result of recent design and operating improvements. Thermal spraying provides resistive coating for wear & corrosion in different industries like automobile, structural, power generation etc. The AISI 1045 medium carbon steel has emerged as a suitable structural material. When normalised or hot-rolled, AISI 1045 steel has excellent machinability, high strength, weldability, and toughness, and is widely utilised in the transportation, power, construction and automotive fields. A minor drawback of AISI 1045 is that, unlike other medium carbon steels, it does not sustain complete hardening except for short parts of up to 60 mm. Surface treatment technologies have been the subject of extensive investigation for decades for AISI 1045 to enhance mechanical properties by thermal spray coating [1-16],plasma arc cladding[17],electron beam cladding[18],tungsten inert gas arc cladding[19 22], laser beam cladding[23-27] and other surface modification methods.In order to apply the required coating, the HVOF process employs a mix of heat and kinetic energy to melt and accelerate powder particles. In the production of coatings, combustion and gas dynamics are crucial. In the following two ways, they influence the coating quality: [28]

(1) To strike the target at great velocity, the particles introduced into the gas stream must be accelerated.

(2) Heat must be transferred to the particles from the gas stream in order for them to melt prior to colliding.

High Velocity Oxygen fuel (HVOF) thermal spraying

HVOF spray is among the finest alternatives for hard, less porous, and wear-resistant coatings among the coating techniques by thermal spray [29]. Low temperatures and high velocity (>1000 m/s) are employed in this technique, resulting in semi-molten particles interlocking with one another and increasing the cohesive strength

Content from this work may be used under the terms of the Creative Commons Attribution 3.0 licence. Any further distribution of this work must maintain attribution to the author(s) and the title of the work, journal citation and DOI.
Published under licence by IOP Publishing Ltd

of the particles. Low permeability and great bond strength are achieved as a result of this process. The adherence of an HVOF spray coated substrate to its substrate can be ten times greater than that of conventional flame spraying techniques. A spray weapon with a powder supply unit, the flow metre, air and gas supply unit are part of the HVOF spraying systems. Schematically shown perspective of HVOF spray gun Figure I. In this process, oxygen and gas or liquid fuel in combustion-chamber are mixed, continuously inflamed and combusted. Hot gases are ejected at high pressure as a result of combustion, and these gases pass through a nozzle before passing through nozzle. Fuels may consist of hydrogen, acetylene, methane, kerosene, natural gas etc. The speed of sound is exceeded by the gas jet that emerges from the barrel's exit (>1000 m/s). The powder feedstock is then fed into a high-velocity gas stream and it is further speedup to 800 metres per second. This accelerated powder is further impacted on the substrate where coating has to be deposited. The powder partly melts in the flow and partly melts on the deposited surface because of high-speed collisions of powder grains on the base metal. Particles become totally molten due to heat generated upon particle impact on the substrate, resulting in high-quality HVOF coatings. As a result, they have become intertwined and have created high-quality coatings. HVOF coatings have an excellent bond strength, a low porosity, and a high density, which results in a material with high mechanical strength, toughness and hardness, making it suitable for use in worst conditions such as elevated temperature, corrosive and erosive environments. The thickness and characteristics of HVOF sprayed depositions are also affected by spray variables such as compressed air flow rate, spray distance, rate of powder feed and oxygen to fuel flow-rate ratio.

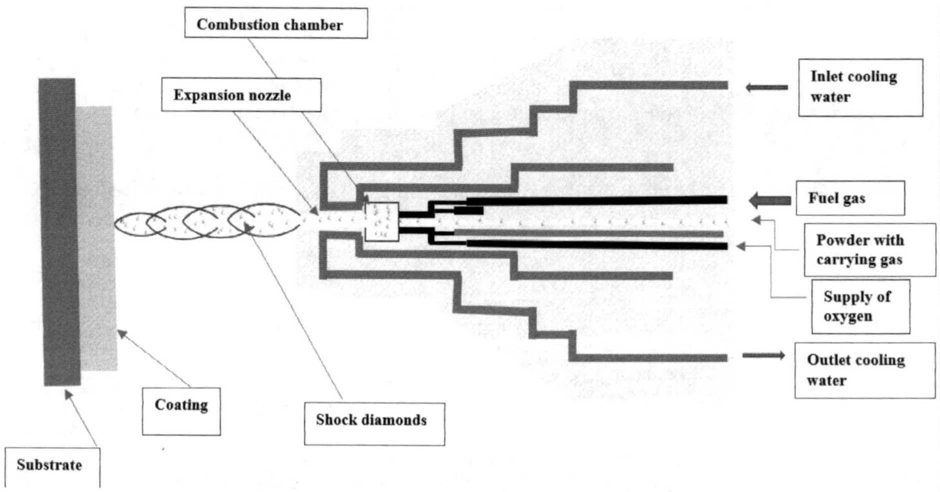

FIGURE 1. Schematic representation of HVOF system

2. RESEARCH PROGRESS IN HVOF SPRAY COATING ON AISI 1045

2.1 Mechanical and Physical characteristics of Coatings

The cladding materials and the processing parameters has a significant impact in obtaining the microstructure & characteristics of the coatings. Pulido-Gonzalez et al. [30] used HVOF thermal Spray to deposit titanium coatings on AISI 1045 and AISI 316L steel substrates. According to the findings, using a high number of layers and speed of gun can result in strong and well-adhered coatings with medium to high permeability and intermediate thickness. For biological purposes, these coatings would be sufficient.

Nguyen et al. [31] studied effect on life of fatigue due to thickness of coating for steel shaft bending sample coated with WC-10Co-4Cr HVOF coatings and hard chrome electroplated coatings. In samples with hardened chrome plated coatings, the fatigue crack originates towards the surface of the coating, propagates through the coating, and then develops in the thickness direction of the substrate to cause ultimate breakage. The fatigue strength of coated sample rises as the coating thickness grows, which may be related to the amount of the compressive residual stress along the substrate-HVOF coating interface increasing as increases in thickness of coating. When the coating thickness of chrome hardened plated samples is raised, the fatigue strength decreases. Because the microcrack density of hard chromium coatings rises with coating thickness, fatigue strength may be reduced.

Vinh et al. [32] examined the effect on the fatigue-strength of substrate made of C45 steel on covering layers 30 μm and 60 μm thick (with chromium and HVOF techniques). The results of research into microstructure indicated that the residual stress of the surface exposed impacted the fatigue strength. Tungsten carbide film has a fatigue resistance of 12.5% above 30 μm and 17.9% above 30 μm and 60 μm over chromium filming.
Castro et al. [33] used an HVOF spray to deposit WC-CoCr powder on AISI 1045 steel. The WC-CoCr coating morphology revealed a distinct oxide film between the layers with low porosity (1.0 percent). Furthermore, the sprayed coating had a high hardness of 1250 $HV_{0.01/15}$.
In HVOF thermal spraying method, the influence of WC-12Co particle temperature on coating thickness was investigated by Azizpour and Majd [34]. The quantitative findings demonstrate that particle morphology and height of HVOF spray coatings are influenced by spray distance and oxygen-to-fuel ratio.
Fu et al. [35] utilised a high-velocity oxy-fuel system to coat coarse feedstock powders with Al-Cu-Cr quasi-crystalline (QC) coatings. The contrast studies were carried out with the identical 65Al20Cu15Cr QC powders utilising low power atmospheric plasma spray (LPAPS) and HVOF spraying. It has been shown that, in comparison with those deposited with LPAs using the same feedstock, the preferable feedstock powder dimensions in HVOF spraying are 5-45 μm, HVOF spread coatings from such coarse (61-74 μm) QC powder had a far lower porosity and considerably denser microstructure.
Hong et al. [36] employed the approach of HVOF thermal spraying on an AISI 1045 steel substrate with WC-10Co–4Cr coating. The goal of this research was to investigate the precise microstructures, thermal stability, and microhardness distribution of WC–10Co–4Cr coatings sprayed with HVOF. The coating is densely packed and adheres effectively to the substrate. The WC and W2C phases make up the majority of the coating. The fast-cooling rates of melted droplets and the multi - component feedstock powder system result in the formation of an amorphous phase throughout the coating. The Vickers microhardness of the coating has a bimodal distribution under 300 g indentation force and a monomodal distribution under 100 g indentation load, according to the Weibull analysis. The existence of weakly bound splats and holes inside the covering is responsible for this dispersion.
The effect of the key control parameters of HVOF on the properties of particles and residual stresses of surface of WC12Co coatings had been investigated by Nourouzi et al. [37]. According to the outcomes, the O/F ratio and the distance from spray were the most significant impacts on particle residual stresses & characteristics.
Garcia et al. [38] utilised AISI 1045 steel for tungsten carbide coatings to analyse the impact of fatigue on thin coating of WC. Tungsten carbide coverings sprayed thermally by HVOF were proposed to be used to produce very hard surfaces and restore shafts or components with worn out areas, such as bearing supports, ring restraint, etc. Results indicate that the fatigue life and the enormous variations between materials may be predicted by this coating and treatments.
Gonzalez et al. [39] used the HVOF spraying method to deposit a cermet coating composed of WC10Co4Cr on a C45 steel substrate to assess fatigue life. The findings show that after grit blasting, Because of the cermet coating, fatigue fractures at deep notches and alumina particles lodged in the steel substrate take longer to initiate, as a consequence, the fatigue strength of coated specimens is not significantly reduced as compared to polished specimens. Fine grinding, on other side, prevents the coat from mechanically adhering to the substrate, causing it to dislocate from the surface at high alternativc stress levels.
The WC-Co-Cr coating, according to Casteletti et al. [40] provides satisfactory performance & emerge out as a good option for hard chromium replacement.
Petsas et al. [41] created Mo and Cr–Mn alloyed steel HVOF coatings that could be sprayed to diesel engine components, using both conventional and nanostructured feedstock. With suitable deposition rates, both nano-phased and conventional coatings demonstrated strong adhesion strength, microhardness and minimal porosity. Nano-phased coatings have better tribological performance than traditional coatings.

2.2 Wear Resistance Properties

Castro et al [33] applied a WC-10Co-4Cr coating. Hydraulic cylinders are frequently coated with electrodeposited hard chromium to provide a wear and oxidation resistant surface. However, owing to wear, the bearing surface of this type of coating for the sealant components increases gradually, it interfaces with the lubrication of the hydraulic rod, damages the sealant and thus the oil leaks. The HVOF procedure is now an alternative to Hard Chrome Plating coating method, employing composites, that give low rates of wear and minimal friction. In comparison to the hard plated one, in terms of friction, roughness and wear in particular, the findings showed superior characteristic of WC-Co-Cr HVOF laminated steel.
Hong et al. [42] investigated the wear behavior of nanostructured coating of cermet WC-Co-Cr, sprayed by HVOF versus Alumina in dry friction at different temperatures. The rate of wear for coatings raised as the

increment of test temperature. The primary wear mechanisms of the coatings as temperature grew were extrusion distortion at ambient temperature, adhesive wear, carbide particle knock and oxidation wear at 200 °C, and a combination of binder extrusion and fatigue deformation combined with oxidation wear at 500 °C.

The issue of fracturing equipment failing due to wear in the oil and gas industry, according to Zheng et al. [43], may be addressed by employing the HVOF technique to prepare WC-based coatings. Thermal spray WC based coatings were investigated for their overall wear characteristics. In comparison to the uncoated specimens, according to the results, tungsten carbide composite coatings had a lower coefficient of friction, higher resistance against wear and higher hardness.

Wu et al. [44] used AISI 1045 steel substrate by HVOF spraying technique to create a WC-10Co-4Cr coating. The coating's hardness, microstructure and dry sliding wear behaviour were examined and compared to Cr12MoV, a cold work die steel. The coating exhibited low porosity, excellent microhardness, and a consistent spread of tungsten carbide particles and WC and W2C phases found in coated surface. As the load raised, the decrement in coefficients of friction found in cold work die steel the coating. The coefficients of friction and wear weight lost were less than Cr12MoV steel. Abrasion wear and extrusion deformation were the primary wear mechanism. Oxidation and Plastic deformation were the primary wear mechanisms in steel Cr12MoV.

Zhang et al. [45] evaluated the behaviour of wear of amorphous coatings composed of Fe48Cr15Mo14C15B6Y2 produced by HVOF thermal spraying to wear-resistant coatings such Al_2O_3, hard Cr and conventional steels. Wear resistance of amorphous Fe based coatings is greater. The force acting has no effect on the wear rate, although it grows linearly as the sliding velocity rises. Oxidative wear coupled with deformation wear is the most prevalent wear mechanism for Fe-based amorphous coatings, with the oxidation process being controlled by oxygen inward diffusion.

Zhang et al. [46] used HVOF to spray WC-12Co particles onto a AISI 1045 substrate that had been prepared with electro-spark and grit blasting before spraying. The grit blasting and electro-spark reinforced surfaces had roughness values of 60-70 μm & 40-50 μm. The grit blast & electro-spark prepared coating was thick and adhered effectively to the substrate. The electro-spark prepared specimen's heat impacted zone was bigger than the grit blasted specimen. The coatings prepared with grit blasting and eletro-spark had average adhesion strengths of 35N and 12.48N.

Zhang et al. [47] used HVOF spray processing to manufacture composite coatings of Ni-based sub-micron CaF_2/ WS_2 self-lubricating on an AISI 1045 substrate, coatings mechanical and microstructure characteristics were studied. At 400 ^{0}C, the coating had a minimal coefficient of friction of 0.32-0.38 and fair anti wear characteristics.

Zhang et al [48] utilised an HVOF spraying and laser remelting method to create a coating on AISI1045 steel using a Ni-based alloy nano h-BN powder. HVOF spraying and remelted laser coatings were examined for their frictional and microstructure characteristics. High porosities and reduced bonding strength were produced using the HVOF spraying technique. The interaction between h-BN and Ni happened during the laser remelting process and formed Ni_3B (CrB) a hard phase. In the layer remelted by laser, some carbide was also deposited. Results indicate that the micro-structure of HVOF Ni/h -BN composite coatings may be improved by Laser remelting significantly hard phases of the removable and lubricating coatings are homogeneous and thick, have a strong interface and are wear resistant.

Fedrizzi et al. [49] explored replacement by new HVOF cermet coatings of hard chromium. Coatings with $75Cr_3C_2$–25NiCr composition were sprayed to steel cylinders in earth-moving equipment using conventional and nano-powders. Cermet coatings were compared to traditional hard chromium coatings in terms of behaviour. Under all of the tested working circumstances, coatings with nano powder had a lesser value of weight loss than standard HVOF and hard chromium coatings. Decreased surface roughness and better carbide dispersion in the matrix of metal, as well as the minimal coating's porosity, may all be attributed to the improved performance.

2.3 Erosion and Corrosion Behaviour of Coatings

With AISI1045 steel, Ren et al. [50] suggested an iterative coupling technique for determining the dynamic performance of the substrate while taking the spray torch trajectory into account. The group for renormalization and realisable k-ε models for turbulence were utilised independently in the presented research to evaluate the simulation findings on the dynamic behaviour of the substrate using those two distinct models. For comparison, temperature field simulations of the substrate were chosen considering two different types of operating conditions with varied speed of scanning. The k-ε & RNG model gives very precise dynamic performance results than the realisable k-ε turbulence model, based on the assessments with experimental data.

Hong et al. [51] used the HVOF spraying method to deposit WC-10Co-4Cr cermet powder onto AISI 1045 steel substrates. The results revealed that ultrasonic sealing increased sealant penetration into the as-sprayed coating, increased resistance by an order of magnitude, and decreased corrosion current density.
Thermal spray is a robust approach for improving the cavitation resistance of hydraulic flow of water devices, as per Huang et al. [52], mainly in on site restoration situations. A WC-Co-Cr cermet coating was produced by high velocity oxygen fuel spraying for use in hydraulic components. The cavitation was investigated using an ultrasonic vibratory device, which measured the microstructure and hardness. When compared to martensite stainless steel, the coating has a better anti-cavitation performance in terms of mass weight reduction.
To enhance the resistance TO corrosion of AISI 1045 mild steel, Huang et al. [53] employed an HVAF spraying approach to create Fe-based amorphous alloy coating. Coating beats steel in terms of resistance to corrosion under electrochemical corrosive conditions, and also in a neutral saline mist atmosphere, coating beats steel for long tenure resistance to corrosion.
In an alkaline-sulfide solution, Li et al. [54] investigated the corrosion behaviour of WC-17Co, WC-10Co-4Cr, and Cr_3C_2-25NiCr coatings sprayed with HVOF. The immersion test is performed for 18 days and the corrosion rate analysis indicates that the best corrosion resistance is demonstrated by the low porosity Cr3C2-NiCr coating. The corrosion behaviour of coatings in solutions is influenced by passive film, micro-galvanic and porosity between binder phase and hard phase. SEM shows that the corrosion occurs mostly in the binder phase, while XPS shows WO3, WS_2, and Cr_2S_3. The binder phase Co in WC-17Co coatings converts to Co oxides, resulting in severe corrosion in the binder phase. In WC-10Co-4Cr coatings, galvanic corrosion happens between locations with different tungsten solubilities in the Cobalt binder, generating localised corrosion. The Cr_3C_2-25NiCr coating degrades somewhat due to the formation of Cr_2S_3 from the hard phase and $NiS/Ni_2O_3/Cr_2O_3$ from the binder. The results are supported by polarisation curves, which demonstrate that the Cr_3C_2-25NiCr coating has the least corrosion and largest passive area.
Hong et al. [55] utilised the Taguchi technique, which included the S/N ratio and ANNOVA, to improve the porosity and corrosion resistance of the WC-10Co-4Cr cermet coating by HVOF on steel substrate. The parameters investigated in this paper were distance from spray, kerosene flow and oxygen flow. The results showed that a spray distance>oxygen flow>kerosene flow was a relevant spraying parameter in succession for the coating's porosity and the only significant component was the spray distance. The kerosene flow should be 6.0 gph, oxygen flow should be 1900 scfh and spray distance should be 300 mm for optimum. Another set of findings revealed that the microstructure has a substantial impact on the coatings' resistance to corrosion. The highest corrosion resistance is found in the coating of cermet WC10Co4Cr, which was achieved by using the suitable spraying conditions with least porosity. It appears to be a viable alternative to hard chromium coating.
Verdian [56] mechanically alloyed Ni–10Ti particles were sprayed on AISI 1045 substrate using HVOF method. The study found that the feed powders' phase composition did not change, and the coatings included only super-saturated solution of NiTi. Coatings had a hardness of around 510 HV, according to hardness testing (50 g). Tafel polarisation experiments indicated that the substrate exhibits no noticeable effect the corrosion behaviour of Ni-Ti coatings in HVOF. In comparable circumstances, however, delamination of Ni–10 wt% Ti air plasma sprayed (APS) coatings and substantial substrate corrosion were found. After one day of immersion, polarisation tests on an Ni-Ti coating by HVOF revealed corrosion of the substrate surface but no delamination of the coating.
Hong et al. [57] used the HVOF spraying method to coat on AISI 1045 steel for coating by WC–10Co–4Cr nano-structured powder. The coating contained WC, W2C, nanoclusters and amorphous phase. The coating exhibited a thick structure and high thermostability while having a low porosity of 0.85 percent. 645°C was the temperature at which the amorphous phase transformed into a nanocrystalline structure. Electrochemical measurements were used to examine the coating's corrosion behaviour. The coating outperformed the hard chromium coating in a 3.5 wt.% NaCl solution.
Verdian et al. [58] used the HVOF spray technique to cover AISI 1045 steel with NiTi intermetallic coatings. HVOF coating's current densities are the same as those of NiTi alloys disclosed in the research, as per polarization tests. EIS analysis showed that when the solution penetrates the coating flaws and produces substrate corrosion, the coatings' corrosion performance increases over time as corrosion products close the defects, preventing the substrate from being addressed more aggressively.

3. SUMMARY AND CONCLUSION

Thermal spraying with HVOF is a modern technique of manufacture for adding or improving functionality to surfaces of the component. In a dynamic situation under controlled conditions, the HVOF process effectively creates a super-sonic flame and powder particles in partly melted has been deposited on the surfaces of component. The HVOF spraying technology is extensively used in industry and offers benefits over alternative thermal-spray methods for a range of material deposition, including carbide cermet coatings. One possible replacement for hard chromium, 'cermet' layers (e.g.Cr_3C_2–NiCr and WC–CoCr), that can be coated by the HVOF method, can help to alleviate health concerns associated with hexavalent chromium solutions. The wear characteristics and adhesive strength of carbide coatings produced by HVOF are improved by subsequent heat treatment (HT), as per the findings of the research. Subsequent HT is commonly utilised to modify and adjust the micro-structure and composition of phase after coating done by HVOF Stress reduction is also achieved by heat treating as-sprayed coatings. Different kinds of HVOF coatings have been reported to be altered by remelting using electron beam or high energy laser as treatment of surface methods after HVOF deposition. These subsequent treatments of surface may be employed to mix the material of cladding powder with base metal to create a new dense & hard coating with increased coating/substrate adhesion. New application markets will likely emerge when new generations of HVOF and appropriately selected post-spray methods mature. Nano-phased thermal spray coatings might possibly provide new, more effective reduced friction solutions with cost savings for end-users.

References

[1] Vanat, K. J., Sucharski, G. B., Paredes, R. S. C., & Pukasiewicz, A. G. M. (2021). Deposition of Aluminum Coatings by HVOF Process on Complex Geometries. *Journal of Thermal Spray Technology*, 1-22.

[2] Vats, A., Kumar, A., Patnaik, A., & Meena, M. L. (2021). Influence of deposition parameters on Tribological Performance of HVOF Coating: A review. In *IOP Conference Series: Materials Science and Engineering* (Vol. 1017, No. 1, p. 012015). IOP Publishing.

[3] Rusinov, P., & Blednova, Z. (2020). Structural behavior and mechanical properties of high-entropy coatings synthesized by HVOF. *Surface Innovations*, *9*(2–3), 127-138.

[4] Rusinov, P., & Blednova, Z. (2020). Structure and properties of the CoCuTiZrHf coating obtained by the HVOF method. *Surface Innovations*, *9*(2–3), 120-126.

[5] Tillmann, W., Schaak, C., Hagen, L., Mauer, G., & Matthäus, G. (2019). Internal diameter coating processes for bond coat (HVOF) and thermal barrier coating (APS) systems. *Journal of Thermal Spray Technology*, *28*(1-2), 233-241.

[6] Luo, X. T., Smith, G. M., Wang, Y., Gildersleeve, E., Sampath, S., & Li, C. J. (2019). Cracking induced tribological behavior changes for the HVOF WC-12Co cermet coatings. *Ceramics International*, *45*(4), 4718-4728.

[7] Hong, S., Wu, Y., Wang, B., & Lin, J. (2019). Improvement in tribological properties of Cr12MoV cold work die steel by HVOF sprayed WC-CoCr cermet coatings. *Coatings*, *9*(12), 825.

[8] Zhang, Y., Hong, S., Lin, J., & Zheng, Y. (2019). Influence of ultrasonic excitation sealing on the corrosion resistance of HVOF-sprayed nanostructured WC-CoCr coatings under different corrosive environments. *Coatings*, *9*(11), 724.

[9] Graf, K., & D'Oliveira, A. S. C. M. (2012). PTA hardfacing of Nb/Al coatings. *Soldagem & Inspeção*, *17*, 158-165.

[10] Matthews, S., & James, B. (2010). Review of thermal spray coating applications in the steel industry: part 1—hardware in steel making to the continuous annealing process. *Journal of thermal spray technology*, *19*(6), 1267-1276.

[11] Kawakita, J., Kuroda, S., Fukushima, T., & Kodama, T. (2003). Development of dense corrosion resistant coatings by an improved HVOF spraying process. *Science and Technology of Advanced Materials*, *4*(4), 281-289.

[12] Kong, G., Zhang, D., Brown, P. D., McCartney, D. G., & Harris, S. J. (2003). Microstructural characterisation of high velocity oxyfuel thermally sprayed Stellite 6. *Materials science and technology*, *19*(8), 1003-1011.

[13] Crawmer, D. C. (1992). Coating development for HVOF process using design of experiments. In *Proceedings 13th International Thermal Spray Conference, Orlando* (Vol. 127). ASM International.

[14] Dent, A. H., Horlock, A. J., McCartney, D. G., & Harris, S. J. (2001). Microstructural characterisation of a Ni-Cr-BC based alloy coating produced by high velocity oxy-fuel thermal spraying. *Surface and Coatings Technology*, *139*(2-3), 244-250.

[15] Gil, L., & Staia, M. H. (1999). Microstructure and properties of HVOF thermal sprayed NiWCrBSi coatings. *Surface and Coatings Technology*, *120*, 423-429.

[16] Haman, J. D., Boulware, A. A., Lucas, L. C., & Crawmer, D. E. (1995). High-velocity oxyfuel thermal spray coatings for biomedical applications. *Journal of Thermal Spray Technology*, *4*(2), 179-184.

[17] Graf, K., & D'Oliveira, A. S. C. M. (2012). PTA hardfacing of Nb/Al coatings. *Soldagem & Inspeção*, *17*, 158-165.

[18] Lienert, T. J., Lippold, J. C., & Brandon, E. D. (1993). Laser and electron beam welding of SiC [sub p] reinforced aluminum A-356 metal matrix composite. *Scripta Metallurgica et Materialia;(United States)*, *28*(11).

[19] Bello, K. A., Maleque, M. A., Adebisi, A. A., & Dube, A. (2016). Preparation and characterisation of TIG-alloyed hybrid composite coatings for high-temperature tribological applications. *Transactions of the IMF*, *94*(4), 211-221.

[20] Hojjatzadeh, S. M. H., Halvaee, A., & Sohi, M. H. (2012). Surface alloying of AISI 1045 steel in a nitrogen environment using a gas tungsten arc process. *Journal of Materials Processing Technology*, *212*(11), 2496-2504.

[21] Dyuti, S., Mridha, S., & Shaha, S. K. (2011). Wear behavior of modified surface layer produced by TIG melting of preplaced Ti powder in nitrogen environment. In *Advanced Materials Research* (Vol. 264, pp. 1427-1432). Trans Tech Publications Ltd.

[22] Wang, X. H., Zhang, M., Zou, Z. D., Song, S. L., Han, F., & Qu, S. Y. (2006). In situ production of Fe–TiC surface composite coatings by tungsten-inert gas heat source. *Surface and Coatings Technology*, *200*(20-21), 6117-6122.

[23] Liu, H., Liu, J., Chen, P., Yang, H., Hao, J., & Tian, X. (2019). Microstructure and properties of AlCoCrFeNiTi high-entropy alloy coating on AISI1045 steel Fabricated by laser cladding. *Journal of Materials Engineering and Performance*, *28*(3), 1544-1552.

[24] Hamada, A. S., Järvenpää, A., Ahmed, E., Sahu, P., & Farahat, A. I. Z. (2016). Enhancement in grain-structure and mechanical properties of laser reversion treated metastable austenitic stainless steel. *Materials & Design*, *94*, 345-352.

[25] Veiko, V., Odintsova, G., Gorbunova, E., Ageev, E., Shimko, A., Karlagina, Y., & Andreeva, Y. (2016). Development of complete color palette based on spectrophotometric measurements of steel oxidation results for enhancement of color laser marking technology. *Materials & Design*, *89*, 684-688.

[26] Oliveira, J. P., Fernandes, F. B., Miranda, R. M., Schell, N., & Ocaña, J. L. (2016). Residual stress analysis in laser welded NiTi sheets using synchrotron X-ray diffraction. *Materials & Design*, *100*, 180-187.

[27] Nath, S., Pityana, S., & Majumdar, J. D. (2012). Laser surface alloying of aluminium with WC+ Co+ NiCr for improved wear resistance. *Surface and Coatings Technology*, *206*(15), 3333-3341.

[28] Hackett, C. M., Settles, G. S., & Miller, J. D. (1993). *On the gas dynamics of HVOF thermal sprays.[HVOF (High-Velocity Oxy-Fuel)]* (No. CONF-9306130-3). Pennsylvania State Univ., University Park, PA (United States). Gas Dynamics Lab.

[29] Sidhu, B. S., Puri, D., & Prakash, S. (2005). Mechanical and metallurgical properties of plasma sprayed and laser remelted Ni–20Cr and Stellite-6 coatings. *Journal of Materials Processing Technology*, *159*(3), 347-355.

[30] Pulido-Gonzalez, N., Garcia-Rodriguez, S., Campo, M., Rams, J., & Torres, B. (2020). Application of DOE and ANOVA in Optimization of HVOF Spraying Parameters in the Development of New Ti Coatings. *Journal of Thermal Spray Technology*, *29*(3), 384-399.

[31] Nguyen, V. P., Dang, T. N., Le, C. C., & Wang, D. A. (2020). Effect of coating thickness on

fatigue behavior of AISI 1045 steel with HVOF thermal spray and hard chrome electroplating. *Journal of Thermal Spray Technology*, *29*(8), 1968-1981.

[32] Vinh, P. N., Thien, N. D., & Le Chi, C. (2018, December). Evaluating the effect of HVOF sprayed WC-10Co-4Cr and hard chromium electroplated coatings on fatigue strength of axle-shaped machine parts. In *International Conference on Engineering Research and Applications* (pp. 309-317). Springer, Cham.

[33] Castro, R. D. M., Rocha, A. D. S., Mercado Curi, E. I., & Peruch, F. (2018). A comparison of microstructural, mechanical and tribological properties of WC-10Co4Cr-HVOF coating and hard chrome to use in hydraulic cylinders. *American Journal of Materials Science. Rosemead, CA. Vol. 8, no. 1 (2018), p. 15-26.*

[34] Azizpour, M. J., & Majd, H. M. (2014). The Effect of Impinging WC-12Co Particles Temperature on Thickness of HVOF Thermally Sprayed Coatings. *International Journal of Mechanical and Mechatronics Engineering*, *8*(12), 1281-1284.

[35] Fu, Y., Peng, T., Yang, D., Sun, C., Chen, Y., & Gao, Y. (2014). HVOF sprayed Al–Cu–Cr quasicrystalline coatings from coarse feedstock powders. *Surface and Coatings Technology*, *252*, 29-34.

[36] Hong, S., Wu, Y. P., Gao, W. W., Wang, B., Guo, W. M., & Lin, J. R. (2014). Microstructural characterisation and microhardness distribution of HVOF sprayed WC–10Co–4Cr coating. *Surface Engineering*, *30*(1), 53-58.

[37] Nourouzi, S., Azizpour, M. J., & Salimijazi, H. R. (2014). Parametric study of residual stresses in HVOF thermally sprayed WC–12Co coatings. *Materials and Manufacturing Processes*, *29*(9), 1117-1125.

[38] García, J. R., Fernández, J. E., Cuetos, J. M., & Costales, F. G. (2011). Fatigue effect of WC coatings thermal sprayed by HVOF and laser treated, on medium carbon steel. *Engineering Failure Analysis*, *18*(7), 1750-1760.

[39] Gonzalez-Hermosilla, W. A., Chicot, D., Lesage, J., La Barbera-Sosa, J. G., Gruescu, I. C., Staia, M. H., & Puchi-Cabrera, E. S. (2010). Effect of substrate roughness on the fatigue behavior of a SAE 1045 steel coated with a WC–10Co–4Cr cermet, deposited by HVOF thermal spray. *Materials Science and Engineering: A*, *527*(24-25), 6551-6561.

[40] Casteletti, L. C., Nucci, R., Neto, L. A., Arnoni, E. A. B., & Totten, G. E. (2008). Hard chromium substitution using HVOF coatings. *International Heat Treatment And Surface Engineering*, *2*(1), 27-31.

[41] Petsas, N., Moutsatsou, A., Papapanos, G., Economou, S., Vardavoulias, M., Pantelis, D. I., & Sarafoglou, C. I. (2006). Nanophased thermal sprayed alloyed steel coatings for diesel engine components. *Book-institute of materials*, *807*, 439.

[42] Hong, S., Wu, Y., Wang, B., Zhang, J., Zheng, Y., & Qiao, L. (2017). The effect of temperature on the dry sliding wear behavior of HVOF sprayed nanostructured WC-CoCr coatings. *Ceramics International*, *43*(1), 458-462.

[43] Zheng, C., Liu, Y., Qin, J., Chen, C., & Ji, R. (2017). Wear behavior of HVOF sprayed WC coating under water-in-oil fracturing fluid condition. *Tribology International*, *115*, 28-34.

[44] Wu, Y., Wang, B., Hong, S., Zhang, J., Qin, Y., & Li, G. (2015). Dry sliding wear properties of HVOF sprayed WC–10Co–4Cr coating. *Transactions of the Indian Institute of Metals*, *68*(4), 581-586.

[45] Zhang, C., Liu, L., Chan, K. C., Chen, Q., & Tang, C. Y. (2012). Wear behavior of HVOF-sprayed Fe-based amorphous coatings. *Intermetallics*, *29*, 80-85.

[46] Zhang, M. L., & Zhao, M. W. (2011). Study of Microstructure and Adhesion Strength of WC-Co Coating Sprayed by High Velocity Oxygen Fuel. In *Advanced Materials Research* (Vol. 189, pp. 306-310). Trans Tech Publications Ltd.

[47] Zhang, X. F., Zhang, X. L., Wang, A. H., & Huang, Z. W. (2009). Microstructure and properties of HVOF sprayed Ni-based submicron WS2/CaF2 self-lubricating composite coating. *Transactions of nonferrous metals society of china*, *19*(1), 85-92.

[48] Zhang, X. F., Zhang, X. L., Wang, A. H., Xu, H. G., & Huang, Z. W. (2008, April). Microstructure and properties of laser remelted Ni-based alloy self-lubricating HVOF sprayed coating. In *Pacific International Conference on Applications of Lasers and Optics* (Vol. 2008, No. 1, pp. 640-645). Laser Institute of America.
[49] Fedrizzi, L., Rossi, S., Cristel, R., & Bonora, P. L. (2004). Corrosion and wear behaviour of HVOF cermet coatings used to replace hard chromium. *Electrochimica Acta*, *49*(17-18), 2803-2814.
[50] Ren, J., Rong, Y., & Ma, Y. (2021). Comparison of the renormalization group and the realizable k-ε turbulence models for dynamic performance of hvof process with a coupled two-stage cae method.
[51] Hong, S., Wei, Z., Wang, K., Gao, W., Wu, Y., & Lin, J. (2021). The optimization of microbial influenced corrosion resistance of HVOF sprayed nanostructured WC-10Co-4Cr coatings by ultrasound-assisted sealing. *Ultrasonics Sonochemistry*, *72*, 105438.
[52] Huang, Q., Qin, E., Li, W., Wang, B., Pan, C., & Wu, S. (2020). The Cavitation Resistance of WC-CoCr Cermet Coating Deposited by HVOF for Hydraulic Application. *Journal of Thermal Spray and Engineering*, *3*(1), 68-73.
[53] Huang, F., Kang, J. J., Yue, W., Fu, Z. Q., Zhu, L. N., She, D. S., ... & Wang, C. B. (2019). Corrosion behavior of FeCrMoCBY amorphous coating fabricated by high-velocity air fuel spraying. *Journal of Thermal Spray Technology*, *28*(4), 842-850.
[54] Li, S., Guo, Z., Xiong, J., Lei, Y., Li, Y., Tang, J., ... & Ye, J. (2017). Corrosion behavior of HVOF sprayed hard face coatings in alkaline-sulfide solution. *Applied Surface Science*, *416*, 69-77.
[55] Hong, S., Wu, Y., Zheng, Y., Wang, B., Gao, W., Li, G., ... & Lin, J. (2014). Effect of spray parameters on the corrosion behavior of HVOF sprayed WC-Co-Cr coatings. *Journal of materials engineering and performance*, *23*(4), 1434-1439.
[56] Verdian, M. M. (2014). Characterisation and corrosion performance of HVOF sprayed Ni–10 wt-% Ti coatings. *Surface engineering*, *30*(9), 670-674.
[57] Hong, S., Wu, Y., Zheng, Y., Wang, B., Gao, W., & Lin, J. (2013). Microstructure and electrochemical properties of nanostructured WC–10Co–4Cr coating prepared by HVOF spraying. *Surface and Coatings Technology*, *235*, 582-588.
[58] Verdian, M. M., Raeissi, K., & Salehi, M. (2010). Electrochemical impedance spectroscopy of HVOF-sprayed NiTi intermetallic coatings deposited on AISI 1045 steel. *Journal of alloys and compounds*, *507*(1), 42-46.

Hybridized and Complementary Comparative Analysis of Dissolved Gases in a Power Transformer

Kulraj Kaur
Research Scholar, I.K.G Punjab Technical University, Kapurthala
kulraj22@gmail.com
Deepika Bhalla
I.K.G Punjab Technical University, Kapurthala
deepika.bhalla89@gmail.com
Jashandeep Singh
Bhai Gurdas Polytechnic College, Sangrur
jashanjb@gmail.com

Abstract

Power transformer is considered important equipment both in the transmission and distribution zone of electric power. Safety of power system elements plays a crucial role for enhancing its reliability. Transformer being a major component plays a much vital role and needs to be monitored and maintained at a regular basis. Transformer oil helps in performing the functions of insulation, cooling and helps extinguish arcs. Transformer oil also dissolves the gases which get generated by the degradation of oil, presence of moisture and gas from insulation deterioration of cellulose, gases and moisture from the surrounding, to which the oil is exposed. The dissolved gas present in an oil sample can be helpful in determining the nature of a fault which can vary according to gas types and its concentrations. DGA (Dissolved Gas Analysis) have been used as a suitable tool in enabling the reliability of the system and safety of the transformer for a long time. An analysis of Dissolved gases in oil is analytical method and have been used and interpreted differently by different methods and users. DGA methods using ratios, gas concentration and graphical methods have been presented in this work. The tools use mix of fuzzy based analysis, MATLAB. The basic motivation of the work is to create a platform for DGA analysis so as to provide early and easy detection and diagnosis of Transformer Failures.

Keywords: DGA, Diagnosis, Power Transformer, Transformer faults, MATLAB, Pentagon

1. Introduction

There is an increased interest in diagnosing and determining the condition due to the fault occurrence. Faults in a transformer can interrupt the power supply in the line in which it is connected. In existing electrical power transmission and distribution network the transformers that are present are aged and at times operate under overload conditions. So, it is very important to find out and predict the developing faults in a transformer and prevent it from more damage. Today so many efforts are being made to explore new techniques and approaches of testing, monitoring, diagnosis, life and condition assessment. During working, the transformer is subjected to various stresses, which can cause deterioration of the insulation of a power transformer.

Dissolved gas analysis (DGA) is extensively accepted technique on account of it being sensitive and reliable, it is employed for estimating the health of an oil filled power transformers. It is used for regular up keeping of power transformer. Transformer oil helps in performing the functions of arc suppression, provide cooling and insulation. Transformer oil also help in dissolving the gases which are generated by the oil deterioration, presence of moisture contents and the gas generated from cellulose deterioration, moisture and gas content from the surroundings to which the oil is exposed to [31].

Content from this work may be used under the terms of the Creative Commons Attribution 3.0 licence. Any further distribution of this work must maintain attribution to the author(s) and the title of the work, journal citation and DOI.
Published under licence by IOP Publishing Ltd

This paper is based on development of a comparative analysis between the diagnostic methods using DGA data where actual fault is specified. The fuzzy logic concept is used to predict the fault for ratio methods and hence more detailed results are obtained. There could be a possibility of one or two faults occurring in some region, hence a mapping is done between various techniques so that a more selective result is obtained. Duval triangle as a graphical method is used to predict the nature of fault. Comparative analysis between different methods is done by their relative accuracy levels.

2. Types of fault identified by DGA

The faults that can be detected using DGA method are Partial Discharge, Discharge and Thermal faults. Discharge is sub classified as Low energy Discharge and High energy Discharge, whereas Thermal fault is classified as T1,T2 and T3 depending upon the temperature ranges. Partial Discharge (PD) is a localized fault and leads to bubble formation in transformer oil. It occurs due to high voltage field stress. As the molecules are excited at a temperature between 60-150°C, the bond breakage of hydrocarbons occur in the partial discharges. Partial discharge deteriorates the insulation and hence can completely damage the transformer. Low energy Discharge (D1) is a sparking type of partial discharges which leads to pin holes and small carbonized punctures in paper. As the low energy arcs occur, it results into the formation of small particles of oil. High energy Discharges (D2) occur in paper or oil, which can result in damaging the paper. Thermal fault which occurs at a temperature below 300 °C is classified as T1. These are visible by change in paper colour as it turn brownish at a temperature more than 200 °C or carbonized at a temperature more than 300 °C. Thermal fault which occurs in between 300 °C to 700 °C is classified as T2 [31]. It is indicated by carbon particles in oil and carbonization. High temperature fault which occurs at a temperature of more than 700 °C is classified as T3. It is visible in the form of large scale carbon particle formation in the oil, tint in the metal at a temperature of 800 °C and greater than 1000 °C. Thermal & Electrical fault (DT) occurs when both the electrical and thermal fault occurs at the same time inside a transformer. This fault leads to increase in the decomposition of both dielectric fluid and solid insulation.

3. Fault Diagnosis

Transformer is a crucial device of a Power System. It plays its role in generating station to transmission station and then finally to the distribution area. The ever increasing energy demand makes it much more useful component of an electrical power system. Generally, it is a very reliable component but if some failure occurs, it becomes difficult to rectify it due to unavailability of spare transformer. Hence, some techniques like Sweep frequency response analysis and Furan analysis have been developed to diagnose the fault at an early stage. Among those techniques since many years DGA is one which detects the incipient faults that occur in a transformer. The major parameters for performing DGA are the gases produced from insulation decomposition of transformer. The main gases generated are methane (CH_4), ethane (C_2H_4), hydrogen (H_2), ethylene (C_2H_6), carbon dioxide (CO_2), carbon monoxide (CO). After separating these gases from oil sample, the diagnosis technique of DGA is applied so as to identify the faults or problems like Partial Discharge, Overheating, Arcing, etc. This method of DGA is used not only for detecting the incipient faults but also online condition monitoring. The cause of failure can also be identified before it happens. The tools or techniques to diagnose the faults are Key Gas method, Roger's ratio method, Doernenburg ratio method, Duval Triangle method [32].

IOP Conf. Series: Materials Science and Engineering 1228 (2022) 012003 doi:10.1088/1757-899X/1228/1/012003

3.1 Key gas method analysis

The gases dissolved in a transformer are known as Key Gases. The fundamental of key gases is based on the generation of gases whenever some fault occurs which ultimately lead to rise in the temperature of the Transformer. The temperature and energy resolves the existence of faults as these are the two parameters which break the anatomy of the transformer oil. Key gas analysis basically measure the amount of fault gases which are released from the insulating oil on occurrence of fault. The key gases present in oil are mentioned in table 1[1,2]. CO_2 (carbon dioxide), CO (carbon monoxide), O_2 (oxygen) are the gases which are obtained from cellulose insulation. C_2H_6

Key gases in oil	Symbol
Hydrogen	H_2
Methane	CH_2
Ethane	C_2H_6
Acetylene	C_2H_4
Ethylene	C_2H_4

Table 1: Key Gases in oil

This method is able to determine the fault caused by some predominant gases at different temperature levels. The percentage of combustible gases is used to relate the key gas to different fault type. This method depends upon quantity of fault gases released in mineral oil on the occurrence of fault. Key fault gases are produced due to arcing, corona, discharge, pyrolysis (thermal heating). Key gas method can identify the faults as shown in figure 3. The oil and cellulose decomposition due to thermal effect produces more than 60% of ethylene and 90% of carbon monoxide, while corona produces 80% hydrogen and arcing produces 30% of acetylene with some traces of hydrogen. [2]

3.1.1 Fuzzy Logic Application of Key Gas method

When a transformer has multiple fault, it becomes difficult and complicated to diagnose the relation between different gases. Hence a MATLAB model was designed and developed to automate the evaluation of Key Gas method. Figure 1 shows gas concentration levels of the key gases. The proposed model is shown in figure 2. As per this representation, it indicates that the relative proportion of particular gas leads to the mentioned type of fault i.e. high value of CO indicates overheated cellulose, high value of C_2H_2, medium and lower value of C_2H_6 and CH_4 indicates overheated oil, value of gases in the decreasing order H_2, C_2H_2, CH_4, C_2H_6 indicates arcing in the oil, extremely higher value of H_2, and very low value of CH_4, C_2H_6, C_2H_2 (in the decreasing order) indicates the existence of Corona.

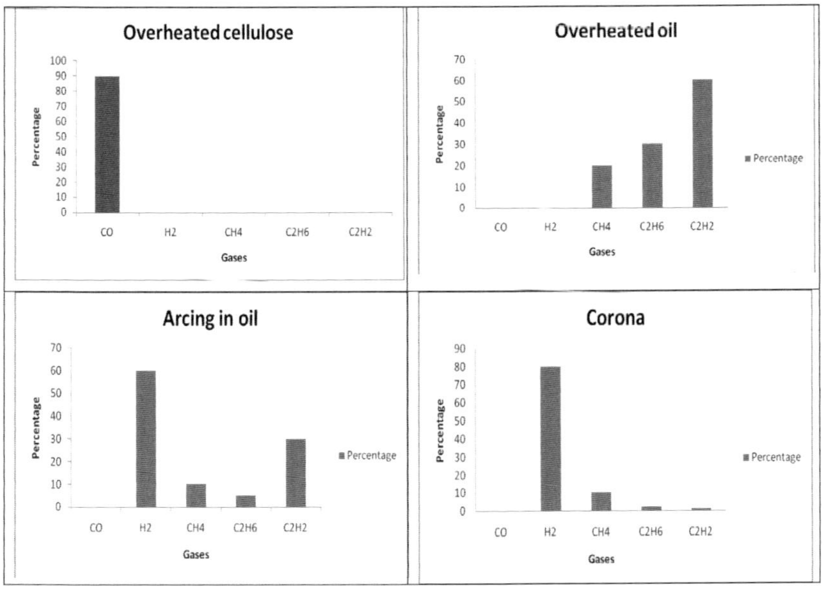

Figure 1. Gas concentration levels

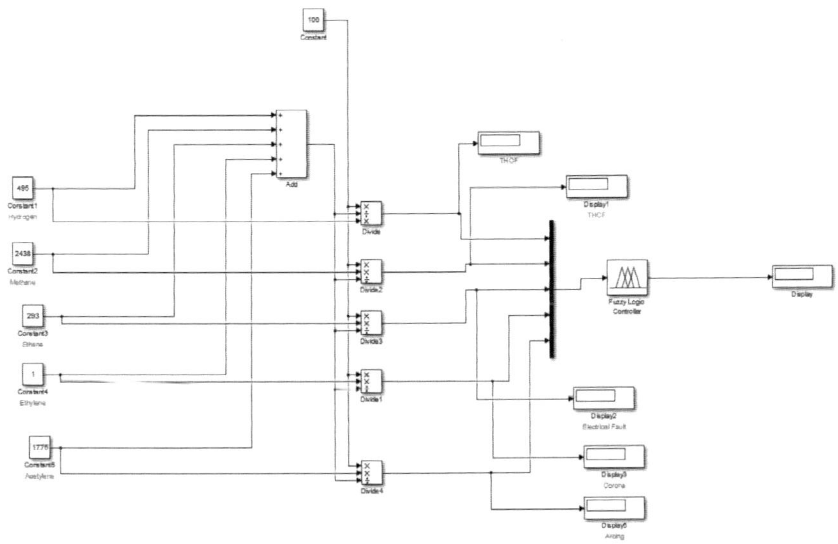

Figure 2. MATLAB Model for Key Gas Fault Diagnosis

3.2 Roger's Ratio method

This method uses 3 gas ratios and identifies the faults as mentioned in table 3. It provides decision for almost all cases but in some cases, it may not lead to any result.

IOP Conf. Series: Materials Science and Engineering 1228 (2022) 012003 doi:10.1088/1757-899X/1228/1/012003

Hence in such case it is called no decision problem. Also in the newer redefined method C_2H_6/CH_4 is not used as this ratio indicates a limited temperature range and donot provide further fault identification [4], [5].

Table 2. Fault range

C_2H_2/C_2H_4	CH_4/H_2	C_2H_4/C_2H_6	Fault type
<0.1	>0.1 to <1.0	<0.1	Normal
<0.1	<0.1	<0.1	Arcing (Low density)
0.1-3.0	0.1 to 1.0	>3.0	Arcing (High density)
<0.1	>0.1 to <1.0	1.0-3.0	Thermal (<700°C)
<0.1	>0.1 to <1.0	>3.0	Thermal (>700°C)

3.2.1 Implementation of Roger's Ratio method

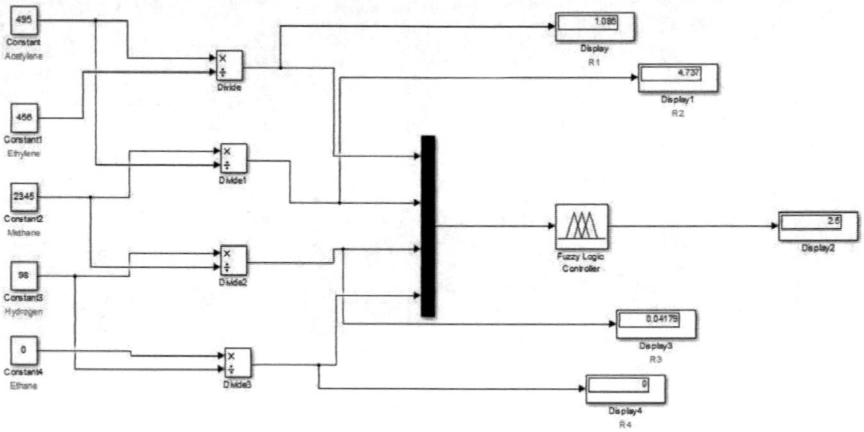

Figure 3: MATLAB model for Roger's Ratio method

3.3. Dornenburg ratio method

The calculation procedure for this method also requires significant gas levels so as to obtain a valid diagnosis. The ratios which are used over here are Ethylene/ Acetylene (C_2H_6/C_2H_2), Acetylene/Ethane (C_2H_2/C_2H_4), Acetylene/Methane (C_2H_2/CH_4), Methane/hydrogen(CH_4/H_2) [6]. Implementation of this method is done in MATLAB coding. Sample data was tested and result interface is shown in figure 5. A flow chart describing the procedure involved in calculation of fault using this method is shown in figure 4.

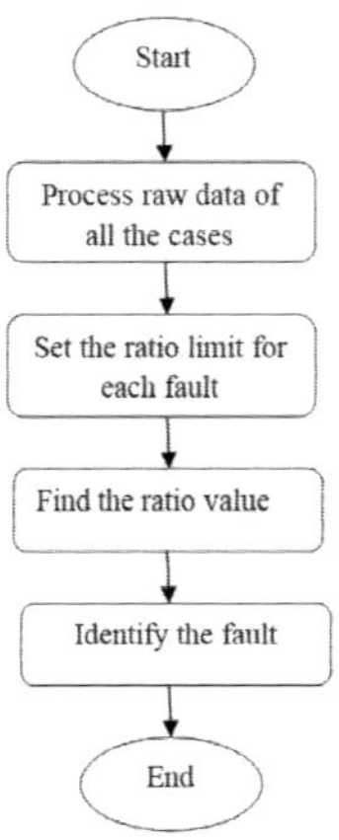

Figure 4. Flow Chart

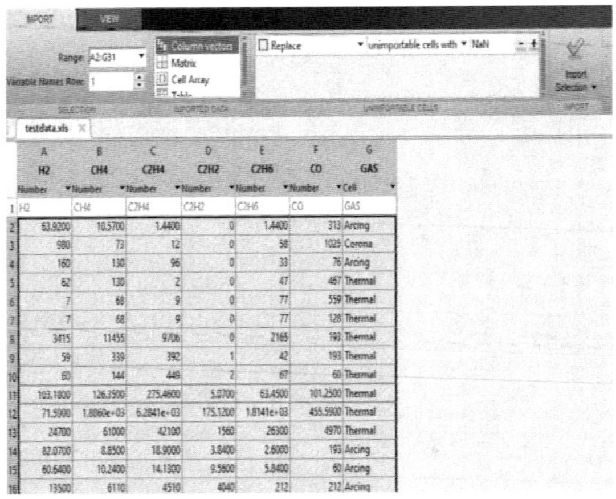

	A H2 Number	B CH4 Number	C C2H4 Number	D C2H2 Number	E C2H6 Number	F CO Number	G GAS Cell
1	H2	CH4	C2H4	C2H2	C2H6	CO	GAS
2	63.9200	10.5700	1.4400	0	1.4400	313	Arcing
3	980	73	12	0	58	1025	Corona
4	160	130	96	0	33	76	Arcing
5	62	130	2	0	47	467	Thermal
6	7	68	9	0	77	559	Thermal
7	7	68	9	0	77	128	Thermal
8	3415	11455	9706	0	2165	193	Thermal
9	59	339	392	1	42	193	Thermal
10	60	144	449	2	67	60	Thermal
11	193.1800	126.3500	275.4600	5.0700	63.4500	101.2500	Thermal
12	71.5900	1.8860e+03	6.2841e+03	175.1200	1.8141e+03	455.5900	Thermal
13	24700	61000	42100	1560	26300	4970	Thermal
14	82.0700	8.8500	18.9000	3.8400	2.6000	193	Arcing
15	60.6400	10.2400	14.1300	9.5600	5.8400	60	Arcing
16	13500	6110	4510	4040	212	212	Arcing

Figure 5. MATLAB Result interface

The pseudo code developed in MATLAB for Doernenburg ratio method of DGA for detection of faults in transformer is given below:

```
function (thermal, corona, arcing)
Define Ratio A,B,C,D
{
if ratio A < 1
 if ratio B < 0.75
  if ratio C < 0.3
   if ratio D > 0.4
display "Thermal"
else
        "corona"
or
        "arcing"
}
else
Print "not defined"
```

3.4 Duval triangle method

Duval Triangle uses a combination of 3 hydrocarbon gases namely H_4, C_2H_4 and C_2H_2.These gases get generated when energy levels in the transformer increases and due to that the fault gases are released while transformer in service. This method is an improved technique out of all previously discussed ones and it provides much more accurate and consistent result as compared to all other techniques. Figure 6 indicates the zones which are divided in the Duval Triangle which shows all the individual faults falling in each zone. Seven type of faults can be identified through these zones which are DT (intermediate), PD (Partial Discharge), T3(Thermal fault), T2 (Thermal fault), T1 (Thermal fault), D2(Discharge), D1 (Discharge) [4],[6],[11].

IOP Conf. Series: Materials Science and Engineering 1228 (2022) 012003 doi:10.1088/1757-899X/1228/1/012003

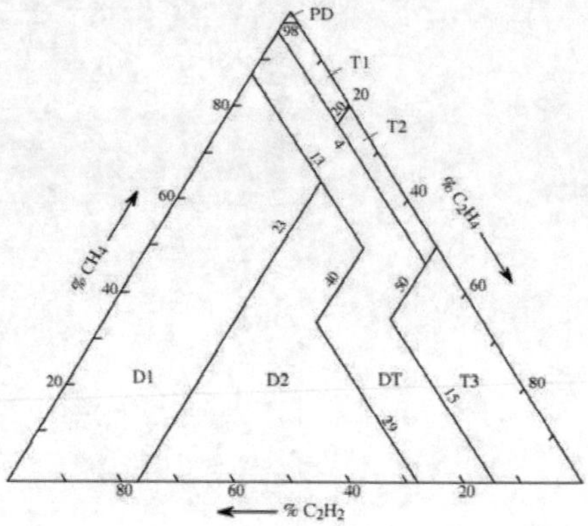

Figure 6. Duval Triangle

3.4.1 Implementation

The fault calculation is achieved in a stepwise procedure as follows:
Step 1: Obtain the gas data
Step 2: The gas concentration is always given in ppm, calculate it in percentage using the formula (% Gas 1= 100*Gas1 ppm/Gas1 ppm+Gas2 ppm+Gas3 ppm)
Step 3: Plot each individual Gas concentration in percentage and find the intersecting point.
Step 4: Identify the region in which the intersecting point falls.

This procedure is followed and an expert system in excel that is used to analyze the data for fault identification shown in figure 7.

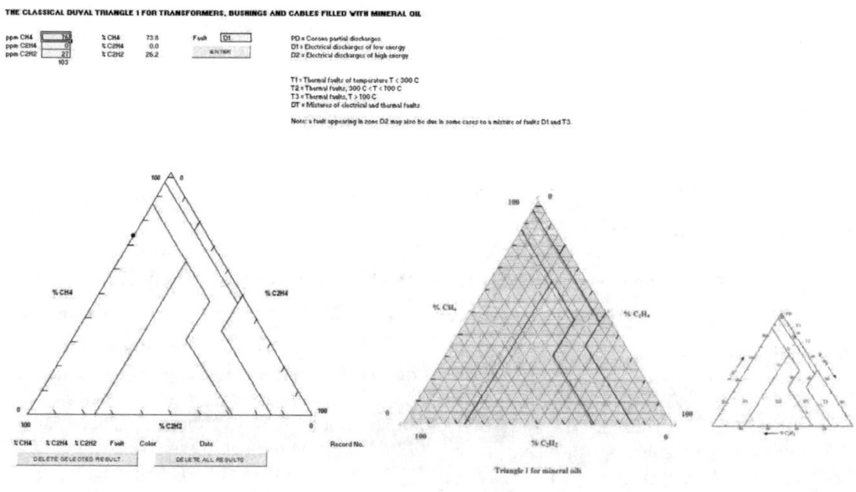

Figure 7: Excel Interface

4. Results and Discussion

So as to vindicate the combination process, a data base of around 452 known DGA test samples from IEC TC 10 Database were used and tested by this system. Out of that 30 data samples are used and shown in table 4. The gas content is provided in the data and it is used as a sample calculation. Prior to applying it on Duval Triangle it is applied to Key gas, Roger's Ratio, Doernenburg Method. For comparison mapping is done among the techniques. It is done by providing the fault codes to all the different type of faults each technique can evaluate as shown in table 3. As discussed earlier, developing a diagnosing and monitoring system for a Power transformer is a complex task. The techniques mentioned are able to take into account those complexities and requirements. After mapping the data using fault codes, comparative result is obtained for 30 oil samples. Looking into the results it is an undeniable fact that Duval Technique are the most appropriates techniques out of all other as the accuracy of fault prediction by DT method is found to be 46.6%, whereas Doernenburg method gives 30% accurate results, Key gas method and Rogers ratio method gives 5% accuracy. Moreover Duval Triangle technique is able to predict 'not known' faults even, which other techniques are lacking.

5. Conclusions

Due to more and more number of available methods, techniques and the obtained samples utilized in the process, a much more pliable system's need arises. The use of diagnosis tools which are software based helps to implement the coherent management policies and strategies for many engineering applications which also include a Power transformer. In this effort a system is implemented by mapping, combining and comparing various DGA diagnostic schemes. This comparison shows that mistaken diagnosis, no diagnosis, available fault; all these three conditions are clearly available in the result and can be easily compared with the result obtained from other technique. Hence this increases the likelihood of identifying the faults before it lead to the transformer failure. Moreover, the Duval triangle technique is found to be favorable approach to support advanced monitoring and diagnosis systems and also beat the complications. 300°C

Method	PD	D1	D2	T1	T2	T3
Key gas method	Sparking	Arcing	Corona	Overheating	NA	NA
Rogers ratio method	Normal Fault	Flashover	Corona	Circulating current	Thermal fault	Overheating
Doernenburg ratio method	Partial discharge	Arcing	Corona	Thermal	Thermal	Thermal
Duval Triangle method	Partial discharge	Discharge (Low energy)	Discharge (High energy)	Thermal (<300°C)	Thermal (>300°C <700°C)	Thermal (>700°C)

Table 3. Mapping of faults among various methods

Sr. No	H_2	CH_4	C_2H_4	C_2H_6	C_2H_2	Actual fault	KG	RR	DR	DT
1	63.92	10.57	1.44	1.44	0	Not known	Not known	Not known	D1	T1
2	980	73	12	58	0	Partial Discharge	D1	D1	D2	T1

3	160	130	96	33	0	Thermal	D1	T1	D1	T2
4	62	130	2	47	0	Thermal	Not known	Not known	T1	PD
5	7	68	9	77	0	Thermal	PD	Not known	T1	T1
6	3415	11455	9706	2165	1	Thermal	Not known	Not known	T1	T2
7	59	339	392	42	1	Thermal	Not known	Not known	T2	T3
8	60	144	449	67	2	Thermal	T1	Not known	T1	T3
9	56	286	928	96	7	Thermal	Not known	Not known	T1	T3
10	103.18	126.35	275.46	63.45	5.07	Not known	Not known	Not known	T1	T3
11	71.59	1886	6284.1	1814	175.1	Not known	Not known	Not known	T1	T3
12	24700	61000	42100	26300	1560	Not known	Not known	T1	T2	T2
13	82.07	8.85	18.9	2.6	3.84	Not known	T1	D2	D1	T3
14	60.64	10.24	14.13	5.84	9.56	Not known	T1	Not known	D1	DT
15	13500	6110	4510	212	4040	Not known	Not known	D2	D1	D2
16	32.4	5.5	12.6	1.4	13.2	Not known	T1	D2	D1	D2
17	8200	3790	4620	250	5830	Not known	Not known	D2	D1	D2
18	545	130	153	16	239	Discharge	D1	D2	D1	D2
19	20000	13000	29000	1850	57000	Discharge	Not known	D2	D1	D2
20	535	160	305	16	680	Not known	D1	D2	D1	D2
21	645	86	110	13	317	Discharge	D1	D2	D1	D1
22	106.69	1.71	1.56	3.97	0	Not known	Not known	D1	D2	T2
23	77.31	5.01	1.42	3.14	0	Not known	Not known	D1	D2	T2
24	55.04	43.28	46.25	17.82	0	Not known	PD	T1	D1	T3
25	416	695	867	74	0	Not known	D1	Not known	Not known	T3
26	80	619	2480	326	0	Thermal	Not known	Not known	Not known	T3
27	425	17424	37043	7299	0	Thermal	D1	Not known	Not known	T3

IOP Conf. Series: Materials Science and Engineering 1228 (2022) 012003 doi:10.1088/1757-899X/1228/1/012003

28	244	754	1281	172	27	Thermal	D2	Not known	Not known	T3
29	6709	10500	17700	1400	75	Thermal	Not known	Not known	Not known	T3
30	212	8712	18521	3649	79	Thermal	D2	Not known	Not known	T3

Table 4. Comparative Fault Diagnosis

References

[1] IEEE Standards C57.104-1991. IEEE Guide for the Interpretation of gases generated in Oil-Immersed Transformers, 1991.

[2] IEC Standards 60599. Mineral oil-impregnated electrical equipment in service - Guide to the interpretation of dissolved and free gases analysis, second edition, Mars 1999.

[3] R. R. Rogers, "IEEE and IEC Codes to Interpret Incipient Faults in Transformers, Using Gas in Oil Analysis," IEEE Trans. on Electrical Insulation, vol. 13, no. 5, pp. 349-354, 1978.

[4] M. Duval, "Dissolved Gas Analysis: It Can Save Your Transformer," IEEE Electrical Insulation Magazine, vol. 5, no. 6, pp. 22-27, 1989.

[5] R.D.Stebbins, J.J.Kelly. S.D.Myers,"Power Hansformer Fault Diagnosis', 1997 IEEE PESWM, Panel Session, New York, Feb.6, 1997.

[6] M. Duval, "Fault Gases Formed in Oil-Filled Breathing EHV Power Transformers-The Interpretation Of Gas Analysis Data", IEEE PAS Conf. Paper No C 74 476-8, 1974.

[7] Duval, Michel Lamarre, Laurent, "The duval pentagon-a new complementary tool for the interpretation of dissolved gas analysis in transformers", IEEE Electrical Insulation Magazine, Year 2014, Volume 30, Issue 6, Pages 9-12.

[8] A. Akbari, A. Setayeshmehr, H. Borsi, E. Gockenbach, I. Fofana "Intelligent Agent-Based System Using Dissolved Gas Analysis to Detect Incipient Faults in Power Transformers", IEEE Electrical Insulation Magazine · January 2011, Vol. 26, No. 6.

[9] Khiar, Mohd Shahril Ahmad Thayoob, Yasmin Hanum Md Ghazali, Young Zaidey Yang Ghani, Sharin Ab Chairul, Imran Sutan, Procedia Engineering, 2013,volume 68, pages 477-483

[10] Muhamad, N. A. Phung, B. T. Blackburn, T. R. Lai, K. X., "Comparative study and analysis of DGA methods for transformer mineral oil", 2007 IEEE Lausanne POWERTECH, Proceedings, Year 2007, Issue August, pages 45-50.

[11] Duval, Michel, "Dissolved Gas Analysis and the Duval Triangle", TechCon Asia Pacific, Sydney, Australia, Year 2006, pages 1-20.

[12] Saranya, S Mageswari, Uma Roy, Natalya Sudha, R, "Dissolved Gas Analysis Methods To Diagnose Transformer Faults", International Journal of Engineering Research and Applications (IJERA), Year 2013, Volume 3, Issue 3, Pages 592-595.

[13] Duval, Michel, "New versions of the Duval Triangle for equipment filled with mineral oil are proposed for load tap changers of the oil type, equipment filled with non-mineral oils, and for low-temperature faults where stray gassing of oils may interfere with diagnostics. The Duval Triangle for Load Tap Changers, Non-Mineral Oils and Low Temperature Faults in Transformers", Pages 22-29.

[14] Duval, Michel Lamarre, Laurent, "The duval pentagon-a new complementary tool for the interpretation of dissolved gas analysis in transformers", IEEE Electrical Insulation Magazine, Year 2014, Volume 30, Issue 6, Pages 9-12.

[15] Pattanadech, N, "Application of Duval Pentagon Compared with Other DGA Interpretation Techniques: Case Studies for Actual Transformer Inspections Including Experience from Power Plants in Thailand", 2019 5th International Conference on Engineering, Applied Sciences and Technology (ICEAST), Year 2019, Pages 1-4.

[16] Lakehal, Abdelaziz Tachi, Fouad, "Bayesian Duval Triangle Method for Fault Prediction and Assessment of Oil Immersed Transformers", Measurement and Control (United Kingdom), Year 2017, Volume 50, Issue 4, Pages 103-109.
[17] Soni, Jatinkumar Suthar, Dhaval, "An Experimental Analysis to Check Accuracy of DGA", Year 2018, Volume 1, Pages 394-385.
[18] Geetha, R, "Fault Diagnosis of Power Transformer using Duval Triangle Based Artificial Intelligence Techniques", Year 2016, Pages 78-88.
[19] Gedam, Abhishek V Swami, P S Thosar, Archana, "A Comparative Analysis of DGA Method for the Incipient Fault Diagnosis in Power Transformer Using ANN Approach", International Journal of Scientific & Engineering Research, Year 2015, Volume 6, Issue 5, Pages 387-391.
[20] Duval, Michel DePablo, Alfonso, "Interpretation of gas-in-oil analysis using new IEC publication 60599 and IEC TC 10 databases", IEEE Electrical Insulation Magazine, Year 2001, Volume 17, Issue 2, Pages 31-41.
[21] Singh, Sukhbir Joshi, Dheeraj Bandyopadhyay, M N, "Software Implementation of Duval Triangle Technique for DGA in Power Transformers", International Journal of Electrical Engineering, Year 2011, Volume 4, Issue 5, Pages 529-540.
[22] Mineral Oil-Impregnated Electrical Equipment in Service - Guide to the Interpretation of Dissolved and Free Gases Analysis. IEC Publication 60599 (1999-03), Mar. 1999.
[23] Guide for the Interpretation of Gases Generated in Oil-Immersed Transformers, IEEE Std C57.104-1991, 1991.
[24] M. Duval, "A review of faults detectable by gas-in-oil analysis in transformers," IEEE Electr. Insul. Mag., vol. 18, no. 3, pp. 8–17, 2002.
[25] P. Griffin, L. Lewand, et al., "Load tap changer diagnostics using oil tests – A key to condition-based maintenance," Minutes of Doble Engineering Conf., 2005.
[26] S. Girard, "Protective maintenance at Chugach Electrical Association, Anchorage, Alaska", TechCon 2004 North America Conference, San Antonio, TX, 2004.
[27] M. Duval et al., Final Report of CIGRE TF D1-01-15, to be published in 2009.
[28] Cooper Power Systems, Envirotemp FR3 Fluid – Dissolved Gas Guide, Technical Brochure R900-20-19, 2006.
[29] P. Griffin, "Analysis for combustible gases in transformer silicone fluids", Minutes of Doble Engineering Conf., 1985, Sec.10-701/ 707.
[30] H. Kuwahara, et al., Study of Explosion and Fire Hazards of Silicone Liquid Under Arc Conditions, Mitsubishi Technical Brochure, 1987.
[31] K. Chatterjee, V. K. Jadoun and R. K. Jarial, "Emerging trends for determining incipient faults by dissolved gas analysis," *2017 3rd International Conference on Condition Assessment Techniques in Electrical Systems (CATCON)*, 2017, pp. 63-68, doi: 10.1109/CATCON.2017.8280185.
[32] Chatterjee, K., Dawn, S., Jadoun, V.K. and Jarial, R. (2019), Novel prediction-reliability based graphical DGA technique using multi-layer perceptron network & gas ratio combination algorithm. IET Sci. Meas. Technol., 13: 836-842. https://doi.org/10.1049/iet-smt.2018.5397

CFD analysis of Effect of journal bearing lubricated with non-Newtonian fluid

Jai Prakash sharma[1], M R Tyagi[1], G D Thakre[2]

[1]Department of Mechanical Engineering, Manav Rachna University, Faridabad, India
[2]Tribology & Combustion Division, CSIR-Indian Institute of Petroleum, Dehradun, India

Abstract: The present paper deals with the CFD analysis of hydrodynamic bearing lubricated with a non-Newtonian fluid and the non-Newtonian behaviour of lubricant is represented by a power-law. Bearing parameters such as fluid film pressure and load-carrying capacity are determined by varying the power-law index (n) in terms of its viscosity from 0.8 to 1.2 under steady-state conditions through the CFX 17.2 software module and the flow model opted as Laminar based on Reynolds number calculation as per the geometry developed for the simulation. The relationship between the pressure and load-carrying capacity are also obtained for different relative clearance, revolution per minute and L/D ratio. Fluid film pressure profile was drawn by varying input parameters at different power-law index for understanding the performance of HDB under different regimes i.e. dilatant, Newtonian and pseudoplastic fluids. Plots were also created between Pressure, load carrying capacity and power law index to study the performance of HDB. This has been observed that the bearings lubricated with dilatant fluid (n >1) have higher pressure and load carrying capacity as compared to those lubricated with pseudoplastic fluid (n < 1). Occurrences of cavitation were also seen at lower angle with increases in power law index for 0.0006 relative clearance but increases at 0.0008.

Keywords: Hydrodynamic Bearing, CFD, Power-law fluid, fluid film pressure, load-carrying capacity.

1. Introduction

Journal bearings are used to support the rotating shafts and heavy loads applied to the mechanical system. The lubricant used dictates the bearing performance while supporting the loads. The lubricants are often considered Newtonian fluids for the sake of simplicity. However, the lubricants are blended products comprising polymeric and organic/inorganic compounds as additives. The additives used to the lubricants render non-Newtonian behaviour. Hence, to have a realistic analysis of the bearings, it is essential to consider lubricant behaviour as non-Newtonian.

The flow behaviour differs for the Newtonian, pseudoplastic and dilatant fluid [1]. Power-law fluid is a simplistic representation of the non-Newtonian fluid. For power-law fluids, the relation between shear stress and shear strain is represented by a consistency factor (n). The consistency factor may be < 1 or > 1. Mathematically the power-law fluids are represented by the Ostwald-de Waele power-law model [2], [3]. Researchers have used non-Newtonian models to solve the lubrication problems [4]–[12]. They have used various analytical approaches to determine the performance parameters of bearings lubricated with power-law fluids [6], [13], [14]. Studies have revealed that load capacity, pressure, temperature, and frictional power increases when n>1, and vice versa for n<1 [15].

In the hydrodynamic bearings (HDBs), as the lubricant flow from the convergent to divergent region, the lubricant properties get influenced. The lubricant film gets stressed. As a result, the viscosity changes with an increase in pressure and temperature. The HDBs may also experience cavitation [16–19]. In such cases the lubricant rheology plays a pivotal role in predicting the bearing performance. In this context, mathematical models are developed by researchers to study the

Content from this work may be used under the terms of the Creative Commons Attribution 3.0 licence. Any further distribution of this work must maintain attribution to the author(s) and the title of the work, journal citation and DOI.
Published under licence by IOP Publishing Ltd

rheological properties of lubricant for slider and journal bearings [20]. Studies have shown that the fluid film thickness is larger than the shear zone thickness. In such cases, the additives like nanofluids or polymers [21, 23] can help increase the load-carrying capacity (LCC) with a decrease in friction coefficient.

The journal bearing performance is also affected by the geometric and operating parameters like speed (N), L/D ratio, load, temperature (T), eccentricity ratio (ε), radial clearance (c), fluid dynamic viscosity (μ) etc. The load is usually determined at the beginning of the bearing design. Also, by using a viscous lubricant, LCC can be increased. The eccentricity ratio is directly related to LCC as per the relation which affects the performance. While bearing operation, cavitation occurs due to pressure difference between low and high-pressure region. This can create a condition of metal to metal contact causing rise in temperature and wear of journal and bearing.

This paper investigates the performance of HDB lubricated with pseudoplastic, Newtonian and dilatant fluids, which was described by a power-law model and by taking cavitation into account. The bearing performance (LCC due to pressure variation) is affected by the type of lubricant used. Hence, it is essential to study the effect of non-Newtonian fluid on bearing performance by varying the power-law index (n) between 0.8 to 1.2. Simulations were performed on Ansys 17.2 CFX to determine the influence of speed, L/D ratio, relative clearance (Ψ) and power-law index (n) on journal bearing of fixed diameter (50 mm).

2. Physical and mathematical models

The bearing geometry was created using 3D modelling software and then imported to CFX to perform the simulation. The lubricant density is 871 kg/m^3 at 15^{0}C and kinematic viscosity of 46.3 cSt at 40^{0}C was taken for simulations. Table 1 shows the design parameters used to create the geometry and the working model.

Table 1: Bearing design parameters

Design Parameter	Value
D (mm)	50
Ψ(%)	0.06, 0.08
L/D	0.5, 1, 2
N(rpm)	10,100, 1000

2.1 Reynolds's equation

For solving the fluid flow problem, Reynolds's equation forms the basic governing equation. The 3D Reynolds's equation can be written as,

$$\frac{\partial}{\partial x}\left[\frac{h^3}{\mu}\frac{\partial p}{\partial x}\right]+\frac{\partial}{\partial z}\left[\frac{h^3}{\mu}\frac{\partial p}{\partial z}\right]=6\left(U\frac{\partial h}{\partial x}+V\frac{\partial h}{\partial z}\right)+12(w_h+w_o) \quad (1)$$

Where U and V are circumferential and linear velocities, w_h and w_{ho} are the velocities of top to the bottom column, h is fluid film thickness and μ is dynamic viscosity. It is too difficult to solve the 3D Reynolds's equation, and therefore it is simplified using various approximations. By applying the unidirectional velocity, steady film thickness and isoviscous approximations, the equation can be written as:

$$\frac{\partial}{\partial x}\left[h^3\frac{\partial p}{\partial x}\right]+\frac{\partial}{\partial z}\left[h^3\frac{\partial p}{\partial z}\right]=6\mu U\frac{\partial h}{\partial x} \quad (2)$$

With the short bearing approximation, the above equation is simplified to evaluate the pressure analytically with boundary conditions p = 0 at y = ±L/2 and dp/dy = 0 at y = 0. By integrating the pressure over the journal surface, the LCC can be calculated as.

$$W_1=\int_0^n\int_{-\frac{L}{2}}^{\frac{L}{2}}PR\cos\theta\, d\theta\, dy$$

$$W_1 = \int_0^{n} \int_{-\frac{L}{2}}^{\frac{L}{2}} PR \sin\theta \, d\theta \, dy \qquad (3)$$

To solve the flow problem, the conservation of mass and momentum equations used by CFX are,

$$\frac{\partial \rho}{\partial t} + \nabla(\rho \vec{v}) = 0 \qquad (4)$$

Where ρ and $\vec{v}$ are the fluid density and fluid velocity vector. Pressure (P) can be evaluated from the one dimensional Reynolds equation, but the value of Ψ & ϵ depends on the configuration, loading condition and lubricant used.

2.2 Momentum conservation equation

For an inertial (non-accelerating) reference frame, the momentum conservation equation is:

$$\frac{\partial}{\partial t} (\overrightarrow{\rho v}) + \nabla (\rho \, \vec{v} \, \vec{v}) = -\nabla p + \nabla (\bar{\bar{\tau}}) + \rho \vec{g} + \vec{F} \qquad (5)$$

Where $\bar{\bar{\tau}}$, p, $\vec{F}$ and $\rho\vec{v}$ is the stress tensor, static pressure, external body force and gravitational body force, respectively. The $\bar{\bar{\tau}}$ is given by

$$\bar{\bar{\tau}} = \mu \left[(\nabla \vec{v} + \nabla \vec{v}^T) - \left(\frac{2}{3}\right) \nabla . \vec{v} \, \mathrm{I} \right] \qquad (6)$$

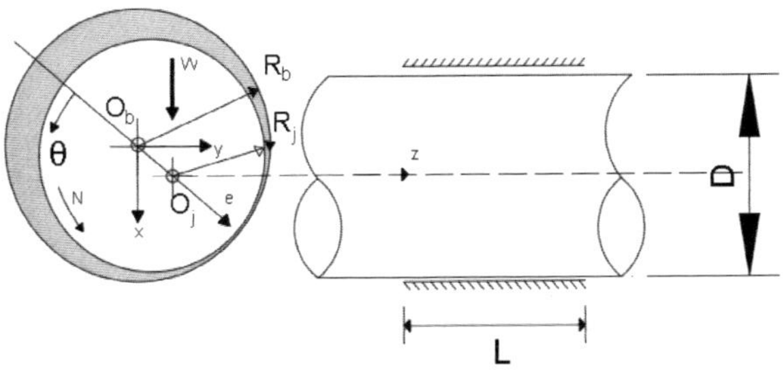

Fig. 1: Journal bearing geometry

The bearing configuration used as figure 1, where O_j is the journal's origin, O_b is the origin of bearing, R_b and R_j is the radius of the bearing and journal. Under loaded condition, eccentricity is the distance between O_B and O_j. The expression used to evaluate fluid film thickness for the configuration was h = c(1+εcos θ) when the bearing radius is fixed.

2.1. Viscosity for Non-Newtonian model

The working fluid used in this simulation is VG 46 to study its effect on journal bearing performance. The lubricant behaviour as non-Newtonian is modelled through Power-Law. The n for pseudoplastic fluid is n<1, Newtonian fluid n=1 and dilatant fluid n>1 [24, 25]. The power-law equation describes the rheology of fluid

$$\mu = K(\lambda\gamma)^{n-1} \qquad (7)$$

Where μ, k, ϒ, λ & n is the apparent viscosity, flow consistency index, shear strain rate, normalizing time constant and flow behaviour index. As n varies, the non-Newtonian behaviour of fluid increases or decreases. The upper and lower limit of the power-law function was set for simulation. The value used for the setup of the non-Newtonian model with power-law in CFX is given in table 2:

Table 2: CFX setup parameters

Parameter	values
Consistency index (k) (Pa-s)	0.0403273

Power law index (n)	0.8 - 1.2
Lower limit (s^{-1})	0.0001
Upper limit (s^{-1})	1000

2.2. CFD model

The CFD package used for the study is CFX 17.2. The geometric model is created in the design modeller and then meshed in the CFX setup. The model was divided into 20 nodes circumferentially and 200 nodes axially. The inner surface of the geometry was considered rotating and the outer surface as stationary, with both inlet and outlet as symmetrical. Based on mesh independence study, the number of elements used for simulation was 104000, and the minimum and maximum element volume ratio lies between 1.00448 – 1.25438, which is in the prescribed range for normal analysis. The non-Newtonian model was set up with the lubricant properties and power-law index to analyse the effect of eccentricity ratio, L/D ratio, $n < 1$, $n = 1$, $n > 1$ (fluid rheology) and speed on pressure causing cavitation and evaluating the LCC of HDB. Also the residual type was opted as RMS and residual target was 0.0001 under convergence criteria. The governing equation was solved under steady-state conditions with no account of gravity, and operating pressure was set at 101325 Pa.

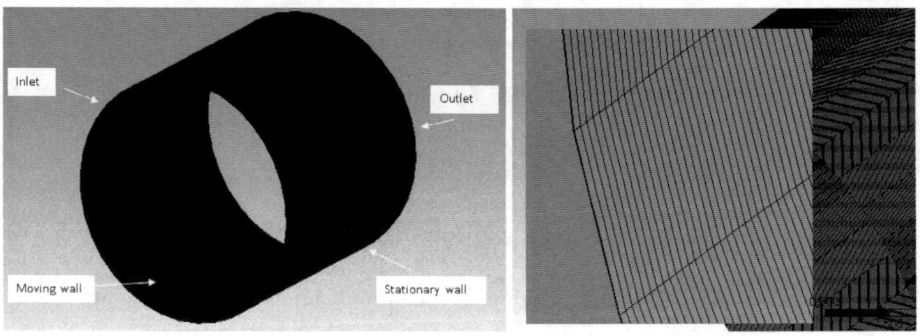

Figure 2: Flow domain with boundary condition

3. Results and discussion

For performance evaluation of HDB, simulation was performed using Ansys CFX 17.2 by varying L/D ratio, eccentricity and rpm with power-law index $n = 0.8 – 1.2$. The influence of input parameters is discussed through plots between pressure and theta. Also its performance in terms of pressure attainment and LCC was discussed.

3.1. Effect of speed, eccentricity and power-law index

Simulated results show peak pressure distribution attained at $n = 0.8$ was $2 * 10^6$ Pa with different eccentricities and speed through figure 3. At the start of the initial run, the pressure was high for a 0.00075 eccentricity value; this may be due to lower speed as the system tried to achieve the hydrodynamic condition. However, with an increase in eccentricity, the load-carrying capacity decreased. Negative pressure was observed at 10rpm with 0.0008 relative clearance between 0^0 to 42^0, but positive pressure was achieved after this. It is also perceived, negative pressure increases with the increase in rpm and relative clearance for different eccentricities. With maximum eccentricity, the system is not stable resulting in contact between journal and bearing as the pressure is almost zero in some cases.

IOP Conf. Series: Materials Science and Engineering 1228 (2022) 012004 doi:10.1088/1757-899X/1228/1/012004

<table>
<tr><th>Ψ
rpm</th><th>0.0006</th><th>0.0008</th></tr>
<tr><td>10</td><td>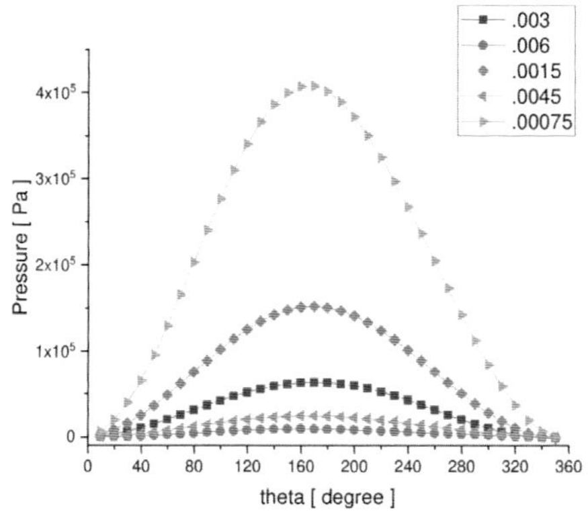</td><td>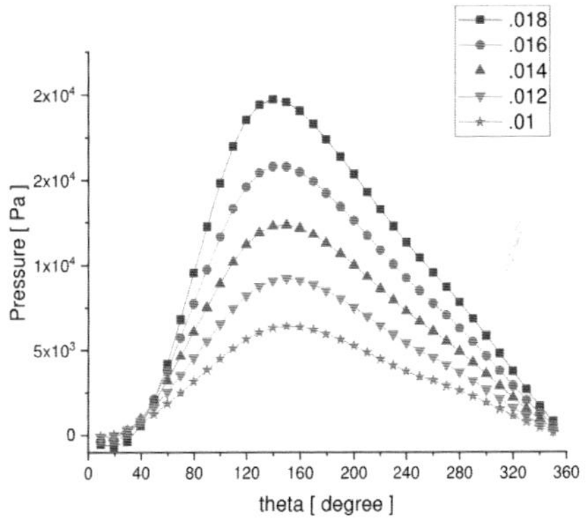</td></tr>
<tr><td>100</td><td>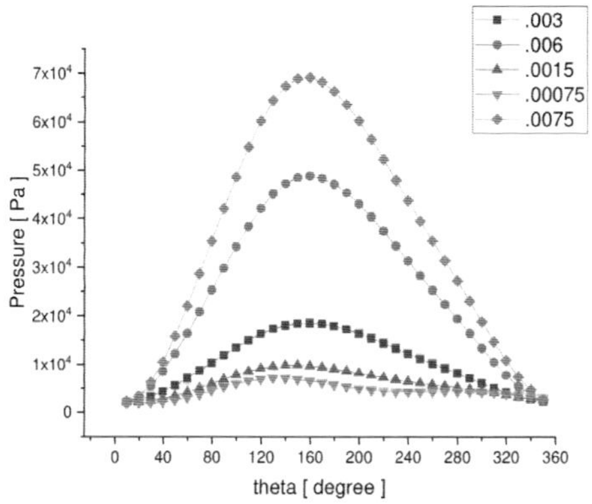</td><td>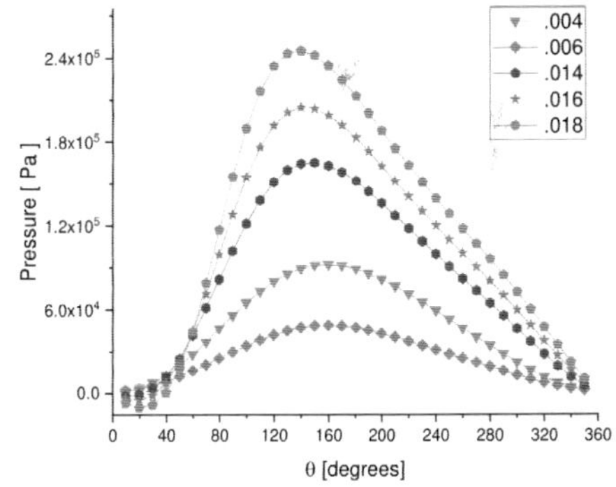</td></tr>
<tr><td>1000</td><td>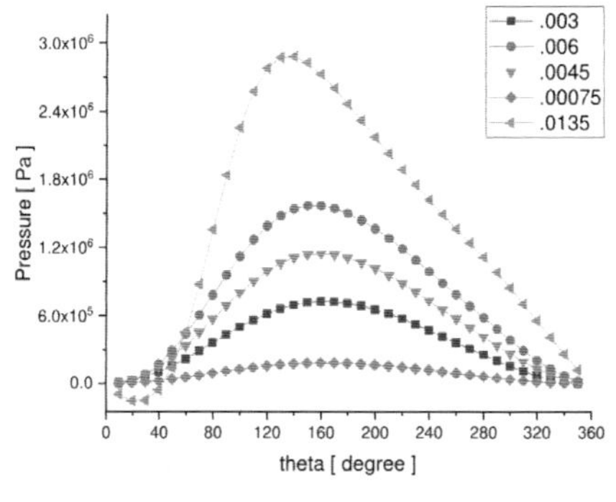</td><td>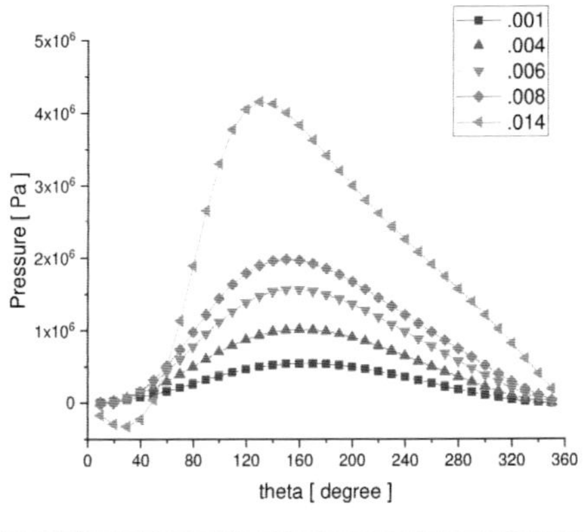</td></tr>
</table>

Figure 3: Pressure vs. theta (L/D = 0.5, n = 0.8)

Figure 4 illustrates the simulation results for pressure distribution on the midplane for n = 0.9. The peak pressure achieved was $8*10^6$ Pa. Pressure distribution plot shows that the LCC raises with a rise in rpm & eccentricity.

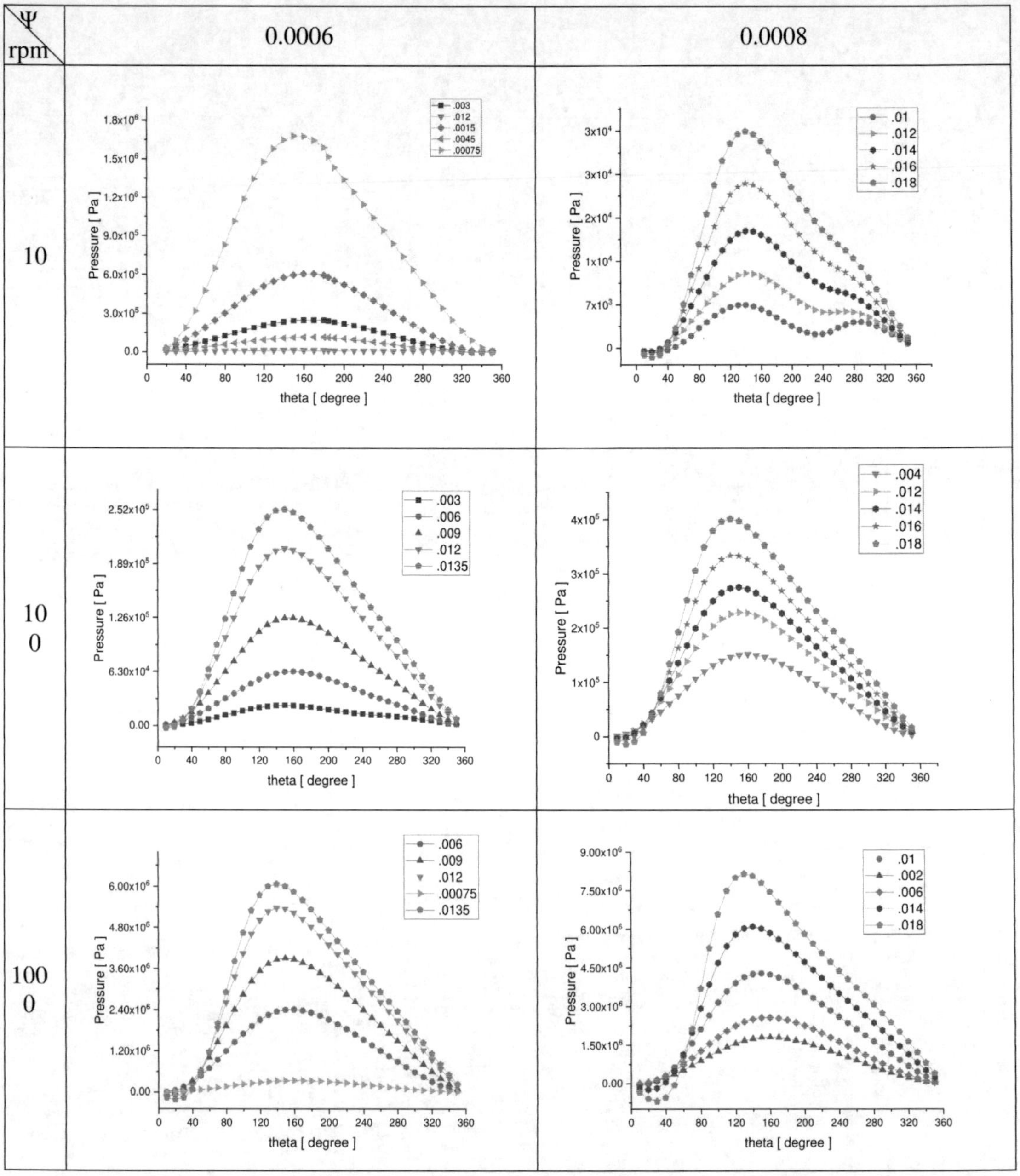

Figure 4: Pressure vs. theta (L/D = 0.5, n = 0.9)

Similarly, the pressure plots for the HDB lubricated with Newtonian fluid are plotted in figure 5. For n =1, the peak pressure achieved was $1.2*10^7$ Pa. As Ψ & rpm increases, the pressure profile first increases and then decreases from 140°. Also, the pressure profile with slight variation was seen in few runs for lower relative clearance. At 10 rpm for 0.0006 (Ψ), positive pressure regions occurred for all eccentricities. Still, negative pressure occurred for max eccentricity at 100 & 1000 rpm with different Ψ (0.0006 & 0.0008) till 42°, and then positive pressure region was achieved till 360°.

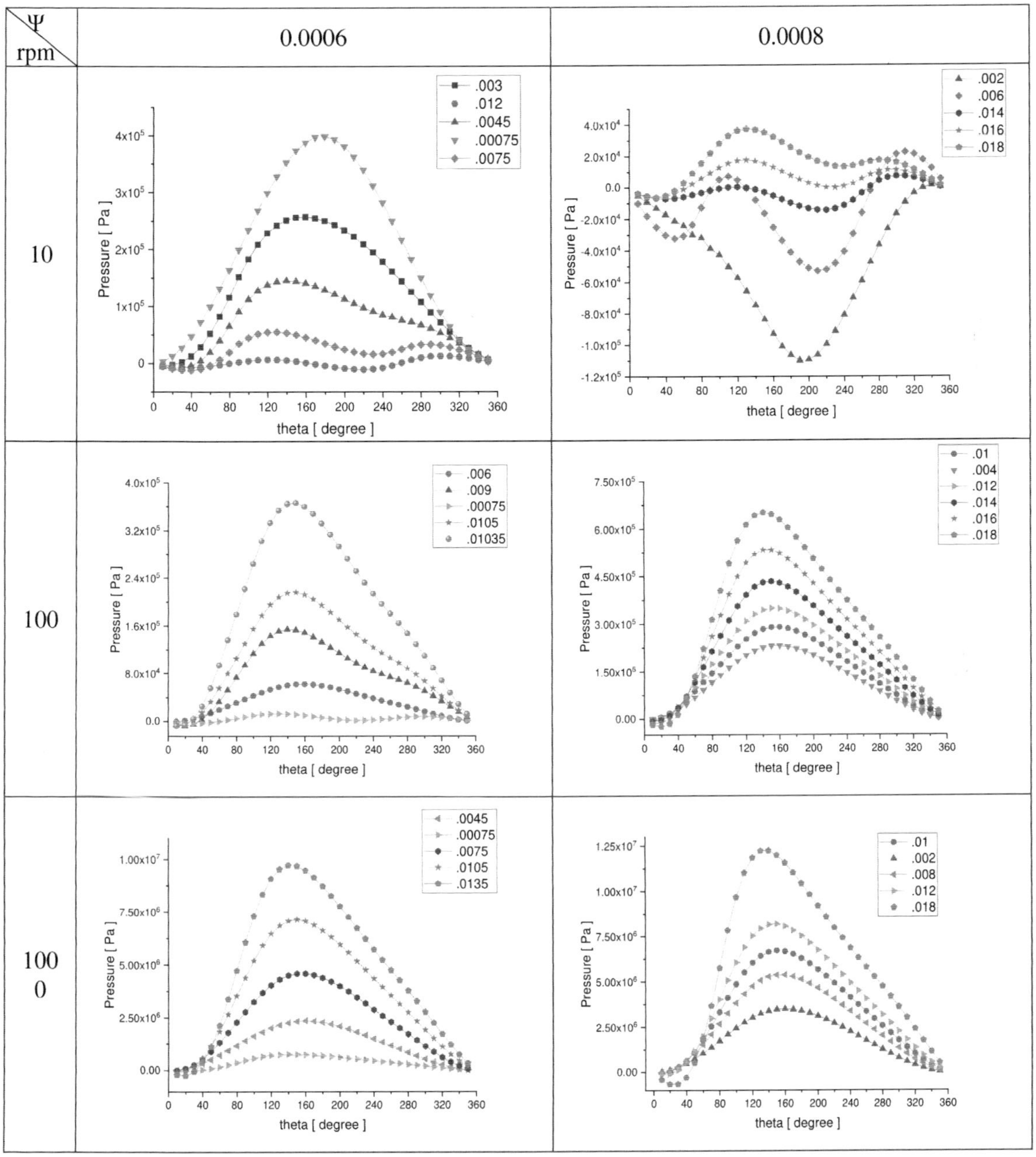

Figure 5: Pressure vs. theta (L/D = 0.5, n = 1)

Figure 6 presents the pressure profile for the bearing at n=1.1. The maximum pressure attained was 2.2 * 10^7 Pa at max eccentricity. It was also seen that the negative pressure region does not exist for 0.0006 (Ψ) at 10, 100 & 1000 rpm. However, the pressure profile was too disturbing with negative pressure for 0.0008 (Ψ) at 10 pm and then stabilised. The pressure profile seems reasonable as the rpm increases, with the negative pressure range lying between 40^0 - 50^0.

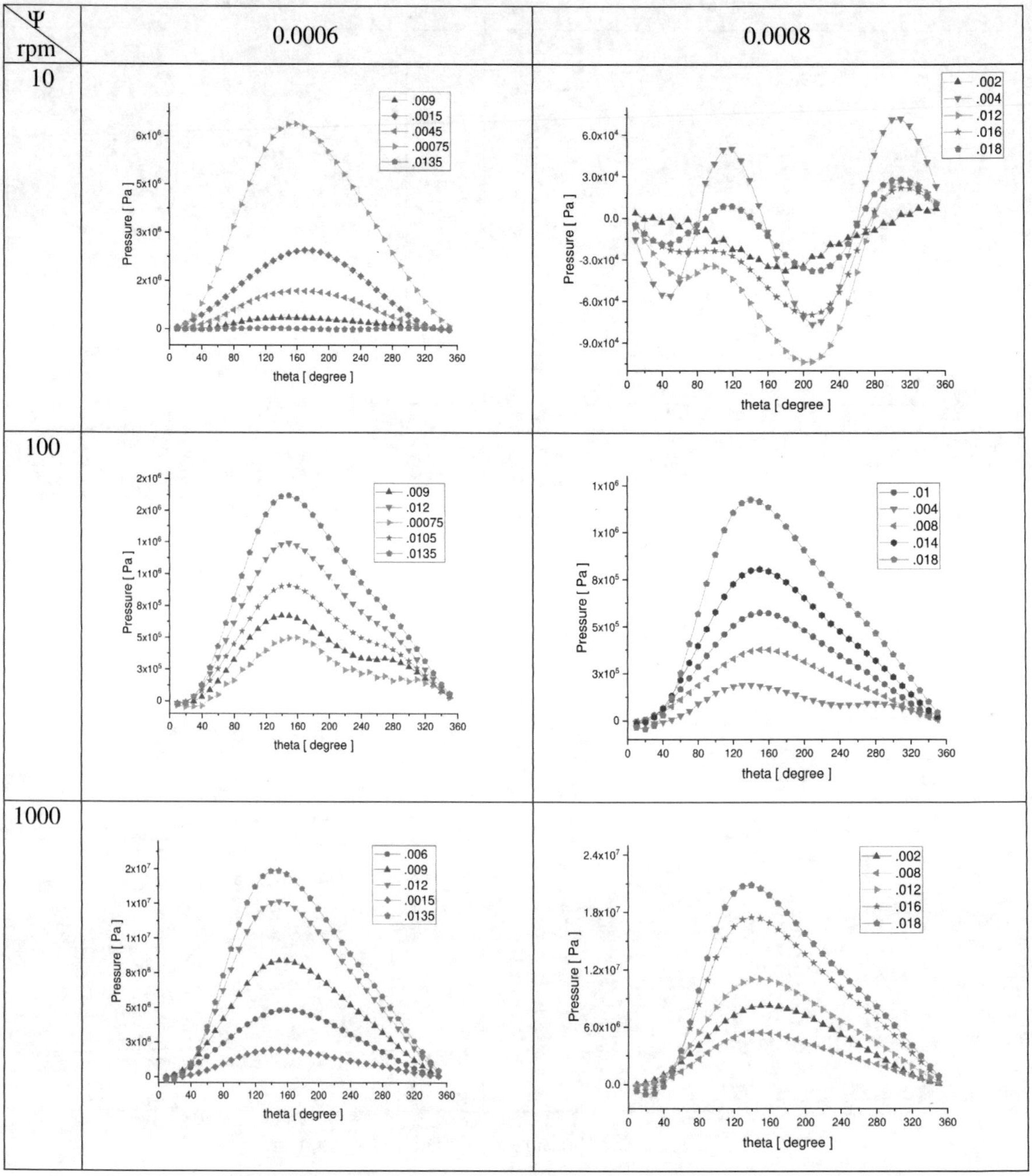

Figure 6: Pressure vs. theta (L/D = 0.5, n = 1.1)

IOP Conf. Series: Materials Science and Engineering 1228 (2022) 012004 doi:10.1088/1757-899X/1228/1/012004

As observed from Figure 3 - 6, the pressure rise increases with an increase in Ψ and rpm. On the similar lines an increase in pressure was observed for n=1.2 as seen in figure 7. The pressure increases and reaches to $3.5*10^7$ Pa. Also, the occurrence of negative pressure with maximum eccentricity (e) at 1000 rpm decreases as compared with earlier simulation results. This can be seen from the comparison of all pressure vs. theta plots for all power-law indexes given in Table 3.

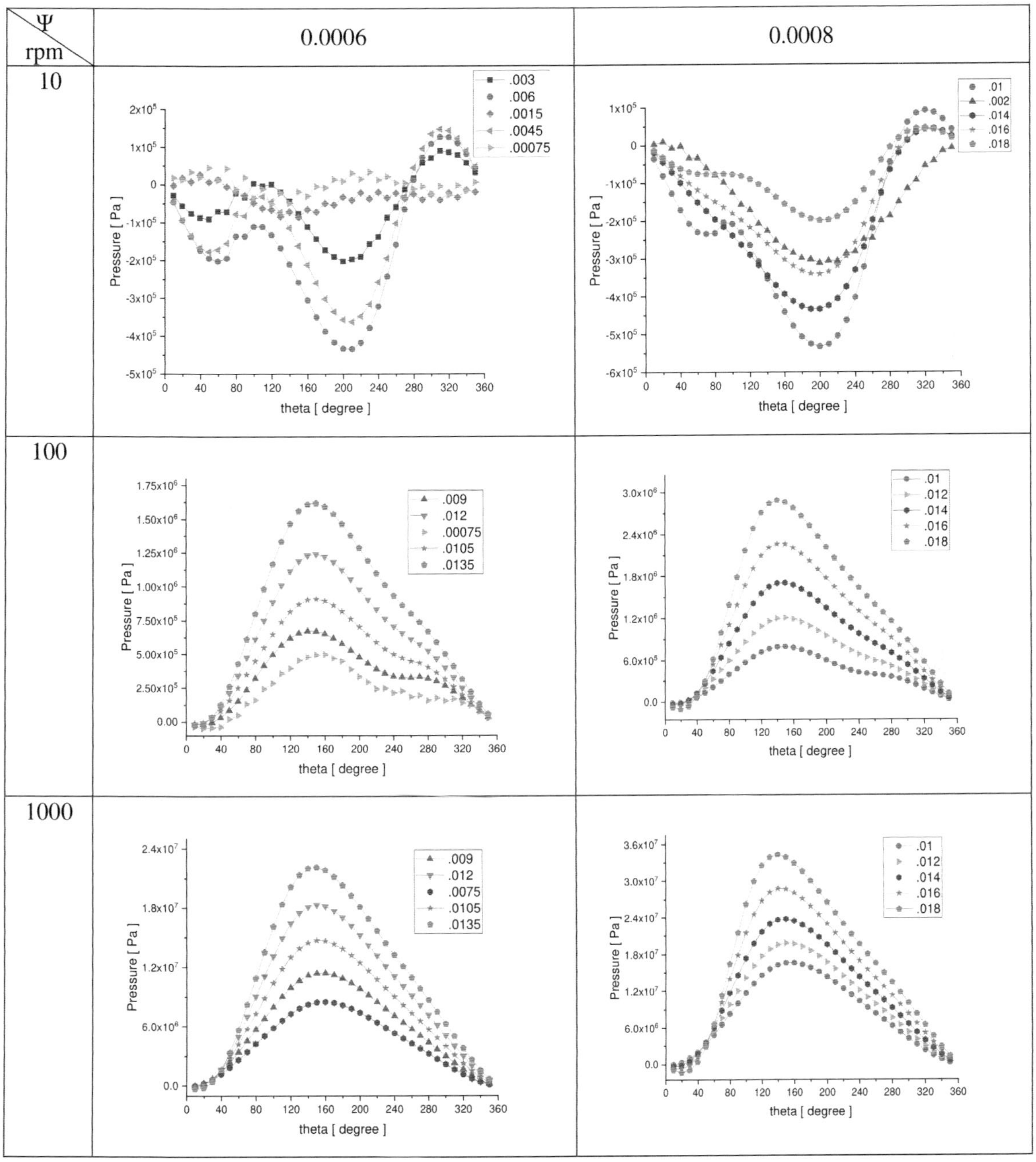

Figure 7: Pressure vs. theta (L/D = 0.5, n = 1.2)

IOP Conf. Series: Materials Science and Engineering 1228 (2022) 012004 doi:10.1088/1757-899X/1228/1/012004

Table 3: Max pressure and cavitation at different rpm and Ψ

Sr. No.	Power Law Index (n)	Max Pressure (Pa)	θ for Cavitation (Ψ =0.0006) in degree	θ for Cavitation (Ψ =0.0008) in degree
1.	0.8	$4.2*10^6$	45	51
2.	0.9	$8*10^6$	40	46
3.	1	$1.2*10^7$	32	42
4.	1.1	$2.2*10^7$	25	40
5.	1.2	$3.5*10^7$	20	38

It is noticed from Table 3 as n and rpm rises, the pressure increases and reaches to $3.5*10^7$ Pa. Furthermore, with increase in speed and relative clearance, the occurrence of cavitation decreases for both cases (0.0006 & 0.0008). The maximum pressure attained is $3.5*10^7$ Pa for n = 1.2, so the load-carrying capacity will also be higher at higher n and L/D ratio values. Also, the cavitation region occurs at a lower value of theta, reducing the negative pressure region compared to the shear thinning fluids, which is higher.

Figure 8 presents the variation in pressure distribution for the L/D = 1 and n = 0.9 for different rpm. It was observed that with an increase in L/D ratio at lower speeds, the pressure profile is not stable. However, as the relative clearance increases with e, a more stable pressure profile can be achieved. Maximum value of theta for occurrence of cavitation was seen at 35^0 (Ψ = 0.0006). From figure 8, it is concluded that the maximum pressure achieved was $8 * 10^6$ for L/D = 1. Hence, the LCC was observed to increase with the increase in L/D ratio for the same rpm as pressure increases.

Ψ / rpm	0.0006	0.0008
10	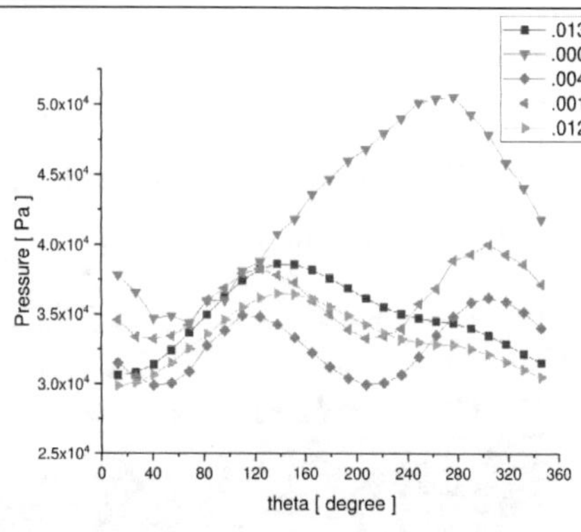	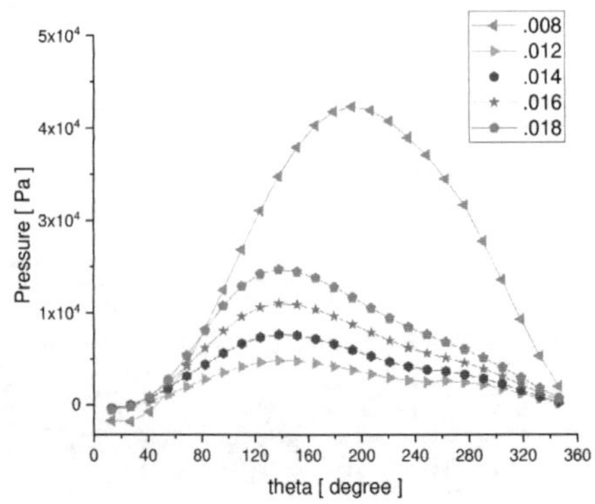

100	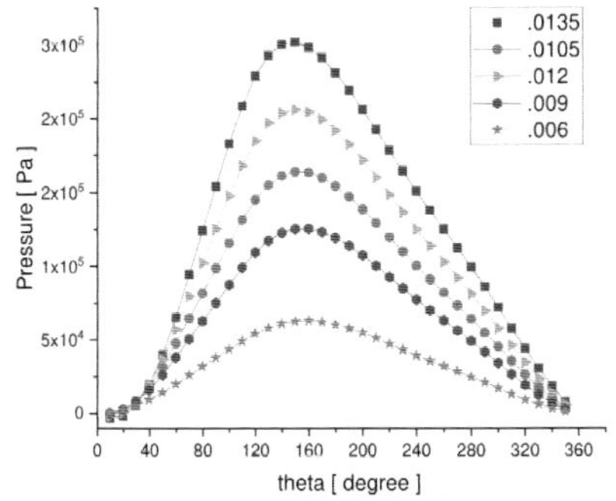	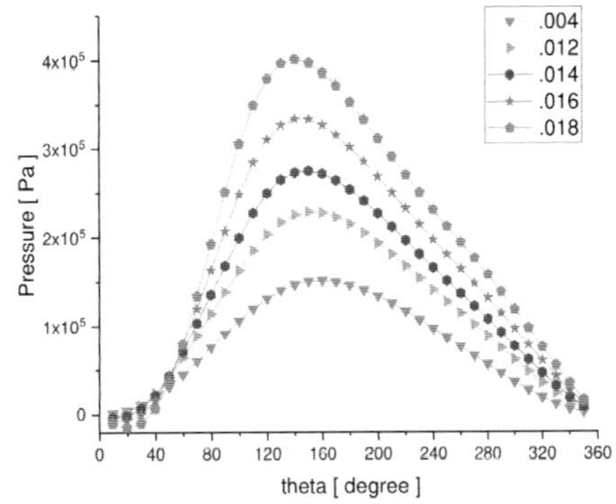
1000	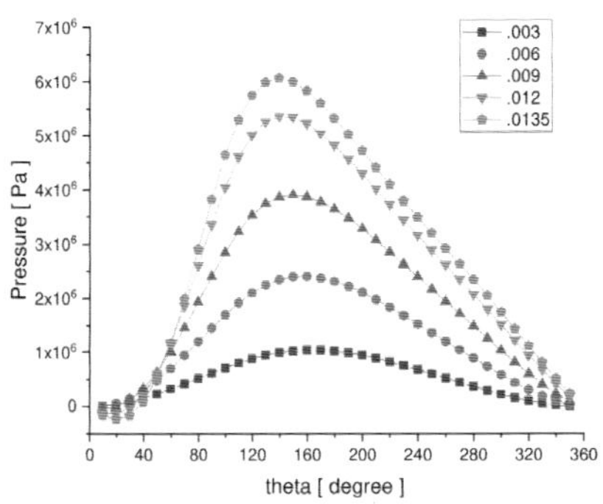	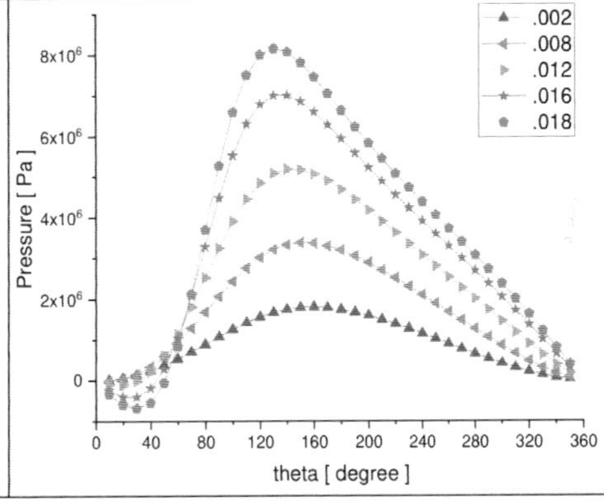

Figure 8: Pressure vs. theta (L/D = 1, n = 0.9)

Similarly from figure 9 it is observed that at the start, the pressure rises is almost equal for Ψ = 0.0006 and gets better for Ψ = 0.0008 at 10 pm. With the increase in rpm and e, the pressure increases and reaches 8.1*10^6 Pa. The cavitation region lies till 58^0 which is maximum in all runs compared to lower values of n and L/D ratio.

Ψ / rpm	0.0006	0.0008
10	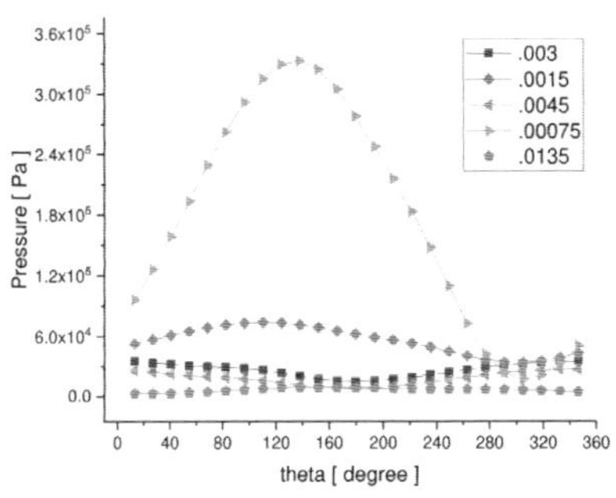	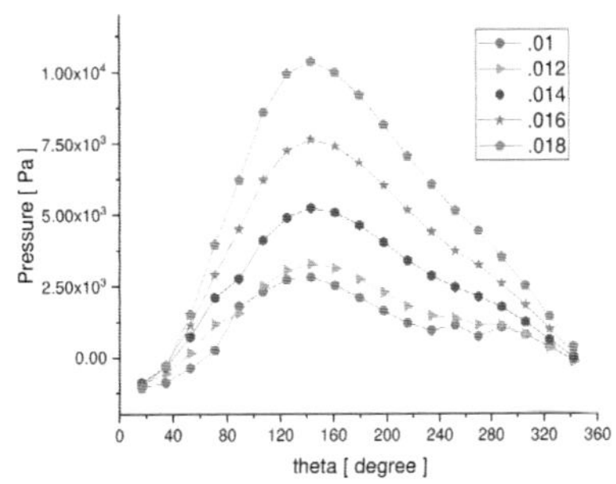

100

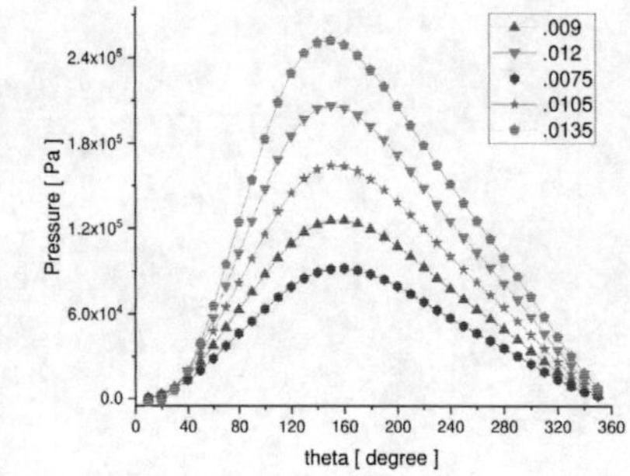

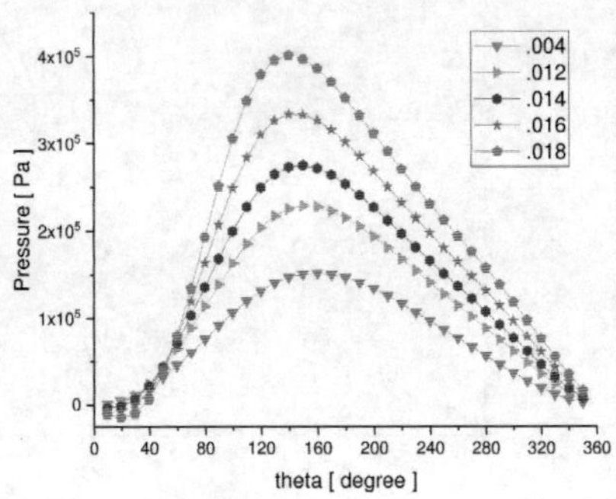

1000

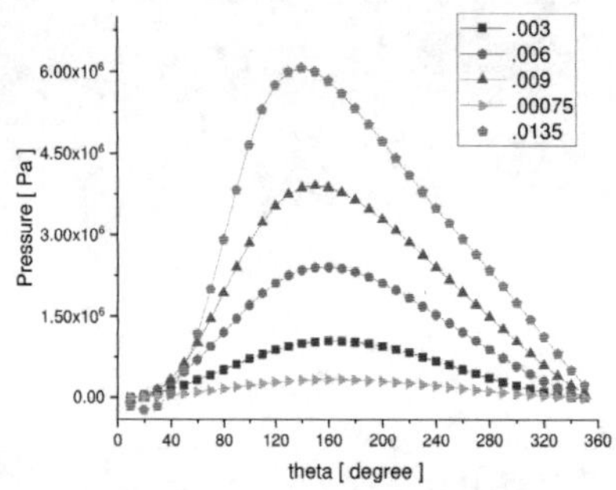

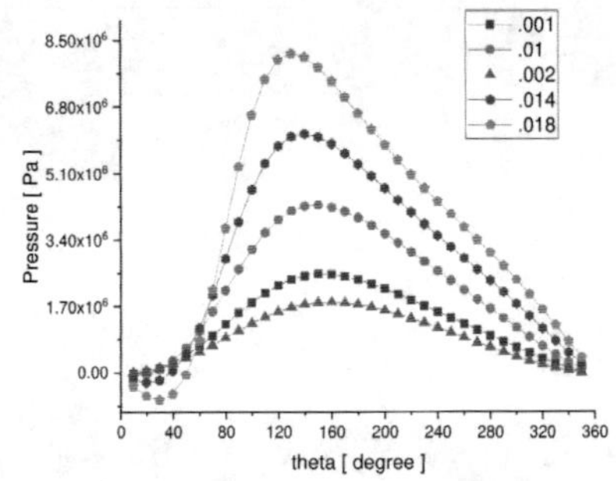

Figure 9: Pressure vs. theta (L/D = 2, n = 0.9)

The present study reveals that the pressure constantly rises with increase in rpm and n, as shown in figure 10. For rpm, the minimum pressure is achieved for pseudoplastic fluid (n<1) and maximum pressure ($3.25*10^7$ Pa) for dilatant fluid (n>1). As the pressure increases with rpm and n, the load-carrying capacity will also rise for HDB. A maximum pressure region was seen when dilatant fluid is used during the simulations. For N= 10 rpm with L/D = 0.5,1 & 2, the initial pressure rises $0.8*10^4$ Pa, but with reference to N = 100 & 1000 rpm, the rise in initial pressure is $1.5*10^5$ Pa, so the load-bearing capacity for this will be more. Also, the pressure variation for n = 0.8 to 1.2 is almost similar with few deviations at 10 rpm.

0.5

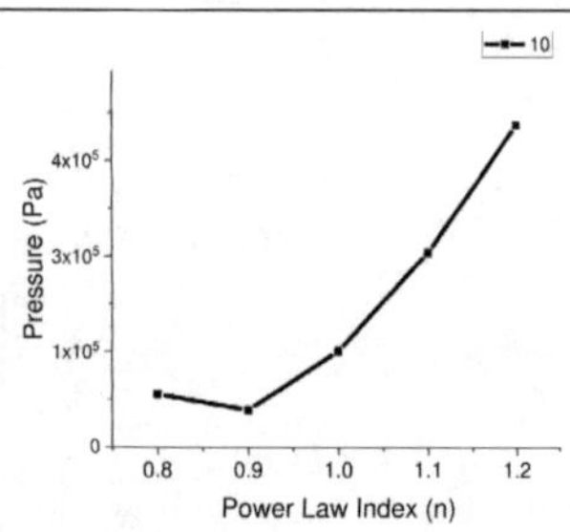

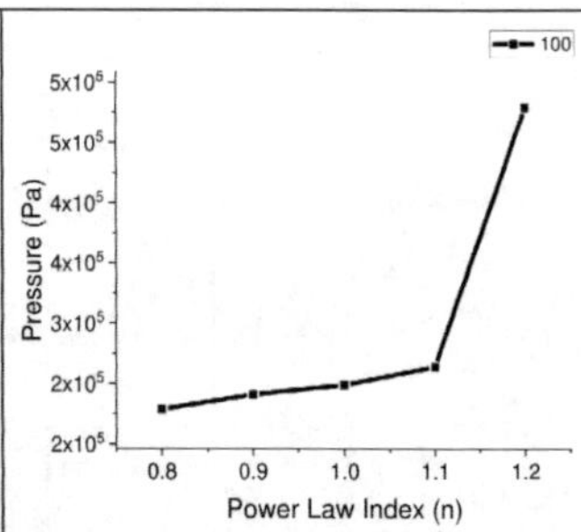

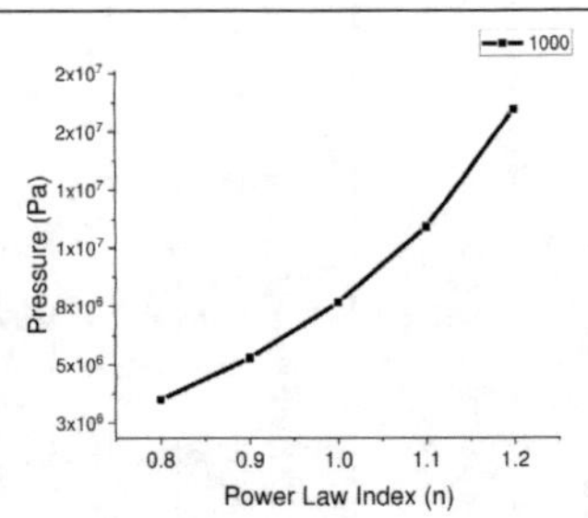

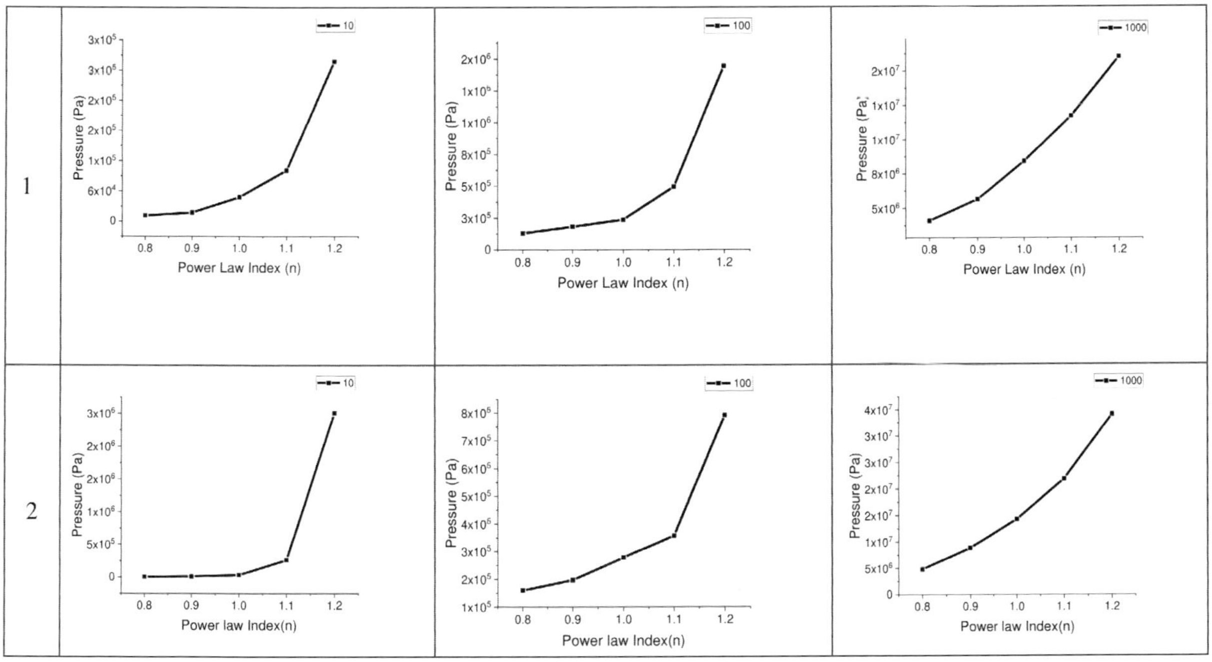

Figure 10: Pressure vs. Power-law index (n)

It is percieved from figure 11 that with increase in the n, the LCC increases as speed increases. The maximum LCC is $6*10^5$ N for L/D =2 at 1000 rpm. There is a slight difference in the plot for L/D =0.5, 1 & 2 at 10 and 100 rpm. However, when HDB is running at 1000 rpm, the run was smooth as n increases for all three cases.

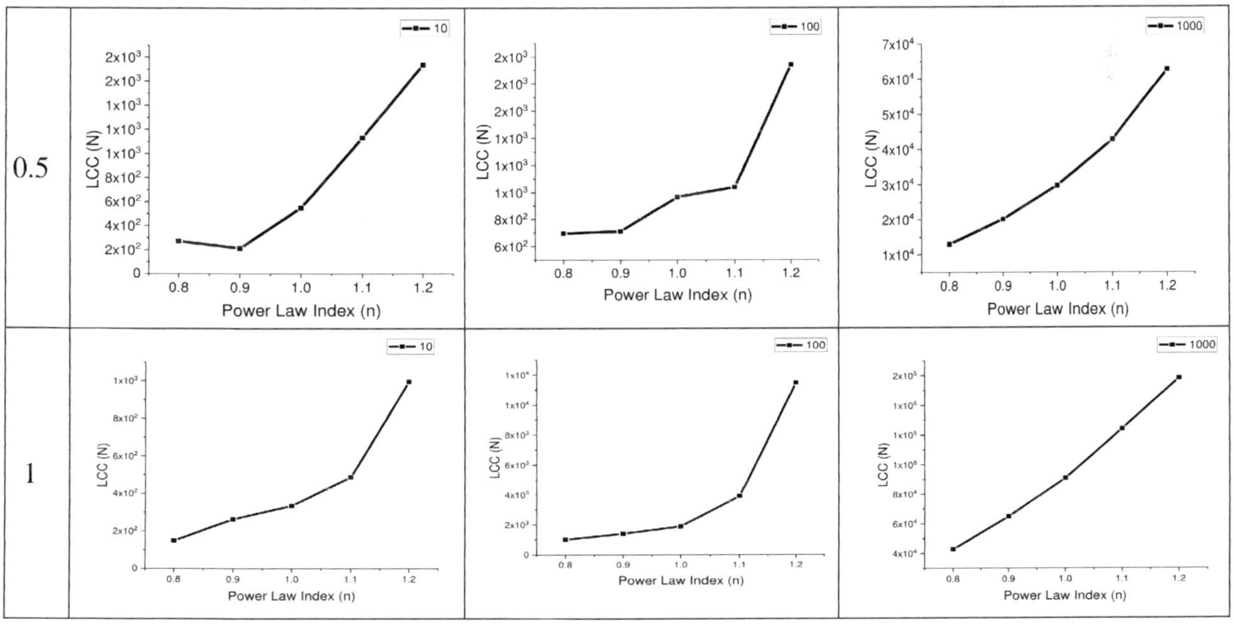

2

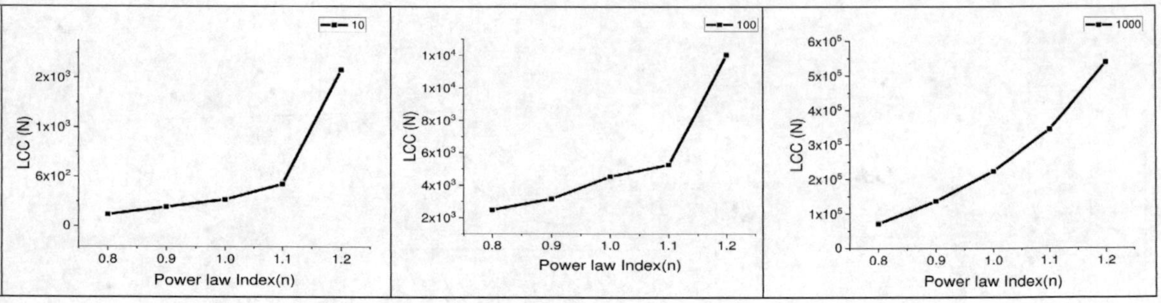

Figure 11: Load-carrying capacity vs. power-law index

4. Conclusions

The simulation was performed to investigate the fluid film pressure distribution for an HDB through Ansys CFX 17.2. This investigation was executed to find the influence of the Power-law index (n) on pressure distribution and LCC. Simulation outcome show the relationship between the power-law index for Newtonian and non-Newtonian fluid used in HDB. The following conclusion was made from the results:

1. With increase in power-law index (n) pressure also increases
2. The cavitation region decreases with an increase in rpm, n and eccentricity
3. The pressure profile is more disturbing at a lower value of n and improves as n increases with relative clearance and rpm
4. With maximum eccentricity, negative pressure was seen at the start and then diminishes with positive pressure. It was also observed that the occurrence of cavitation at theta decreases as rpm and n increases for different Ψ and L/D ratios
5. The LCC also increases with the n and rpm
6. The performance of HDB at 1000 rpm was better than 10 and 100 rpm for both L/D =0.5, 1 and 2

5. References

[1] P. Sinha and C. Singh, “Lubrication of a Cylinder on a Plane With a Non-Newtonian Fluid Considering Cavitation.,” *Am. Soc. Mech. Eng.*, vol. 104, no. 81-LUB-2, 1981.
[2] W. E. S. and E. N. L. R. B. Bird, *Transport Phenomena*. 1960.
[3] P. A. Longwell, *Mechanics of Fluid Flow, McGraw-Hill, New York, NY*. 1966.
[4] Z. S. Safar and G. S. A. Shawki, “Performance of thrust bearings operating with non-Newtonian lubricating films,” *Tribol. Int.*, vol. 12, no. 1, pp. 31–33, Feb. 1979.
[5] C. W. Ng and E. Saibel, “Nonlinear viscosity effects in slider bearing lubrication,” *J. Fluids Eng. Trans. ASME*, vol. 84, no. 1, pp. 192–195, Mar. 1962.
[6] R. I. Tanner, “Non-newtonian lubrication theory and its application to the short journal bearing,” *Amt. J. Appl. Sci.*, no. 129, 1963.
[7] C. S. Prawal Sinha, J. B. Shukla, K. R. Prasad, “Non-Newtonian power law fluid lubrication of lightly loaded cylinder wiht normal and rolling motion.,” vol. 89, pp. 313–322, 1983.
[8] J. B. Shukla, “Load capacity and time relation for squeeze films in conical bearings,” *Wear*, vol. 7, no. 4, pp. 368–371, Jul. 1964.
[9] A. F. Elkouh, N. J. Nigro, and A. Glowacz, “A Generalized squeezing flow of a power-law fluid between two plane annuli,” *J. Tribol.*, vol. 108, no. 1, pp. 80–85, 1986.
[10] D. Prasad, P. Singh, and P. Sinha, “Thermal and inertia effects in hydrodynamic lubrication of rollers by a power law fluid considering cavitation,” *J. Tribol.*, vol. 115, no. 2, pp. 319–326, 1993.
[11] M. W. Johnson and S. Mangkoesoebroto, “Analysis of lubrication theory for the power law fluid,” *J. Tribol.*, vol. 115, no. 1, pp. 71–77, 1993.
[12] R. H. Buckholz, “Effects of power—law, non-newtonian lubricants on load capacity and

friction for plane slider bearings," *J. Tribol.*, vol. 108, no. 1, pp. 86–91, 1986.
[13] R. A. Burton, "An analytical investigation of visco-elastic effects in the lubrication of rolling contact," *ASLE Trans.*, vol. 3, no. 1, pp. 1–10, 1960.
[14] Z. S. Safar and G. S. A. Shawki, "Performance of thrust bearing operating with non- newtonian lubricating fluids," *Tribal. Ink*, vol. 12, no. 31, 1979.
[15] A. Nessil, S. Larbi, H. Belhaneche, and M. Malki, "Journal bearing lubricaiton aspect analysis using non newtionian fluids," vol. 2013, 2013.
[16] D. C. Sun, D. E. Brewe, and P. B. Abel, "Simultaneous pressure measurement and high-speed photography study of cavitation in a dynamically loaded journal bearing," *J. Tribol.*, vol. 115, no. 1, pp. 88–95, 1993.
[17] D. Siano, E. Frosina, and A. Senatore, "Diagnostic Process by Using Vibrational Sensors for Monitoring Cavitation Phenomena in a Getoror Pump Used for Automotive Applications," *Energy Procedia*, vol. 126, pp. 1115–1122, 2017.
[18] P. A. S. F. Da Silva, L. D. Shinomiya, T. F. De Oliveira, J. R. P. Vaz, A. L. A. Mesquita, and A. C. P. B. Junior, "Design of Hydrokinetic Turbine Blades Considering Cavitation," *Energy Procedia*, vol. 75, pp. 277–282, 2015.
[19] E. Van Emden, C. H. Venner, and G. E. Morales-Espejel, "Aspects of flow and cavitation around an EHL contact," *Tribol. Int.*, vol. 95, pp. 435–448, 2016.
[20] M. Hlaváček, "A central film thickness formula for elastohydrodynamic lubrication of cylinders with soft incompressible coatings and a non-Newtonian piecewise power-law lubricant in steady rolling motion," *Wear*, vol. 205, no. 1–2, pp. 20–27, 1997.
[21] N. D. S. A. Santos, V. R. Roso, and M. T. C. Faria, "Review of engine journal bearing tribology in start-stop applications," *Eng. Fail. Anal.*, vol. 108, p. 104344, 2020.
[22] X. L. Wang, K. Q. Zhu, and S. Z. Wen, "Thermohydrodynamic analysis of journal bearings lubricated with couple stress fluids," *Tribol. Int.*, vol. 34, no. 5, pp. 335–343, 2001.
[23] T. P. Gundarneeya and D. P. Vakharia, "Performance analysis of journal bearing operating on nanolubricants with TiO2, CuO and Al2O3 nanoparticles as lubricant additives," *Mater. Today Proc.*, Mar. 2021.
[24] P. Roy, S. London, P. R. Soc, P. R. Soc, P. Trans, and R. Soc, "The flow of cream through narrow glass tubes," no. 341, 1953.
[25] D. Zhang *et al.*, "Electrospun polyacrylonitrile nanocomposite fibers reinforced with Fe3O4 nanoparticles: Fabrication and property analysis," *Polymer (Guildf).*, vol. 50, no. 17, pp. 4189–4198, 2009.

Face Recognition Method Based on Residual Convolution Neural Network

Arshi Husain[1] and Virendra P. Vishvakarma[2]
USICT, Guru Gobind Singh Indraprastha University
arshihusaincdac@gmail.com, virendravishwa@rediffmail.com, vpv@ipu.ac.in

Abstract. With the advancement of information technology and societal growth, social security has become more important than ever. Face recognition, as compared to other traditional recognition methods like fingerprint recognition, palm recognition, etc, has the benefit that it is contact less, and now it is becoming one among the most prominent technologies in development. Although there are numerous recognition systems that use DNNs in the field of facial expression recognition, their accuracy and practicality are still insufficient for real-world applications. A facial recognition approach based on Resnet 152 v2 has been proposed in this work. In this paper, a residual learning approach is presented to make the training of networks that are far deeper than previously employed networks easier. The proposed method, employs the AT&T face dataset, and supposing that normalization and segmentation are complete, we concentrate on the subtask of person verification and recognition, demonstrating performance using a testing database comprising illumination, pose, expression and occlusion variations. SoftMax is the activation function that has been used, which adjusts the output sum up to one allowing it to be understood as probabilities. Then, the model would generate a judgment depending on which option has a strong likelihood. This system employs Adam as an optimizer to control the learning rate through training and categorical cross entropy as its loss function. The proposed approach has a 97 percent face recognition accuracy on AT&T dataset, showing its efficacy after a significant number of analyses and experimental verification.

Keywords: Deep convolutional neural network, Face recognition, ResNet.

1. Introduction

In many computer visual identification applications, the CNN model has outperformed standard machine learning algorithms, thanks to the advancement of deep learning.
Face recognition algorithms are now being improved in three areas: face pre- processing (comprising face alignment and detection), extraction of features (mostly developing ANN structures), and classification of features.
Face confirmation and face verification are generally two subtasks in face recognition. The former categorizes faces into distinct identities, whilst the latter assesses if 2 face picture pairs are associated with the same identity.
Automatic face recognition is a multi-step process which includes face detection and localization in a messed-up environment, normalization, recognition, and verification. Some of the subtasks may be difficult to complete depending on the nature of the application, such as the size of the training and testing databases, background variability, occlusion, noise, and speed requirements. Supposing that normalization and segmentation are complete, we concentrate on the subtask of person verification and recognition, demonstrating performance using a testing database of about 400 pictures.
In the realm of image identification, CNN is the most used machine learning algorithm. Visual Geometry Group (VGG-15) network contains nineteen layers whereas LeNet comprises five layers of network. The 100-layer barrier was not broken until the development of networks like ResNet in 2016. By constructing a short connecting channel from the front layer to the rear layer, the signal is sent immediately from the one layer to the next via ResNet.
Although there are numerous recognition systems that use Neural Networks in the field of facial expression recognition, their accuracy and practicality are still insufficient for real-world applications. A facial recognition approach based on Resnet 152 v2 has been proposed in this work which has better accuracy than the existing ones. Since we have used the deep Neural networks in our system, hence the features need not to be extracted manually.

Content from this work may be used under the terms of the Creative Commons Attribution 3.0 licence. Any further distribution of this work must maintain attribution to the author(s) and the title of the work, journal citation and DOI.
Published under licence by IOP Publishing Ltd

2. Related Work

In the past, various classification approaches were used to address the problem of facial recognition. Trunk-Branch Ensemble CNN model was given by Changxing Dingand Dacheng Tao.Onetrunk network and numerous branch networks are contained by Trunk-Branch Ensemble CNN. The trunk network was trained so that it can learn face representations for holistic face pictures, whereas the branch networks were programmed in order to make it learned face representations for image patches sliced from 1 facial component. GoogLeNet has been used to implement the trunk network. They also present a technique to generate video like training picture data in order to achieve blur-resistant face identification. They tested their findings using PaSC and YouTube Faces, two publicly available large-scale video face datasets. On the YouTube Faces Database, their mean verification accuracy using the TBE-CNN technique is 94.96 percent, and on the PaSC dataset, their verification rate using the TBE-CNN method is 95.83 and 94.80 on the control set and hand held set, respectively[1]. Vallimeena P, Uma Gopalakrishnan, Bhavana B Nair, and Sethuraman N Rao used the FERET Database to perform ethnicity-based face categorization on 447 samples (357 for training, 90 for testing) Extraction of skin color and computation of the normalised forehead area utilising the Sobel Edge Detection technique, with a 94 percent experimental accuracy [2].

Karthikeyan Shanmugasundaram, Sathees Kumar Ramasamy and Sharma S deployed a FAREC - CNN Based Efficient Facial Recognition Technique using Dlib. Face Recognition Grand Challenge dataset has been used in their paper [3]. Face detection is the first method employed in this study to detect human frontal faces from digital pictures or videos. For this purpose, computer vision technology was used to detect frontal faces using facial landmark identification of the nose and upper lips. Following that, the Dlib is used to align the frontal faces. Following that,face cropping is used to cut off the face with varying resolution depending on the distance between the face and the camera. Finally, CNN is used to extract features from the face images. So finally, FAREC takes 20 epochs and produces 96% accuracy for FRGC [3]. Yuxiang Zhou and Hongjun Ni developed a Face and Gender Recognition System Based on Convolutional Neural networks. The datasets used were Labeled Faces in the Wild (LFW), YouTube Face (YTF) and VGGFace2 and the highest achieved accuracy was 93.22% [4].

Ying Wen, Pengfei Shi developed a face recognition approach using PCA, 2DPCA (2D)2, PCA, IPCA as the feature extraction techniques on ORL and Yale datasets. They used Nearest Neighbour classifiers for the purpose of classification and the highest recognition rate was 90%. [5] Divya Meena, Ravi Sharan proposed a method for face detection and face recognition. In their work, viola Jones algorithm was used to detect face and principal component analysis for face recognition. The highest accuracy achieved was 90% [6].

A method proposed by W. Nan, Z. Zhigang, M. et al used ResNet and enhanced edge cosine loss transform for the purpose of face recognition and highest recognition accuracy achieved was 72.602% and verification accuracy was 85.420% [7]. P. Shamna, C. Tripti et al investigated the Difference Component Analysis (DCA) Approach on AT & T dataset for face recognition and 73.3% correct recognition rate was acquired [8].

W. Wang and W. F. Wang proposed a Grayscale Face Recognition approach, in which feature extraction was done by deriving the whole expression of gray conversion; the face classifier was trained by back propagation learning algorithm and achieved 93.33% recognition rate [9]. H. Hatimi, M. Fakir et al developed an approach for face Recognition Using Fuzzy and Multi-agent systems from Video Sequences. Face detection was done using texture color and geometrical face, multi-agent system and fuzzy approach are used in the recognition process and achieved 95% recognition rate [10].

C. Lu and X. Tang. investigated human-level face verification performance on LFW dataset using Gaussian Face model and achieved 93.73% accuracy [11]. D. Wang, H. Yu, D. Wang and G. Li explored different activation functions in CNN for Face Recognition System Based and the highest achieved accuracy was 88.41% [12].

S. Ergin and M. Bilginer Gulmezoglu developed a Face Recognition method using Face Partitions with Common Vector Approach and achieved 93.3% recognition rate [13]. Y. Wen and P. Shi proposed a method for the purpose of Face Recognition using PCA and achieved 90% accuracy [14] S. M. S. Hossain, A. Yousuf and M. S. Sadi, proposed Eigen vector and Covariance matrix-based approach for face recognition and achieved 96.25% accuracy [15].

3. Dataset and Proposed Methodology

3.1. Image Dataset

This method employs the AT&T face dataset, formerly referred as the ORL Database Of Faces. It is a collection of 400 face photos, with 10 different images of each of 40 different subjects. The pictures for people had been taken at varying conditions like different facial emotions (e.g smiling / not smiling etc), varied lighting and different facial features (with glasses / without glasses) were being used.

The people were snapped in a frontal, upright position against a background that was black and uniform. Pictures in the dataset were saved in **Portable Gray Map (pgm)** format before being converted to jpg. Each picture is 92x112 pixels in size and has 256 grey levels per pixel. A preview image of the Database of Faces is shown in Fig.1.

Fig.1. Dataset Images

3.2. Proposed Methodology

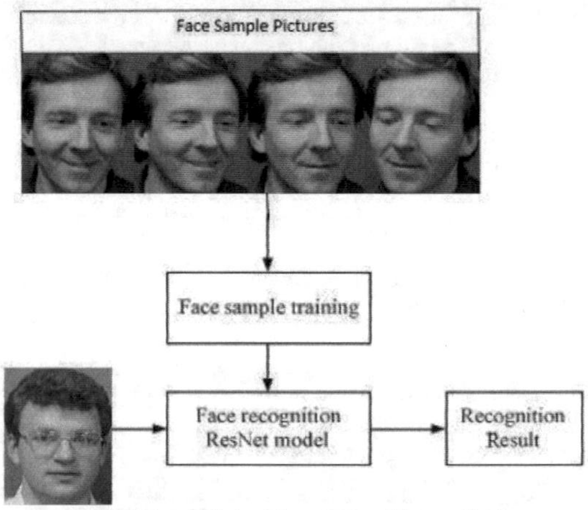

Fig.2. Resnet based Face recognition method

The proposed system mentioned below uses residual networks for the face recognition problem. Figure 2 depicts the concept of a *ResNet*-based face recognition algorithm. The *ResNet* based face recognition method is used in this work.

Step1: The grayscale images of dataset which are in pgm format are loaded into the system and then they are converted into jpg format.

Step 2: We split the original dataset which contain 400 images in the ratio of 70:30, i.e, into two sub datasets known as training dataset and testing dataset, as a result, 280 pictures are used to train the system and 120 pictures are used to test it.

Step 3: We import the required libraries, add a preprocessing layer to the front of *RESTNET152V2*.

Step 4: Model Compilation has been done using the three parameters that are Optimizer, Loss function and metrics

Step 5: The model has been trained by utilising the 'fit()' function with the following three parameters:

(A) number of epochs used (B) validation data (C) training data target data.

RESIDUAL NETWORK MODEL

Lately, Deep convolutional networks have led to tremendous breakthroughs for image generation, classification and so on. With development of Residual networks, the difficulty of training incredibly deep networks has been alleviated and these *Resnets* are made up from Residual Blocks. The residual learning module allows to train layers with hundreds or even thousands of them and still get great results. The *resNet* learning fundamental idea is to preserve a portion of the original input data throughout CNN unit training in order to prevent classification accuracy saturation generated by numerous convolutional layers. Simultaneously, it is not necessary for the residual module to memorize the entire output; instead, it only has to learn the differences between the input and output, which optimizes the learning objectives and complexity.

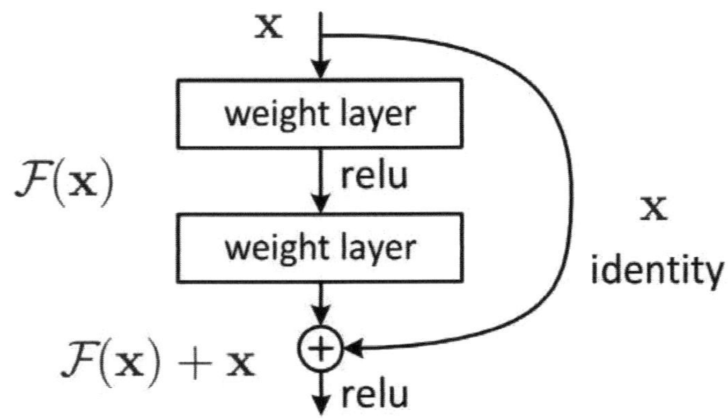

Fig.3. Residual Learning: A building block

There is a direct link in *ResNet* that skips certain layers (which may vary depending on the model) in between. The core of residual blocks is a link known as the 'skip connection.' That adds previous layer outputs to stacked layer outputs. By including skip connections to our network, we are permitting the network to leave out training for the layers that are not relevant and do not contribute value to overall accuracy, rather than using the number of layers as an essential hyperparameter to tune. Skip connections, in a sense, make our neural networks dynamic, so that it can tune the number of layers appropriately during training

The residual block shown in Fig.3. is mathematically represented as F(x)

$$y = F(x, \{Wi\}) + x \quad (1)$$

Here the input and output vectors of the layers in concern are x and y. The residual mapping to be learned is represented by the function F(x, Wi).
The residual block consists of various weight layers which are represented as Wi.The number of weight layers should be greater than one. The two layers contained in the residual block can be illustrated using the following equation:

$$F\,(x, \{Wi\}) = W_2\sigma(W_1x) \quad (2)$$

where σ represents the Rectified Linear Unit activation function. so

$$H(x) = \sigma\,(y) \quad (3)$$

IOP Conf. Series: Materials Science and Engineering 1228 (2022) 012005 doi:10.1088/1757-899X/1228/1/012005

In this research, our system utilises Resnet152V2, which employs batch normalisation (BN) processing to speed up the training pace of the entire training network in the proposed module which is based on residual learning. The *ResNet V2* contains the non-linearity which refers to an identity mapping. This mapping can be considered as the

output of an addition of the residual mapping and the identity that is to be passed on for the processing in the next block. In ResNet V1, the result of the addition operation is transmitted to the next block as the input via *ReLU activation*.

4. Experimental Result and Discussion

The TASK was performed using a 64-bit operating system laptop. The machine had 16 GB of RAM installed. The CPU utilized was an Intel i5-9300H with a clock speed of 2.5 GHz. The operations were performed on anaconda software running on python 3.8 and jupyter notebook version2.1 with libraries used for image classification like *tensorflow*. Initially, the grayscale images in pgm format are loaded into the system and then they are converted into jpg format. We have split the original dataset containing 400 images into two sub datasets known as training dataset and testing dataset in a ratio of 70:30, as a result, 280 pictures are used to train the system and 120 pictures are used to test it.

We import the required libraries, add a preprocessing layer to the front of *RESTNET152V2*. SoftMax is the activation function that has been used, which adjusts the output sum up to one allowing it to be understood as probabilities.

Maximum Entropy (MaxEnt) Classifier is another term for *Softmax Regression* and the formula for *Softmax function* is as follows:

$$\sigma(\vec{z})_i = \frac{e^{z_i}}{\sum_{j=1}^{K} e^{z_j}} \tag{4}$$

where,

$\vec{z}$	The input vector to the softmax function, made up of (z0, ...zK)
z_i	All the Z_i values are the elements of the input vector to the softmax function, and they can take any real value, positive, zero or negative.
e^{z_i}	The standard exponential function is applied to each element of the input vector, this gives a positive value above 0, which will be very small if the input was negative, and very large if the input was large.
$\sum_{j=1}^{K} e^{z_j}$	It is the normalization term. It ensures that all the output values of the function will sum to 1 and each be in the range (0,1), thus constituting a valid probability distribution.
K	The number of classes in the multi-class classifier.

Table.1. Symbols used in the Softmax Formula

Then, the model would generate a judgment depending on which option has a strong likelihood.

4.1. Model Compilation

Loss, Optimizer and metrics are the parameters using which the model was then compiled. The system utilizes Adam Optimizer which governs the rate of learning during the training phase.

The impact of learning rate affects the speed of computation of optimum weights. Although the lower speed increases the computation time but yields better results. For our loss function, this system uses *categorical cross entropy* and a lower score means our model is doing better. The system uses 'accuracy' metric to view accuracy score on validation set when we train the model which will make things much easier to comprehend.

4.2. Training the model

In this system, the 'fit()' function has been utilized with the following parameters to train our model: 1. number of epochs used 2. validation data 3. training data target data. The test set created for us for the AT & T dataset was used for validation data. In this model, 25 epochs the no.of epochs is set 25. After 25 epochs, we have gotten to 97% accuracy on our validation set. Fig.4. and Fig.5. shows the comparison graph between Training Accuracy and Validation Accuracy respectively on AT&T Dataset.

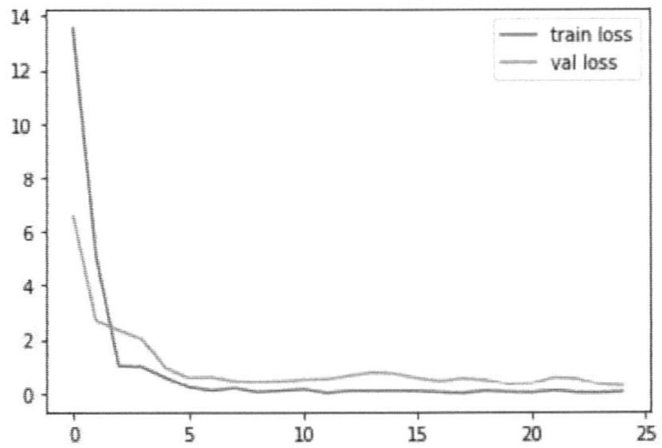

Fig 4. *Comparison of Training & Validation Loss*

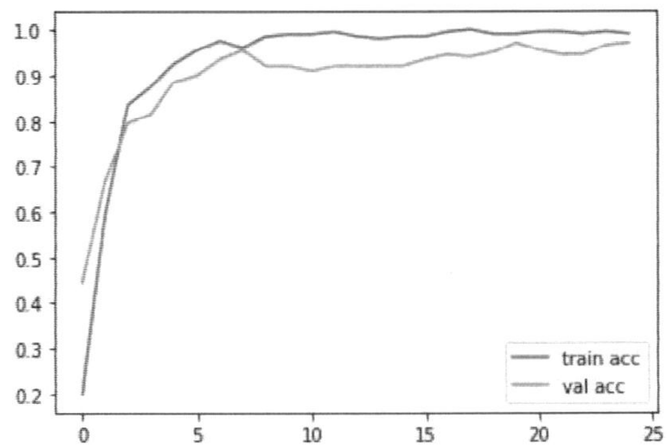

Fig. 5. *Comparison of Training & Validation Accuracy*

5. Conclusion

A facial recognition model using ResNet is proposed in this paper. Resnet 152v2 is the residual network variation that we selected since, despite the additional depth, the 152-layer ResNet has lesser complexity and is more accurate than other networks such as VGG-16/19 networks. The softmax activation function, adam optimizer and categorical cross entropy as a loss function have been employed in our system. We haven't noticed any degradation issue; therefore, we get large accuracy benefits from the greater depth. This paper's facial recognition algorithm has been trained and tested on the AT&T dataset. The findings show our suggested algorithm is superior and has a lot of promise in the open face recognition problem. In our future work, larger size databases will be explored for the investigation of proposed methods.

References

[1] r C. Ding and D. Tao, "Trunk-Branch Ensemble Convolutional Neural Networks for Video-Based Face Recognition," in IEEE Transactions on Pattern Analysis and Machine Intelligence, vol. 40, no. 4, pp. 1002-1014, 1 April 2018

[2] P. Vallimeena, U. Gopalakrishnan, B. B. Nair and S. N. Rao, "CNN Algorithms for Detection of Human Face Attributes – A Survey," 2019 International Conference on Intelligent Computing and Control Systems (ICCS), 2019, pp. 576-581

[3] S. Sharma, K. Shanmugasundaram and S. K. Ramasamy, "FAREC — CNN based efficient face recognition technique using Dlib," 2016 International Conference on Advanced Communication Control and Computing Technologies (ICACCCT), 2016, pp.192-195

[4] Y. Zhou, H. Ni, F. Ren and X. Kang, "Face and Gender Recognition System Based on Convolutional Neural networks," 2019 IEEE International Conference on Mechatronics and Automation (ICMA), 2019, pp. 1091-1095

[5] Y. Wen and P. Shi, "Image PCA: A New Approach for Face Recognition," 2007 IEEE International Conference on Acoustics, Speech and Signal Processing - ICASSP '07, 2007, pp. I-1241-I-1244

[6] D. Meena and R. Sharan, "An approach to face detection and recognition," 2016 International Conference on Recent Advances and Innovations in Engineering (ICRAIE), 2016, pp. 1-6

[7] W. Nan, Z. Zhigang, M. Jingqi, L. Huan, L. Junyi and Z. Zhenyu, "Face Recognition Method Based on Enhanced Edge Cosine Loss Function and Residual Network," 2019 Chinese Control and Decision Conference (CCDC), 2019, pp. 3320-3324

[8] P. Shamna, C. Tripti and P. Augustine, "DCA: An Approach for Face Recognition through Component Analysis," 2013 Third International Conference on Advances in Computing and Communications, 2013, pp. 34-38

[9] W. Wang and W. F. Wang, "A GrayScale Face Recognition Approach," 2008 Second International Symposium on Intelligent Information Technology Application, 2008, pp. 395-398

[10] H. Hatimi, M. Fakir and M. Chabi, "Face Recognition Using a Fuzzy Approach and a Multi-agent System from Video Sequences," 2016 13th International Conference on Computer Graphics, Imaging and Visualization (CGiV), 2016, pp. 442-447

[11] C. Lu and X. Tang. Surpassing human-level face verification performance on LFW with GaussianFace. arXiv Preprint arXiv:1404.3840, 2014.

[12] D. Wang, H. Yu, D. Wang and G. Li, "Face Recognition System Based on CNN," 2020 International Conference on Computer Information and Big Data Applications (CIBDA), 2020, pp. 470-473

[13] S. Ergin and M. Bilginer Gulmezoglu, "Face Recognition Based on Face Partitions Using Common Vector Approach," 2008 3rd International Symposium on Communications, Control and Signal Processing, 2008, pp. 624-628

[14] n S. M. S. Hossain, A. Yousuf and M. S. Sadi, "Towards an efficient face recognition approach," 2015 International Conference on Electrical Engineering and Information Communication Technology (ICEEICT), 2015, pp. 1-5

[15] Bhawna Ahuja, Virendra P Vishwakarma, "Deterministic Multi-kernel based extreme learning machine for pattern classification," Expert System with Applications,2021, pp. 115308.

[16] Sahil Dalal, Virendra P Vishwakarma, "Classification of ECG signals using multi-cumulants based evolutionary hybrid classifier", Scientific Reports, 2021

[17] Virendra P Vishwakarma, Varsha Sisaudia, "Self-adjustive DE and KELM-based image watermarking in DCT domain using fuzzy entropy", International Journal of Embedded Systems, 13 (1), 74-84

[18] Sahil Dalal, Virendra P Vishwakarma, "A Novel Approach of Face Recognition Using Optimized Adaptive Illumination–Normalization and KELM", Arabian Journal for Science & Engineering (Springer Science & Business Media BV), 2020, Vol.45, issue 12

[19] VP Vishwakarma, S Dalal, "A novel non-linear modifier for adaptive illumination normalization for robust face recognition", Multimedia Tools and Applications, 2020, 79 (17), 11503-11529

Smart Harvest Diseases Identification using Image Processing

Kanaklata Roshan[1], Ritam Dutta[2], Bhaskar Roy[1] and Chiranjib Goswami[1]

[1]Department of Electronics & Communication Engineering, Asansol Engineering College, Asansol – 713305, India, West Bengal
[2]Center of Intelligent Systems and Robotics, ITER, Siksha 'O' Anusandhan (Deemed to be University), Bhubaneswar – 751030, India, Odisha

bhaskar.ec11@gmail.com

Abstract. Manual identification of leaf or fruit diseases is nearly impossible when it comes to large scale production, and it is a very time-consuming task that also requires a large amount of manpower. Hence, the existing technology is used where the disease identification is done within minutes. When a code is developed and deployed in any of the surveillance drones, a large scale area can be covered in minutes and with less manpower.Here, in this proposal, there are two methods proposed. One is using the keywords and the other method is where a classifier is trained with the database available and the classifier with the highest accuracy is tested against the test set and a result is obtained. When there is a huge area of land, manually identifying the leaf diseases may require a lot of manpower and also be time-consuming. Hence, deploying this code in any sprayer drone will help in identifying the diseases within minutes. Along with disease identification, the particular longitude and latitude of the disease affected leaves' location will be saved due to the fact that when the land is huge, the farmer need not go in search of the location; instead, a drone will fly to the exact location and the respective fertiliser will be sprayed. Before identification of leaf disease, a photo of the diseased leaves will be captured, due to which, by looking at the picture, the farmer can decide whether medication is necessary or not.

1. Introduction

Manually identifying disease of any fruits or leaves can be done for small scale. For a large scale it may take long time and it is nearly impossible. Hence, by using smart farm technology within less time large area can be covered. This is done with the combination of image processing technique. Here first we acquire the image of diseased leaf and then through image processing we identify the image by which it has been affected. Image processing is a method used to perform image enhancement technique and also get features extracted. It is a type of signal processing technique in which we supply input as an image and output can be extracted from enhanced image or features. However, to get an optimized work flow and to avoid losing time, it is important to process images after the capture, in a post-processing step. Computer vision, remote sensing, feature extraction, face detection, prediction, optical character recognition, finger-print detection, optical sorting, claim fact,

Content from this work may be used under the terms of the Creative Commons Attribution 3.0 licence. Any further distribution of this work must maintain attribution to the author(s) and the title of the work, journal citation and DOI.
Published under licence by IOP Publishing Ltd

microscope imaging, lane departure caution are essential advantages of image processing. Extract the size, scale, or number of objects in a scene. [5]

Common plant diseases are fungal, infectious and bacterial. The processing of images is used to measure the affected region of the disease and to detect the disease. Image segmentation is the process of segmenting the picture in various sections. There are several different ways to perform image segmentation at present, ranging from the basic thresholding approach to sophisticated methods of segmentation of color images. Clustering of K-means and Otsu segmentation approaches are most widely used. The segmentation mechanism is based on various characteristics found in the image. In agriculture, segmentation plays an important role, as having diseases in plants is very normal. If due care is not taken during segmentation, then serious effects on plants are caused and productivity is impaired because of this. A disease called small leaf disease, for example, is a harmful disease found in US pine trees. It is useful to detect plant disease through any automated technique as it eliminates a significant amount of monitoring work in significant crop farms and detects the symptoms of diseases at a very early stage, i.e. when they appear on plant leaves. This paper introduces an image segmentation process algorithm that is used for the identification of diseases and classification of plant leaf diseases. It also contains surveys on various leaf diseases, techniques of classification that can be used to diagnose plant leaf disease. Picture segmentation, an important component of disease detection in plant leaf disease, is performed using the technique of clustering K-means. [1]

The challenging task of manufacturing agricultural products comes as numerous micro-organisms invade the plant, causing bacterial diseases and pests. The signs of the attacks are generally described by the leaves, steam, or fruits checked. The present article addresses the processing of photographs and the methods used in the early detection of plant diseases. This work is aimed at the target of introducing techniques for image processing and classification for leaf disease extraction and classification. The picture of the leaf is to assess the status of each leaf, captured and then processed. The proposed paradigm is a four-part image preprocessing model.

In order to preserve the plant's health, the infection must be detected and given due consideration. Physically, it is impossible to do so because the minute differences in the infected portion of the leaf cannot be seen by the human eye. In this way using MATLAB, we have created a system programmed to differentiate plant leaf diseases by using image handling procedures. The programmed is designed so that infected leaves can be identified effectively by an individual who does not have prior knowledge of the plants and their ailments. We have used the clustering of k-means to differentiate the contaminated area of the plant leaf. [2] Picture acquisition, image pre-processing, segmentation and feature extraction and SVM classification are included in the diseased recognition section. The images are taken through a high-resolution digital camera and are subjected to k-means clustering after pre-processing to obtain the diseased portion of the leaf. These are then run through the various algorithms of machine learning and categorized based on their characteristics of color and texture. In order to achieve maximum accuracy, a precision-based comparison is made between different machine learning algorithms: K-Nearest Neighbors, Multi-Class Support Vector Machine, Naive Bays and Multinomial Logistic Regression. Using MATLAB, the implementation was completed.

A study on the examination of leaf parameters, the identification of healthy, sick or damaged leaf regions and the classification of leaf diseases using various methods. The identification of the exact form of leaf disease by naked eyes is important and difficult for human eyes. Each leaf of a plant has various symptoms of different diseases. The algorithm developed for one plant, with the leaves of another plant, does not work correctly. To detect leaf diseases along with the leaf parameter analyzer, specialized algorithms for the any plant are needed. Picture processing and machine learning methods are helpful in the precise detection of any leaf diseases.

Image processing is a method of performing any operations on an image. Image processing is among the fast-growing technologies nowadays. It also forms the central research area within the fields of engineering and computer science. Basically, image processing requires the following three steps:

1. Importing the image using tools for image acquisition.
2. Analyzing the image and manipulating it.
3. The performance that results in an image or report that is based on image processing may be changed.

2. Methodology

It asks for the image to be fed as soon as we execute the code. After that, the picture given will be converted from RGB to LAB if we provide the image. This conversion is done because the test image given is divided into clusters using the clustering technique k-means, which depends on the values of L, A, and B, where L corresponds to black (+) and white (-) (while A corresponds to green (+) and red (-)), and B is equivalent to blue (+) and yellow-). Then we need to include the number of the cluster that has the region of interest (ROI) as well. If we provide the cluster number, the code will take the corresponding image and create a grey level co-occurrence matrix (GLCM) that can be extracted from different characteristics of that particular image. Contrast, correlation, energy, entropy, homogeneity, kurtosis, mean, RMS, skewness, and standard deviation are the characteristics that can be extracted. Both of these features are used when using the excel sheet to build a database. To train the classifiers, the generated database is then used. In this classification method, various methods, such as the neural network method, were attempted, but the efficiency was not satisfactory. [1],[2],[3],[4],[5]

2.1 Image capture

Image acquisition is conducted with a standard digital camera and the disease is detected within minutes with the aid of the required image processing technique. The camera is going to be attached to the drone. Therefore, if any part of the region is affected while operating in a wide field, then appropriate fertilizer is sprayed for that region by entering the drone's latitude and longitude values for that area. [6]

2.2 Preprocessing

Image preprocessing to remove unnecessary and noisy material from the image using a variety of filtering techniques. [1]

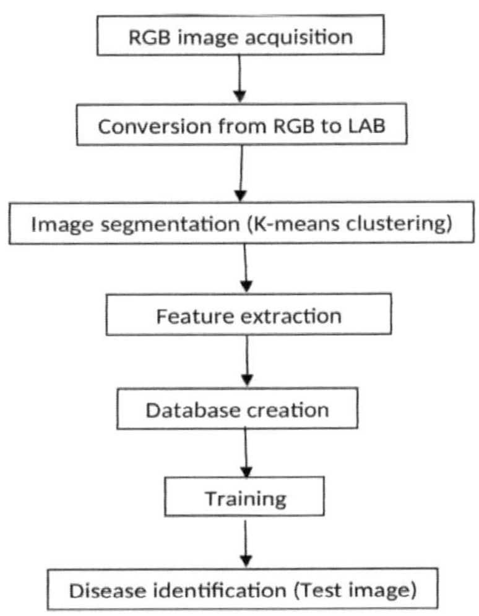

Figure 1: Flowchart of Image pre-processing

IOP Conf. Series: Materials Science and Engineering 1228 (2022) 012006 doi:10.1088/1757-899X/1228/1/012006

3. Image Segmentation

Segmentation is then performed on the preprocessed image, so it is possible to isolate the intersected sections from the original image. The preprocessed image is added to the threshold and k-mean clustering here. For the detection of grey-level images, k-mean clustering is used. For feature extraction, the disease-affected area is taken as the ROI. [7]

4. Feature Extraction

From the image collected, characteristics such as texture, form, color, etc. are extracted. Using GLCM (Gray Level Co-occurrence Matrix), the texture, shape, and pixel value are extracted. One of the extracted features is taken into account and correlated with the value of the database and then made on the basis of this judgment [12]. Any of the principal characteristics are:

1. Contrasting
2. Correlation
3. Energy
4. Homogeneousness
5. Standard Deviation

4.1 Disease recognition

The available database of about 2000 images is used to train the classification learner, and the well-trained classifier is used to cross-validate with the available test collection. There will be two approaches described in this section on how to classify any picture of the disease. Both approaches rely on the test image. If there is only one test image, then Method A must be selected; if there is a set of images to be evaluated, then Method B must be selected.

4.2 Method A: Keywords method

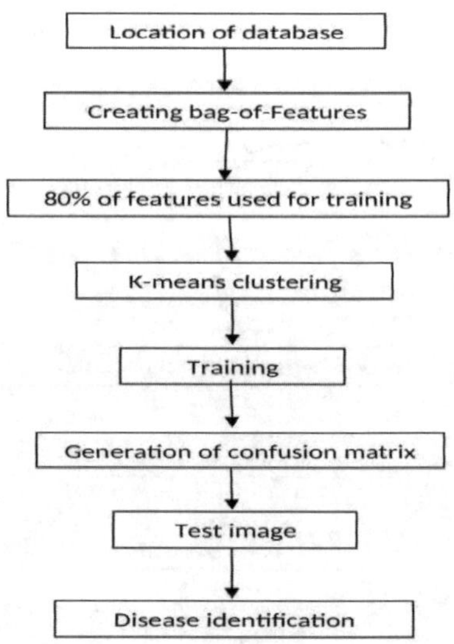

Figure 2: Flowchart describing the Method A

At the start, we give two main categories of input for training purposes. In this method, the main thing that comes into the picture is keywords. The list of keywords used in this method is given below.

4.2.1 Set directory

This keyword is used to access the location of the images where they are stored.

4.2.2 Image data store

Image Data store is one of the primary keyword used to read each and every images in the location provided.

4.2.3. Splitting of data

The provided amount of data is divided into training set and test set. By default, the splitting is done as 0.3 for training and 0.7 for testing purpose.

4.2.4 Feature collection

From each of the images in the database, a bag of features is being created. A proprietary algorithm often used in computer vision tasks and related to object detection purposes is S.U.R.F or Speeded Up Robust Functionality. The algorithm works as follows: it seeks characteristics / key-points that are likely to be present in the same object's various images. Keeping 80% of the strongest features from each category sufficient amount of cluster centres have been created

4.2.5 Training

Train an SVM multiclass classifier for 2 categories. After training evaluate the performance of the classifier on a test set.

4.2.6 Matrix of Confusion

A confusion matrix is a table frequently used to define the output of a classification model (or "classifier") on a collection of test data that are considered to be true values. Using this matrix, we can evaluate the performance of the algorithm. The average of the diagonal elements gives the accuracy of the algorithm.

4.2.7 Evaluation

If we enter any random image related to the two classifiers that we used in training, then the classifier will recognise the image accurately and the result will be displayed with an accuracy of around 1.

4.3 Method B: Classification learner method

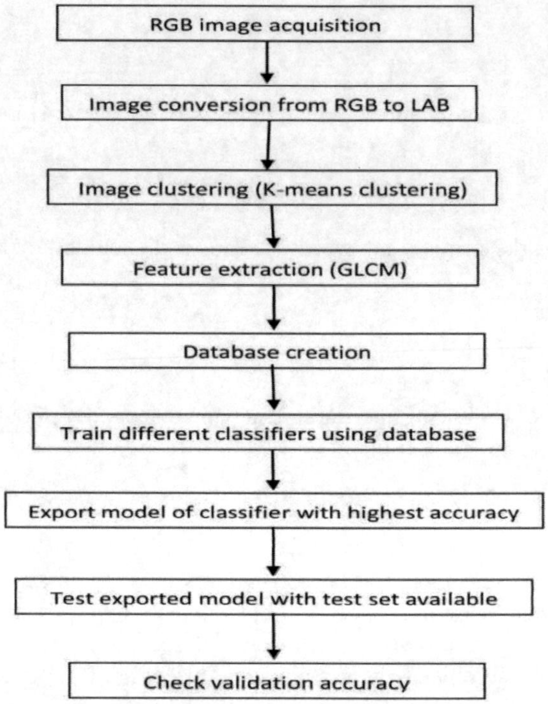

Figure 3: Flowchart describing Method B

Primary features such as contrast, correlation, energy, homogeneity, and standard deviation are extracted from each of about 2000 images in this process, and an excel sheet is used to construct a database.

4.3.1 Development of a database

In this method, first we will extract primary features such as Contrast, Correlation. Energy, Homogeneity and Standard deviation from around 2000 images and database is created using excel sheet.

New to MATLAB? See resources for Getting Started.

VarName1	VarName2	VarName3	VarName4	VarName5
0.14645	0.86681	0.55422	0.95545	29.593

ans =

1×5 table

VarName1	VarName2	VarName3	VarName4	VarName5
0.14491	0.86803	0.55457	0.95546	29.555

ans =

1×5 table

VarName1	VarName2	VarName3	VarName4	VarName5
0.14161	0.91866	0.35425	0.95474	36.581

Figure 4: Feature extraction parameters

In the above figure, VarName1 denotes contrast, VarName2 denotes correlation, VarName3 denotes energy, VarName4 denotes homogeneity, and VarName5 denotes standard deviation.

4.3.2 Learner of classification

MATLAB classification Learner is one of the apps that is available on MATLAB. This classification learner is used to classify or distinguish between things and has quite a good number of datasets. There are a good number of classifiers available, out of which, depending on our requirements and the database available, different classifiers are trained. In the same way, the available database is used to train different classifiers with different validation schemes available at the start of the training.

Here we prefer to go with the "no validation" scheme because all the data will be used for training purposes, whereas the remaining two of the validation schemes will divide the given database data into training set and test set, due to which a sufficient amount of data will not be available for the classifiers to get trained during the training set. Here there are quite a sufficient number of classifiers, but during this classification preference was given to the Ensemble classifier (Bagged trees) because it had a good performance during the training period compared to other classifiers.

4.3.3 Training

The training of different classifiers is done using the available database and the trained model is exported corresponding to the model that has the highest accuracy. Various classifiers are trained with the database created and, at last, the trained model of the classifier with the highest accuracy is exported and validated with the test set available.

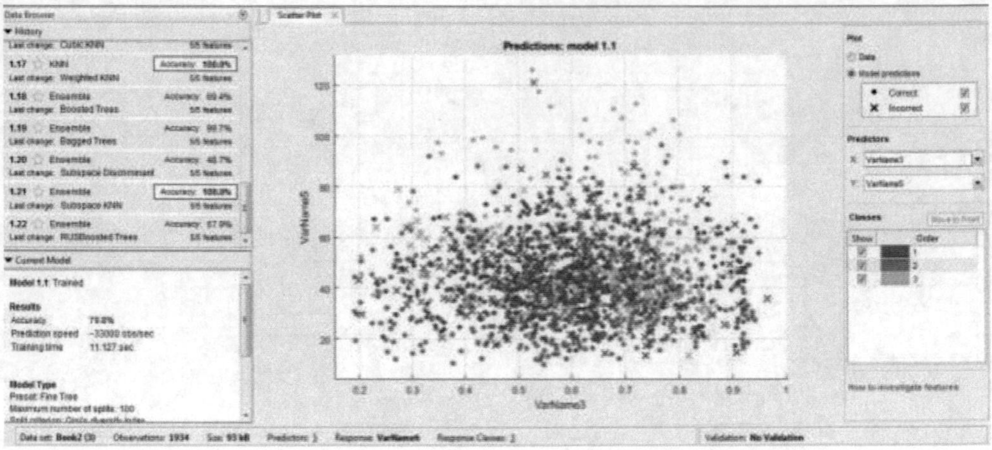

Figure 5: Training graph of different classifiers

4.3.4 Validation

The trained model is exported with the default name "trained Model" and is tested against the test set available using the command yfit = trained Model. Predict Fcn (testset) where in place of the test set, the name of the training set has to be given.

Then the result will be produced in the form of a table and saved in the workspace under the name yfit.

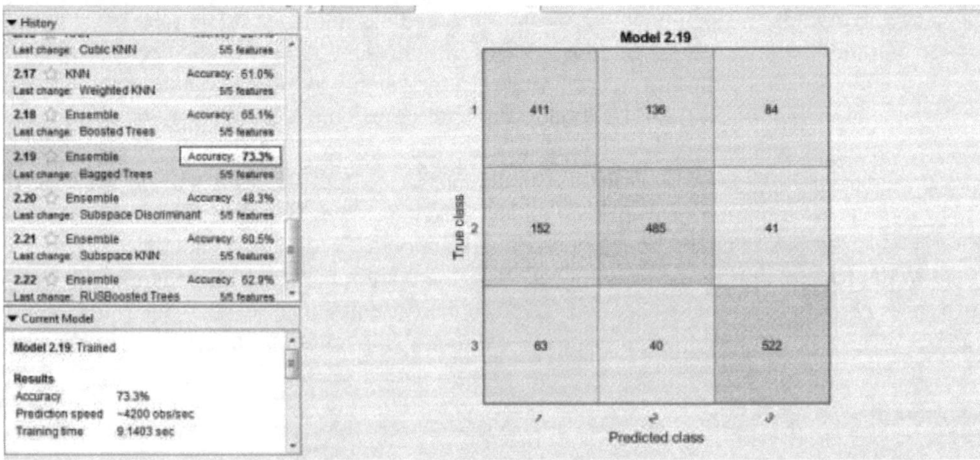

Figure 6: Training accuracy with confusion matrix

4.3.5 Validation precision

A code has been developed and the accuracy of the exported model is being tested against the already predetermined result. Here, the different models have been trained using the database available, and the one performing the best during the training period will have its model exported, while the performance of the different classifiers during the training period depends on the validation scheme which has to be selected at the start of the training.

IOP Conf. Series: Materials Science and Engineering 1228 (2022) 012006 doi:10.1088/1757-899X/1228/1/012006

4.4 Extracted characteristics

4.4.1 Contrast

The human visual system is more sensitive to contrast than absolute luminance; we can interpret the environment equally, regardless of the vast variations in lighting during the day or from location to location. The overall contrast of an image is determined by the contrast ratio, or dynamic range. The formula to calculate the contrast of an image is as shown below:

$$Contrast = \sum_{i=1}^{M} \sum_{j=1}^{N} (i-j)^2 P_{ij} \qquad (1)$$

Where P denotes the grey level co-occurrence matrix.

i = the number of rows (1–M)

j = the number of columns (1–N)

4.4.2 Correlation

Correlation is the process of moving a filter mask over an image and calculating the number of products at each position, also known as a kernel.The role of the philtre's displacement is correlation. In other words, the first value of the Identification of Crop Diseases by Image Processing correlation corresponds to zero philtre displacement, the second value corresponds to one displacement unit, and so on.

$$Correlation = \frac{\sum_{i=1}^{M} \sum_{j=1}^{N} (ij) P_{ij} - \mu}{\sigma} \qquad (2)$$

Where P = grey level co-occurrence matrix
μ = mean of the matrix
σ = standard deviation of the matrix

4.4.3 Energy

Energy is a calculation of uniformity in the GLCM with the summation of square elements. Between 0 and 1 is the range. For a constant picture, energy is 1.

$$Energy = \sum_{i=1}^{M} \sum_{j=1}^{N} P(i,j)^2 \qquad (3)$$

Where P = grey level co-occurrence matrix

4.4.4 Homogeneity

The homogeneity of an image region is related to the intensity changes that occur in that region. In this sense, whether we function with the true image or its negative, it is not necessary because the changes are the same and, thus, homogeneous.

$$Homogeneity = \sum_{i=1}^{M} \sum_{j=1}^{N} \frac{P(i,j)}{1+|i-j|} \qquad (4)$$

Where P = grey level co-occurrence matrix

IOP Conf. Series: Materials Science and Engineering 1228 (2022) 012006 doi:10.1088/1757-899X/1228/1/012006

4.4.5 Standard deviation

A picture is a set of light intensity data points, STD picture variance implies a gross measure of the imprecision / variation at each such data point about the target value of light intensity. Normally, the symbol σ signifies standard deviation.

$$\sigma = \sqrt{\left(\frac{1}{MN}\right)\sum_{i=1}^{M}\sum_{j=1}^{N}(P(i,j) - \mu)^2} \tag{5}$$

Where P = grey level co-occurrence matrix
σ = mean of the matrix

4.5 GLCM

Haralick introduced the co-occurrence matrix and texture features in 1973, which today are the most common statistical characteristics of the second order. For texture feature extraction, Haralick proposed two steps. The first step is to calculate the matrix of co-occurrence, and the second step is to calculate the texture function based on the matrix of co-occurrence. This method is useful in a broad range of image processing applications, from biomedical to remote sensing techniques. Using Grey-Level Co-Occurrence Matrix (GLCM) Texture Analysis The GLCM functions describe an image's texture by measuring how frequently pixel pairs of unique values occur in an image and in a given spatial relationship, generating a GLCM, and then extracting statistical measurements from that matrix.

4.5.1 Diseases

1. Alternaria leaf spot
2. Apple Mosaic
3. Marssonina leaf blotch
4. Powdery mildew
5. Anthracnose
6. Die back
7. Phoma blight
8. Leaf spot
9. Bacterial leaf spot
10. Early blight
11. Septoria leaf spot
12. Healthy
13. Apple Scab
14. Cedar Apple Rust

5. Results

5.1 Method A

In this method, the predefined keywords have been used to extract the features from the images provided, and these features are used for training purposes, with the result displayed in a cell array with around 100 percent accuracy. Also, along with disease detection, the fertiliser which can be used will be suggested depending on the value of levelIndex.

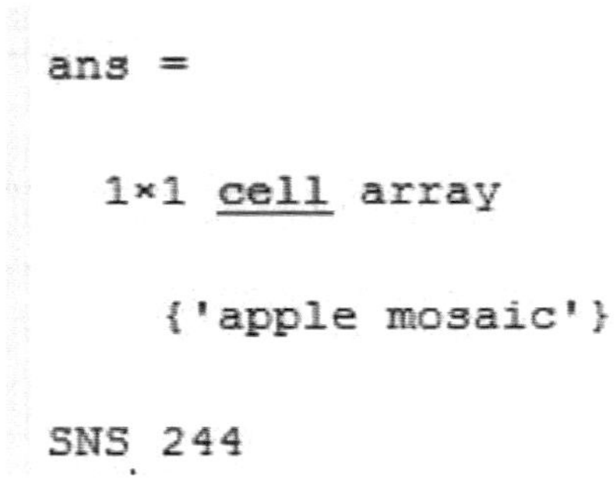

Figure 7: Disease identification with remedy

But in this method, there is a drawback. If there are two categories to be identified, the accuracy will be around 1, that is 100 percent. But if we increase the number of categories to be identified, then there will be a gradual decrease in the accuracy.

5.2 Method B

In this method, there are three different parts. The first one is a MATLAB code that is issued to extract features from the images available, and a database is created using the extracted features. The second part is where all the classifiers that are available in the classification learner are being trained using the database. The final section is where the classifier model with the highest accuracy during the training period is exported and tested against the available test set database, yielding an accuracy of 65 to 70%.

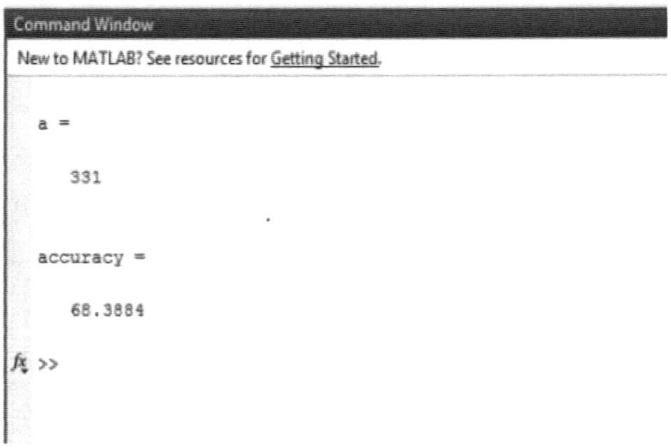

Figure 8: Validation of accuracy

6. Conclusion

In normal environments, plant diseases occur and their symptoms vary greatly at various stages of the disease. Segmentation of the image acquired is segmented accurately and only the region of interest is taken for consideration where the features such as contrast, correlation, energy, homogeneity, and standard deviation are extracted using MATLAB code and a database is created. Then, depending on the number of disease categories provided during training, either of the two methods may have been selected. In the first method, an accuracy of 100 percent is achieved only with two categories of diseases provided for training. When the category for recognition is increased, the accuracy decreases gradually.

So, in the future, even with more than two categories, it has to be seen that accuracy near to 100 percent is achieved. In the second method, the database provided for training purposes of different classifiers has to be increased to achieve greater accuracy. Finally, comparing both the methods, we can conclude that, depending on the number of diseases to be identified, the method has to be decided. If the number of diseases is less than or equal to two, then the keyword has to be considered because the accuracy will also be high, that is, around 100 percent. But when the diseases to be identified are more than two, the classification learner method is more suitable as it has good accuracy and the user will have a higher probability of getting accurate results.

References

[1] K. R. Gavhale, U. Gawande, and K. O. Hajari, "*Unhealthy region of citrus leafdetection using image processing techniques*," *International Conference forConvergence for Technology*-2014. IEEE 2014.

[2] M. Jhuria, A. Kumar, and R. Borse, "*Image processing for smart farming: Detection of disease and fruit grading*," *IEEE Second International Conference on Image Information Processing* (ICIIP-2013). IEEE 2013.

[3] V. Singh, A. Misra et al., "*Detection of unhealthy region of plant leaves usingimage processing and genetic algorithm,*" *International Conference onAdvances in Computer Engineering and Applications.* IEEE, 2015.

[4] Esmaeel et al., "*A novel approach to classify and detect bean diseases basedon image processing*," *IEEE Symposium on Computer Applications &Industrial Electronics* (ISCAIE). IEEE, 2018.

[5] L. C. Dorado, J. I. C. Aguila, and R. B. Caldo, "*Smart farm: automated classifying and grading system of tomatoes using fuzzy logic*," *Laguna J. Eng. Comput.Stud*, vol. *3*.

[6] N. G. Kurale and M. V. Vaidya,"*Classification of leaf disease using texturefeature and neural network classifier*,"*International Conference on Inventive Research in Computing* Applications (ICIRCA). IEEE, 2018.

[7] R. Anand, S. Veni, and J. Aravinth, "*An application of image processingtechniques for detection of diseases on brinjal leaves using k-means clustering*" *international conference on recent trends in information technology* (ICRTIT). IEEE, 2016.

[8] Devaraj, K. Rathan, S. Jaahnavi, and K. Indira, "*Identification of plantdisease using image processing technique*," *International Conference onCommunication and Signal Processing* (ICCSP). IEEE, 2019.

[9] S. Weizheng, W. Yachun, C. Zhanliang, and W. Hongda, "*Grading method ofleaf spot disease based on image processing*," *International conferenceon computer science and software engineering*, vol. 6. IEEE, 2008.

[10] L. S. Pinto, A. Ray, M. U. Reddy, P. Perumal, and P. Aishwarya, "*Crop disease classification using texture analysis*," *IEEE International Conference on Recent Trends in Electronics, Information & Communication Technology* (RTEICT). IEEE, 2016.

[11] H. Krishnan, K. Priyadharshini, M. Gowsic, N. Mahesh, S. Vijayananth, and P. Sudhakar, "*Plant disease analysis using image processing in matlab*,"*IEEE International Conference on System, Computation, Automation and Networking* (ICSCAN). IEEE, 2019.

[12] J. N. Reddy, K. Vinod, and R. A. AS, "*Analysis of classification algorithms forplant leaf disease detection,*" *IEEE International Conference on Electrical, Computer and Communication Technologies* (ICECCT). IEEE, 2019.

A Review on Electrical and Electronics Part of 3D Printer

Kiranlal S[1], Brathikan V M[2]Anandh B[3], Vikash S[4]
[1] Assistant Professor of Mechanical Department, Kumaraguru College of Technology, Coimbatore
[2] Incharge ré- KCT's Platform for Research and Exploration, Kumaraguru College of Technology, Coimbatore
[3] Department of Electrical and Electronics Engineering, Kumaraguru College of Technology, Coimbatore
[4] Department of Electrical and Electronics Engineering, Kumaraguru College of Technology, Coimbatore

Abstract—3D printing is one of the additive manufacturing technologies in the everyday field where three-dimensional objects are created by laying down successive layers. The electrical parts hold nearly 80% process of the printer. The major electrical components are the controller, motor, sensors, and extruders. In present years, FDM technology (Fused Deposition Modelling) is one of the most widely used 3D printers. This paper will give detailed information on the electrical and electronic parts of the 3D printer and its types. It gives the information of each controller with the respective data. The division of each electrical part in a printer is analyzed.

Index Terms – 3D printer, FDM, Electric and Electronics, Controller, Motor.

1. Introduction

The journey of the 3D printing started in 1940 with the very small bit to the very large extent of constructing the house. 3D printing is an additive manufacturing in construction of a three-dimensional object from a CAD model or a digital 3D model. Various materials are fused, and an 3D model is created by additive manufacturing by fusing the material layer by layer. Many different working mechanisms like axial and delta type of printing have been evolved. Different types of nozzles, extruders and many different materials are used for building the 3D material. The types of the 3D printer are: (TABLE I)

a. Fused deposition modelling (FDM)
b. Directly ink writing (DIW)
c. Stereo lithography apparatus (SLA) & (Digital light procession) DLP
d. Laminated object manufacturing (LOM)
e. Selective Laser Sintering (SLS) & Selective Laser Melting (SLM)
f. Photopolymer Jetting (Ployjet)
g. 3D Powder Binder Jetting (3DP)

Content from this work may be used under the terms of the Creative Commons Attribution 3.0 licence. Any further distribution of this work must maintain attribution to the author(s) and the title of the work, journal citation and DOI.
Published under licence by IOP Publishing Ltd

2. FDM Printer

Fused Deposition modelling is most preferred by consumer because of its fast-paced manufacturing and low-cost fabrication. The reason for more preference of this printer than others, based on working is because of vast usage of materials in FDM printer. Materials like thermoplastic, chocolate can be used to manufacture the model. The FDM printer is classified on major two types, they are 'Cartesian' and 'Delta' model of printing. Each has an advantage over the other in different terms. It depends on the company or the consumer and the manufacturer of the printer. [1]

There are 7 major parts in a 3D printer. These parts contribute a well and good outcome of a model after printing. They are frame, extruder, cooling fan, nozzles, power supply, display screen, motherboard (controller board), printer bed.

A. Frame:

A frame is a mechanical component which is made of a steel or any other suitable material for a rigid support to hold the components. Choosing a right material for a frame is very important and this decides the rigidity of a entire setup.

B. Extruder:

A extruder is responsible for drawing in, melting, and pushing out the filament. It is broadly classified into two types as Bowden type extrude and Direct drive Extrude. [7]

C. Cooling fan:

It cools the layers of plastic when the hot end puts down the 3D model. It is very important part of a 3D printer. When one layer is melted down, before printing another layer, the fan cools the previous printed layer.

D. Nozzle:

It conducts the thermal energy provided by the heating cartridge and block the filament, melting it. The bigger the nozzle, the more mass and surface area available for transferring heat to the filament making the process efficient.

E. Power Supply:

It consists of a set of transformers, which steps down the voltage to 12 or 24V accordingly. The rated voltage: 110V @60Hz and 240V @50Hz AC

F. Display:

As the name suggest it is used to display output. It is used it display the time taken for printing or the left-over time for printing. It can be programmed accordingly to display various information like speed of the fan and few other things according to our needs.

Table 1. Types Of 3D Printers and Specification

TYPE	TECHNIQUE	MATERIALS	ADVANTAGES	Ref.
Fused deposition modelling - (FDM)	Extrusion Based Principle	ABS, PLA, PC, PA, Glass, Ceramics, Eutectic metal, etc.,	Economical, easily accessible and maintaining and multi-material structures	[1]
Directly ink writing (DIW)	Extrusion Based Principle	Ceramics, Biodegradable plastics, Living Cells, Food, Composites	Adaptable to convenient printing mode	[2]
Stereo lithography apparatus (SLA) & Digital light procession (DLP)	Photocuring Principle	Photopolymers, Plastics	Accessible and high accuracy	[3]
Laminated Object Manufacturing (LOM)	Lamination Principle	Paper, Metal sheets, Cellulose, Plastic Film, Flexible sheets, etc.,	Economical, Versatile	[4]
Selective Laser Sintering (SLS) & Selective Laser Melting (SLM)	Powder based laser curing Principle	ABS wax, acrylic styrene, PC, PVC, Powdered Plastics, Metal, Ceramics, etc.,	High accuracy, Increases the strength of printed material and wide adaptation of materials	[5]
Photopolymer Jetting (Ployjet)	Inkjet-based Principle	Liquid photopolymers	High accuracy	[6]
3D Powder Binder Jetting (3DP)	Inkjet-based Principle	Plaster, ceramics, sugar, etc.,	Economical, used for multicolour printing, versatile	[6]

3. Stepper Motor

As we know there are many different types of motors. Specific motor can be made for specific purpose. Motors like AC, DC, Shunt motors are the most famous motors. With computer control stepping can be achieved to a greater extent and to a precise manner. They can also move in interval and then hold the position. (Fig. 1)

3D printer requires a specific type of motor, which should be stepped accordingly to the CAD model. The purpose is achieved is by the 'Stepper Motor'. These motors are the most important part of an extruder as a whole 3D printer. Since they can move in discreate steps. These are used to move the extruder or the build platform alone the x, y, z axis. They play a major role in extruders. As they are responsible for the amount of filament to be pushed to the hot end. They play a significant role in direct drive extruders, since they are flexible the amount of material coming out can be controlled by stepper motor.

3.1 ADVANTAGES OF USING STEPPER MOTOR IN 3D PRINTER:

Step angle of stepper motor is proportional to pulse rate. Stepper motor outputs its maximum torque at the moment it stops running (when winding excitation). As stepper motors are brushless ones, the lifetime of stepper motor is determined by the bearings' lifespan only. As t0he speed is proportional to the pulse rate, a wide speed range can be available.

Fig 1. Stepper Motor

4. End Stops

They are little electronic pieces at the end axes that connects to the main board. These sensors ensures that a machine keeps on a object on axis (in case of 3D printing). There are different end stops on different mechanism. Like, optical and magnetic end stops. Mostly electronic is preferred when compared to other types. [9]

These are broadly classified into three major types:

a. Endstop switches.

Endstop switches work using a simple touch sensor that functions as a switch. When the switch is touched by an object, it signals to the mainboard that the object is at the end of the path. These are easy to install since they are a switch. (Fig. 2. A)

b. Optical Endstops.

Optical endstops consist of a U-shaped device, called a photo-interrupter, with a light emitter on one arm of the U and a light sensor on the other. Optical endstops work by sensing when the light beam, cast by the light emitter, is disrupted. These are little trickier to install and set up. (Fig. 2. b)

c. Magnetic Endstops

These endstops, as the name implies, use magnets to determine when an object has reached the end of its path. Magnetic endstops use two parts: a Hall effect sensor and a magnet. The Hall effect sensor is attached to the end of an axis and detects the magnetic field of the magnet, which is attached to the object in motion.Since this consist of magnet, they must have a right place to install. (Fig. 2. c)

Fig 2. a. Endstop Switch, **b.** Optical Endstops, **c.** Magnetic Endstops

5. Extruder

The advancement in these extruders is the multiprinting and using of multiple filaments at the same time. This is achieved by using multiple filaments and nozzles. There are different applications of this method, it depends on the consumer. Multiple extruders with multiple and multiple nozzles with single extruder is available. [7] The primary reason for using multiple extruders is to decrease the time of the 3D printing. By implementing multiple extruders with several nozzle, it is possible to feed multiple material at the same time. These multipoinding extruders must have a strong controller system and a well programmed code. (Fig. 3)

The hot end must be designed accordingly to have a nozzle for different filament working at the same time. Filament changing procedure is a time-consuming procedure and it should be coded accordingly. As there are many filaments used the power consumption will be very high when compared to using one nozzle. The power consumption must be reduced accordingly. For using a multifilament extruder, the CAD design must be made accordingly, as it must consist of motor to rotate the filament for printing and for pushing the filament for printing. [8]

The calculations for the flow of filament must be calculated with respective formulas and fed it into the code.

Fig 3. Extruder

6. Controller

The brain of the 3D printer is the mother board or controller board. This board is responsible for all electronic functioning of a 3D printer. The printer would not be able to do much complex things. Choosing a right controller is very important for a 3D printer. As the load and programming is fully depends on the controller unit. There are specific pins in the controller which is used to control the extruders and motors. The controllers are the most important component of a 3D printer. There are various controllers available in the market. Those controllers can be integrated accordingly to work for the need of the consumer. There are specific controllers available in the market for 3D printer. The Firmware in these controllers are coded accordingly for the usage of multiple-axis and single-axis printing. The code in the firmware cannot be modified as it is predefined. The Arduino UNO ATmega328 and ATmega16U2. Where ATmega328 is a used coded micro controller and Atmega16U2 is a firmware, which is coded by the manufacturer itself. As the bit rate increases the speed and other processing increases. [8] (Fig. 4)

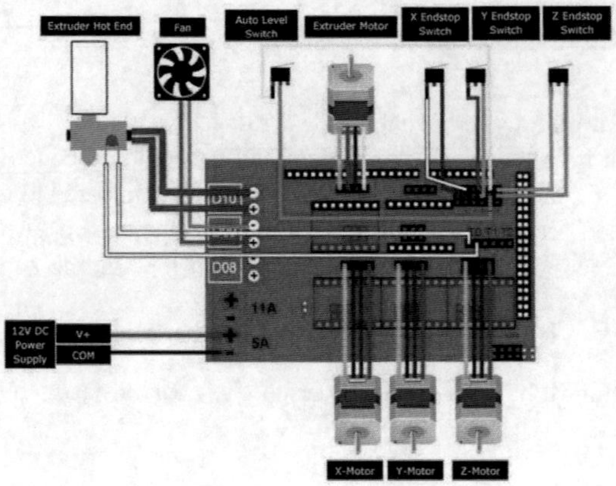

Fig. 4. Controller Wire Connection

Types Of Controllers For 3D Printer:

A. Smoothie Board V1

It is 32-bit micro-controller which has 120-MHz processor and a clock speed of 8-MHz. It has a built-in stepper driver, so there is no need of placing external motor shields. It has a micro-controller LPC1793. This has a built-in shield A5984. This also has a USB, Ethernet, and a micro-SD slot. The input power is 12 to 24V.

B. SKR Mini E3

This board is an upgradation of reality 3D printing. It has a 32-bit micro-controller and an onboard driver TMC2209, which is a sensor less homing. Another advantage of this board is the firmware flashing is far simpler. It has a dedicated 5V circuits for auto- bed leveling sensor, TFT screen, and LED lights. It has a STM32F103RCT6 micro-controller. The input power is of 12V to 24V.

C. Duet 2 Wifi

This board also has 32-bit ARM chip, great drivers, connectivity options and many more. The big advantage of this is the built in Wi-Fi connection and 1/256th micro-stepping. It has TCM2660 driver which can produce 2.8A motor current. This firm ware supports array of movements system including the multi-independent axis. It has Atmel SAM4E8E micro-controller. And a input power of 12V to 24V.

D. MSK SBASE V1.3

It is also a 32-bit-based controller, which has five integrated DRV8825 drivers. It has a LPC1768 micro-controller. Also, it has a USB jack, Ethernet jack and a micro-SD slot. It has an input power of 10V to 24V.

E. Re-ARM

It is a 32-bit controller which has a LPC1768 controller. It is a good replacement for the Arduino Mega an 8-bit controller. The Re- ARM can bypass documentation reading, software configuration and also even the connector's compatibility. We can choose between Marlin 2.0 and Smoothieware firm ware. The input power is of 10V to 30V.

F. MKS GEN L

It is an 8-bit controller for more conservative users. This is the most preferred board for desktop and DIY 3D printers. This is compatible with three drivers (TMC2100, TMX2209, TMC2130). It also provides jumper pins for better connections. It has Arduino Mega compatible. The input power is of 12V to 24V.

G. Archim2

This board is named after the Greek scientist Archimedes. It is highly capable of multitasking with Atmel SAM3X8E 32-bit processor and a built in Tarinamic TMC2130 drivers. It supports up to five motors, which means dual extrusion is possible also capable of 1/256th stepping. It can support different firmware. The input power is of 10V to 24V.

H. Revolve (TO BE RELEASED)

Beaglebone and Replicape are the two most popular boards for 3D printing which has a inbuilt wi-fi and an ethernet jack. Revolve is the combination of those two boards. The most fascinating thig about this controller is that it has 1-GHz processor, which can do multi-axis and multi-filament printing more easily. This is 8-times faster than a 32-bit controller. This power enables it to run a fully-fledged version of Linux. It has a OSD3358 SIP controller, also a USB and a HDMI port. It has a TMX2130 built in driver.

I. SRK 1.4 Turbo

It is 32-bit board with 120-Hz processor. It is meant to accommodate a dual Z axis motors and dual extrusion. It has 6 stepper motor ports also; extra ports can be integrated accordingly. It also has couple of safety features and thermistor. It has 6 motor ports, 3 thermistor, 3 end-stops, 4 fans and a RGB port. Also the liner advance depends on the motor driver (TMC2209). It has a inbuilt USB Type-B, wi-fi ports.

J. MKS Robin E3D

These company board are specially meant for FDM and SLA printer. This board has a 32-bit controller, which suits best for upgrades. It has 5 stepper motor ports and 5 corresponding stepper motor ports. The abundance of port is used for dual Z-axis motors. It also has automatic bed levelling ports. It also has 2 thermistor, 3 end-stops, four fans and RBG ports, also a filament runout sensor. The linear advance depends on the stepper motor driver (TMC2209).

The main thing to look before buying a controller is the processor. The processor speed must match the size of a 3D printer. It can start from 8-bit to 1-GHz controller. Next thing to loom after is the stepper driver. The controller must have a inbuilt driver or else an external driver must be placed which takes up the pins. Next is up with the ports. The controller must have a enough ports for End-stops sensor, fans, RGB light, Filament Runout sensor, Thermistor, TFT display. The connectivity is more important. Weather it is connected by a USB port, a inbuilt wi-fi jack or a micro SD slot. (TABLE II)

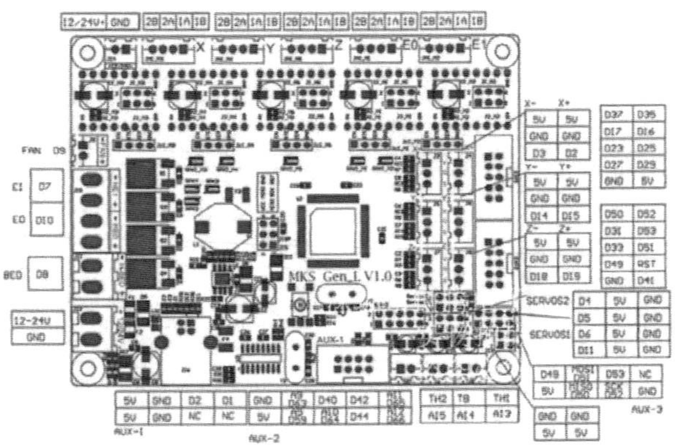

Fig 5. Controller Board Layout

IOP Conf. Series: Materials Science and Engineering 1228 (2022) 012007 doi:10.1088/1757-899X/1228/1/012007

Table 2. Types Of Controllers

S.No	CONTROLLERS	MICROCONTROLLER / PROCESSOR	INPUT POWER	DRIVERS
1	Smoothie Board V1	32-bit microcontroller (LPC1793) / 120-MHz Processor	12 to 24 V	A5984
2	SKR Mini E3	32- bit microcontroller (STM32F103RCT6)	12 to 24 V	TMC2209
3	Duet 2 Wifi	32- bit ARM (SAM4E8E)	12 to 24 V	TMC2660
4	MSK SBASE V1.3	32- bit (LPC1768) microcontroller	12 to 24 V	DRV8825
5	Re-ARM	32 – bit microcontroller (LPC1768) with 8 bit Mega controller	10 to 30 V	DRV8990
6	MKS GEN L	8- bit controller with Arduino Mega controller	12 to 24 V	3 Drivers – TMC2100, TMX2209, TMC2130
7	Archim2	32- bit microcontroller (SAM3X8E)	12 to 24 V	TMC2130
8	Revolve	32- bit microcontroller (OSD3358 SIP) / 1-GHz Processor	13 to 30 V	TMX2130
9	SRK 1.4 Turbo	32- bit microcontroller / 120Hz Processor	12 to 30 V	TMC2209
10	MKS Robin E3D	32- bit microcontroller (STM1793)	12 to 30 V	TMC2209

7. Sensors

A. Strain Sensor:

Th0ese sensors are used to detect an electric shift upon mechanical deformation. Stretchable electronics is a new class of electronics devices which offers many opportunities. This sensor converts pressure, tension, weight, etc. into electrical resistance which can be measured. Few sensors have transduction mechanism of resistance few has capacitance.[9]

B. Pressure Sensor:

This sensor work by converting pressure into analogue electrical signal. We can also measure pressure electronically using pressure transducers and pressure switches. This sensor delivers measured pressure and shear stress. The test result showed that it can measure pressure and shear at stress up to 350kPa and 80kPa. [10]

C. Tactile Sensor

These sensors have been implemented for a variety of capacitance sensing applications like, strain sensor, humidity sensor and tactile sensor. These sensors are mostly used in touch devices on mobile phones and computing. [11]

D. Displacement sensor

These sensors are generally called ass eddy current sensors. It produces a high-frequency magnetic field by applying high-frequency current to the coil inside the sensor head. This sensor measures distance based on the change of oscillation caused by this phenomenon. The systems consisting of these sensors are expected to enable the construction of measurement. [12]

E. Accelerometers.

This is a kind oof device that measures coordinate acceleration, which is the rate of change oof velocity. The have various application in industry and science. These sensors are made in highly sensitivity, those are applied in aircraft, medical instruments etc.. this sensor measures the acceleration force acting on an object in order to determine the object's position in space and also to monitor the object's movement. [13]

F. Angular sensor

Angular and linear position sensor, which is used to measure angle and linear positional changes relative to a reference position. There are sensors to detect only angle or linear position. The angular acceleration is exposed to inertia-driven fluid flows, which can be used too rotational acceleration sensing. [14]

G. Electromagnetic sensor

These sensors are non-destructive evaluation technologies. They are widely used in health monitoring and damage detection for infrastructure. The main use of this sensor is in stress monitoring for steel cables. [15]

H. EEG sensor

Electroencep0halogram sensor, this detects electrical activity in brain, using metal discs attached to scalp.[16] (TABLE III)

Table 3. Information on 3D Printer Sensors

S.No	SENSORS	MATERIAL	TRANSDUCTION MECHANISM	3D – PRINTED PARTS	Ref.
1	Strain Sensor	Silicone rubber, Verolue RGD840, Graphene aerogel, Carbon-based ink	Capacitance, Resistance	Sensing Part	[9]
2	Pressure Sensor	ABS, PVDF, ABS-based material, TangoBlack polymer	Capacitance, Optical absorbance	Mechanical Frame, Sensing Part, Bionic Sensing Part	[10]
3	Tactile Sensor	PC, Photopolymers	Capacitance, Piezo resistance	Origami package, Sensing part	[11]
4	Displacement sensor	SunTronic Jet Silver U6503, Nd-Fe-B Magnets	Inductance, Hall Effect	Sensing Part	[12]
5	Accelerometers	Silver nanoparticles, Thermoplastics	Capacitance, Gravity	Sensing Part	[13]

6	Angular sensor	Suntronic Jet Silver U6503, Plastic	Capacitance, Hall effect	Electrodes, Sensing Part	[14]
7	Electromagnetic sensor	zp®150, VisiJet® EX200, VisiJet® S100	Resonance frequency, Inductor-Capacitor(LC)	Sensing Part	[15]
8	EEG sensor	ABS, PLA	Resistance	EEG electrode	[16]

8. Discussion

The paper gives the detailed structural types of all electrical parts of the 3D printers. The most preferred and economical FDM (Fused Deposition Modelling) printers specification are given with the respective electrical phenomenal types. For a large-scale 3D printer, we need a micro-controller board of minimum of 8BG RAM to perform the operation in a smooth and healthier manner. Also, the type of material plays a crucial role in the model. The type of extruder is very important for the detailing of the model. The type of motor is also very important for the printer. Because it determines the stability and other factors of the model. Also, different types of sensors used for the detect various functions. Finally, the advantages of using a particular motor for 3D printing.
In this paper we have discussed about the types of 3D printer available in the market. Also, the materials required for the process of printing. All the electronics have been discussed in this paper along with the important components used in 3D printer.

9. Conclusion

The paper has been discussed on the basics of electrical and electronics and parts of an 3D printer. Also, it has a brief about the various materials, extruders and sensors used in the process of 3D printing. This paper will give detailed information on the electrical and electronic parts of the 3D printer and its types. It gives the information of each controller with the respective data. The division of each electrical part in a printer is analyzed.

10. References

[1]. Crump, S.S. Apparatus and Method for Creating Three Dimensional Objects Google Patents. U.S. Patent 5,121,329, 9 June 1992.
[2]. Wegst, U.G.K.; Bai, H.; Saiz, E.; Tomsia, A.P.; Ritchie, R.O. Bioinspired structural materials. Nat. Mater. 2015, 14, 23–36.
[3]. Hull, C.W. Apparatus for Production of Three-Dimensional Objects by Stereolithography. U.S. Patent 4575330,
11 March 1986.
[4]. Feygin, M. Apparatus and Method for Forming an Integral Object from Laminations. U.S. Patent 4752352, 11 October 1988.
[5]. Deckard, C.R. Method and Apparatus for Producing Parts by Selective Sintering. U.S. Patent 4863538, 5 September 1989.
[6]. GothaitHApparatus and Method for Three-Dimensional Model Printing 2001. U.S. Patent 6259962.
[7]. Landry, T. (2017). Extruders 101: A crash course on an essential component of your 3D printer | MatterHackers. [online] MatterHackers. Available at:
https://www.matterhackers.com/articles/extruders-101:-a-crash-course-on-an-essentialcomponent-of-your-3d-printer
[8]. Scott, C. (2017). New DeltaWasp Pellet 3D Printer Prints Big Without Big Cost. [online]
3DPrint.com | The Voice of 3D Printing / Additive Manufacturing.
[9]. Muth, J.T.; Vogt, D.M.; Truby, R.L.; Mengüç, Y.; Kolesky, D.B.; Wood, R.J.; Lewis, J.A. 3D Printing: Embedded
3D Printing of Strain Sensors within Highly Stretchable Elastomers. Adv. Mater. 2014, 26, 6307–6312

[10]. Saari, M.; Xia, B.; Cox, B.; Krueger, P.S.; Cohen, A.L.; Richer, E. Fabrication and Analysis of a Composite 3D
Printed Capacitive Force Sensor. 3D Print. Addit. Manuf. 2016, 3, 136–141.
[11]. Shemelya, C.; Cedillos, F.; Aguilera, E.; Espalin, D.; Muse, D.; Wicker, R.; MacDonald, E. Encapsulated copper
wire and copper mesh capacitive sensing for 3-D printing applications. IEEE Sens. J. 2015, 15, 1280–1286.
[12]. Bodnicki, M.; Pakula, P.; Zowade, M. Miniature Displacement Sensor; Advances in Intelligent Systems and Computing 393 (AISC); Springer: Cham, Switzerland, 2016.
[13]. MacDonald, E.; Salas, R.; Espalin, D.; Perez, M.; Aguilera, E.; Muse, D.; Wicker, R.B. 3D printing for the rapid
prototyping ofstructural electronics. IEEE Access 2014, 2, 234–242.
[14]. Tiem, J.V.; Jarno, G.; Sanders, R.; Krijnen, G. 3D printed bio-inspired angular acceleration sensor. In Proceedings of the 2015 IEEE Sensors, Busan, Korea, 1–4 November 2015; pp. 1430–1433.
[15]. Wu, S.Y.; Yang, C.; Hsu, W.; Lin, L. RF wireless LC tank sensors fabricated by 3D additive manufacturing. In
Proceedings of the 2015 Transducers—2015 18th International Conference on Solid-State Sensors, Actuators and Microsystems (Transducers), Anchorage, AK, USA, 21–25 June 2015; pp. 2208–2211.
[16]. Krachunov, S.; Casson, A.J. Casson.3D Printed Dry EEG Electrodes. Sensors 2016, 16, 1635.

Acknowledgements

We express our heart- felt gratitude to Kumaraguru College of Technology for providing us such an excellent platform to discover ourselves and to showcase all our talents. A special note to mention ré- KCT's Platform for Research and Exploration without whom this research work would have stood stagnant. We would like to thank all for help and constant support.

IOP Conf. Series: Materials Science and Engineering 1228 (2022) 012008 doi:10.1088/1757-899X/1228/1/012008

Water absorption study and characterization of polymer composites developed from discarded nylon carpet

Rajesh Kumar Verma[1]*, Balram Jaiswal[1], Rahul Vishwakarma[1#], Kuldeep Kumar[1], and Kaushlendra Kumar[1]

1Materials & Morphology Laboratory, Department of Mechanical Engineering, Madan Mohan Malaviya University Technology, Gorakhpur, Uttar Pradesh 273010

rkvme@mmmut.ac.in

Abstract. The scientific objective of this work is to reuse the polymer materials generated from carpet waste. The nylon obtained from the discarded carpet was used as reinforcement material into the epoxy matrix phase. A modified approach of Vacuum-assisted resin transfer molding (VARTM) was used to infuse the mixture of epoxy and hardener into the nylon carpet polymer composites. This investigates the water absorption behavior of carpet waste composite. It will justify the application of proposed composites for lightweight structural applications in a moisture environment. The nylon fiber-based-epoxy composite was dipped in water as per ASTM standard for a different duration, namely, 24, 48, 72, 144 hours. The findings demonstrate that it can effectively withstand in moisture environment, and a maximum 4.5% weight of samples increased among all the samples for a different duration. Also, the developed composite showed a lower water diffusion co-efficient than the nail head structure and linden wood. Also, the carpet waste composite samples were investigated for thermal degradation and chemical behavior using X-Ray diffraction (XRD) and Fourier transform infrared (FTIR) spectroscopy, respectively. The finding shows that the proposed composite could be used for lightweight components such as dashboard panels, sound absorbers, panel sheets, wall tiles, etc.

1. Introduction

Nowadays, waste management and recycling are widely recognized as major issues to balance environmental concerns. A large amount of plastic is generated from various resources involving industries, domestic, medical, etc. The waste generated from the carpet and textile sector consists of a substantial amount of pre-consumer and post-consumer materials. There is an extreme need for the recycling of waste generated from such types of industries. It helps to optimize energy consumption and maximize the profit in the sectors. The textile industry is essential in the lives of every human being because it is attributed to basic needs for living. Carpets are a vital part of the textile sector manufactured by hand or by automation [1]. Generally, carpet has 4 to 6 years; after this duration, it is converted into waste. Then the waste was disposed of in landfills [2], energy production through incineration. Carpet waste could be effectively used to fabricate composite for applications such as furniture, subflooring, roof panels, exterior support, interior panels of the vehicle, etc. Among all of the nylon are the most significant produced synthetic textiles [3]. Therefore, a considerable quantity of nylon fibers is available. Various pioneer scholars reused the carpet waste and converted it into some useful form. In this series, Venkateshwaran et al. [4] reported that the water permeability depends on the absorption of water by the jute hybrid composite. Jain et al. [5] developed a structural composite from the discarded carpet of

Content from this work may be used under the terms of the Creative Commons Attribution 3.0 licence. Any further distribution of this work must maintain attribution to the author(s) and the title of the work, journal citation and DOI.
Published under licence by IOP Publishing Ltd

nylon and olefin material in a different configuration such as Back- Fiber-Fiber-Back (BFFB) and Fiber-Back-Back-Fiber (FBBF) through the VARTM method. Furthermore, they applied a tensile test on both the composite material and obtained nylon with better tensile strength. Mishra et al. [6] fabricate a carpet composite material from the VARTM method. Later, they tested the hardness and compressive test of the composite materials. Karmaker [7] studied the moisture effect on jute/polypropylene composite. They conclude that water absorption was linearly proportional to the fiber content in the composite. Gassan and Bledzki [8] performed mechanical testing with the moisture effect on the silanized jute/ epoxy composite. They conclude that the moisture content degrades the tensile and flexural strength of the developed sample. Onal et al. [9] examined the behavior of water absorption of polymer composite, which was comprised of carpet-based jute yarn and thermoset matrices. The impact properties of the composite were explored by aging in distilled water. The water absorption behavior of composites was investigated by treating jute yarn with NaOH. It was found that when alkali-treated jute yarn composite reduces, the water absorption of the composite results in interface improvement. Mohanty et al. [10] modified hessian cloth and carpet backing cloths, which were the two varieties of jute fibers, and done involvement of alkali treatment and cyanoethylation. The characterization of the fabrics treated chemically was done by X-ray diffraction analysis (XRD) and thermogravimetric analysis (TGA). Mechanical properties were found to be increased in comparison with the parent material.

The present research article contains the recycling method of post carpet (nylon) waste. The moisture absorption of composite has been used to calculate the amount of water consumed under defined conditions. Factors influencing water absorption include the type of plastic, chemicals used, temperature, and exposure time. Evidence highlights the durability of products in damp or moisture conditions. The findings of XRD and TGA demonstrate the usableness of the composite for advance structural applications. This work aims to reutilize the discarded nylon carpet to expand the field of polymer composites.

2. Materials and methods

2.1. Sample preparation

The waste carpet (nylon) was used as reinforcement and epoxy as a matrix material to fabricate composite samples. Primarily, the nylon-based carpet waste is treated by detergent for cleaning and drying in direct sun light. In this process, GY-257, Diglycidyl ether bisphenol-A (DGEBA) based resin. It was modified with aromatic glycidyl ether hardener (Araldite® HY-951) as a curing agent in the proportion of 10:1. This blend shows properties of low viscous 500 – 700 mPa-s (at 25 ℃). The samples were fabricated in two configurations: BFFB and FBBF.

The VARTM technique is used in this study to fabricate the composite material. Fig. 1 represents the schematic of the VARTM Setup. The setup's upper mold and spiral wrap are present to ensure a homogeneously flow of resin over the carpet pile.

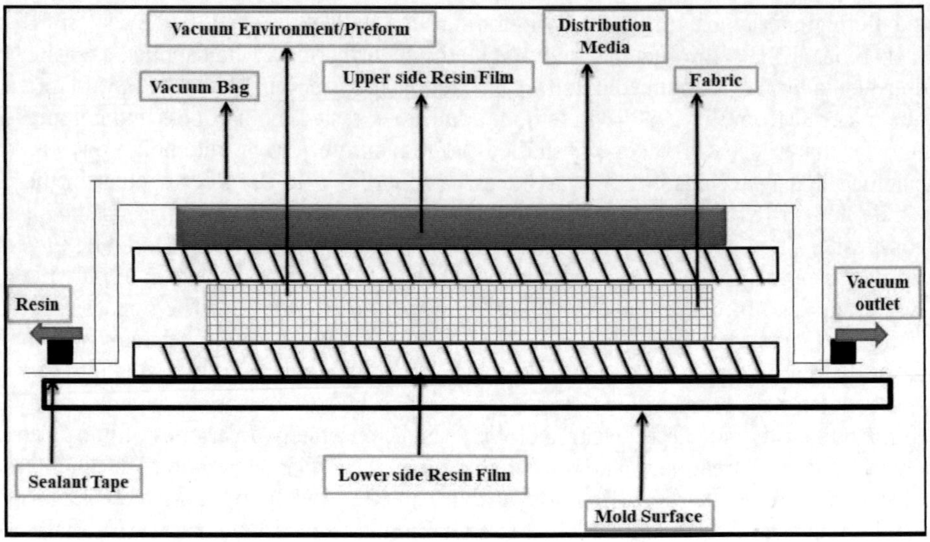

Fig. 1 Setup of VARTM for Nylon carpet waste Composite

The carpets piles were put in the dimension of 250×300 mm2 in the BFFB and FBBF configuration separately. The polythene sheet and tacky tape created an airtight mold with the inlet and outlet valve. The matrix material flow to infuse the carpet pile through the vacuum pressure applied by the vacuum pump. A peel ply and transfer media are used to infuse matrix material into the carpet ply properly. After the matrix has been wholly infused, the inlet and outlet valves are tightly closed for 30 hours and cured.

2.2. *Water absorption behavior of waste carpet composites*

The water absorption test was conducted according to an ASTM-D570 (as displayed in Fig.2). Generally, three concept which regulates the moisture dispersion in the material. The first one is water diffusion into the micro-voids among adjoining polymer chains. Further, with the help of capillary action arises in the polymer composite, and the last one is diffusion over the micro holes presented in the composite. After absorbing moisture, the interfacial adhesion force between the reinforcement and matrix is reduced [11]–[13].

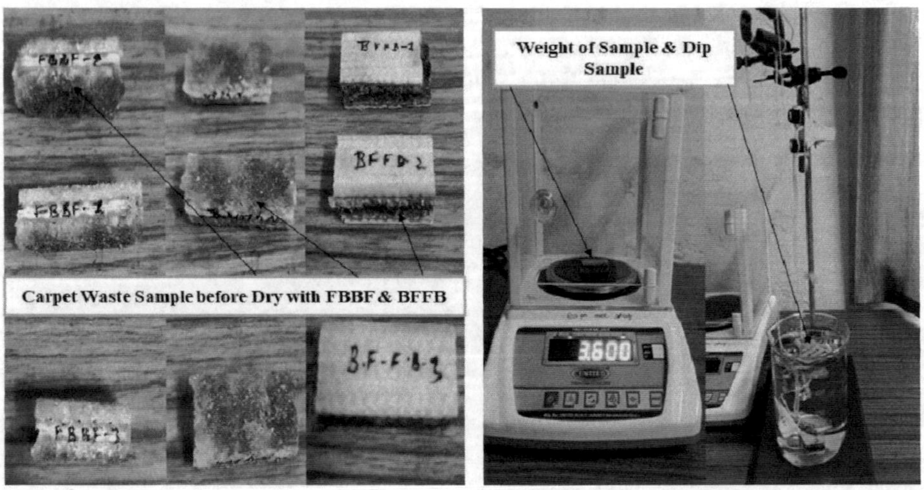

Fig. 2 Water absorption test

The following procedure has opted for the water absorption test

1. Firstly, the specimens are entirely dipped in water for several hours.
2. Then, the samples are taken out from the water and allowed to drain for one minute.
3. The specimens after draining are wiped off and weighed immediately, and the weight of each sample is noted.
4. The dry weight of each specimen has been recorded.
5. The percentage of water absorption is calculated by the water absorption percentage (M) given by the following formula.

$$M = \frac{Wet\ weight - Dry\ weight}{Dry\ weight} \times 100 \quad (1)$$

2.3. Morphology analysis

X-Ray Diffraction (XRD) tests were used in the present study to analyze the crystallographic structure of the waste carpet composite material. The XRD research is used to classify crystalline phases of a substance utilizing a crystal structure review and expose details on the chemical composition. X-ray diffraction curves of the studied samples are characterized by diffraction peaks of the nylon carpet and epoxy matrix crystalline structures. XRD test was done on the Ultima IV diffractometer. XRD pattern has been obtained with the help of Bragg's equation:

$$n\lambda = 2dsin\theta \quad (2)$$

where,
n= Integral number of wavelength
λ= wavelength of the electron
d= Spacing of the crystal
θ= Reflected angle

Thermogravimetric analysis (TGA) of the waste carpet composite (TGA 55) was performed from room temperature to 800 °C with a 10 °C/min heating rate. In this test, the sample's thermal stability was investigated regarding weight loss as a function of temperature in a nitrogen atmosphere.

3. Result and discussion

3.1. Water absorption test

The water absorbed tests on the nylon-based composite material. The test was performed for 216 hours at the iteration of 24 hrs. The amount of moisture absorbed in the sample is shown in Table 1, and in graphical form, in Fig. 3, the test is carried out on BFFB and FBBF, both configuration and two samples of each configuration. The result shows that composite samples are absorbed a maximum of 4.22% at the 144 hours after that negligible water absorb. Also, similar results are confirmed by the research [14], [15].

Table 1 water absorption of different samples

Time	% water absorbed			
Hrs	BFFB configuration		FBBF configuration	
	BFFB-1	BFFB-2	FBBF-1	FBBF-2
24	3.04	2.76	3.3	2.59
48	3.63	3.03	3.55	2.99
72	3.76	3.25	3.74	3.16
96	4.06	3.53	4.36	3.3
120	4.1	3.59	4.41	3.52
144	4.14	3.68	4.53	3.56
168	4.22	3.78	4.73	3.67
192	4.22	3.78	4.73	3.67
216	4.22	3.78	4.73	3.67

IOP Conf. Series: Materials Science and Engineering 1228 (2022) 012008 doi:10.1088/1757-899X/1228/1/012008

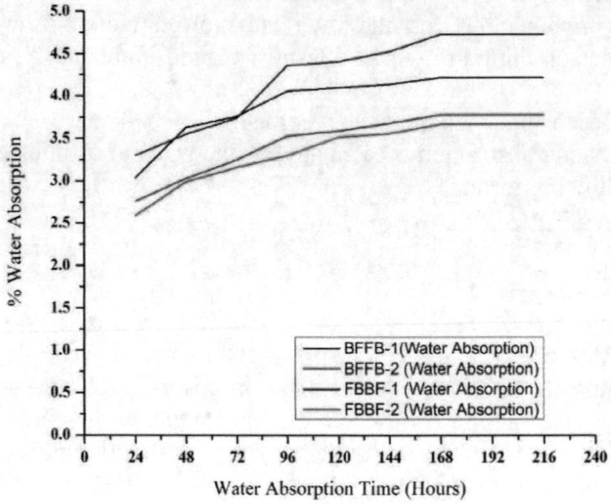

Fig. 3 Graph between %water absorption vs. time

3.2. X-Ray diffraction analysis (XRD)

The results of XRD of the carpet waste Above graph of XRD such that of carpet waste nylon reinforcement with epoxy materials. X-ray diffraction is a non-destructive research tool for studying the crystalline material composition [16], abbreviated as maxima point position peaks at certain angles of incidence theta. XRD two-point atoms more highly count par second 16500 and 14000 of intensity. In epoxy, the XRD characterization method is primarily used to determine the degree of epoxy and nylon reinforcement. The crystalline characteristics of nylon/epoxy composite are visible in the diffractogram at 2θ =31°, shown in Fig. 4. The peaks of epoxy are also exhibiting at 20°.

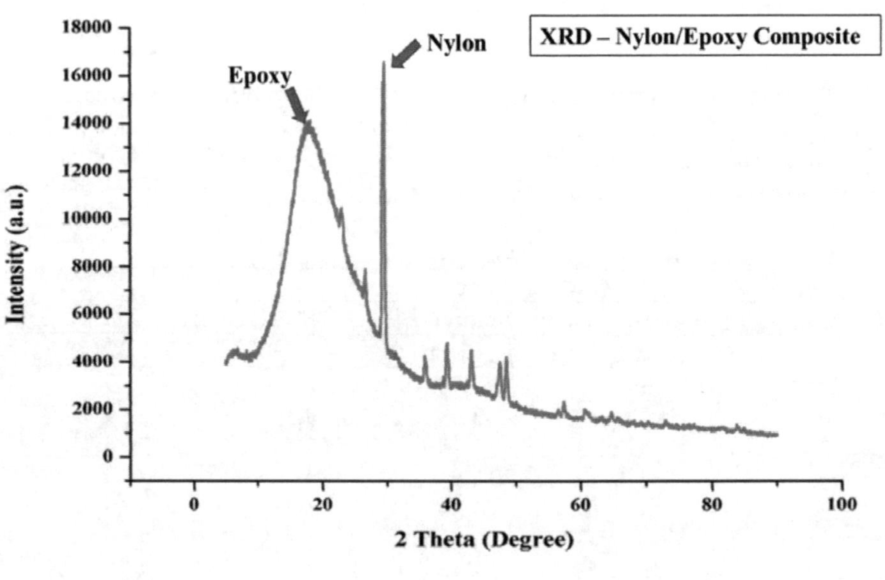

Fig. 4 XRD analysis of Nylon/Epoxy composite

3.3. Thermogravimetric analysis

Fig. 5 shows the TGA of a nylon-based carpet composite. There are three primary reaction stages of degradation. The first stage shown at 240 °C was caused by the decomposition of impurities and volatile components. The loss of weight fraction was around 15.6%. The second degradation step is associated with the polymeric fraction's decomposition. At this point, there was massive thermal degradation related to the deterioration of the polymeric. A temperature rises of about 300 °C is equivalent to the highest weight loss resin was decomposed. Then in the third stage, the nylon is decomposed. The decay of the nylon ended approximately at 680 °C, with a 70 % weight loss. Finally, around 800 °C, it degrades about 75% weight loss of the whole sample [17].

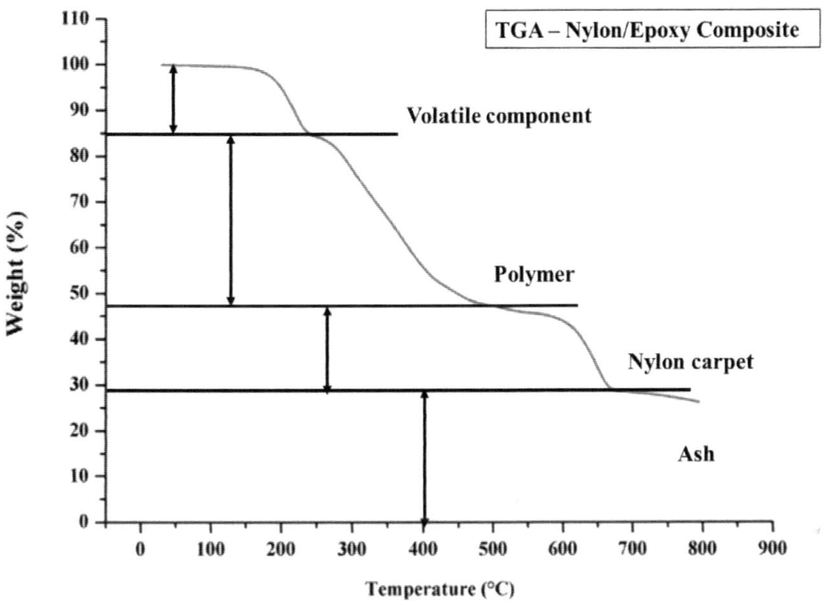

Fig. 5 TGA analysis of Nylon/Epoxy composite

4. Conclusion

The discarded carpet was successfully converted into a structural composite material with the help of epoxy matrix polymer. The fabricated sample was tested for water absorption for 210 hours. Form the experimental analysis conclusion can be summarized as followings points:

1. The composite was developed from a modified vacuum environment to transfer epoxy resin into the carpet pile structure. The vacuum environment is remarked as a useful method for developing composites with uniform dispersion and infusion.
2. The results show that composite materials absorb less than 4.8% of their weight, then after, the sample absorbs negligible moisture. It demonstrates the higher feasibility of the proposed composite in the moisture area.
3. An XRD pattern confirms the crystalline structure of a nylon-based composite produced at 31o. This structure is highly desired for structural composite materials.
4. The thermogravimetry analysis results demonstrate that the produced composite has good thermal stability. The graph indicates that since the temperature rises to 700°C, the material degrades by over 75%.

The finding shows that carpet composite could be transformed for lightweight components such as dashboard panels, sound absorbers, panel sheets, wall tiles, etc. In the future, other types of carpet waste such as polyester, polypropylene, wool can be used to manufacture polymer composites. This approach could be efficiently used for solid waste management of the carpet and textile sector.

Acknowledgment
The authors would like to acknowledge the kind support of O/o DC (Handicrafts), Ministry of Textiles, Govt. of India, New Delhi, India, to directly extend all possible help in this research work.

Declaration of Conflicting Interests
The authors declared no potential conflicts of interest.

Funding
This research is by O/o DC (Handicrafts), Ministry of Textiles, Govt. of India, Under Project ID: K-12012/4/19/2020-21/R&D/ST.

References

[1] R. Vaidyanathan, R. P. Singh, and T. Ley, "Recycled carpet materials for infrastructure application," p. 88, 2013.
[2] T. Domina and K. Koch, "The Textile Waste Lifecycle," *Cloth. Text. Res. J.*, vol. 15, no. 2, pp. 96–102, Mar. 1997, doi: 10.1177/0887302X9701500204.
[3] F. Puertas, T. Amat, A. Fernández-Jiménez, and T. Vázquez, "Mechanical and durable behaviour of alkaline cement mortars reinforced with polypropylene fibres," *Cem. Concr. Res.*, vol. 33, no. 12, pp. 2031–2036, 2003, doi: 10.1016/S0008-8846(03)00222-9.
[4] N. Venkateshwaran and A. ElayaPerumal, "Mechanical and water absorption properties of woven jute/banana hybrid composites," *Fibers Polym.*, vol. 13, no. 7, pp. 907–914, 2012, doi: 10.1007/s12221-012-0907-0.
[5] A. Jain, G. Pandey, A. K. SINGH, V. Rajagopalan, R. Vaidyanathan, and R. P. Singh, "Fabrication of structural composites from waste carpet," *Adv. Polym. Technol.*, vol. 31, no. 4, pp. 380–389, 2012, doi: 10.1002/adv.20261.
[6] K. Mishra, S. Das, and R. Vaidyanathan, "The use of recycled carpet in low-cost composite tooling materials," *Recycling*, vol. 4, no. 1, pp. 1–8, 2019, doi: 10.3390/recycling4010012.
[7] A. C. Karmaker, "Effect of water absorption on dimensional stability and impact energy of jute fibre reinforced polypropylene," *J. Mater. Sci. Lett.*, vol. 16, no. 6, pp. 462–464, 1997, doi: 10.1023/A:1018508209022.
[8] J. Gassan and A. K. Bledzki, "Effect of moisture content on the properties of silanized jute-epoxy composites," *Polym. Compos.*, vol. 18, no. 2, pp. 179–184, 1997, doi: 10.1002/pc.10272.
[9] L. Onal and Y. Karaduman, "Mechanical characterization of carpet waste natural fiber-reinforced polymer composites," *J. Compos. Mater.*, vol. 43, no. 16, pp. 1751–1768, 2009, doi: 10.1177/0021998309339635.
[10] D. M. Paleri, A. Rodriguez-Uribe, M. Misra, and A. K. Mohanty, "Pyrolyzed biomass from corn ethanol industry coproduct and their polypropylene-based composites: Effect of heat treatment temperature on performance of the biocomposites," *Compos. Part B Eng.*, vol. 215, no. November 2020, p. 108714, 2021, doi: 10.1016/j.compositesb.2021.108714.
[11] W. Wang, M. Sain, and P. A. Cooper, "Study of moisture absorption in natural fiber plastic composites," *Compos. Sci. Technol.*, vol. 66, no. 3–4, pp. 379–386, 2006, doi: 10.1016/j.compscitech.2005.07.027.
[12] M. S. Bodur, K. Englund, and M. Bakkal, "Water absorption behavior and kinetics of glass fiber/waste cotton fabric hybrid composites," *J. Appl. Polym. Sci.*, vol. 134, no. 47, pp. 1–9, 2017, doi: 10.1002/app.45506.
[13] E. Muñoz and J. A. García-Manrique, "Water Absorption Behaviour and Its Effect on the

Mechanical Properties of Flax Fibre Reinforced Bioepoxy Composites," *Int. J. Polym. Sci.*, vol. 2015, p. 390275, 2015, doi: 10.1155/2015/390275.
[14] H. Mohammadhosseini, N. Hasanah, A. Shukor, A. Rahman, M. Sam, and M. Samadi, "Effects of Elevated Temperatures on Residual Properties of Concrete Reinforced with Waste Polypropylene Carpet Fibres," *Arab. J. Sci. Eng.*, vol. 43, no. 4, pp. 1673–1686, 2018, doi: 10.1007/s13369-017-2681-1.
[15] Z. Kamble and B. K. Behera, "Mechanical properties and water absorption characteristics of composites reinforced with cotton fibres recovered from textile waste," *J. Eng. Fiber. Fabr.*, vol. 15, pp. 0–7, 2020, doi: 10.1177/1558925020901530.
[16] Y.-T. Hsieh, S. Nozaki, M. Kido, K. Kamitani, K. Kojio, and A. Takahara, "Crystal polymorphism of polylactide and its composites by X-ray diffraction study," *Polym. J.*, vol. 52, no. 7, pp. 755–763, 2020, doi: 10.1038/s41428-020-0343-8.
[17] K. R. Chow, N. Mohamad, and S. C. Koay, "Mechanical properties of palm fiber reinforced polyester composite based on different weight ratio of matrix and fiber," *AIP Conf. Proc.*, vol. 2233, no. May, 2020, doi: 10.1063/5.0003792.

IOP Conf. Series: Materials Science and Engineering 1228 (2022) 012009 doi:10.1088/1757-899X/1228/1/012009

Defect detection in rotating machine elements by using an improved image segmentation technique

V.K. Chawla and Ekta Yadav*

Indira Gandhi Delhi Technical University For Women, New Delhi, India
ektabhinda@gmail.com

Abstract. In industrial applications rotating machinery is widely used. Therefore it becomes necessary to diagnose the health of rotating machinery and detect any fault during their working. The pre-determination of defects can alert the operators and overall system for any type of incident and loss. Generally, fault in the rotating machine causes a rise in temperature of the rotating elements of the machine during their working. In this work, the image segmentation technique is used on thermal images of rotating elements of a machine to identify and detect the defects in them. The main focus of this investigation is to apply linear transformation and improved hyper smoothing based local binary pattern(LTIHLBP) image segmentation technique for defect detection and compare its performance with other conventional defect detection techniques namely improved hyper-smoothing function based local binary pattern (ILHBP), Sobel and Canny. In this paper, three tests are performed to identify the effectiveness of the proposed method. The first test is performed on synthetic images to identify the accuracy and Jaccard similarity index(JSI) of the proposed segmentation method, the second test is performed on healthy moving element images and the third test is performed on faulty moving element images. The outcome of this investigation revealed that the proposed method of LTIHLBP outperforms the conventional image segmentation-based defect detection techniques and found to be capable and accurate in diagnosing and detecting defects.

Keywords: *Defect detection and diagnosis, Image segmentation, Rotating machine elements, Thermal condition.*

1. Introduction

Rotating machinery performs an important part in manufacturing, transportation, electric-powered, and many more industries [1, 2]. Rotating machinery generally includes some common parts like rolling element bearings, gears, and rotors. During machine operations, it has a lot to suffer from various failures that result in financial losses and unscheduled maintenance costs. In consideration of these losses, it becomes important to perform defect detection of rotating machinery. Defect detection procedure ensures safe operation and prevents catastrophic failure [3, 4]. Generally, 44% of fault in rotating machines is due to bearings [5]. Bearing defects are generally defined as mechanical defects and classified as outer-inner race and ball bearing defects [6, 7, 8]. Sehgal et al. [9] used the fault tree digraph approach to identify failures in tribo-mechanical systems. Angra et al. [10] studied bearings and found elastic effects of hydrodynamic bearings are more significant than any other bearings. Sehgal et al. [11] used graph theory and matrix approach for the selection of bearing for an application.

Content from this work may be used under the terms of the Creative Commons Attribution 3.0 licence. Any further distribution of this work must maintain attribution to the author(s) and the title of the work, journal citation and DOI.
Published under licence by IOP Publishing Ltd

Nowadays the area of defect detection comes under condition monitoring. Researchers are focusing to develop advanced defect detection and classification methods that could overcome the existing condition monitoring techniques [12]. New and advanced defect detection techniques have been proposed and are used in the rotor-bearing system and many other industrial types of equipment for health monitoring [13, 14]. Various defect detection techniques are used in fault diagnosis of rotating machinery like acoustic emission, current, vibration, rotating encoder, torque, and sound signal, etc [15-20]. Along with these techniques, the vibration measurement technique is most commonly used as this technique directly represents the effective performance of rotating machinery [20-23]. But these existing methodologies have some limitations as they are not able to provide noise-free signals. The existing methodologies mostly depend upon time-varying operating conditions. It becomes hard to draw out the fault indications. Pre-defect detection is not possible in the most popular vibration measurement technique [24]. The aforementioned reasons motivated the researchers to develop a new technique that can diagnose the healthy conditions of rotating machines [23]. Nowadays, the temperature signal becomes more popular and gains the attention of researchers. A thermocouple is used to collect temperature signals. This technique is more effective as the fault in machinery cause to temperature increase, advancement in research has proven that rich information is available in temperature signal that becomes a good instrument for robust state inspection in machinery [25-27]. The thermography camera is used to get an infrared thermal (IRT) image that contains rich fault information. In current years, the Infrared Thermography technique is introduced as a new health condition monitoring method, as it is no-contact, no-intrusive, only one sensor-dependent temperature measurement technique [28]. In [12] the induction motors are diagnosed by equating a hotter area of healthy motor image with a hotter area of the faulty motor image. The thermal images captured with a thermal camera contain a bunch of knowledge about motor and these knowledge features are withdrawn using various images processing techniques [29]. Additionally, the easiest way to recognize thermography images is by using image segmentation methods. Pal et al. [30] discussed various segmentation methods and also studied their comparison with each other. The technique depends upon threshold values as this method is instinctive, also easily applied. Researchers used grayscale-based segmentation techniques without paying much attention to limitations related to conversions using grayscale. The aim of using grayscale-based conversion is to divide a gray-level histogram at specific threshold levels. Singh et al. [31] have applied an automatic threshold technique named as Otsu method to extract the region which has the highest temperature in the thermal image that can be applied in various operations. In addition, an algorithm named as infrared thermal anomaly detection algorithm (ITADA) is applied for diagnosing defects inside capacitors, generators that depend upon the Otsu method in which statistical calculations are performed using the gray-level histogram [29-31]. Jadin et al. [32] suggested that IRT can sense the infrared energy emissions emitted from objects and the temperature of any electric device is controlled using IRT. It is infamous for the life of any machinery that life gets reduced when its temperature gets increased. In recent years, machine learning algorithms and artificial intelligence techniques are widely applied in manufacturing automation [33-40]. Vasti and Dev [41] used different types of machine learning algorithms like support vector machine (SVM), K-nearest neighbor (KNN), Naïve Bayes (NB), and random forest (RF) for classification on real-world earthquake dataset. Ramirez et al. (42, 43) proposed induction motor fault detection techniques using thermography images. Thermography technique has many advantages as compared to other techniques like a wide range of analyses, non-invasive and not required to mount on the body.

Literature reveals that most of the fault detection techniques are based on image segmentation technique, vibration, and sound signal. These techniques are unable to detect the pre-fault and proper location of fault. These issues are overcome by using proposed thermal image processing techniques. In this work, a linear transformation and improved hyper-smoothing based local binary pattern (LTIHLBP) is proposed for fault detection in rotating machine elements. The results of LTIHLBP are compared to other segmentation techniques and found that proposed method can detect proper location of a defect.

2. Linear transformation and improved hyper smoothing function based local binary pattern (LTIHLBP) for fault detection

The main limitations of existing fault detection techniques are unable to detect the pre-fault and proper location of fault. These limitations are overcome by using the thermal image and edge-based segmentation techniques. In this technique, linear transformation improved local binary pattern technique and rectangular boxes are used to identify the position of the fault in rotating machines.
The contrast of the thermal image is adjusted with two methods commonly known as linear transformation and non-linear transformation. The more suitable method to process thermal images is a linear transformation which is given below [44].

$$I_{LT}(k) = \begin{cases} 0, & k < P \\ I(k), & P \leq k \leq Q \\ 255, & k > Q \end{cases} \tag{1}$$

Where I_{LT} represents the thermal images obtained from the linear transformation, $I(k)$ represents the gray thermal image and (k) is the kth pixel in the image. P and Q have represented two fixed thresholds.
In this work, an improved hyper smoothing function-based local binary pattern (IHLBP) is applied to obtain the effective edge. The IHLBP method is based on a counting scheme-based threshold value, hyper smoothing functions, and local binary pattern. In this technique, the adaptive threshold value is used to lessen the noise and thin edges. The adaptive threshold value is calculated by comparing the surrounding pixel with the central pixel. If the value of surrounding pixel value is greater than center pixel then the output is 1 else the output is zero [43].

$$T_{th} = \sum_{i=0}^{7} I_{LT}(s_j - s_c), \quad I_{LT}(s_j - s_c) = \begin{cases} 1 & (s_j - s_c) \geq 0 \\ 0 & (s_j - s_c) < 0 \end{cases} \tag{2}$$

Where s_j represent neighboring pixel value and s_c denotes the center pixel value.

$$T_{th\beta} = \beta T_{th} \tag{3}$$

Where $T_{th\beta}$ is the threshold value used for thinning the edges. The value β is a constant whose value is 1.5 [45].
The LBP method has some limitations like the degree of difference between the pixels is not described by this method. This limitation is overcome by the use of the hyper smoothing function in the LBP method. In this work hyper-function is calculated using the equation [45]:

$$HS(s_j - s_c) = \begin{cases} 0 & (s_j - s_c) < T_{th\beta} \\ \frac{1}{2}\left(\frac{1}{\pi} log\left(\frac{(s_j+s_c)^2}{(s_j-s_c)}\right)\right) & Otherwise \end{cases} \tag{4}$$

The neighboring pixels are compared to center pixel. If difference is greater than threshold value then HS (hyper smoothing) provides some value that appears as a white line.
The linear transformation and improved hyper smoothing function-based local binary pattern are calculated [45].

$$LTIHLBP = \begin{cases} 0 & T_{th} = 0 \\ \sum_{i=0}^{7} HS(s_j - s_c)2^i & 1 \leq T_{th} \geq 6 \\ 0 & T_{th} = 7 \end{cases} \tag{5}$$

The LTIHLBP method is used to calculate the accurate edges and to remove the noise from the images without destroying the edges. When the threshold values are 0 and 7 then the LTIHLBP value is zero, otherwise some positive number. The flow graph of LTIHLBP method is shown in Figure 1.
By using the LTIHLBP method the hottest region in the thermography images is calculated and encircled by the rectangular box. So the region of interest is easily identified and calculated by the LTIHLBP method.

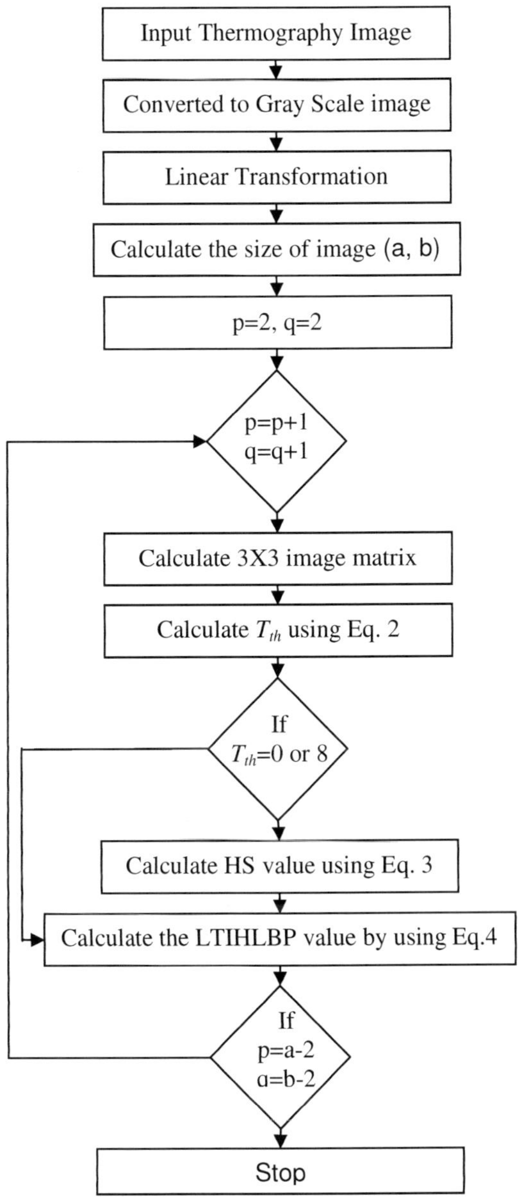

Figure1 Flowchart of LTIHLBP

3. Result and discussion

The experiments are performed in MATLAB 2018A on Intel Core i5-370M 3.1GHz 8 GB RAM Laptop. The experiment is evaluated on thermography images of rotating machines to calculate the effectiveness of the LTIHLBP technique. The result of the defect detection depends upon various parameters like geometric structure, noise in the images, and optical properties. The result of the LTIHLBP technique is compared with various segmentation methods. The thermal images are divided into three test benches. The first one is for synthetic images, the second for healthy images, and the third one for faulty images.

3.1. Test 1

Edges of synthetic images are drawn out by using various edge detection methods. The edge charts acquired with proposed method, IHLBP, Canny and Sobel techniques for a synthetic image of the star, are shown in Figure 2. The outcomes of objective differentiation methods unveiled that the Proposed method and IHLBP have detected proper edge maps. However, Sobel and Canny method is not providing the proper edge on the corner of the star. It is also shown that proposed and IHLBP methods contribute to narrow and continual boundaries.

The quantifiable investigation of proposed and other edge detection methods on the basis of accuracy and Jaccard Similarity Index is provided in Table 1. Results reveal that proposed and IHLBP method efficiently extracts the edges of an image. Furthermore, Figure 3 presents the graphical representation of noisy synthetic star images. It is concluded by outcomes that the proposed and IHLBP method provide large performance measures in comparison with other edge detection techniques. Hence the proposed and IHLBP methods reduce the limitations of Canny and Sobel method.

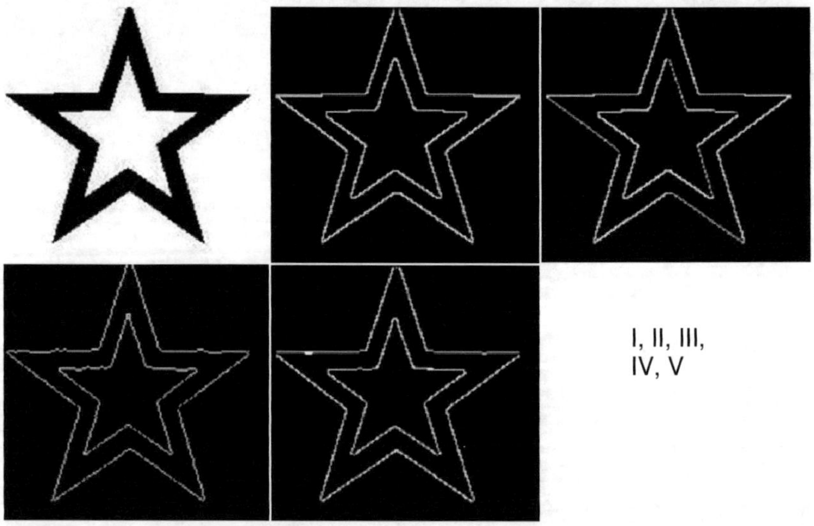

Figure 2 Edge detection of star image (synthetic image) using (I) input image, (II) Proposed method (III) IHLBP (IV) Sobel (V) Canny Method

Table 1 Performance measurement of various segmentation techniques

	Proposed Method	IHLBP	HLBP	Sobel	Canny
Accuracy	96.94	96.94	90.87	95.10	96.78
Jaccard Similarity	85.56	85.56	69.98	45.90	58.32

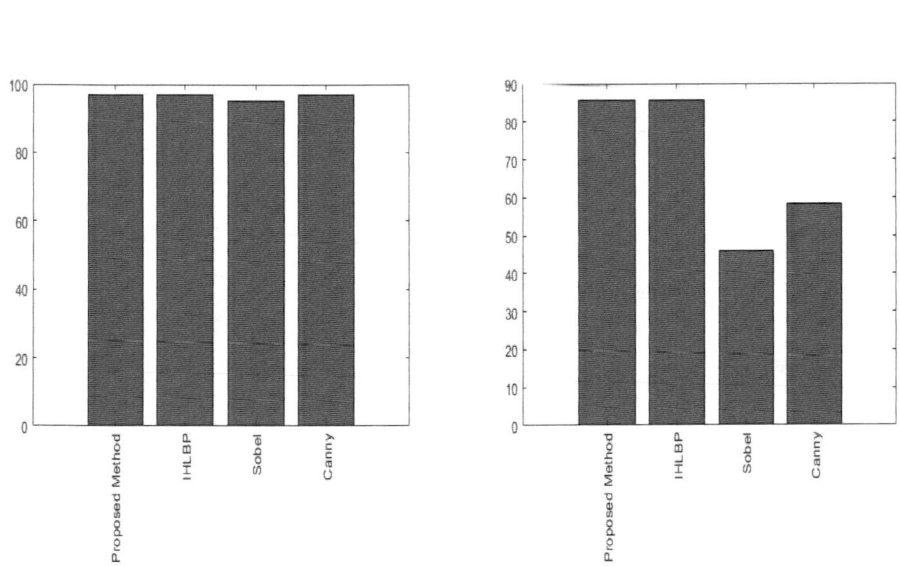

Figure 3 Graphical representation of performance measurement (a) Accuracy (b) Jaccard Similarity Index

3.2 Test 2

Thermography images have a complex background and contain all information about the object. In this test, 20 various thermography images of the healthy rotating object from the open sources are taken for testing purposes however the result of three images of the shaft with bearing house, a motor coupled with the shaft, and electric motor are shown in Figure 4. The proposed linear transformation and improved hyper smoothing function-based local binary pattern techniques successfully identified that all these testing machines are healthy. In these images there is no fault existing, so the LTIHLBP method has not detected any fault. Later the other segmentation techniques like IHLBP, Sobel, and Canny methods have detected faults in these healthy machines. These segmentation techniques are unable to differentiate the faulty and unfaulty parts in thermal image captured from rotating body whereby the proposed method can differentiate the healthy and faulty parts.

3.3 Test 3

In this experiment, The Linear transformation and improved hyper smoothing function based local binary pattern technique, improved hyper smoothing function based local binary pattern technique, Sobel and Canny segmentation techniques are utilized for fault detection in thermal images of rotation body. The test is performed on 20 faulty images out of this; the results of three images fan belt with a pulley system, the motor with coupling shaft, and shaft with bearing house are shown in Figure 5. The fault location in the three images is adjustable pulley in Fan belt and pulley system, motor bearing fault in motor coupling with a shaft, and bearing house fault in the shaft with bearing house. All fault detection method has produced a different pattern for all images and fault location. The proposed method is able to detect the proper location of the fault in all testing images. While the IHLBP method, Sobel method, and Canny method are not able to detect proper location of fault. The method selects the complete machine as a faulty portion in place of the proper fault position shown in Figure 5(C). The LTIHLBP method is better than other fault detection techniques as latter techniques have produced false information after image processing. Hence, an LTIHLBP technique has provided as much more information as possible to differentiate the healthy and faulty portions in rotating bodies.

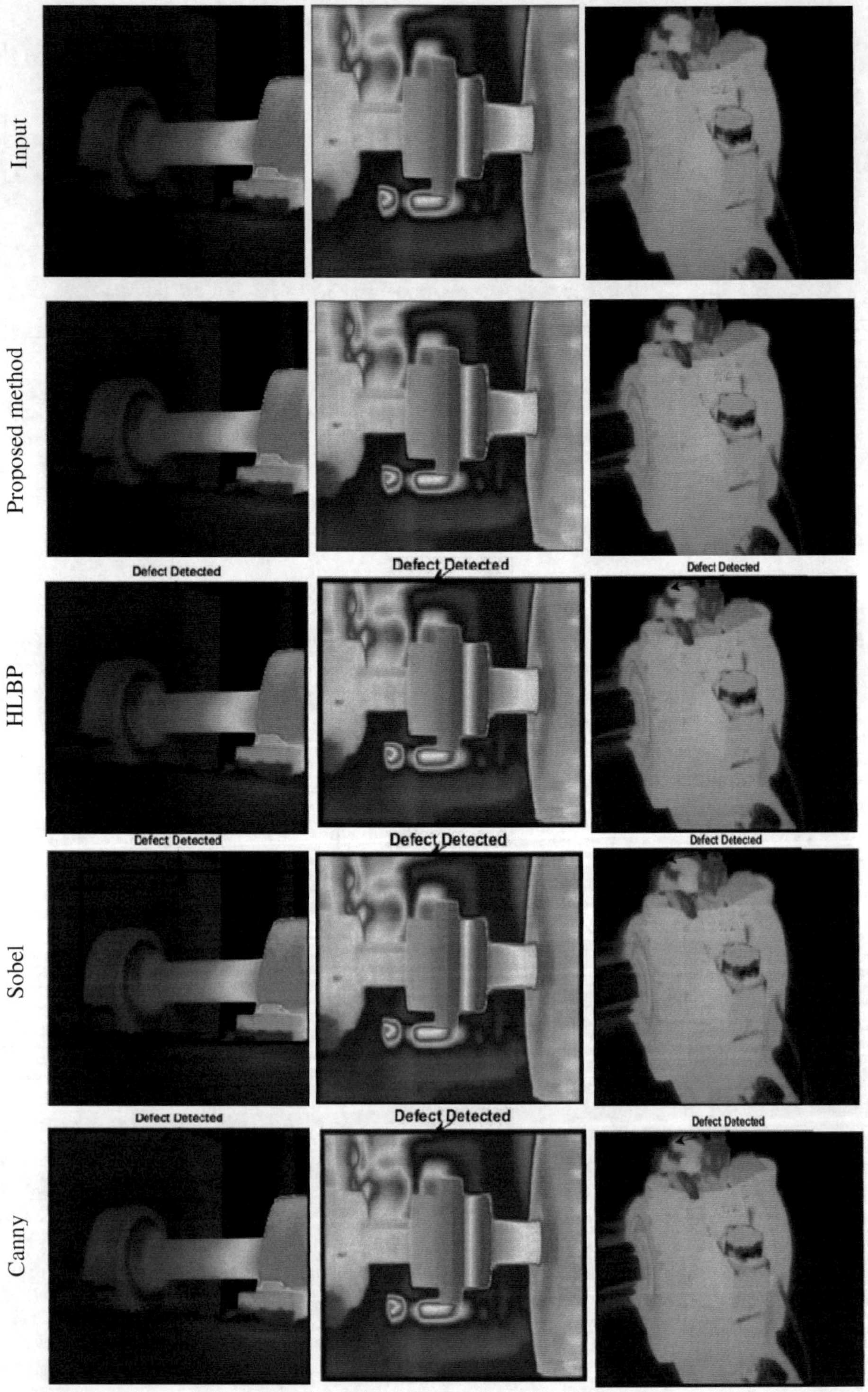

Figure 4 Comparisons of various fault detection techniques

IOP Conf. Series: Materials Science and Engineering 1228 (2022) 012009 doi:10.1088/1757-899X/1228/1/012009

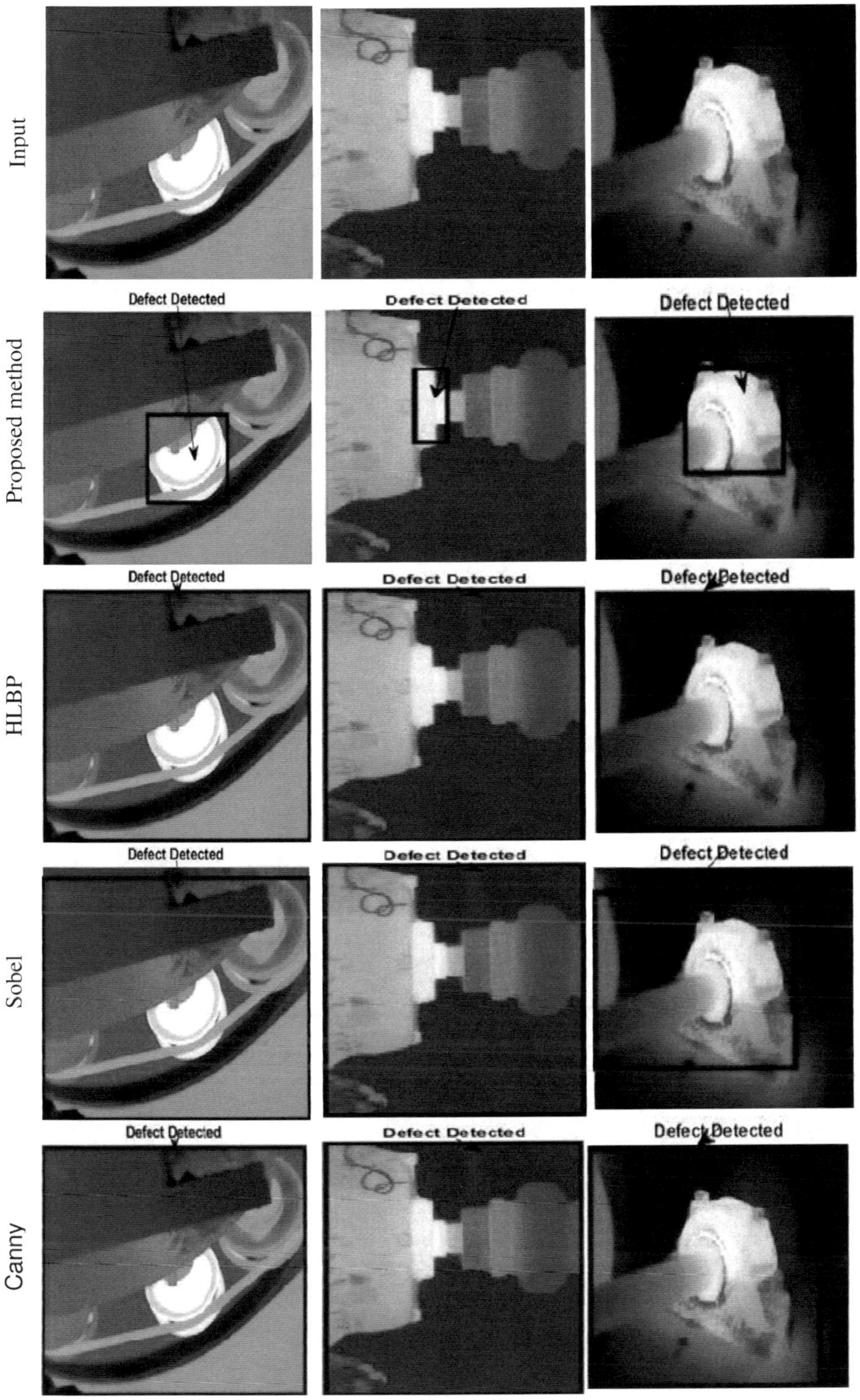

Figure 5 Comparison of various fault detection techniques using thermography images

The results obtained from the segmentation methods that are used in this paper are found to be very helpful in detecting the pre-fault in the rotating body. The proposed segmentation method can find out the proper condition of the rotating body either faulty or healthy. It is also able to find out the proper location of the fault in the moving body. On the other hand, other segmentation methods are producing a false result and not giving the proper location of a fault. Thus the suggested segmentation technique provides better result than other segmentation techniques.

4. Conclusion

Thermography-based image condition monitoring is applied in rotating elements to detect the faults based upon linear transformation and improved hyper smoothing based local binary pattern. Various methods based on image segmentation such as ILHBP, Sobel, and Canny have been compared with the LTILHBP method. The thermal image with LTIHLBP is used to diagnose the defects as it is capable to diagnose the hottest region of an image. Further, various image matrices are applied to thermal images for extracting the best fault information. To test the ability of the proposed techniques in detecting faults, forty various images of rotating elements are considered for investigation. Less time and computational complexity are two important benefits of the proposed method as they can diagnose the faults with more accuracy. Hence, the LTIHLBP technique proposed in this paper is successful in detecting the faults in the rotating machines by extracting the prime characteristic that might be promising tool for future works. In future, these features can be used by connecting them with different classifications system so that defects can be classified accordingly.

References

1. C. Ding, M. Zhao, J. Lin, J. Jiao and K. Liang, "Sparsity based Algorithm for Condition Assessment of Rotating Machinery Using Internal Encoder Data", *IEEE Trans. Ind. Electron.*, 2019, DOI 10.1109/TIE.2019.2941172.
2. H. Shao, H. Jiang, H. Zhang, and T. Liang, "Electric locomotive bearing fault diagnosis using a novel convolutional deep belief network", *IEEE Trans. Ind. Electron.*, vol. 65, no. 3, pp. 2727-2736, Mar. 2018.
3. Zhang, C., Yuling, L. I. U., Fangyi, W. A. N., Binqiang, C. H. E. N., Jie, L. I. U., & Bingbing, H. U. (2020). Multi-faults diagnosis of rolling bearings via adaptive customization of flexible analytical wavelet bases. *Chinese Journal of Aeronautics*, *33*(2), 407-417.
4. Li, Y., Li, G., Yang, Y., Liang, X., & Xu, M. (2018). A fault diagnosis scheme for planetary gearboxes using adaptive multi-scale morphology filter and modified hierarchical permutation entropy. *Mechanical Systems and Signal Processing*, *105*, 319-337.
5. Pandarakone, S. E., Mizuno, Y., & Nakamura, H. (2018). Evaluating the progression and orientation of scratches on outer-raceway bearing using a pattern recognition method. *IEEE Transactions on Industrial Electronics*, *66*(2), 1307-1314.
6. Younus, A. M., & Yang, B. S. (2012). Intelligent fault diagnosis of rotating machinery using infrared thermal image. *Expert Systems with Applications*, *39*(2), 2082-2091.
7. Prieto, M. D., Cirrincione, G., Espinosa, A. G., Ortega, J. A., & Henao, H. (2012). Bearing fault detection by a novel condition-monitoring scheme based on statistical-time features and neural networks. *IEEE Transactions on Industrial Electronics*, *60*(8), 3398-3407.
8. Kumar, H. S., Pai, P. S., Sriram, N. S., & Vijay, G. S. (2013). ANN based evaluation of performance of wavelet transform for condition monitoring of rolling element bearing. *Procedia engineering*, *64*, 805-814.
9. Sehgal, R., Gandhi, O. P., & Angra, S. (2003). Failure cause identification of tribo-mechanical systems using fault tree—a digraph approach. *Tribology international*, *36*(12), 889-901.
10. Angra, S., Mehta, N. P., & Rattan, S. S. (1996). Effects of elastic deformation of a bearing liner on performance of finite offset-halves pressure dam bearings. *Tribology Letters*, *2*(3), 273-285.
11. Sehgal, R., Gandhi, O. P., & Angra, S. (2000). Reliability evaluation and selection of rolling element bearings. *Reliability Engineering & System Safety*, *68*(1), 39-52.
12. Al-Musawi, A. K., Anayi, F., & Packianather, M. (2020). Three-phase induction motor fault detection based on thermal image segmentation. *Infrared Physics & Technology*, *104*, 103140.

13. Zhao, M., Jiao, J., & Lin, J. (2018). A data-driven monitoring scheme for rotating machinery via self-comparison approach. *IEEE Transactions on Industrial Informatics*, *15*(4), 2435-2445.
14. Wan, L., Han, G., Shu, L., Chan, S., & Feng, N. (2016). PD source diagnosis and localization in industrial high-voltage insulation system via multimodal joint sparse representation. *IEEE Transactions on Industrial Electronics*, *63*(4), 2506-2516.
15. Liang, X., Zuo, M. J., & Feng, Z. (2018). Dynamic modeling of gearbox faults: A review. *Mechanical Systems and Signal Processing*, *98*, 852-876.
16. Naha, A., Samanta, A. K., Routray, A., & Deb, A. K. (2017). Low complexity motor current signature analysis using sub-Nyquist strategy with reduced data length. *IEEE Transactions on Instrumentation and Measurement*, *66*(12), 3249-3259.
17. Wang, Y., Xue, C., Jia, X., & Peng, X. (2015). Fault diagnosis of reciprocating compressor valve with the method integrating acoustic emission signal and simulated valve motion. *Mechanical systems and signal Processing*, *56*, 197-212.
18. Shibata, K., Takahashi, A., & Shirai, T. (2000). Fault diagnosis of rotating machinery through visualisation of sound signals. *Mechanical Systems and Signal Processing*, *14*(2), 229-241.
19. Han, C. F., He, H. Q., Wei, C. C., Horng, J. H., Chiu, Y. L., Hwang, Y. C., & Lin, J. F. (2018). Techniques developed for fault diagnosis of long-range running ball screw drive machine to evaluate lubrication condition. *Measurement*, *126*, 274-288.
20. Zhao, M., & Lin, J. (2017). Health assessment of rotating machinery using a rotary encoder. *IEEE Transactions on Industrial Electronics*, *65*(3), 2548-2556.
21. Ciabattoni, L., Ferracuti, F., Freddi, A., & Monteriu, A. (2017). Statistical spectral analysis for fault diagnosis of rotating machines. *IEEE Transactions on Industrial Electronics*, *65*(5), 4301-4310.
22. Samuel, P. D., & Pines, D. J. (2005). A review of vibration-based techniques for helicopter transmission diagnostics. *Journal of sound and vibration*, *282*(1-2), 475-508.
23. Dalvand, F., Dalvand, S., Sharafi, F., & Pecht, M. (2017). Current noise cancellation for bearing fault diagnosis using time shifting. *IEEE Transactions on Industrial Electronics*, *64*(10), 8138-8147.
24. Yongbo, L. I., Xiaoqiang, D. U., Fangyi, W. A. N., Xianzhi, W. A. N. G., & Huangchao, Y. U. (2020). Rotating machinery fault diagnosis based on convolutional neural network and infrared thermal imaging. *Chinese Journal of Aeronautics*, *33*(2), 427-438.
25. Ciampa, F., Mahmoodi, P., Pinto, F., & Meo, M. (2018). Recent advances in active infrared thermography for non-destructive testing of aerospace components. *Sensors*, *18*(2), 609.
26. Laborda, A., Robinson, A., Wang, S., Zhang, Y., & Reed, P. (2018). Fatigue assessment of multilayer coatings using lock-in thermography. *Materials & Design*, *141*, 361-373.
27. Grammatikos, S. A., Kordatos, E. Z., Matikas, T. E., & Paipetis, A. S. (2018). On the fatigue response of a bonded repaired aerospace composite using thermography. *Composite Structures*, *188*, 461-469.
28. Bagavathiappan, S., Lahiri, B. B., Saravanan, T., Philip, J., & Jayakumar, T. (2013). Infrared thermography for condition monitoring–A review. *Infrared Physics & Technology*, *60*, 35-55.
29. Chou, Y. C., & Yao, L. (2009, December). Automatic diagnostic system of electrical equipment using infrared thermography. In *2009 international conference of soft computing and pattern recognition* (pp. 155-160). IEEE.
30. Pal, N. R., & Pal, S. K. (1993). A review on image segmentation techniques. *Pattern recognition*, *26*(9), 1277-1294.
31. Singh, G., Kumar, T. C. A., & Naikan, V. N. A. (2016). Induction motor inter turn fault detection using infrared thermographic analysis. *Infrared Physics & Technology*, *77*, 277-282.
32. Jadin, M. S., & Taib, S. (2012). Recent progress in diagnosing the reliability of electrical equipment by using infrared thermography. *Infrared Physics & Technology*, *55*(4), 236-245.
33. Chawla, V. K., Chanda, A. K., Angra, S., & Rani, S. (2019). Effect of nature-inspired algorithms and hybrid dispatching rules on the performance of automatic guided vehicles in the flexible manufacturing system. *Journal of the Brazilian Society of Mechanical Sciences and Engineering*, *41*(10), 1-17.
34. Chawla, V. K., Chanda, A. K., & Angra, S. (2019). Automatic guided vehicle systems in flexible manufacturing system–A review. *International Journal of Industrial Engineering: Theory, Applications and Practice*, *26*(5).
35. Chawla, V., Chanda, A. K., Angra, S., & Bonyadi Naeini, A. (2020). Coexistent scheduling in the tandem flow path configuration of a flexible manufacturing system by using an advanced grey wolf optimizer. *Scientia Iranica*.

36. Sharma, N., Chawla, V., & Ram, N. (2020). Comparison of machine learning algorithms for the automatic programming of computer numerical control machine. *International Journal of Data and Network Science*, *4*(1), 1-14.
37. Chawla, V., Chanda, A., & Angra, S. (2019). Material handling robots fleet size optimization by a heuristic. *Journal of Project Management*, *4*(3), 177-184.
38. Chawla, V., Angra, S., Suri, S., & Kalra, R. (2020). A synergic framework for cyber-physical production systems in the context of Industry 4.0 and beyond. *International Journal of Data and Network Science*, *4*(2), 237-244.
39. Solanki, A., & Kumar, E. (2017). Design and development of web enabled fuzzy expert system using rule advancement strategy. *International Journal of Intelligent Systems Design and Computing*, *1*(1-2), 3-27.
40. Palia, N., Kant, S., & Dev, A. (2019). Performance evaluation of speaker recognition system. *Journal of Discrete Mathematical Sciences and Cryptography*, *22*(2), 203-218.
41. Vasti, M., & Dev, A. (2020). Classification and Analysis of Real-World Earthquake Data Using Various Machine Learning Algorithms. In *Advances in Data Sciences, Security and Applications* (pp. 1-14). Springer, Singapore.
42. Garcia-Ramirez, A. G., Morales-Hernandez, L. A., Osornio-Rios, R. A., Benitez-Rangel, J. P., Garcia-Perez, A., & de Jesus Romero-Troncoso, R. (2014). Fault detection in induction motors and the impact on the kinematic chain through thermographic analysis. Electric Power Systems Research, 114, 1-9.
43. Garcia-Ramirez, A. G., Morales-Hernandez, L. A., Osornio-Rios, R. A., Garcia-Perez, A., & Romero-Troncoso, R. J. (2014, September). Thermographic technique as a complement for MCSA in induction motor fault detection. In 2014 International Conference on Electrical Machines (ICEM) (pp. 1940-1945). IEEE.
44. Navdeep, Singh, V., Rani, A., & Goyal, S. (2020). An improved hyper smoothing function based edge detection algorithm for noisy images. Journal of Intelligent & Fuzzy Systems, 38(5), 6325-6335.
45. Navdeep, Singh, V., Rani, A., & Goyal, S. (2020, February). Improved depth local binary pattern for edge detection of depth Image. In 2020 7th International Conference on Signal Processing and Integrated Networks (SPIN) (pp. 447-452). IEEE.

IOP Conf. Series: Materials Science and Engineering 1228 (2022) 012010 doi:10.1088/1757-899X/1228/1/012010

Design and implementation of RESTFUL API based model for vulnerability detection and mitigation

Bhavya Modi, Uday Chourasia, Rajeev Pandey

University Institute of technology, RGPV
Bhopal

Modi.bhavya88@gmail.com, udaychourasia@rgtu.net, rajeevpandey@rgtu.net

Abstract. With the rapid rise of online applications in the contemporary digital world, security concerns are also developing day by day. A web application is a collection of several web pages that are kept on a server. It is stated that as the economy grows, so will the security risks. There are numerous current web vulnerabilities that are still a significant concern, and with time, there will be much more new vulnerability that emerges as technology advances. In this paper, There is a detailed study of various web threats i.e., vulnerabilities that are emerging and proposed a system which detects those vulnerabilities, resolves them for the user and generates a user log for the administrator to monitor the type of vulnerabilities. In the proposed model we have generated the system in the format of API. In the testing phase we've detected various web vulnerabilities such as SQL Injections, Cross Site Scripting, Session Hijacking, Unauthorized access, Middle man attack etc.

1. Introduction

Our dependency on web applications is increasing with each passing day. Whether it be railway ticket booking, online movie tickets, Hospital reports or patient data, or financial sector everything around us is dependent on web applications. With this rapid growth the frequency of attacks is also increasing. The resources available and ease of use of the internet is the main factor which contributes to the instantaneous growth of web vulnerabilities.
In the last one decade there's a significant rise in cyber attacks in which the most common attacks are Structured Query Language (SQL) Injection Attack, a Man-in-the-middle (MitM), Phishing Attack, Denial of Service Attack (Dos).
The following records shows the recent statistical data related with the web application attack:

- Form-jacking code compromises 4,800 websites per month on average (Symantec).
- 329 days and 233 days, respectively, were spent in the data breach lifecycle by the healthcare and banking industries (IBM).
- Between 2017 and2018, the global number of web assaults banned each day climbed by 56.1 percent (Statista).
- Office apps were the most often used applications globally in the third quarter of 2018. (Statista).
- Between 2017 and2019, the number of persons affected by health data breaches increased by 80%. (Statista)

Content from this work may be used under the terms of the Creative Commons Attribution 3.0 licence. Any further distribution of this work must maintain attribution to the author(s) and the title of the work, journal citation and DOI.
Published under licence by IOP Publishing Ltd

- Form jacking assaults may net cyber criminals up to $2.2 million by collecting only 10 credit cards each website (Symantec).
- Hackers may target users in around 9 out of 10 web applications. Some frequent types of attacks include sending users to a particularly harmful link, phishing attempts for credential theft, and infecting machines with malware, adware, and spyware.
- Unauthorized access to the application is feasible on 39% of the sites. According to 2019 statistics, for 16% of online applications, complete system control may be acquired. On 8% of the systems, complete control of the web application allowed for local network attacks.
- In 68 percent of online apps, sensitive data breaches posed a risk. According to some assessments, there is a 47 percent probability of a breach for personal data and a 31 percent chance for Credentials.
- Approximately 82 percent of the problems were discovered in the application code itself.
- When compared to 2018, the average number of vulnerabilities per online application increased by one-third.
- According to studies, each system had an average of 22 vulnerabilities, four of which were extremely dangerous.
- In 2019, one out of five vulnerabilities was classified as very vulnerable.

By reviewing these stats we can say that the security of web applications is the most important for the safety of the web application owner and the application users as well. A web application is critical software which contains complex structure and data flow. Any loophole or data breach in a web application can result in heavy data loss which can result in quite a large amount of financial loss to the web app users.

In the current scenario the web application development and the testing phase contains no security assistance and hence there are web applications which contain many vulnerabilities. Developing a secure web application requires a significant amount of time, employees or developers with good knowledge of development and security about the process to develop an application. And due to lack of these basic skills there remains security vulnerability in the web application. So to fill this knowledge gap there is a middle solution to the problem which is Web Application Firewall (WAF). A WAF is a firewall which helps in monitoring and filtering of traffic between the web application server and the internet i.e., between the client - side and server side of an application. When an application firewall is deployed in the front of the web application it creates a barrier between the web application and the internet. The WAF acts as a reverse - proxy that provides a security shield for the server and the server from exposure by requiring the client to pass through the security shield before accessing the server.

2. Related Researches in the field of Web Application Firewall

2.1. A Firewall engine for web application

The The Web Application Firewall's core component is the WAF Processor [10]. The Web Application Firewall and the Web Server are both installed on the same computer. Only the Apache HTTP Web Server is secured by the Web Application Firewall. The Packet Analyzer Module and the Configuration Module are the two components that make up the WAF Engine. In this research, 2 injection vulnerabilities are used for testing of the Web Application Firewalls that are SQL injection and XSS injection vulnerability. Distinct types of SQL and XSS injection vulnerability are used during the testing phase. Packets are queued at the kernel level in the packet analyzer module, where they are examined. These packets are analyzed using the command line instructions or file shell commands. Iptables is a command line instruction that queues packets at the kernel level and sends them via the Packet Analyzer first. The Packet Analyzer intercepts packets that approach the Apache HTTP Web Server for data analysis. The data from the packets that were examined is analyzed to measure whether the packets should be allowed or rejected. The information about the rejected packets is logged so that it may be analysed later, in the stored logs there are details about the user socket details and the vulnerability detected.

2.2. A SQL Injection & DDOS

The SOA platform allows a group of web services applications to be unified and coordinated over a network. The XML communications framework designed for SOA-based applications has a visible security wall. Bypassing the HTTP and SMTP protocols, SOAP messages may be read and other sorts of attacks can be carried out. This issue can be resolved by implementing the information in the particular SOAP message. This will authenticate the user every time when a request is passed to the server [12]. Passing over a firewall causes security issues that can disrupt business networks and result in DDOS assaults Web services also use digital signatures in order to provide security which lies in non-repudiation techniques. These techniques allow various structures or formats for the digital signature, also availability of multiple encryption algorithms is provided in this method. The digital signature used in this case contains the information about the service holder of the corresponding private key. The '*Integrated Application and Protocol Framework*' is being developed particularly for DDOS and SQL Injection Attacks [12]. Anti-attack protections are implemented using the UDDI protocol, WSDL protocol, and SOAP protocol inside this framework.

2.3. A SQL Injection & DDOS

An attacker can inject malicious input into a web application and modify the program's functionality by compelling it to execute particular instructions during an injection attack.

An injection attack can expose or destroy data, cause a denial of service, or compromise the entire web server. Such attacks are feasible because of flaws in an application's programming that allow for invalidated user input. If an application performs run-time X-Path query formation, injecting unsafe user input into the query, the attacker will be able to enter the data into the query, causing the freshly created request to be processed differently than just the developer expected. It is said that, filtering user-supplied/controlled inputs to the best of your abilities is always preferred. GET/POST parameters are among some of the request: Parameters for URLs, Cookie parameters, HTTP header parameters

. To identify assaults on Web services, a programme named CIVS-WS (*Command Injection Vulnerability Scanner for Web Services*) [3] was created. It aims to identify similar assaults by training the system query types. Ultimately, the scanner tokenizes the query and fetches the request parameter. Various types of queries could also be tokenized in the manner that user requests are entered. After acquiring the request parameters, a probable vulnerability should be identified. In order to identify any form of suspected injection, this approach must be general and effective.

2.4. A CDATA Field exploit, XML Injection, SOAP attacks, Mitnick attack

Kevin Mitnick was the first to carry out the Mitnick assault in the United States. The technique makes use of the flaws discovered in TCP's three-way handshake mechanism. Assuming that two devices called H1 and H2 are connected over TCP, the attacker attempts to impersonate one of the devices by disrupting the connection between X1 and X2. If it acts like X1 here, it will assault the X1 device with a Syn / Flood attack. It also sends multiple TCP/IP packets to X2 in order to determine the TCP/IP sequence number X2 produces for X1. By spoofing X1's IP address, the attacker impersonates X1 and delivers the SYNC packet to X2.The handshake procedure begins with this step. The objective is to maintain a continuous TCP connection between the attacker and the X2. H1 will wish to terminate the connection since he is vulnerable to a flood assault. X2 sends a SYNC / ACK message to X1, but X1 is unable to send a REST packet to terminate the connection since it is under SYNC flood attack. In this scenario, it is presumed that a secure connection is created between the X2 device and the X1, and that X2 is unaware that it is really linked to a hostile individual based on the TCP/IP sequence number and handshake process.

2.5. DDOS, XDOS

There are two modes in the system that has been created for this study. Normal and assault modes are available. A verification request is not sent if the system is not in regular mode. If we put the system into attack mode, the services that server requests have to be validated before the processing. Only one of the requests made for the attack will be handled if any of them are verified. As a result, the server does not waste resources as a result of resource hacking. Additionally, the server uses resources to do these validations. The request that results must be authenticated.

By making a high number of requests that demand verification and authentication, the exploiter is not able to drain the user side resources. A technique known as the operations provider has been created to prevent this. All the incoming requests for the authentication are routed through the service provider through the name of the authenticator of the operation provider.

2.6. DOS-DDOS

Distributed Denial Of service (DDOS) assaults is split into two categories in this study: *Semantic & Flood attacks*. Flood assaults are explained as - when a service does not reply to other packages by delivering a large number of normal packets to the web service server. Semantic attacks are those in which an attacker sends a large number of WS-Security headers with nested encrypted objects. In the semantic attack the necessity of responding to a request requires a significant amount of resource and time to react to the request, the service will no longer respond to other packages. To avoid these two scenarios, the author devised the Client Puzzle technique. When a client application requests to communicate with another node, both machines exchange a series of texts simultaneously. This is referred to as the 3 Way Handshaking mechanisms. The client first sends a SYNC (synchronization) message to the server. The server then responds with SYNC - ACNK (acknowledgement) message to acknowledge the client's SYNC message. The client then responds with an ACNK message to conclude the setup process. After that, the connection has been established, and flow of data between both the server and the client proceeds. A malicious user faked source IP addresses to send a series of SYN packets to the victim in order to exploit this method. As a result, the victim responds with SYNC - ACNK message and then waits for a predetermined period of time for an ACNK to arrive in order to complete the connection creation. Here discrimination of DOS and DDOS assaults are possible under IPv4 and IPv6.

Before beginning a complex or expensive process in terms of incoming requests and calculating the response, the server requests the user to answer various intermediaries to tough problems. The final result is to differentiate between the person who makes the call for the request and the user who is the automated bots in this case.

2.7. XSS (Cross site scripting) Attacks

The system that is developed detects the XSS susceptible web apps that are used to evaluate our technology that is widely available on the Internet. We collected a number of XSS susceptible websites and manually added test scripts to HTTP requests before sending them to the vulnerable websites. The manually inserted test scripts are only the most basic type of XSS attack scripts for demonstration purposes; more complex and complicated scripts (JavaScript, VBScript, etc.,) exist, but they are outside the scope of our discussion; additionally, XSS vulnerabilities exist in web applications, not scripts. Actually, HTML tags such as "", ">", and "The main objective in this of a XSS attack is to trigger malicious Scripts in the user's web browser, though there are a few fundamentally different strategies to achieve those goals. [18]

As a result, the known solutions for these vulnerabilities are classified using the Sdlc, and their effectiveness is validated using a debugging tool. The conclusions of a security flaw before and after the execution of appropriate solutions illustrate that if systems are established with protection in order to move ahead of the SDLC, many threats on web applications could be mitigated basically without any significant work or expense

IOP Conf. Series: Materials Science and Engineering 1228 (2022) 012010 doi:10.1088/1757-899X/1228/1/012010

2.8. The technology utilised by many academics and scholars in the subject of online security and web vulnerabilities is shown in this table.

S.no	Topic	Objective of the research	Technique used
1	Firewall engine for web application	Data packets are examined for acceptance and rejection of data.	The Packet Analyzer Module and the Configuration Module are the two components that make up the WAF Engine
2	D-Dos Attack And SQL Injection	Detection of exploitation and attack types.	Anti-attack protections are implemented using the UDDI, WSDL, and SOAP protocols.
3	X-PATH Injection And SQL Injection	Identification of X-Path and SQL injection vulnerabilities.	Two phase single pass incremental k-means algorithm. CIVS-WS (Command Injection Vulnerability Scanner for Web Services) was created
4	DDOS, XDOS	Prevent hack attacks	Created Normal and attack mode, where in normal mode the system will not perform the validation checks, whereas in attack mode the client request is verified before sending it to the server.
5	CDATA Field exploit, DDOS attacks, XML Injection, SOAP attacks, Mitnick attack	Vulnerabilities are detected automatically.	Various different types of rules are defined in order to detect various vulnerabilities in the web application.
6	DOS-DDOS	Attack Scenarios	To acquire analysis data, the Simple Network Management Protocol (SNMP) and a network interface card are utilized.
7	XSS (Cross site scripting) Attacks	Penetration testing, security weakness, and hacking prevention	The research includes data from the injection process analysis.

Table No - 1

3. Problem Statement

Application layer vulnerabilities have been attacked with significant repercussions in recent years: the exploits have tried various attacks like faking multi vendor websites for sending commodities for free, collected login credentials, and released various personal information (such as addresses and credit-card details). Researchers are beginning to look at new tools and approaches that attack the challenge of application layer web security from many angles: before, during, and after the fact. Web application firewalls appear to be a viable method of preventing application attacks but they have their own setbacks. When we setup an existing firewall into our systems it works as a middleware which is imported into the system as a library but it impacts the performance of the application as these tools analyse all incoming and outgoing information at the application layer. However, when the firewall examines and evaluates each packet, it must use CPU cycles. As a result, the inspection process takes longer than conventional packet-filtering firewalls, significantly limiting network performance.

IOP Conf. Series: Materials Science and Engineering 1228 (2022) 012010 doi:10.1088/1757-899X/1228/1/012010

4. **Proposed Work**

The relevance of online services is evident from the earlier study on web services. With this level of significance, it's understandable that web service vulnerabilities might create severe issues. Our main motive in this research work is to study various web application vulnerabilities and implement the web application firewall into our existing system to detect these vulnerabilities. Our system monitors every request that has been sent to the server and filters the data before it is communicated with the database. If any vulnerability is detected in the client request then the request is filtered out and a user log is generated in the Json format which contains the type of vulnerability detected and the user details through which the request is generated. With the help of this system we try to achieve multiple goals. In this setup of web app firewall an algorithm is proposed or developed to handle the existing web application vulnerabilities and the system also creates the client record for the easy verification of attackers log or details. Steps involved in the development of the application firewall :
Step - 1: Initially we've set up an environment for the testing of the Web App Firewall that we've created. The environment setup step consists of creating a request from client side with and without the implementation of firewall Api that we've created. This authentication step works as a double security layer for our firewall where only authenticated users are allowed to send requests.

Sample request looks like:

Name: username

Data: Some Data

Step - 2: When the request is sent from the client - side to server - side then our firewall comes into action. Every request has to pass through the web application firewall for the filtration of data and checking of web application vulnerabilities exists in the request.

Here the web application firewall is called as an API from the client request before sending it to communicate with the database. So for the API authentication we have created some API keys that are stored in encrypted form using MD5 algorithm. These API Keys decide whether the request that is coming from the client is authenticated or unauthenticated. If the system catches any unauthenticated request from the client then it will block the request and the request will not be processed further.

Step - 3: If the API Authentication step is successfully authenticated then the request is detected for the method for which it is requesting data. For example POST or GET methods. The client has requested for which method. Here we have differentiated the POST and GET method requests via CSRF token authentication.

In POST request the request must contain a CSRF unique token which will be stored in the session. This CSRF token is validated when a POST request is sent to the server. If the Client request does not contain a CSRF Token then the request will be considered as Unauthenticated Requests. This step provides the security check for **Unauthorized Requests, Middle man attack and Session hijacking attacks**.

In GET requests the request headers must contain the CSRF Token. It will also work the same as the POST method requests but the basic difference between both the methods is that in POST it will find the CSRF Token in the client request whereas in the GET method it will find the Token in the request header.

Step - 4: Now the request is checked for the next vulnerability i.e., **SQL Injection.** We have researched various conditions for the detection of SQL Injection from the previous studies and research works. These studies helped us to create a set of rules that detected the SQL Injection vulnerability. Our firewall detects the vulnerability and then filters the data for the vulnerability. The data is processed and then sent to the server for the processing. Here we have detected the problem and diagnosed it accordingly. The user log is generated for the administrator as well that SQL Injection is reported in the client request but adequate measure has been taken.

Step - 5: Now the next step is to check requests for another vulnerability i.e., **Cross Site Scripting (XSS).** In Cross Site Scripting attacks the request that is sent by the client contains some HTML tags or the scripting language which when submitted to the database can result in the access of the database or even the personal information of users stored on the server. We detect the scripting tags and the html tags and filter them out. The request is filtered and the processed data is sent to the server for further processing. In this condition the user log is generated for the administrator as well that XSS Attack is reported in the client request but adequate measure has been taken.

Step - 6: Now the next step is to check for the file type that has been uploaded by the user. These files are included in the multipart / form-data type of request where the client can send any images to the server for uploading. We have created a valid MIME file type data for the validation of authentic file types that must be uploaded to the server. The file that is given by the client is first checked by its extension. If some malicious extension is detected by the firewall it will block the file from uploading to the server as the malicious file can cause damage to the server. It can either lead to the crashing of the server. This step detects the **unauthorized file uploading** attacks. The file updating is blocked and the user log is generated stating the unauthorized file uploading attempt.

5. Results

In the previous studies the Web App Firewall is implemented as libraries in the system or the SDK are created that are included in the web applications. But we have implemented the firewall in the form of REST Api using the curl request because REST Api's are platform independent i.e., the firewall we have created can be integrated in any of the system and not only in Php Applications. It is one of the biggest advantages of using REST Api's.

We have secured the API calling with the use of API authentication key which will be unique for all the users. Without the api key or with invalid api key neither the user logs can be accessed nor the web firewall be used. We tried to cover most of the web vulnerabilities that can be found in most of the web systems. These vulnerabilities are SQL injection, Cross Site Scripting Attacks (XSS), Session Hijacking, Business Logic attacks, Unrestricted file upload, Unauthenticated requests, Middle man attack, unauthorized users blocking. It makes our system more efficient for detection of various exploits because the previous models that are created provide detection of limited vulnerabilities.

6. Conclusion

On studying different web vulnerabilities and their implementation and detection in the research study we can conclude that if we implement the web app firewall in the form of REST Api then it can help reduce the load on the client server as the data is processed on a mirror server where the Api is stored. Also the client requests are always safe because we filter the vulnerability and process the data then send the request back to the client server, So the client data is never lost as other firewalls block the requests. For further research, various more sets of rules can be implemented and the client logs can be more detailed so it will help in expanding the firewall that we have created.

On the other hand we can also create an alert system for the administrator or the client so when any vulnerability is detected then it will automatically inform the administrator.

IOP Conf. Series: Materials Science and Engineering 1228 (2022) 012010 doi:10.1088/1757-899X/1228/1/012010

References

[1]. Yogesh Kumari, "Survey on Web Application Vulnerabilities," © 2019, IRJET | Impact Factor value: 7.34 | ISO 9001:2008 Certified Journal | Page 922 - 925.
[2]. http://www.owasp.org/index.php/Top-10-2010-A1-Injection retrieve on 13/01/2010.
[3]. Tajpour,Atefeh, Maslin Masrom, Mohammad Zaman Heydari, and Suhaimi Ibrahim. "SQL injection detection and prevention tools assessment" In Computer Science andInformation Technology (ICCSIT), 2010 3rd IEEE International Conference on, vol. 9, pp. 518-522. IEEE,2010
[4]. Shar, L. S. Tan, H. B. K. and Briand, L. C. "Mining SQL injection and cross site scripting vulnerabilities using hybrid program analysis," Proc. of Int. Conf. on Software Engineering (ICSE '13) IEEE Press, pp 642- 651, 2013.
[5]. Web Application Security Consortium: Thread Conclusion version 1.0.0 by webappsec.org
[6]. Classification of Web Application Vulnerabilities, Savita B. Chavan and Dr. B. B. Meshram, V.J.T.I, Mumbai, India V.J.T.I, Mumbai,India
[7]. A Survey on Server-side Approaches to Securing Web Applications XIAOWEI LI Vanderbilt University YUAN XUE VanderbiltUniversity
[8]. Web Application Vulnerabilities: A Survey, Vandana Dwivedi, Himanshu Yadav and Anurag Jain RITS, Bhopal (India)
[9]. Z. Su and G. Wassermann. The essence of command injection attacks in Web applications. In Proc. POPL, 2006
[10]. Bin Wang1 ,Lu Liu2 ,Feng Li1 ,Jianye Zhang3,Tao Chen3 ,Zhenwan Zou1, "Research on Web Application Security Vulnerability Scanning Technology",2019 IEEE 4th Advanced Information Technology, Electronic and Automation Control Conference (IAEAC 2019).
[11]. S. Prandl, M. Lazarescu, Duc-Son Pham, "A Study of Web Application Firewall Solutions", ICISS 2015: Proceedings of the 11th International Conference on Information Systems Security - Volume 9478 December 2015 Pages 501–510 , https://doi.org/10.1007/978-3-319-26961-0_29.
[12]. Aruna S, "Security in Web Services- Issues and Challenges",International Journal of Engineering Research & Technology (IJERT) http://www.ijert.org ISSN: 2278-0181 IJERTV5IS090245 (This work is licensed under a Creative Commons Attribution 4.0 International License.) Published by : www.ijert.org Vol. 5 Issue 09, September-2016
[13]. R. A. Muzaki, O. C. Briliyant, M. A. Hasditama and H. Ritchi, "Improving Security of Web-Based Application Using ModSecurity and Reverse Proxy in Web Application Firewall," *2020 International Workshop on Big Data and Information Security (IWBIS)*, 2020, pp. 85-90, doi: 10.1109/IWBIS50925.2020.9255601.
[14]. V. Clincy and H. Shahriar, "Web Application Firewall: Network Security Models and Configuration", *42nd IEEE International Conference on Computer Software & Applications*, 2018.
[15]. https://blog.sagipl.com/top-web-app-development-stats/
[16]. https://www.giac.org/certifications/certified-web-application-defender-gweb/
[17]. S. Prandl, M. Lazarescu and D.-S. Pham, "A Study of Web Application Firewall Solutions", *ICISS Perth*, 2015.
[18]. Ms. Daljit Kaur, Dr. Parminder Kaur, "Cross-Site-Scripting Attacks and Their Prevention during Development", 2017 IJEDR | Volume 5, Issue 3 | ISSN: 2321-9939.

Comparative study of Nano-fluids as Coolants in a Car Radiator

Aditi [1], Zoya Farooque [2] and Nathi Ram Chauhan [3]

[1] Surface Sustaining Engineer, Cameron A Schlumberger Company, Coimbatore, Tamil Nadu, India-641014
[2] Assistant Manager, Maruti Suzuki India Limited, Gurugram, Haryana, India-122015
[3] Professor, Head of Department, MAE, Indira Gandhi Delhi Technical University for Women, Delhi, India-110006
E-mail:aditikumar911@gmail.com [1], zoya.farooque@gmail.com [2], nramchauhan@gmail.com [3]

Abstract. The automobile industry has been researching and developing ways to maximize the efficiency of the engine while ensuring the cars remain compact. One of the ways to achieve this result is by decreasing the frontal area of the car, which not only makes the car compact but will also provide the enhanced aerodynamic effect. The efficiency of a car radiator greatly depends upon the rate of heat transfer of the radiator that in turn depends on the properties of the coolant used, a coolant with higher thermal conductivity has the ability to transfer heat at a faster rate and cool down the engine quickly which results in better engine efficiency and a decrease in need of coolant volume required, pumping power and eventually the size of the radiator. That will result in a smaller frontal area which leads to engine efficiency, compact cars, and better fuel economy, and lesser emissions. Through our research, we have demonstrated that with the use of nanofluids, by virtue of their enhanced thermal conductivity property, they offer a better heat dissipation than conventional coolants thereby having a potential practical application.

Keywords: Coolant, CFD, Nanofluids, Radiator

1. Introduction

The performance of engines depends upon many systems present in the automobile, and one of the systems that adversely affect the performance is the inadequacy of the cooling system. As we know, the major component of the cooling system of an automobile is the radiator, the function of which is to cool the engine down by heat transfer. Conventional coolants consist mainly of water and ethylene glycol, present in fifty-fifty proportion by volume. But, with the advancement of technology, a need was identified to develop new coolants because the conventional coolants came with many disadvantages and a new class of coolants were developed called Nanofluids. Since the development of nanofluid coolant is in the extensive research phase. The concept of nanoparticles was first developed by Choi [1]. He reported an improvement in thermal conductivity when the nanoparticles are dispersed in typical heat transfer fluids. On further development in this area, Choi [2] also stated the shortcomings of the typical cooling system, which are as follows:

(i) Conventional coolants and oils have characteristically inferior heat transfer properties.

Content from this work may be used under the terms of the Creative Commons Attribution 3.0 licence. Any further distribution of this work must maintain attribution to the author(s) and the title of the work, journal citation and DOI.
Published under licence by IOP Publishing Ltd

IOP Conf. Series: Materials Science and Engineering 1228 (2022) 012011 doi:10.1088/1757-899X/1228/1/012011

(ii) In order to compensate for the inferior heat transfer properties, design modifications are required to be made in the radiator fins.

Kulkarni et al. [3] explored the effect of aluminum oxide and water-based nanofluids in diesel-electric generators and discovered that there will be a reduction in cogeneration efficiency. Leong et al. [4] analyzed and stated the effect of Ethylene Glycol and Copper nanofluids in the cooling system of an automobile and found that the co-efficient of heat transfer was raised by 3.8%. Peyghambarzadeh et al. [5] found that the application of several fluids such as pure ethylene glycol, pure water, and the binary blends of pure water and pure ethylene glycol with Al_2O_3 nanoparticles. He noticed that the car radiator's cooling performance was improved by 40% when nanofluids (pure water and Al_2O_3 nanofluid and Pure ethylene glycol and Al_2O_3 nanofluid) were used as coolants. Naraki et al. [6], were able to obtain the overall coefficient of heat transfer of CuO-water nanofluids which was supported by experimental investigation

While conducting the experiments on convection heat transfer [9-14], Putnam et al. [7-8] found that the effective static thermal conductivities of nanofluids composed of Silver. Significant improvements were also observed in the results of the experiment. In a project started in 2008, Roubert et al. which involved industrial cooling using nanofluids resulted in energy savings and reduction in emissions.

Singh et al. [16] performed an investigation in which they used nanofluid with enhanced thermal conductivity in radiators, which could show a decrease of about 10% in the radiator's frontal area. Therefore, there was also an increase in the energy efficiency and performance of automobiles. Utilizing fluid alumina as a coolant on a vehicle's plain fin compact heat exchanger, Vasu et al. [17] project reached the conclusion that there will be a decrease in the rate of heat transfer by raising the temperature of the air inlet. In the investigation done by Tzeng et al. [18], the team deduced that there was a clear advantage in the employment of nanofluids in transmission from the viewpoint of thermal performance. By using the nanofluids in the radiator, Ravikanth et al. [19] studied the performance of heat transfer. In the experiment, Ravikanth and his team used CuO and Al_2O_3, they inferred that with particle volumetric concentration and Reynolds number, the average coefficient of heat transfer enhanced. In the study of the performance of diesel-electric generators, Kulkarni et al. [20] spread the Aluminum Oxide in a fifty-fifty mixture of inhibited Ethylene glycol and water. This led to the inference that with the increase in the concentration of particles, the efficiency of the heat exchanger also increases due to the elevated coefficients of heat transfer of nanofluids.

An improvement in the power of computing has been observed in recent times, which has led to an increase in the interest of engineers and researchers towards simulating their problems with the use of computational and numerical methods. A large number of developments in computational tools and methods in the last decade have been made for analyzing the fluid dynamics, different modes of heat transfer, and combustion. In the investigation carried out by Moraveji et al. [21] numerically of the turbulent heat transfer and the factor of friction in a tube passage that uses the CFD method for Fe_3O_4 nanofluid. A maximum error of 10% was found in this project.

Sunder et al. [22,23] in their research where the team investigated the viscosity and thermal conductivity, for varying concentrations of nanofluids ranging from 0.0% to 2.0% with the temperature range from 293K to 333K. Vajjha and Das [24], in their experiments on thermal conductivity for three different nanofluids and proposed correlations. In their study, Park and Cho [25], experimentally explored the effect of the factor of friction and the behavior of suspended particles where they used two separate metal oxide particles of Al_2O_3 and TiO_2.

Rajashekhar and others [26] studied the mixture of alumina, silica, and CuO nanoparticles with water-ethylene glycol in 60:40 ratio and found that the coefficient of heat transfer was directly proportional to the load of the particle and showed higher rates of thermal efficiency.

2. Methodology

2.1. *Selection of Nanofluid Properties*

In order to subject Aluminum oxide and Silver nanofluids to CFD analysis, their thermo-physical properties are calculated by the following relations:

2.1.1. *Density*

Pak and Cho developed a relationship which is the equation for density of mixture of solid particles in two-phase, in liquids for a micrometer and nanometer sized is expressed by the formula

$$\rho_{nf} = \rho_s\varphi + \rho_{bf}(1-\varphi)$$

Where,
$\rho_{nf} = density\ of\ nanofluid, \rho_s\varphi = density\ of\ the\ particle, \rho_{bf} = density\ of\ the\ base\ fluid,$
$\varphi = volume\ fraction$

2.1.2. *Specific Heat*

The equation for two-phase mixture is a function of the concentration of particles. The correlation was developed by Pak and Cho [25]

$$C_{pnf} = (1-\varphi)C_{pbf} + \varphi C_{ps}$$

Buongiorno [9] improved the above equation (2) for dilute the concentration of nanofluids by asserting thermal equilibrium between the liquid phase and nanoscale solid particles.

$$C_{pnf} = \frac{\varphi C_{ps}\rho_s + (1-\varphi)\rho_{bf}C_{pbf}}{\rho_{nf}}$$

Where,$C_{pnf} = specific\ heat\ of\ nanofluids, C_{ps} = specific\ heat\ of\ the\ particle,\ fluid$
$C_{pbf} = specific\ heat\ of\ the\ base$ fluid

2.1.3. *Thermal Conductivity*

This relation for effective thermal conductivity for nanofluids was given by Hamilton-Crosser.

$$\frac{k_{nf}}{k_{bf}} = \frac{\frac{k_p}{k_{bf}} + (n-1) - (n-1)(1-\frac{k_p}{k_{bf}})\varphi}{\frac{k_p}{k_{bf}} + (n-1) + (1-\frac{k_p}{k_{bf}})\varphi}$$

Where, n is the empirical factor of shape and ψ is sphericity; $n = 3/\psi$ and $\psi = 1$ for spherical shaped particles and $\psi = 0.5$ for cylindrical particles.

2.2. *Calculation of Nanofluid Properties*

Table 1: Properties of Aluminium and water, Aluminium and Silver at 27°C

Properties	ρ, $\boldsymbol{Density}$ $(\boldsymbol{kg/m^3})$	$C, \boldsymbol{Specific\ Heat}$ $(J/kg-K)$	$k, \boldsymbol{Thermal\ Conductivity}$ $(W/m-K)$
Water	997.5	4182	0.600
Aluminium	3980	765	237
Silver	10500	240	406

The optimal SBDS (Sodium dodecylbenzenesulphonate) concentration for 50 nm size was found to be 0.01wt % with a fixed pH at 5.7 and the maximum increase in the thermal conductivity for 50 nm size for a fixed nanoparticle concentration of 2%.[24]

φ = Volume fraction = 2% = 0.02, for both the nanoparticles

Table 2: Properties of Aluminium Oxide Nanofluid

ρ, $\boldsymbol{Density}$ $(\boldsymbol{kg/m^3})$	$C, \boldsymbol{Specific\ Heat}$ $(J/kg-K)$	$k, \boldsymbol{Thermal\ Conductivity}$ $(W/m-K)$
1057.83	3922.20	1.036

Table 3: Properties of Silver Nanofluid

ρ, $\boldsymbol{Density}$ $(\boldsymbol{kg/m^3})$	$C, \boldsymbol{Specific\ Heat}$ $(J/kg-K)$	$k, \boldsymbol{Thermal\ Conductivity}$ $(W/m-K)$
1187.5	3485.06	1.061

3. CFD Analysis

We have used Computational Fluid Dynamics (CFD) as the means of verifying our results. CFD is a branch of fluid mechanics that solves algorithms by analysing the problems. CFD analysis is constructed on finding the solution of governing equations of fluid dynamics problems. By making use of mathematical modeling, CFD helps us to foresee fluid flow behaviour. Using this approach, temperature and pressure graphs are generated using specific boundary conditions.

CFD simulation was done on ANSYS FLUENT with a radiator model with four fluids as coolant, namely-

1. Water
2. Ethylene Glycol
3. Aluminium Oxide nanofluid
4. Silver Nanofluid

3.1. Procedure for CFD Analysis

Under the DESIGN MODULE on ANSYS workbench, the radiator model was uploaded.

The diameter of the inlet and outlet tubes is 8 mm and the width of the radiator is 120 mm. The material is aluminium.

Figure 1. Radiator Model in Solidworks

3.1.1. Choice of physical model

The choice of the physical model is based on the Reynolds number ($\rho v d/\mu$). Where, ρ is the fluid's density, d is the pipe's diameter and μ is the fluid's viscosity.

Viscous standard k-ε model was used. The standard k-ε model is used for the turbulent flow of single-phase fluid in a circular tube pipe.

3.1.2. Boundary Conditions

Different boundary conditions such as velocity inlet/pressure outlet/wall are defined.

Table 4: Boundary Conditions

Boundary Type	**Parameter Value**
Inlet Temperature of Coolant **(K)**	385
Velocity Magnitude **(m/s)**	2
Heat Transfer Coefficient (For convection wall) **(W/m²-K)**	80
Inlet Temperature of Air **(K)**	300

3.1.3. Mesh Generation

Meshing is the division of the continuous body into no of elements. A mesh is generated prior to solving the solution. More is the no of elements; more is the accuracy of the result. But increasing the number of elements also increases the processing time. So, for this model, a coarse mesh was created. After that names were given to all the faces.

3.1.4. CFD Simulation

After the model meshes and the inlet, outlet, and the convection name are created, the radiator model is imported to FLUENT. In the FLUENT launcher settings, the solver selected was a Pressure-based type with 3D double precision and parallel processing. Next, the Energy Equation is turned on because of heat transfer and the physical model chosen is the viscous standard k-ε model. Next, selection of the

IOP Conf. Series: Materials Science and Engineering 1228 (2022) 012011 doi:10.1088/1757-899X/1228/1/012011

suitable material is done. Aluminium is the chosen material for solid radiator and water, ethylene glycol, and nanofluids for fluid respectively. The values of nanofluids properties are assigned, and the coolant is defined. Then, suitable boundary conditions are applied.

Under the Solution step, the Solution Method chosen is Pressure-Velocity Coupling with SIMPLEC scheme. In spatial discretization, the gradient used is Least Square cell-based, Pressure was set to Linear. Pressure, momentum, and turbulent kinetic energy are First order upwind. In the next step, the solution is initialized by selecting Hybrid Initialization with 10 iterations. The results were checked and studied by post-processing involving the computed fluid flow with contours of temperature and pressure in the form of graphs, as shown in the results below.

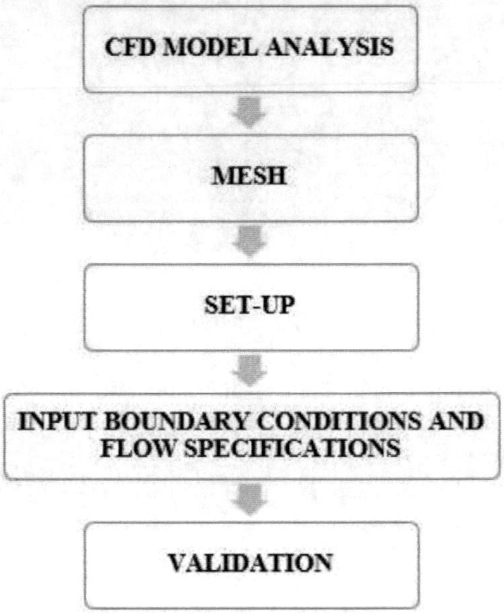

Figure 2: Flow Chart of the Procedure

4. Results and Discussion

Individual results of temperature gradient contour of each of the coolant fluids under study is shown in Figure 3.

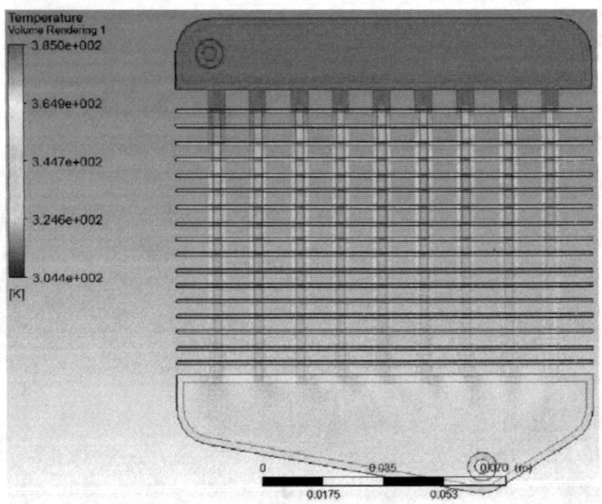

Figure 3. Temperature gradient developed in post processing of Water as coolant

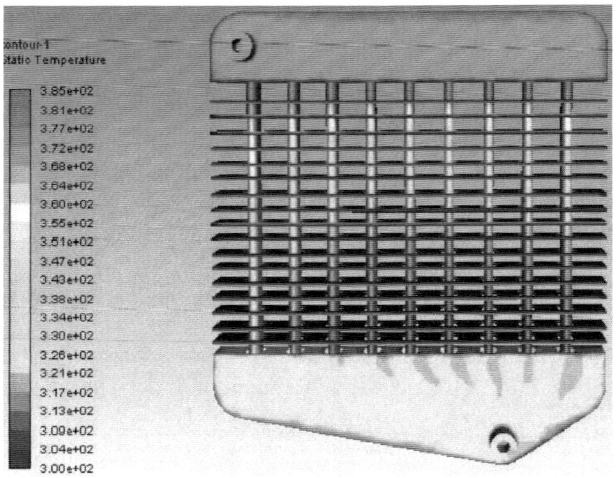

Figure 4. Temperature gradient developed in post processing of Ethylene glycol nanofluid as coolant

Figure 5. Temperature gradient developed in post processing of Aluminium oxide nanofluid as coolant

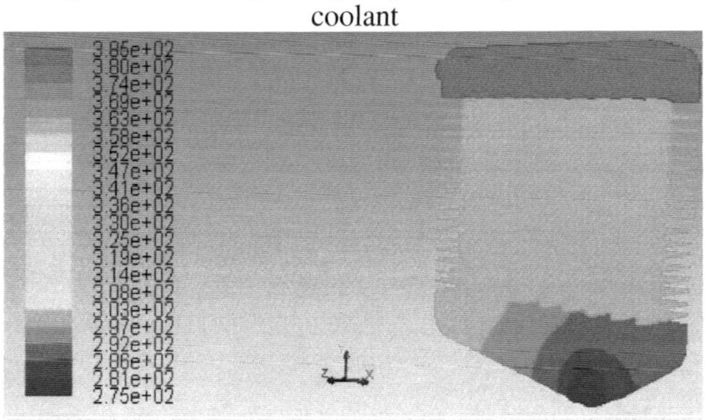

Figure 6. Temperature gradient developed in post processing of Silver nanofluid as coolant

Table 5: Inlet-Outlet Temperature using different coolants

Coolant Used	Inlet Temperature (K)	Outlet Temperature (K)
Water	385.00	367.04
Ethylene Glycol	385.00	315.47
Aluminium Oxide Nanofluid	385.00	300.55
Silver Nanofluid	385.00	275.25

Results are reported in Figures 7 and 8 temperature and pressure from the inlet to the outlet of the radiator.

This was obtained by plotting results for temperature drop and pressure drop for the nanofluids under simulation. In all cases, the volume concentration and nanoparticle size were considered the same i.e., 2% by volume and 50 nm particle size.

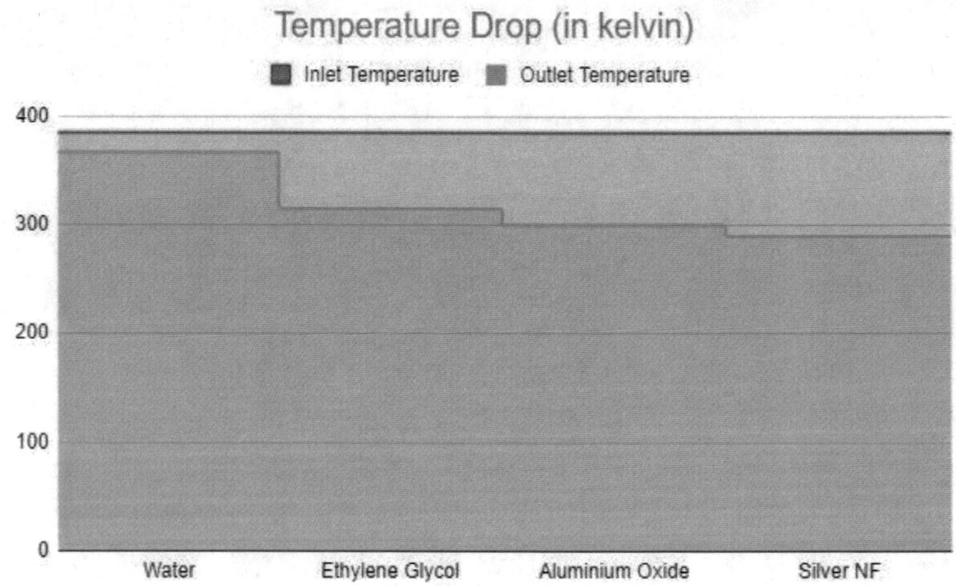

Figure 7. Temperature Drop from radiator

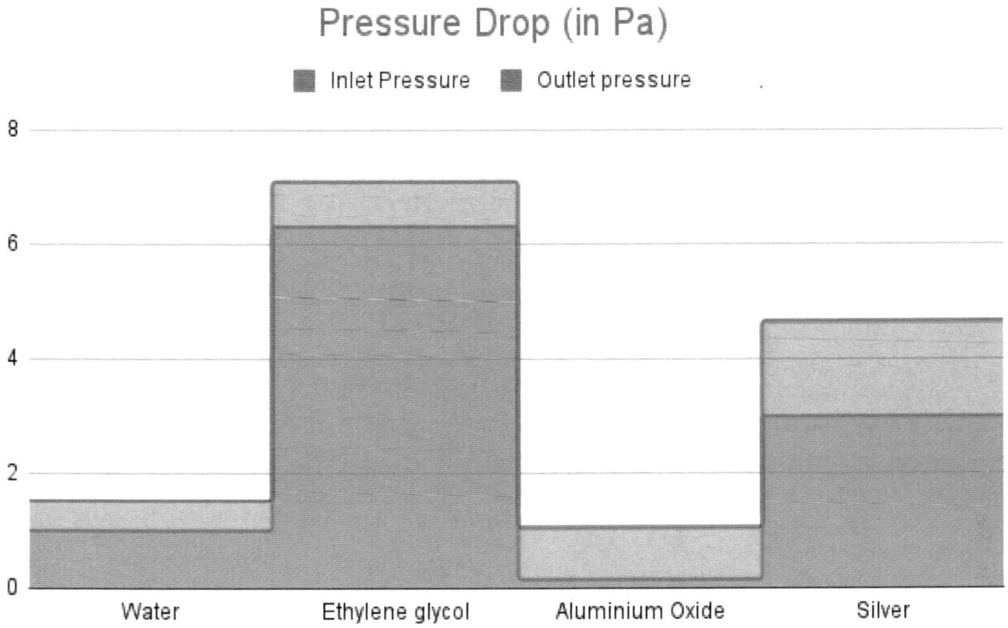

Figure 8. Pressure Drop from radiator

From Figure 8, it can be observed that the drop was maximum in the case of silver nanofluid followed by aluminium oxide nanofluid, ethylene glycol, and lastly water. This drop signifies that as the coolant is moving along the tubes of the radiator, it gets cooled down and signifies the heat transfer capacity of the fluid.

Since ethylene glycol is always used with water in the ratio of 50:50 hence; we can say that silver nanofluid can be considered with maximum pressure drop. The pressure drop of the coolant signifies the efficiency of the nanofluids application.

5. Conclusion

As seen from the simulation results, the overall outlet temperature of the radiator, with the nanofluids as coolant was lesser than that of conventional coolants. This shows that enhanced thermal conductivity of nanofluids has potential practical application in the automotive industry as radiator coolant.

The thermal conductivity of nanofluid is temperature-dependent. At higher rpm and load, with an increase in temperature, a simultaneous increase in the heat-carrying capacity is seen. This is an advantage when the engine is running at high rpm with the load. Nanofluid-based coolants encourage a size reduction of radiator by decreasing the frontal area which allows for better aerodynamic design. Eventually, due to a decrease in the need for coolant circulation, the weight will be reduced which leads to less requirement of pumping power. The overall engine efficiency will increase.

References

[1] Choi U. S., Singer D. A., and Wang H. P., 1995, Development and Application of Non-Newtonian Flows, Vol. 231, ASME, New York, pp. 99–105.

[2] Choi S., 2006, "Nanofluids for Improved Efficiency in Cooling Systems," Heavy Vehicle Systems Review, Argonne National Laboratory, Argonne, Illinois.

[3] Kulkarni, D. P., Vajjha, R. S., Das, D. K., and Oliva, D., 2011 "Application of Aluminium Oxide Nanofluids in Diesel Electric Generators Jacket Water Coolant," Appl. Therm. Eng., 28, pp.1774 1781.

[4] Leong, K. Y., Saidur, R., Kazi, S. N., and Mamun, A. H., 2010, "Performance Investigation of an Automotive Car Radiator Operated with Nanofluid-Based Coolants (Nanofluid as a Coolant in a Radiator)," Appl. Therm. Eng., 30, pp. 2685–2692.

[5] Peyghambarzadeh, S. M., Hashemabadi, S. H., Hoseini, S. H., and Jamnani,M. S., 2011, "Experimental Study of Heat Transfer Enhancement Using Water/Ethylene Glycol Based Nanofluids as a New Coolant in the Car Radiator," Int. Common. Heat Mass Transfer, 38(9), pp.

[6] Naraki, M., Peyghambarzadeh, S. M., Hashemabadi, S. H., and Vermahmoudi,Y., 2013, Parametric Study of Overall Heat Transfer Coefficient of CuO / Water Nanofluids in a Car Radiator," Int. J. Therm. Sci., 66, pp. 82–90.

[7] S. A. Putanm, D. G. Cahill and P. V. Braun, J. Appl. Phys. 99, 084308 (2006). doi:10.1063/1.2189933

[8] R. Rusconi, E. Rodari and R. Piazza, Appl. Phys. Lett. 89, 261916 (2006). doi:10.1063/1.2425015

[9] J. Buongiorno, J. Heat Transf. 128, 240 (2006). doi:10.1115/1.2150834

[10] R. Chein and G. Huang, Appl. Therm. Eng. 25, 3104 (2005). doi: 10.1016/j.applthermaleng.2005.03.008

[11] S. G. Etemad, S. Z. Heris and M. N. Esfahany, Int. J. Heat Mass Transfer33, 529 (2006). doi: 10.1016/j.icheatmasstransfer.2006.01.005

[12] M. A. M. Said and R. K. Agarwal, 38th AIAA Thermo-physics Conference, Toronto, Ontario, Canada 1(2005).

[13] Y.Xuan and Q. Li, J. Heat Transf. 125, 151 (2003). doi:10.1115/1.1532008

[14] J. Kim, Y. T. Kang and C. K. Choi, Phys. Fluids 16, 2395 (2004a). doi:10.1063/1.1739247

[15] W. Yu, D. M. France, J. L. Routbort and S. U. S. Choi, Heat Transfer Eng. 29, 432 (2008).

[16] D. Singh, J. Toutbourt, G. Chen et al, Annual report Argonee National lab (2006).

[17] V. Vasu, K. Ramakrishna and A. C. S. kumar, Int. J. Nano Technol. Appl. 2, 75 (2008).

[18] S. T. Zeng, C. Lin and K. Huang, Acta. Mechanica. 179, 11 (2005). doi:10.1007/s00707-005-0248-9

[19] S. V. Ravikanth, K. D. Debendra and K. N. Praveen, Int. J. Heat Fluid Flow 31, 613 (2010). doi: 10.1016/j.ijheatfluidflow.2010.02.016

[20] P. K. Devdatta, S. V. Ravikanth, K. D. Debendra and O. Daniel, Appl. Therm. Eng. 28, 1774 (2008). doi: 10.1016/j.applthermaleng.2007.11.017

[21] M.K. Moraveji, MajidHejazian, Modelling of turbulent forced convective heat transfer and friction factor in a tube for Fe3O4 magnetic nanofluid with computational fluid dynamics, International communications in Heat and Mass transfer, 39 (2012) 1293-1296.6

[22] L. SyamSundar, M.K. Singh, A.C.M. Sousa, Investigation of thermal conductivity and viscosity of Fe3O4 nanofluid for heat transfer applications, International Communications in Heat Mass Transfer, 44 (2013) 7-14.

[23] L. SyamSundar, M.K. Singh, A.C.M. Sousa, Thermal conductivity of ethylene glycol and water mixture based Fe3O4 nanofluid, International Communications in Heat and Mass Transfer, 49 (2013) 17-24.

[24] R.S. Vajjha, D.K. Das, Experimental determination of thermal conductivity of three nanofluids and development of new correlations, International Journal of Heat Mass Transfer, 52 (2009) 4675-4682.
[25] B.C. Pak, Y.I. Cho, Hydrodynamic and heat transfer study of dispersed fluids with submicron metallic oxide nanoparticles, Experimental Heat transfer, 11 (1998) 150-170.
[26] Pendyala, R., Chong, J. L., & Ilyas, S. U. (2015). CFD analysis of heat transfer performance in a car radiator with nanofluids as coolants. Chemical Engineering Transactions, 45, 1261-1266.

Machining of Polymethyl-methacrylate (PMMA) bone cement-based nanocomposite using Combined compromise solution (CoCoSo) theory

Virat Mani Vidyasagar[1], Kuldeep Kumar[1], Umang Dubey[1], Shivi Kesarwani[1], and Rajesh Kumar Verma[1*]

1Materials & Morphology Laboratory, Department of Mechanical Engineering, Madan Mohan Malaviya University Technology, Gorakhpur, Uttar Pradesh 273010

rkvme@mmmut.ac.in

Abstract. Polymethyl-methacrylate (PMMA) bone cement has become an effective biomaterial in orthopedic applications due to its outstanding biocompatibility and mechanical properties. A machining test is essential required while using PMMA bone cement nanocomposite in implant applications. Drilling on the prosthetic product is mainly needed when using PMMA to insert bolts for safety and fixation in the human body. The drilling on bone cement-based prostheses is the most often used machining method during bone-grafting and implant interface surgical procedures. Henceforth, this article focus on the relatively new Combined compromise solution (CoCoSo) theory to study the influence of varying process variables on the drilling performance of Graphene Nanoplatelets (GNP) infused PMMA bone cement nanocomposites. The control of process constraints, namely, GNP weight percentage (Wt.%), spindle speed (SPEED), and drill bit tool material (TOOL) such as HSS, Carbide, and TiAlN, is mainly done to achieve the desired value of Surface roughness (SR) and Material removal rate (MRR) during the drilling experimentation. Using the CoCoSo technique, the optimum setting was attained as Wt.% = 4, SPEED = 1428 rpm, and a Carbide drill bit tool. The synthesized sample was also subjected to an XRD examination to validate the appropriate infusion of nanomaterials in PMMA bone cement. The flexural analysis of the developed nanocomposite sample indicates that mechanical property improves significantly due to the supplement of GNP in PMMA. The findings of the proposed nanocomposites material show its feasibility in biomedical functions. Also, the drilling optimization results demonstrate new criteria to control the quality and productivity indices. It can be recommended for multicriteria optimization case studies.

Keywords: Bone cement, Graphene, CoCoSo, Surface roughness, MRR

1. Introduction

Poly-methyl methacrylate (PMMA) is widely used for various orthopedic applications like implant fixation, trauma surgeries, cranial implants, and treating congenital bone defects [1], [2]. The improvement in mechanical features of PMMA is possible with the supplement of reinforcement such as carbon nanotubes, [3] titanium oxide, [4] zirconium oxide, [5] polymeric nanofibers. The findings of the studies demonstrated that carbon-based reinforcement has emerged as an effective reinforcement in biomedical functions. It improves the mechanical and chemical properties of the biocomposite. [6], [7].

Content from this work may be used under the terms of the Creative Commons Attribution 3.0 licence. Any further distribution of this work must maintain attribution to the author(s) and the title of the work, journal citation and DOI.
Published under licence by IOP Publishing Ltd

In a study, incorporating 0.04 wt.% graphene oxide (GO) into the bone cement enhanced flexural and compressive strength compared to control cement [8]. Singh et al. [9] fabricated a composite containing multi-walled carbon nanotubes (MWCNTs), PMMA, and HA and remarked it as a superior material for biomedical applications. In comparison to commercially available PMMA bone cement, adding 0.5% by weight GO to PMMA bone cement increased compressive strength by 8.4% [10]. These investigations validate the utilization and potential of carbon nanomaterials like graphene, GO, and MWCNT in boosting the mechanical properties of PMMA bone cement. It has been observed that very limited investigations exist on the machining characteristics of graphene nanoplatelets impregnated bone cement.

Drilling on the prosthesis is frequently required when using PMMA bone cement to place bolts for protection and fixation [11]. The optimal parametric condition are needed to maintain the biometric property during the machining of bio nano composite. Drilling in bone and artificial bones is usually done by hand since the quantity of feed required is not predetermined, although various spindle speeds can be used [12]. As the PMMA-GNP nanocomposite can be used in the form of artificial bone as it provides good mechanical strength. The drilling operation must be performed manually to apply the findings to real-world orthopedic drilling situations. The current work uses PUSCO manually operated vertical drilling machine to facilitate hand drilling. For orthopedic procedures, the most commonly used drill bit is a twist drill [13]. In this article, the drilling operation of GNP incorporated PMMA bone cement nanocomposite has been performed. The aggregated value of the output response such as material removal rate (MRR) and surface roughness (Ra) was considered to estimate the machining performance. The optimum solutions for ideal drilling conditions were determined using Combined Compromise Solution (CoCoSo). The drilling parameters such as spindle speed, GNP Wt.%, and tool material were optimized and analysis of variance (ANOVA) was used to evaluate the most affecting factor. The crystal structure analysis and flexural property was examined before the machining test. This work shows the drilling performance analysis of modified PMMA nanocomposites using CoCoSo hybrid optimization module for quality and productivity indices optimization.

2. Materials and Methods

2.1 Materials

or the fabrication of PMMA bone cement bio-nano composite, the Simplex© PMMA bone cement with monomer was used, provided by M/s. Global Corporation, Ahmedabad, India. The incorporating Graphene Nanoplatelets (GNP) was supplied by M/s. A.D Nano Technology, Bangalore India.

2.2 Fabrication of Composite Materials

The GNP incorporated PMMA bio-nanocomposite is fabricated by the conventional hand blending method with different wt. % of GNP (1, 2.5, and 4 %) as shown in Table 1. For the bonding of composite, PMMA and GNP mixture was mixed with liquid monomer in 2:1 ratio and then blended the mixture manually in a ceramic bowl till the dough state is achieved. The dough state of the mixture is put to the prepared mold and left for curing for 1 hour.

Table 1. Composition of GNP incorporated PMMA composite.

Sample	PMMA	GNP	Wt. %
P-NEAT	40.00 gm	0 gm	0 wt.%
PG1	39.60 gm	0.40 gm	1 wt.%
PG2	39.00 gm	1 gm	2.5 wt.%
PG3	38.40 gm	1.6 gm	4 wt.%

1.7 Crystal Structure Analysis

The chemical composition and homogeneous dispersion of nanomaterials in the resulting composite was validated using X-ray diffraction (XRD). Using a Raguku Model D/max III diffractometer, the crystal structure and phase composition of the developed nanocomposite, Graphene nanoplatelets, and PMMA bone cement were studied. Nickel-filtering was used on the X-ray beam. Cu Ka1 (lZ0.1504 nm) was irradiated at 40 kV and 30 mA using a nickel-filter Cu Ka1 filter (lZ0.1504 nm). At a scanning rate of 28 per minute, data from 1 to 508 were collected. In a combined graph of GNP, PMMA bone cement, and PMMA-GNP nanocomposite, the resulting peaks are presented in Fig. 1. After examining the graph and comparing it with similar research, it was discovered that PMMA bone cement exhibits the required peak values [14]. The diffraction peak of GNP was discovered to be at ($2\theta = 25.64°$), which is similar to previous research [15], [16]. The peak at 25.64° corresponds to roughly 0.336 nm of interlayer spacing, which is extremely near to graphite d-spacing [17]. The GNP nanoparticles are completely distributed in the produced PMMA-GNP nanocomposite, according to the XRD graph and comparable prior research [18][19].

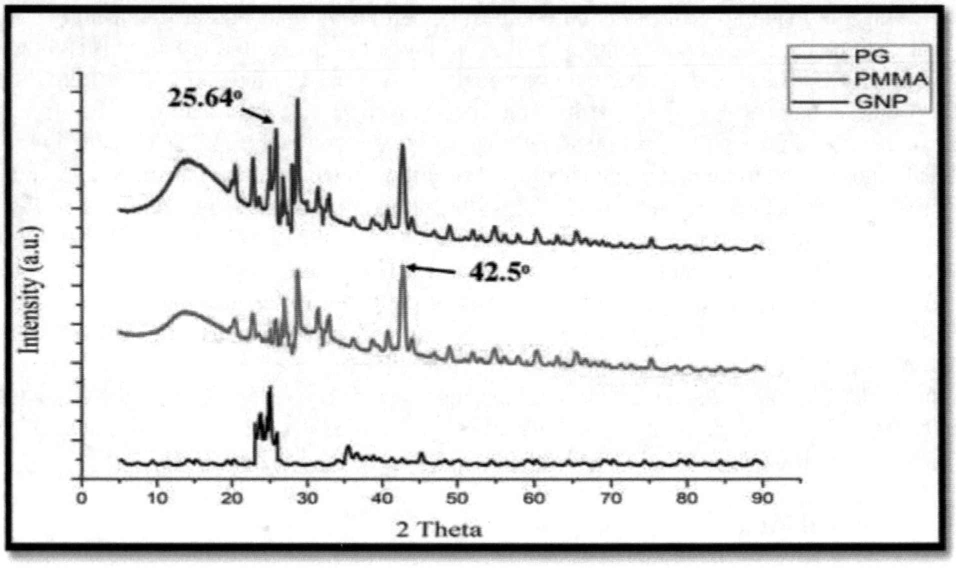

Figure 1. X-Ray Diffraction Peaks.

1.8 Mechanical Property Analysis

Flexural strength becomes a crucial characteristic for applications such as complete hip joint replacement and spine bolt fixation. As a result, flexural property investigation of the produced PMMA-GNP nanocomposite is highly required. The flexural test was carried out according to ISO5833-2002 [20]. Equation 1 gives the flexural strength of a specimen in a three-point bend test [21].

$$S = \frac{3*P_{max}*L}{2*b*h^2} \quad (1)$$

where Pmax = maximum load faced,
b = width,
d = thickness,
and a = between inner loading points gap (20 mm).

Adding different amounts of nanomaterial can considerably improve flexural strength, as seen in Fig.2. In comparison to the pristine sample, the flexural strength of the PG-2 sample raised by 12.8 % (from 63.3 ± 0.4 to 72.4 ± 0.5 MPa). A decrease in flexural strength is also observed when a large amount of GNP is used. When GNP is present in small numbers and is consistently distributed, it acts as load carriers. In turn, it assists in the improvement of mechanical characteristics. Nonuniform distributions of GNP particles can occur as the proportion of GNP particles rises, resulting in particle aggregation, poor matrix adherence, and decreased mechanical characteristics [22]. The large specific area and the wrinkled surface of the graphene sheets contribute to improved filler incorporation in the cement matrix as well as strong matrix adhesion [23]. According to the current findings, GNP should be used up to 2.5% by weight. It is the strongest in terms of flexural strength.

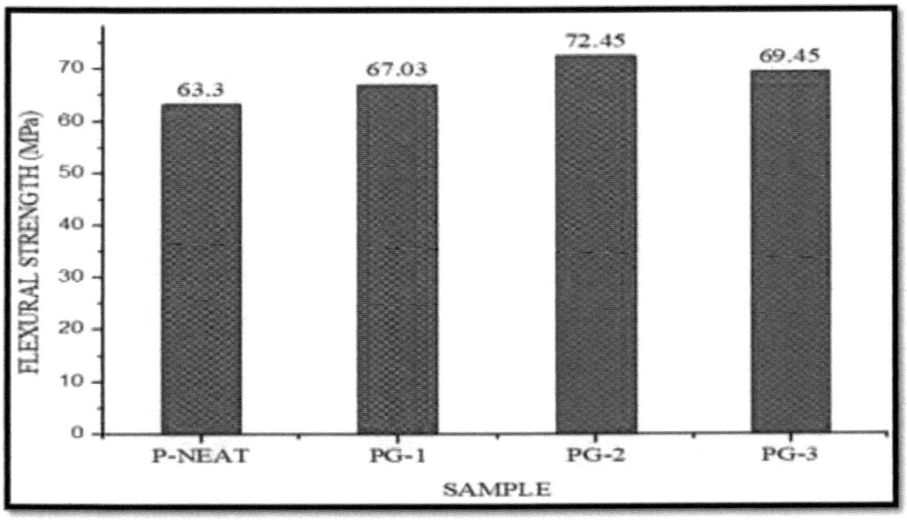

Figure 2. Flexural Strength of Prepared Samples.

2.5 Experimental Design

The design of experiments (DOE) was used to identify control parameters and maximize their effect on machining performance for drilling GNP incorporated PMMA bio-nano composite. As stated in Table 2, different levels were assigned to working range parameters. Taguchi orthogonal array (OA) design consisting of nine experiments (L_9) by considering three varying factors such as speed (rpm), GNP wt.%, and drill bit material. The three different drill bits (HSS-1, Carbide-2, and TiAlN-3) were used to perform the drilling operation. The experiment was carried out on a PUSCO vertical drilling machine shown in Fig. 3.

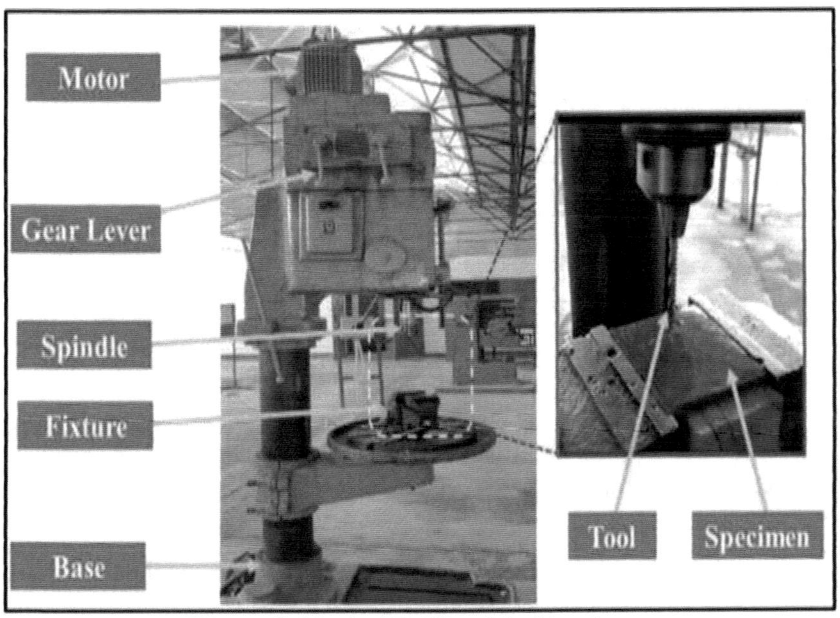

Figure 3. Vertical Drilling Machine Setup.

Table 2. Drilling parameters and level

Parameters	Symbol	Levels			
		Unit	1	2	3
Speed	S	RPM	318	865	1428
GNP wt.%	Wt. (%)	-	1	2.5	4
Tool	T	-	HSS (1)	Carbide (2)	TiAlN (3)

2.6 Measurement and Calculation of Experiment Data

Drilling operation performances such as MRR and Ra of drilled holes were calculated. The images of the drilled holes on composite material is shown in Fig. 4.

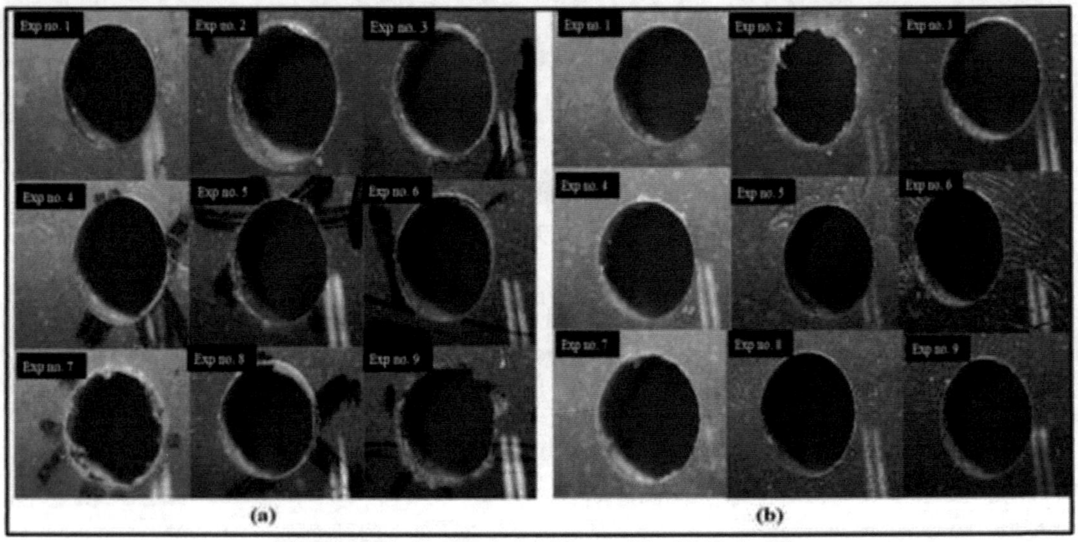

Figure 4. Drilled hole (a) Entry side (b) Exit side

2.6.1 Material removal rate (MRR)

The material removal rate is the measurement of material removed per unit time during the machining process. It can mathematically be calculated from the difference in the volume of material after machining operation or the difference in the weights from the initial sample to the final sample per unit time. MRR was calculated using equation (1).

$$MRR = \frac{Initial\ weight - Final\ weight}{\rho \times Time\ taken}\ mm^3/sec \quad (1)$$

2.6.2 Surface roughness (Ra)

The most significant performance of the drilling operation is surface roughness. The influence of drilling process parameters on the surface roughness of drilled GNP incorporated PMMA bone cement nanocomposite was investigated in this study by using model no. Surtronic S-128 setup (Fig. 5) at three different locations. The surface roughness value has been evaluated with the help of following Equation 2.

$$Ra = \frac{1}{L}\int_0^L |Y(x)|dx \quad (2)$$

Ra is the arithmetic mean deviation from the mean line, L is the sampling length, and Y is the profile curve ordinate.

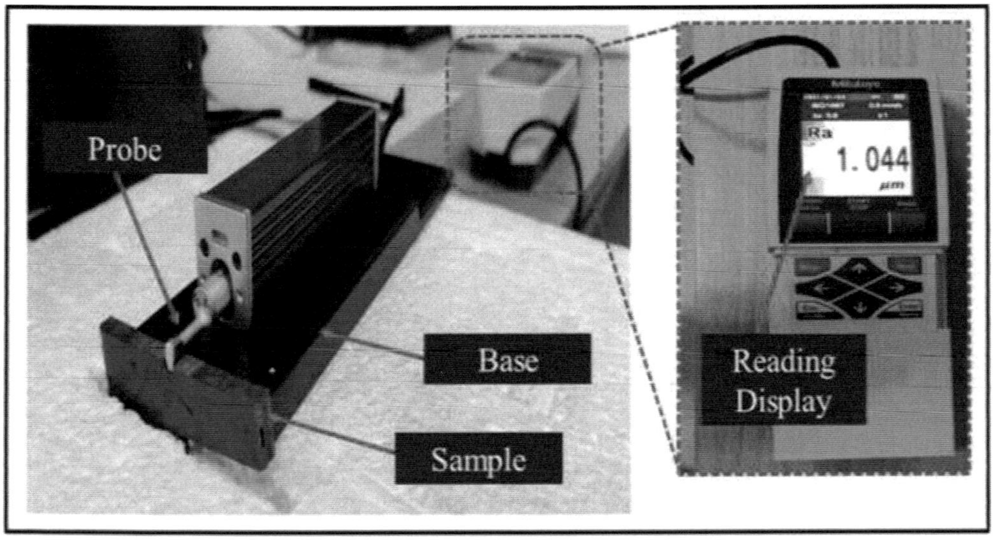

Figure 5. Surface Roughness Tester

2.7 Variance Analysis

To identify significant process boundaries and analyze the impacts of process variables on responses using a mathematical model, analysis of variance (ANOVA) was required. Each study was done under the assumption of a 95% confidence level in order to obtain the best results. The MRR and the surface roughness ANOVA results are shown in Table 3.

Table 3. ANOVA results

Response	Equation	Summary
Material Removal Rate	6.5567 + 0.017 S_318 - 0.157 S_865 + 0.140 S_1428 - 0.940 GNP wt.%_1.0 - 0.070 GNP wt. %_2.5 + 1.010 GNP wt.%_4.0 - 0.160 Tool_1 + 0.217 Tool_2 - 0.057 Tool_3	R-sq = 97.53%, R-sq(adj) = 90.14%
Surface Roughness	1.1311 - 0.0611 S_318 - 0.0511 S_865 + 0.1122 S_1428 + 0.5756 GNP wt.%_1.0 - 0.1811 GNP wt.%_2.5 - 0.3944 GNP wt.%_4.0 - 0.0111 Tool_1 - 0.0544 Tool_2 + 0.0656 Tool_3	R-sq = 99.16%, R-sq(adj) = 96.62%

3. Parametric optimization approach using CoCoSo method

The Combined Compromise Solution (CoCoSo) approach is an optimization approach that combines two famous Multi-Criteria Decision Modelling (MCDM) techniques, namely Simple Additive Weighting (SAW) and Exponentially Weighted Product (EWP). After getting the criteria weights, this technique comprises three mathematical stages that are implemented. It employs a two-step comparison procedure (usual multiplication rule and weighted power of the distance). The aggregated simple additive weighting (SAW) and exponentially weighted product (EWP) approaches are used in this method. The following steps are used to solve the CoCoSo decision problem:

Step-1 First step is determining the initial decision-making matrix, which is represented as:

IOP Conf. Series: Materials Science and Engineering 1228 (2022) 012012 doi:10.1088/1757-899X/1228/1/012012

$$X_{ij} = \begin{bmatrix} a_{11} & a_{12} & \cdot & a_{1n} \\ a_{21} & a_{22} & \cdot & a_{2n} \\ \cdot & \cdot & \cdot & \cdot \\ a_{m1} & a_{m2} & \cdot & a_{mn} \end{bmatrix} \quad (3)$$

Step-2 Normalized values of variables are calculated using the below formula:

$$r_{ij} = \frac{a_{ij} - \min a_{ij}}{\max a_{ij} - \min a_{ij}}; \text{ for benefit criterion;} \quad (4)$$

$$r_{ij} = \frac{\max a_{ij} - a_{ij}}{\max a_{ij} - \min a_{ij}}; \text{ for cost criterion;} \quad (5)$$

Step-3 Determining weighted sequence and power of weight of comparability sequences U_i and V_i,

$$U_i = \sum_{j=1}^{n} (w_j r_{ij}) \quad (6)$$

$$V_i = \sum_{j=1}^{n} (r_{ij})^{\wedge} w_j \quad (7)$$

Step-4 Three different appraisal score for weighted sequence are derived using equation (8), (9), (10);

$$K1 = \frac{V_i + U_i}{\sum_{i=1}^{m} (V_i + U_i)} \quad (8)$$

$$K2 = \frac{U_i}{\min U_i} + \frac{V_i}{\min V} \quad (9)$$

$$K3 = \frac{\lambda(U_i) + (1-\lambda)(V_i)}{\lambda max U_i + (1-\lambda) max V_i} \quad (10)$$

Step-5 Ranking of alternatives determined from higher to lower according to formula 11;

$$k_i = (K1K2K2)^{\frac{1}{3}} + (\frac{1}{3})(K1 + K2 + K3) \quad (11)$$

For this study, the values of weights considered are equal weightage for each response as 0.3, and the value of λ is generally taken as 0.5.

4. Result and discussion

The CoCoSo module is used to optimize the drilling parameters for desired machining performance. The CoCoSo selection index is derived using Eqs. 2 to 8 and then ordered from lowest to the greatest value. Table 4 lists the CoCoSo selection index values for all 9 experiments.

Table 4. Process parameters, corresponding response, and CoCoSo selection index

Exp. no	Input Parameters			Response		K_i
	Speed (rpm)	Wt.% (gm)	Tool	MRR (mm^3 /sec)	Ra (μ)	
1	318.00	1.00	1.00	5.42	1.69	1.688
2	318.00	2.50	2.00	6.9	0.81	12.448
3	318.00	4.00	3.00	7.4	0.71	15.477
4	865.00	1.00	2.00	5.55	1.57	2.863
5	865.00	2.50	3.00	6.22	1.02	8.117
6	865.00	4.00	1.00	7.43	0.65	15.774
7	1428.00	1.00	3.00	5.88	1.86	2.808
8	1428.00	2.50	1.00	6.34	1.02	8.805
9	1428.00	4.00	2.00	7.87	0.85	17.802

4.1 Discussion on the optimal parametric settings

The function parameter rank derived using the CoCoSo module is shown in Table 4. A combination of factors such as GNP (4wt. %, Level-2), Spindle Speed (1428 RPM, Level-3), and tool material (Carbide, i.e., Level-1) gives the optimum MRR and Ra as Ki value is highest (17.802). While drilling PMMA, a high spindle speed produces the best results [24]. Surface smoothness gets better at drilling at high speeds [25]. A polymer composite with a high GNP concentration provides optimal drilling performance [26].

5. Confirmatory test

The confirmation test has been performed on GNP incorporated PMMA bio-nano composites, which showed to be the most influential parameter environment for determining behavior response. According to both optimization approaches, experiment no. 9 obtained the best set of process parameters. The optimal parametric setting is S-1428/GNP wt.%-4/ Tool-2, which provides the best machining characteristics among the completed tests. S-865/GNP wt. % -4/ Tool-2 is the best solution obtained using Taguchi L_9 orthogonal series. Table 5 shows the comparative research outcomes for both the ideal condition and the optimization case. Utilizing the optimization approach, it has been proven that process reaction is substantially enhanced in Ra and MRR. While the Ra decreases from 0.98 to 0.85 m, the MRR increases from 7.15 to 7.87 mm^3/sec.

Table 5. Taguchi design versus optimization result

Optimal setting	Taguchi orthogonal array design S-865/GNP wt. % -4/ Tool-2	Optimization design S-1428/GNP wt.%-4/ Tool-2	Improvement in %
MRR	7.15	7.87	10
Ra	0.98	0.85	13.26

6. Conclusion

The current research focuses on the machinability of PMMA-based bio-nanocomposites when subjected to various process constraints. The nanoparticles were mixed in with the matrix material to enhance the PMMA's mechanical properties and production capabilities. CoCoSo optimization technique was utilized in this study to drill PMMA bone cement bio-nanocomposites. CoCoSo approach effectively determined control parameters and achieved the highest machining efficiency, i.e., a high MRR and a low Ra. From the research, the following findings can be concluded as:

- The peaks in the XRD examination demonstrate that the developed composite crystal structure has GNP dispersed evenly. When the acquired peak was compared to prior research, the results indicated that the data was identical.
- The flexural strength analysis reveals that GNP reinforcement increases flexural strength, with the greatest value being at 2.5% wt.
- CoCoSo has effectively chosen control variables, levels, and the optimum machining environment for decreased surface roughness (Ra) and increased material removal rate (MRR).
- The optimum machining condition for the given experimental plan is to obtain at 4 wt.% of GNP in PMMA bone cement with spindle speed at 1428 rpm and Carbide tool material.
- The confirmatory test reveals that the proposed optimization module enhances the desired response compared to Taguchi's orthogonal array settings.
- The ANOVA analysis shows wt.% of GNP as the most significant parameter for MRR and Ra in the performed experiment.

In terms of the future of this study, numerous significant inspections are still necessary, which will necessitate further research and testing. Different reinforcement weight percentages and other input variables, such as feed and tool type, impact drilling PMMA nanocomposite, which will need to be investigated. Drilling analysis necessitates further investigation of responses like torque and thrust analysis.

References

[1] M. J. Peltola, P. K. Vallittu, V. Vuorinen, A. A. J. Aho, A. Puntala, and K. M. J. Aitasalo, "Novel composite implant in craniofacial bone reconstruction," *European Archives of Oto-Rhino-Laryngology*, vol. 269, no. 2, pp. 623–628, 2012.

[2] J. A. Goldstein, J. T. Paliga, and S. P. Bartlett, "Cranioplasty: Indications and advances," *Current Opinion in Otolaryngology and Head and Neck Surgery*, vol. 21, no. 4, pp. 400–409, 2013.

[3] B. Marrs, R. Andrews, T. Rantell, and D. Pienkowski, "Augmentation of acrylic bone cement with multiwall carbon nanotubes," *Journal of Biomedical Materials Research - Part A*, vol. 77, no. 2, pp. 269–276, 2006.

[4] S. M. Z. Khaled, P. A. Charpentier, and A. S. Rizkalla, "Physical and mechanical properties of PMMA bone cement reinforced with nano-sized titania fibers," *Journal of Biomaterials Applications*, vol. 25, no. 6, pp. 515–537, 2011.

[5] R. J. Kane, W. Yue, J. J. Mason, and R. K. Roeder, "Improved fatigue life of acrylic bone cements reinforced with zirconia fibers," *Journal of the Mechanical Behavior of Biomedical Materials*, vol. 3, no. 7, pp. 504–511, 2010.

[6] E. Paz, F. Forriol, J. C. del Real, and N. Dunne, "Graphene oxide versus graphene for optimisation of PMMA bone cement for orthopaedic applications," *Materials Science and Engineering C*, vol. 77, no. 2016, pp. 1003–1011, 2017.

[7] M. Naffakh, A. M. Díez-Pascual, C. Marco, G. J. Ellis, and M. A. Gómez-Fatou, "Opportunities and challenges in the use of inorganic fullerene-like nanoparticles to produce advanced polymer nanocomposites," *Progress in Polymer Science*, vol. 38, no. 8, pp. 1163–1231, 2013.

[8] Khan *et al.*, "Physical, mechanical, chemical and thermal properties of nanoscale graphene oxide-poly methylmethacrylate composites," *Journal of Composite Materials*, vol. 52, no. 20, pp. 2803–2813, Aug. 2018.

[9] M. K. Singh *et al.*, "Hydroxyapatite modified with carbon-nanotube-reinforced poly(methyl methacrylate): A nanocomposite material for biomedical applications," *Advanced Functional Materials*, vol. 18, no. 5, pp. 694–700, 2008.

[10] E. Paz, F. Forriol, J. C. del Real, and N. Dunne, "Graphene oxide versus graphene for optimisation of PMMA bone cement for orthopaedic applications," *Materials Science and Engineering C*, vol. 77, pp. 1003–1011, Aug. 2017.

[11] C. Foot, "Use of Polymethyl Methacrylate as prosthetic replacement of destroyed foot bones – clinical audit Authors : Abstract :," 2014.

[12] N. Bertollo and W. Robert, "Drilling of Bone: Practicality, Limitations and Complications Associated with Surgical Drill-Bits," in *Biomechanics in Applications*, InTech, 2011.

[13] C. H. Jacob, J. T. Berry, M. H. Pope, and F. T. Hoaglund, "A study of the bone machining process—Drilling," *Journal of Biomechanics*, vol. 9, no. 5, Jan. 1976.

[14] H. Tan, S. Guo, S. Yang, X. Xu, and T. Tang, "Physical characterization and osteogenic activity of the quaternized chitosan-loaded PMMA bone cement," *Acta Biomaterialia*, vol. 8, no. 6, pp. 2166–2174, Jul. 2012.

[15] E. Aram, M. Ehsani, H. A. Khonakdar, S. H. Jafari, and N. R. Nouri, "Functionalization of graphene nanosheets and its dispersion in PMMA/PEO blend: Thermal, electrical, morphological and rheological analyses," *Fibers and Polymers*, vol. 17, no. 2, pp. 174–180, Feb. 2016.

[16] H. Kim and C. W. Macosko, "Morphology and properties of polyester/exfoliated graphite nanocomposites," *Macromolecules*, vol. 41, no. 9, pp. 3317–3327, May 2008.

[17] B. W. Chieng, N. A. Ibrahim, W. M. Z. W. Yunus, M. Z. Hussein, Y. Y. Then, and Y. Y. Loo, "Effects of graphene nanoplatelets and reduced graphene oxide on poly(lactic acid) and plasticized poly(lactic acid): A comparative study," *Polymers*, vol. 6, no. 8, pp. 2232–2246, 2014.

[18] J. Naveen, M. Jawaid, E. S. Zainudin, M. T. H. Sultan, and R. Yahaya, "Improved Interlaminar Shear Behaviour of a New Hybrid Kevlar/Cocos Nucifera Sheath Composites with Graphene Nanoplatelets Modified Epoxy Matrix," *Fibers and Polymers*, vol. 20, no. 8, Aug. 2019, doi: 10.1007/s12221-019-3127-z.

[19] M. Tavakoli, S. S. E. Bakhtiari, and S. Karbasi, "Incorporation of chitosan/graphene oxide nanocomposite in to the PMMA bone cement: Physical, mechanical and biological evaluation," *International Journal of Biological Macromolecules*, vol. 149, pp. 783–793, Apr. 2020.

[20] "INTERNATIONAL STANDARD ISO 5833 Implants for surgery-Acrylic resin cements Implants chirurgicaux-Ciments à base de résine acrylique ISO 5833:2002(E) ii," 2002.

[21] M. Rao *et al.*, "Preparation and characterization of a poly(methyl methacrylate) based composite bone cement containing poly(acrylate-co-silane) modified hydroxyapatite nanoparticles," *Journal of Applied Polymer Science*, vol. 131, no. 15, Aug. 2014.

[22] E. Paz, F. Forriol, J. C. del Real, and N. Dunne, "Graphene oxide versus graphene for optimisation of PMMA bone cement for orthopaedic applications," *Materials Science and Engineering: C*, vol. 77, Aug. 2017, doi: 10.1016/j.msec.2017.03.269.

[23] G. Gonalves, S. M. A. Cruz, A. Ramalho, J. Grácio, and P. A. A. P. Marques, "Graphene oxide versus functionalized carbon nanotubes as a reinforcing agent in a PMMA/HA bone cement," *Nanoscale*, vol. 4, no. 9, pp. 2937–2945, Apr. 2012.

[24] P. Pant, S. Mishra, and P. Mishra, *Advances in Mechanical Processing and Design*. 2021.

[25] M. Kamaraj, R. Santhanakrishnan, and E. Muthu, "Investigation of surface roughness and MRR in drilling of Al2O3 particle and sisal fibre reinforced epoxy composites using TOPSIS based Taguchi method," *IOP Conference Series: Materials Science and Engineering*, vol. 402, no. 1, 2018.

[26] V. Sridharan, T. Raja, and N. Muthukrishnan, "Study of the Effect of Matrix, Fibre Treatment and Graphene on Delamination by Drilling Jute/Epoxy Nanohybrid Composite," *Arabian Journal for Science and Engineering*, vol. 41, no. 5, pp. 1883–1894, May 2016.

IOP Conf. Series: Materials Science and Engineering 1228 (2022) 012013 doi:10.1088/1757-899X/1228/1/012013

Stability Enhancement of Electric Power Systems by Employing Variable Frequency Transformer

Basit Mushtaq
National Institute of Technology, Srinagar, J&K, India
E-mail: basitm40@gmail.com

Mairaj-Ud-Din Mufti
National Institute of Technology, Srinagar, J&K, India
E-mail: muftimd@yahoo.com

Abstract. Economical and technical advantages inherent in the coordinated operation of power systems resulted in large-scale grid interconnection. The major concern regarding the interconnection of the neighbouring power systems with relatively weak tie-lines is the appearance of poorly damped electromechanical oscillations. Inter-area (IA) oscillations reduce system stability and transmission capacity, and without an efficient damping mechanism, these oscillations can cause system failure and even lead to large-scale blackouts. In this work, the impact of variable frequency transformer (VFT) on the IA oscillations of a two-area power system connected in series with the tie-line is investigated. The VFT is a recent power transmission technology that offers continuous control over bi-directional power flow with no harmonic injection. The VFT is made up of a rotary transformer for continuously regulated phase shift, as well as a drive system and control that regulates the power flow through the VFT by adjusting the angle and speed of the rotary transformer. The VFT is controlled to increase system damping, and consequently suppress inter-area oscillations. A comprehensive model of the interconnected power system including VFT is developed in the MATLAB/Simulink environment to study the efficacy of the proposed scheme.

1. Introduction

The improvement in reliability and dynamic stability of power systems due to large scale grid interconnections is attracting increasing attention. Technical rationales for power system interconnection include lower investment in generating capacity, improved load factors, diversity of generation mix, economic interchange and supply security etc [1]. However, electro-mechanical oscillations are inherent in interconnected power systems which influence the safety of large grids. Interconnected power systems have an inter-area osillation mode with an oscillation frequency in the range of 0.1 Hz to 0.9 Hz [2, 3, 4]. Inter-area oscillations are a complex phenomenon and have increased significantly in recent years due to large-scale power system interconnections [5]. Inter-area modes involve generators from different areas oscillating against each other. These low-frequency oscillations are detrimental to the goals of optimal power flow and are observed in the interconnected power systems with weak ties [6]. Low-frequency oscillations curb the conveying capacities of interconnecting ties which results in sub-optimal power flows and sometimes

Content from this work may be used under the terms of the Creative Commons Attribution 3.0 licence. Any further distribution of this work must maintain attribution to the author(s) and the title of the work, journal citation and DOI.
Published under licence by IOP Publishing Ltd

even lead to power system blackouts. Such oscillations can have a two-fold economic cost. Numerous incidents have been reported in different power grids that caused huge loss to economy [7, 8]. Therefore improvement in damping of these modes is of vital concern and their stability have a significant economic benefit.

Until the development of power system stabilizers (PSSs), initial reaction to these oscillations was to restrict operation of the tie-lines. According to various research studies, PSSs embedded in generator excitation systems were used in damping of inter-area (IA) oscillations from early stages of development of interconnected power systems [9, 10]. PSS regulates the voltage reference input in such a way that the generator generates damping torques in response to speed changes. However, use of PSS is limited as the inter-area oscillations are a common result of multiple units, therefore the pressing question is to determine the ideal location for its installation. One of the obvious location for installing PSS is the generator with highest partcipation in the inter-area mode [11]. Therefore PSS needs to be installed in those generator excitation systems which participate in the inter-area oscillations which inturn make the power system complex and costlier.

Flexible AC transmission (FACTS) devices opened a new avenue in damping of these low-frequency oscillations. Different FACTS devices like thyristor controlled series compensator (TCSC), static VAR compensator (SVC), static synchronous series compensator (SSSC) have been used to suppress oscillations and enhance damping of power systems [12, 13]. Since primary objective of these devices is to assist the voltage regulation and power flow, hence rather than finding best installation locations of these controllers, the real problem is whether the proposed FACTS controller can effectively damp the low-frequemcy oscillations. Therefore, supplementary control needs to be realized for effective damping. Application of SSSC and TCSC with supplementary power oscillation damping control is reported in [14, 15, 16]. However, due to certain technical advantages of a recent flexible AC power transmission technology namely variable frequency transformer (VFT) makes it a new choice for damping the low-frequency oscillations. Technical advantages of VFT over other power flow controllers and interconnection devices is reported in [17].

VFT essentially is a controlled bi-directional AC power transmission device that can transfer active power between neighboring power networks which don't need to be at same frequency. Fundamentally VFT is made of wound-rotor induction machine (WRIM) driven by a DC drive motor [18]. The mechanical torque supplied to the rotor shaft of the WRIM determines the active power transfer through VFT[19]. Both WRIM and DC motor of VFT adds inertia to the power systems and thereby improves the stability during system disturbances.

In this paper, effect of VFT connected in series with tie-line on IA oscillations of a two-area power system is analysed. By controlling the power transmisssion of interconnecting sections, VFT effectively helps to improve the damping of power systems to supress low-frequency oscillations.

2. System configuration and mathematical modeling

2.1. Modeling of synchronous generator

The synchronous generator as a two-axis model is used and the dynamical equations for an ith generator are given by:

Transient emf due to the flux linkages of d-axis and q-axis coils is given by Eq(1) and Eq(2):

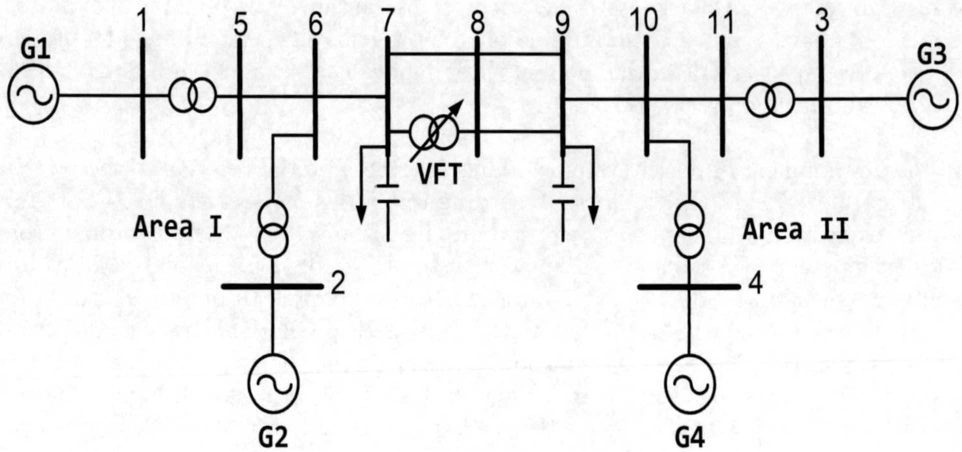

Figure 1. Two area test system with VFT.

$$\frac{dE'_{di}}{dt} = \frac{1}{T'_{qi0}}\left[-E'_{di} + (X_{qi} - X'_{qi})\left(-I_{qi} + \frac{X'_{qi} - X''_{qi}}{(X'_{qi} - X_{lsi})^2}\left\{-\psi_{2qi} - E'_{di} + I_{qi}(X'_{di} - X_{lsi})\right\}\right)\right] \quad (1)$$

$$\frac{dE'_{qi}}{dt} = \frac{1}{T'_{di0}}\left[-E'_{qi} + E_{fdi} + (X_{di} - X'_{di})\left(I_{di} + \frac{X'_{di} - X''_{di}}{(X'_{di} - X_{lsi})^2}\left\{\psi_{1di} - E'_{qi} - I_{di}(X'_{di} - X_{lsi})\right\}\right)\right] \quad (2)$$

Flux linkage in d-axis damper coil and q-axis damper coil results in subtransient emfs Eq(3) and Eq(4)

$$\frac{d\psi_{1di}}{dt} = \frac{1}{T''_{di0}}\left(-\psi_{1di} + E'_q + I_d(X'_{di} - X'_{lsi})\right) \quad (3)$$

$$\frac{d\psi_{2qi}}{dt} = \frac{1}{T''_{qi0}}\left(-\psi_{2qi} - E'_d + I_q(X'_{qi} - X'_{lsi})\right) \quad (4)$$

Eq(5) describes the rotor angular speed variations due to mismatch in mechanical and electrical torques. While as Eq(6) represents the excitation system for generator bus voltage regulation.

$$\frac{d\omega_{ri}}{dt} = \frac{1}{2H_i}\left(T_{mi} - T_{ei} - D_i(\omega_{ri} - \omega_s)\right) \quad (5)$$

$$\frac{dE_{fdi}}{dt} = \frac{1}{T_{ai}}\left(Ka_i * E_{ti} - K_{ai} * V_{refi} - E_{fdi}\right) \quad (6)$$

2.2. Modeling of Variable frequency transformer

VFT as a flexible AC interconnection device comprises of well established doubly-fed induction motor, hydrogenerator and DC drive control. The main body of VFT is a rotary transformer having three-phase winding on both rotor and stator. To adjust the rotational position of the rotor relative to stator, a DC drive system is employed that controls the capacity and direction of active power flowing through the VFT. In order to test the adaptibilty of VFT in suppression of inter-area oscillations, a simplified model of VFT is used in this paper. The ability of VFT to

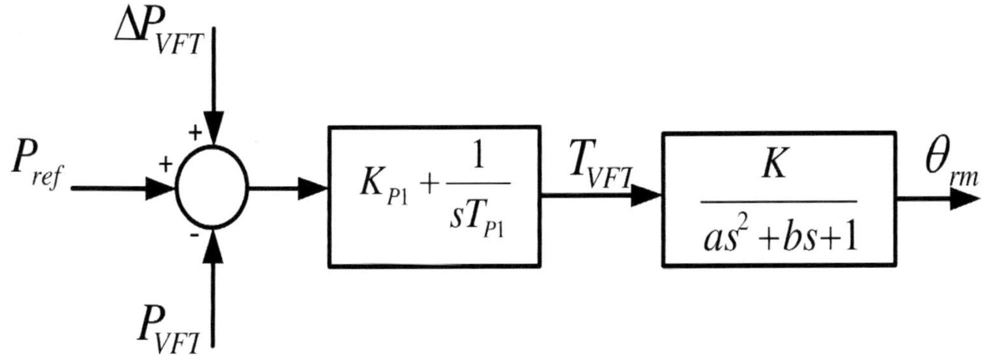

Figure 2. Modeling diagram of VFT.

flexibly regulate the tie power of interconnected systems helps to damp out the low-frequency oscillations. The controller receives an additional input from the active power of the portion where VFT is mounted.

The simulink block diagram representation of VFT is shown in Figure 2., where P_{ref} is the reference power setting of VFT, P_{VFT} is the actual power flowing through the VFT, $\triangle P_{VFT}$ is the given incremental power of VFT. With the PI control action the mechanical torque T_{VFT} is generated from the power input signals. This mechanical torque is applied as input to the VFT so as to adjust the position of rotor θ_{rm} in such a way that the necessary power transmission required for compensating the active power oscillations flows.

3. Simulation Studies and Discussion

A MATLAB/Simulink based model of KRK's two area test system as shown in Figure 1 is developed with VFT incorporated in one of the tie-lines. The performance of the system is analysed when subjected to a three phase fault. The comparative analysis of the various results is presented to show the significant damping of IA oscillations with the help of VFT.

At bus (7) three phase ground fault is applied with a clearing time of 0.1 sec. The response of the system is plotted with and without VFT in figures (3 – 7). Figure 3 presents the active power output of synchronous machines following the disturbance. The increasing oscillations depict that the system is unstable without the application of VFT. When VFT is incorporated the oscillations are well damped and output settles to a particular value.

It is clear from Figure 4 that when speed of G1 and G2 is increasing, speed of G3 and G4 is reducing indicating that generators from different areas participate resulting in inter-area oscillations. Plot of generator speeds with and without VFT is shown in Figure 5 and it is clear that response in presence of VFT is well damped. The rotor angle differences between the generators of interconnected areas is shown in Figure 6, with VFT the damping of inter-area mode is improved. Figure 7 shows the power flow in one of the parallel tie-lines connecting the two areas. It is evident from the figure that VFT effectively damps the power oscillations of tie-line flows.

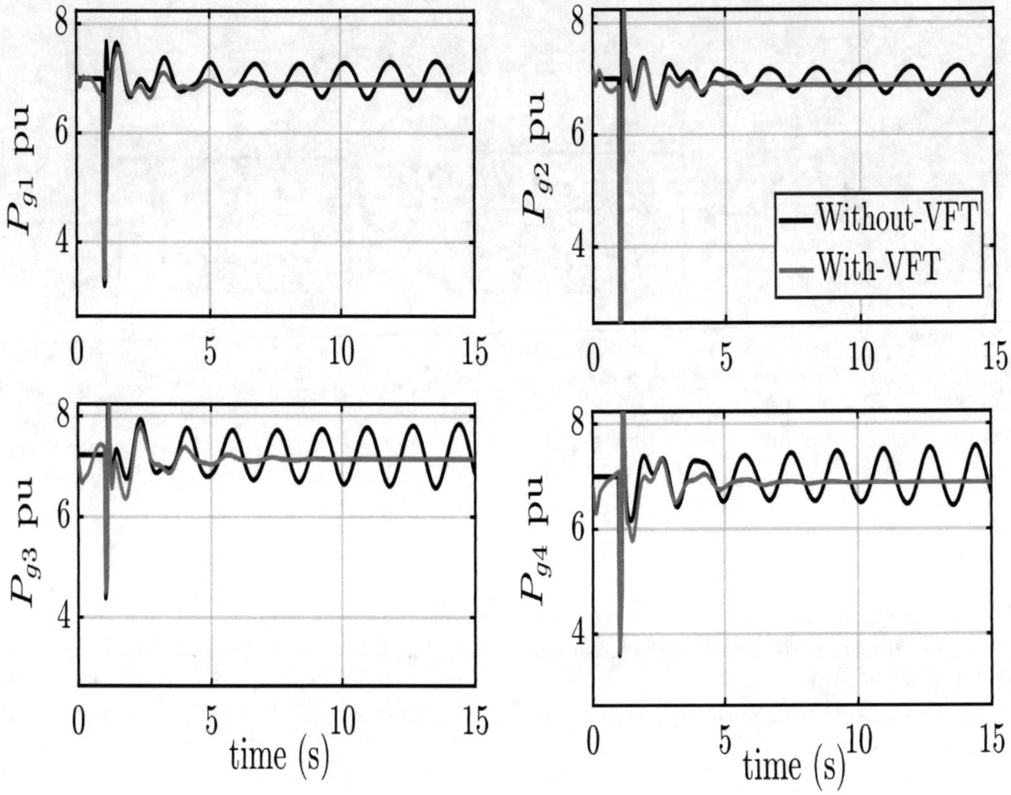

Figure 3. Generator active power outputs.

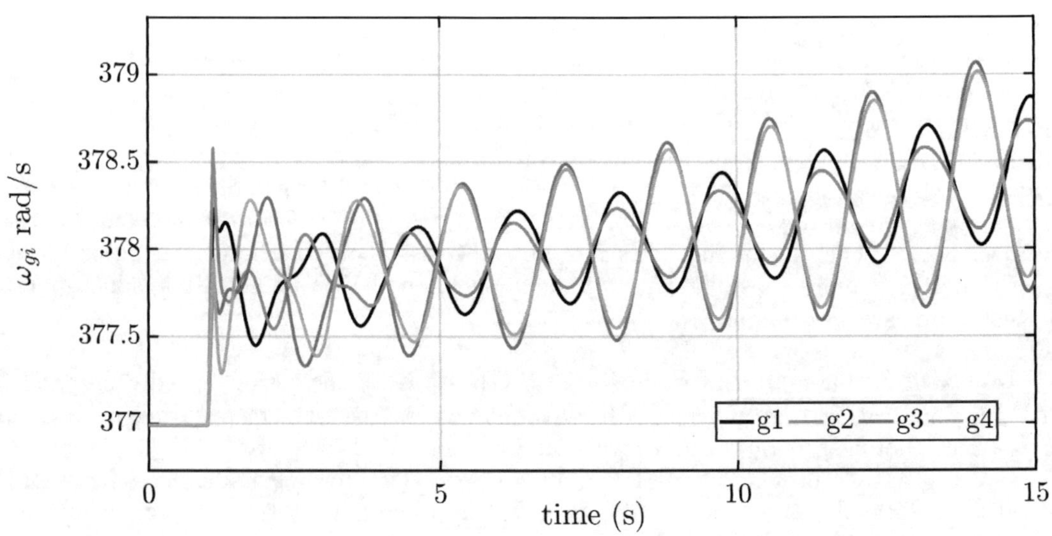

Figure 4. Generator rotor speeds without VFT.

IOP Conf. Series: Materials Science and Engineering 1228 (2022) 012013 doi:10.1088/1757-899X/1228/1/012013

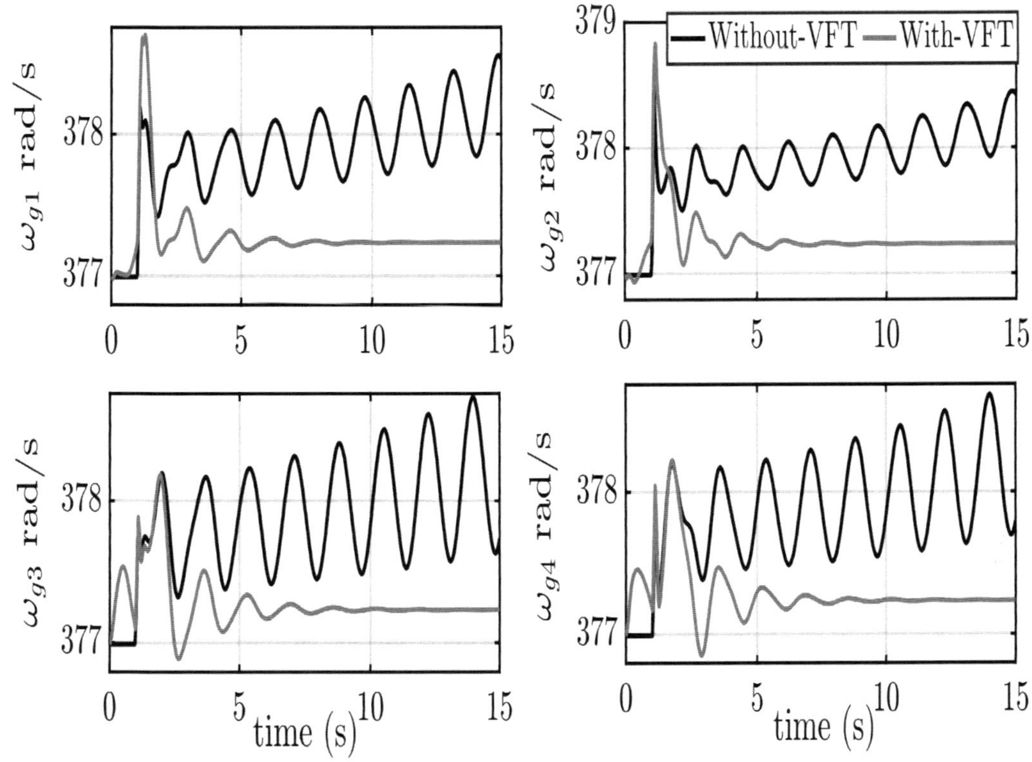

Figure 5. Generator rotor speeds.

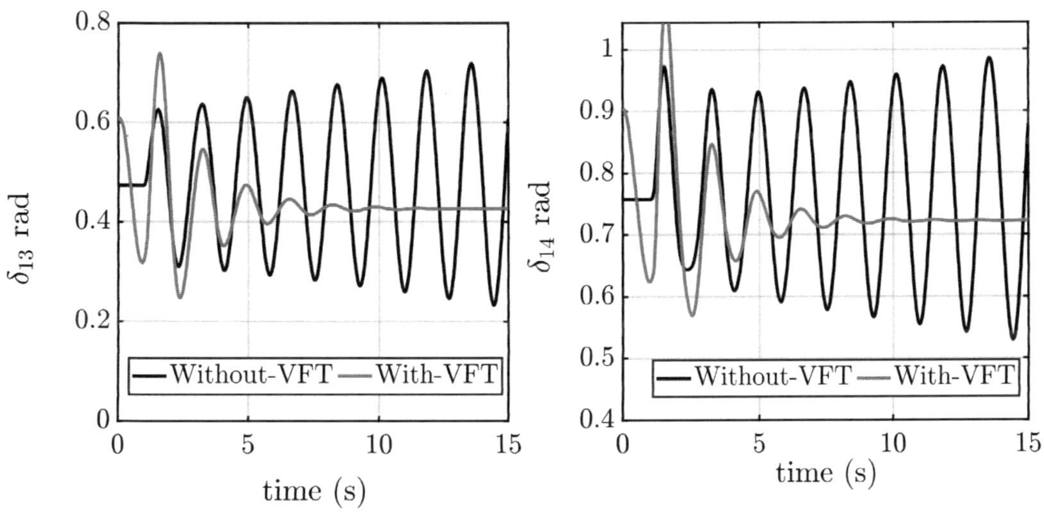

Figure 6. Angle differences between generators.

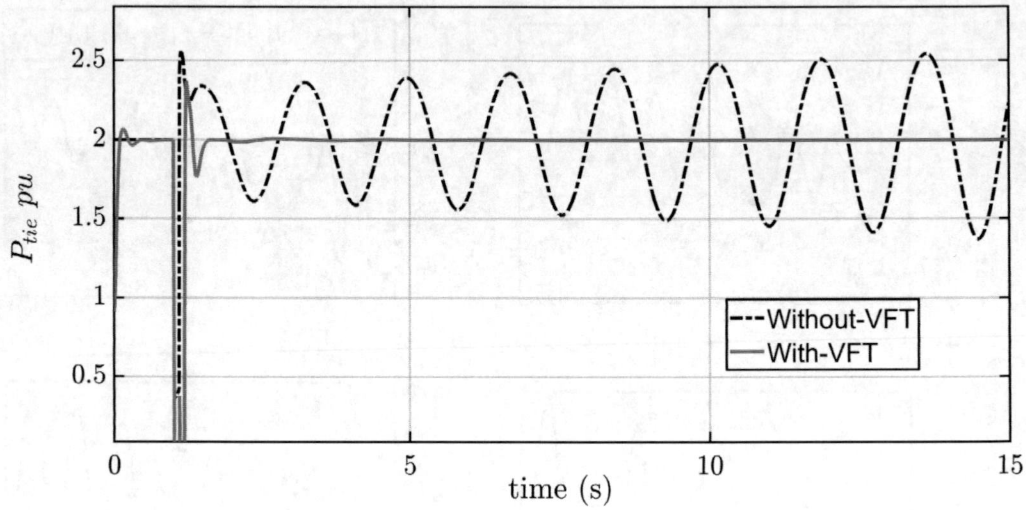

Figure 7. Tie line power oscillations.

4. Conclusion

In this paper, application of variable frequency transformer for damping of inter-area oscillations is analysed. The model verification is tested on Kundurs four machine system with VFT placed in one of the tie-lines. The results obtained show VFT's outstanding capability to adapt to the disturbances in power sytems, thereby suppressing the low-frequency oscillations, thus improving power system stability.

5. References

[1] E. Mariani and S. S. Murthy, *Introduction*, pp. 1–9. London: Springer London, 1997.
[2] L. Rouco, "Eigenvalue-based methods for analysis and control of power system oscillations," in *IEE Colloquium on Power System Dynamics Stabilisation (Digest No. 1998/196 and 1998/278)*, pp. 3/1–3/6, 1998.
[3] M. Klein, G. Rogers, and P. Kundur, "A fundamental study of inter-area oscillations in power systems," *IEEE Transactions on Power Systems*, vol. 6, no. 3, pp. 914–921, 1991.
[4] P. Kundur, *Power system stability and control.* McGraw-Hill, 1998.
[5] A. R. Messina, *Inter-area Oscillations in Power Systems a Nonlinear and Nonstationary Perspective.* Springer US, 2009.
[6] F. R. Schleif and J. H. White, "Damping for the northwest - southwest tieline oscillations - an analog study," *IEEE Transactions on Power Apparatus and Systems*, vol. PAS-85, no. 12, pp. 1239–1247, 1966.
[7] A. Atputharajah and T. K. Saha, "Power system blackouts - literature review," in *2009 International Conference on Industrial and Information Systems (ICIIS)*, pp. 460–465, 2009.
[8] B. Pal and B. Chaudhuri, *Robust control in power systems.* Springer Science Business Media, Inc., 2010.
[9] W. Watson and G. Manchur, "Experience with supplementary damping signals for generator static excitation systems," *IEEE Transactions on Power Apparatus and Systems*, pp. 199–203, 1973.
[10] F. P. de Mello, L. N. Hannett, and J. M. Undrill, "Practical approaches to supplementary stabilizing from accelerating power," *IEEE Transactions on Power Apparatus and Systems*, vol. PAS-97, no. 5, pp. 1515–1522, 1978.
[11] M. Benasla, T. Allaoui, M. Brahami, and B. Hadji, "Frequency response analysis to select the best pss locations in power systems," in *2016 8th International Conference on Modelling, Identification and Control (ICMIC)*, pp. 269–273, 2016.
[12] M. Darabian and A. Jalilvand, "Psss and svc damping controllers design to mitigate low frequency oscillations

problem in a multi-machine power system," *Journal of Electrical Engineering and Technology*, vol. 9, 11 2014.

[13] M. Darabian and A. Jalilvand, "A power control strategy to improve power system stability in the presence of wind farms using facts devices and predictive control," *International Journal of Electrical Power & Energy Systems*, vol. 85, pp. 50–66, 2017.

[14] R. Sadikovic, P. Korba, and G. Andersson, "Self-tuning controller for damping of power system oscillations with facts devices," in *2006 IEEE Power Engineering Society General Meeting*, pp. 6 pp.–, 2006.

[15] M. D. Kumar and P. Sujatha, "Design and development of self tuning controller for tcsc to damp inter harmonic oscillation," *Energy Procedia*, vol. 117, pp. 802–809, 2017. "First International Conference on Power Engineering Computing and CONtrol (PECCON-2017) 2nd -4th March .2017." Organized by School of Electrical Engineering, VIT University, Chennai, Tamil Nadu, India.

[16] A. Siddique, Y. Xu, W. Aslaml, and F. M. Albatsh, "Application of series fact devices sssc and tcsc with pod controller in electrical power system network," in *2018 13th IEEE Conference on Industrial Electronics and Applications (ICIEA)*, pp. 893–899, 2018.

[17] Imdadullah, S. M. Amrr, M. S. Jamil Asghar, I. Ashraf, and M. Meraj, "A comprehensive review of power flow controllers in interconnected power system networks," *IEEE Access*, vol. 8, pp. 18036–18063, 2020.

[18] R. Yuan, Y. Chen, G. Chen, and Y. Sheng, "Simulation model and characteristics of variable frequency transformers used for grid interconnection," in *2009 IEEE Power Energy Society General Meeting*, pp. 1–5, 2009.

[19] P. Truman and N. Stranges, "A direct current torque motor for application on a variable frequency transformer," in *2007 IEEE Power Engineering Society General Meeting*, pp. 1–5, 2007.

A study on engine piston optimization by generative design approach

A Arunkumar[1], A Sooriyamoorthy[1], C Mishael[1] and A Giri[1*]

[1] School of Mechanical Engineering, Vellore Institute of Technology, Vellore, Tamilnadu, India- 632014
*anoj.giri@vit.ac.in

Abstract. Significant enterprises today favor the cooperation of generative plans in the manufacturing and assembling departments. Therefore our aim is to dwell on the more straightforward degrees of generative design to demonstrate the modification of boundary conditions and properties to exhibit modern applications. Generative Design has demonstrated to be a significant upgrade in the machine-producing side of the business because of the simplicity in changes, higher resilience, and reduced number of faults that arise. Nevertheless owing to the added complex shape of structures, the model generation and design are posed with challenges. This paper involves the examination and generative design of a 4-Stroke engine piston. The proposed method is validated through a study which closely resembles a real life prototype. The examination aims to dissect and detail a newly formed design grid for the current cylinder of the F20C 4 - stroke 4 - cylinder engine at the core of the Honda S2000. Our aim to study on piston design, which makes it lighter and more efficient while keeping the similar volume yield in the cylinder. The paper depicts the process of optimizing the component through mass reduction where the mass was reduced by close to twenty three percent.

1. Introduction

Learning outcomes about software aids in the effective practice and performance of generative design in the industry. The most extensively used program amongst others was found to be Autodesk's Generative Design (GD), hosted in Fusion 360. Although there are numerous methods of 3D printing available, only a few can be utilized to model metal bodies with utmost accuracy due to the slicing parameters that have to be followed [1]. A handful of Artificial Intelligence (AI) concepts go into the assembly of a model and 3D printing the same due to the immense growth of digital manufacturing [2, 3]. Nowadays the use of generative design algorithms nowadays is for structural optimization i.e. high strength to weight ratio, creating parts with adequate strength, stiffness fatigue resistance etc. with minimal usage of material [4, 5]. Engineering applications such as internal structural parts for handheld equipment (to improve ergonomics), sports gear (to enhance performance), vehicles and aircraft (to reduce fuel consumption) etc. where the weight is prime consideration [6]. When the material is a key cost driver, superior structural efficiency can lead to significant savings in both from expenditure and sustainability perspective [7]. The 3D printing process carries a lot of weight-age in the article since the major attributes of the component to be tested in going to be determined by the precision and surface finish of 3D printing [8].

The process of manufacturing is slowly starting to evolve, nowadays we want to make products more efficient, cost-effective, and different. Generative Design (GD) is a new rising concept to develop or upgrade manufacturing methods from their conventional methods with the use of AI [9].

Content from this work may be used under the terms of the Creative Commons Attribution 3.0 licence. Any further distribution of this work must maintain attribution to the author(s) and the title of the work, journal citation and DOI.
Published under licence by IOP Publishing Ltd

The term "Generative Design" of tools, is the process of implementing artificial intelligence methods and algorithms, to decipher design problems. Generative Design has proven to be a major development in the machine manufacturing & engineering side of the industry owing to effortless alterations, superior tolerance & fewer errors that can be made [10]. The enhancement of product topology optimization does not always stay up to safety standards. This is the key area where the assistance of the GD process can bring out the minimum compliance weight while not compromising on safety. In GD, lattice orientations in model and combining this concept with additive manufacturing, the complex can be developed that would have been unfeasible to achieve with conventional design & manufacturing methods [11].

Numerous research works were proposed on geometries, materials, and manufacturing techniques, and with these, the piston has also undergone a continuous improvement over the last decades. Based on the studies done in past, there are many reasons for the piston failure. The damaged mechanisms have different origins and are mainly wear, temperature, and fatigue-related [12]. Therefore, there is a necessity to optimize the design of the piston by taking into consideration a variety of parameters. The present work aims to analyse, how generative design can make an enormous difference when applied, where weight and material reduction are main requirement. Along with optimizing the design, our aim is to reduce the weight reduction without compromising the part strength In this research work, the parameters chosen are analysis of piston by applying pressure force acting at the top of the piston in various strokes [13].

2. Part Selection

To study generative design, piston of a specific engine was selected. This research intends to analyze and formulate a new design lattice for the existing piston of the F20C 4-stroke 4-cylinder engine at the heart of the Honda S2000. The specification of engine is shown in Table 1.

Table 1. Engine specification [14]

Parameters	Values
Piston Type	Four stroke piston
Cylinder block alloy	Aluminium
Configuration	Inline 4
Valve train	4 valves per cylinder
Piston Stroke (mm)	84.01
Cylinder bore dia. (mm)	87
Compression Ratio	11.7:1 (JDM), 11.0:1 (USDM)
Number of revolutions per cycle	2
Displacement (cc)	1997
Power output (kW@rpm)	184@8600 (247 HP)
Torque output (Nm@rpm)	203@7500
Redline (rpm)	8900
Weight (g)	348
Fuel type	Gasoline

Aluminium Silicon alloys are widely employed to produce pistons due to their low density, high thermal conductivity, workability & resistance strength [15]. The material used is the A4032

IOP Conf. Series: Materials Science and Engineering 1228 (2022) 012014 doi:10.1088/1757-899X/1228/1/012014

(AlSiMg0.6) alloy which has been used in making the pistons for the F20C engine. This silicon content improves alloy's durability, again making it a great street engine choice when subjected to thousands of repetitions. The addition of silicon to aluminium reduces the melting point while improving the fluidity of the element for usage.

3. Model & Design Approach

To design a piston head with a given engine specification using Generative Design, piston head dimensions were calculated mathematically for a given set of engine specifications and then CAD model is made. Following which the piston CAD model is imported for the Generative Design analysis.

3.1. *Piston Design Considerations*

The design of the piston is made keeping in mind the procedure and specifications given in the machine design and data books. The dimensions are prepared based on two cases, one with the average load on the engine the other with the engine running at maximum load. The parameters like thickness of piston head, piston ribs thickness, piston rings, piston barrels and piston skirt are determined (calculations shown in Appendix.) The figure 1 shows the CAD model for piston.

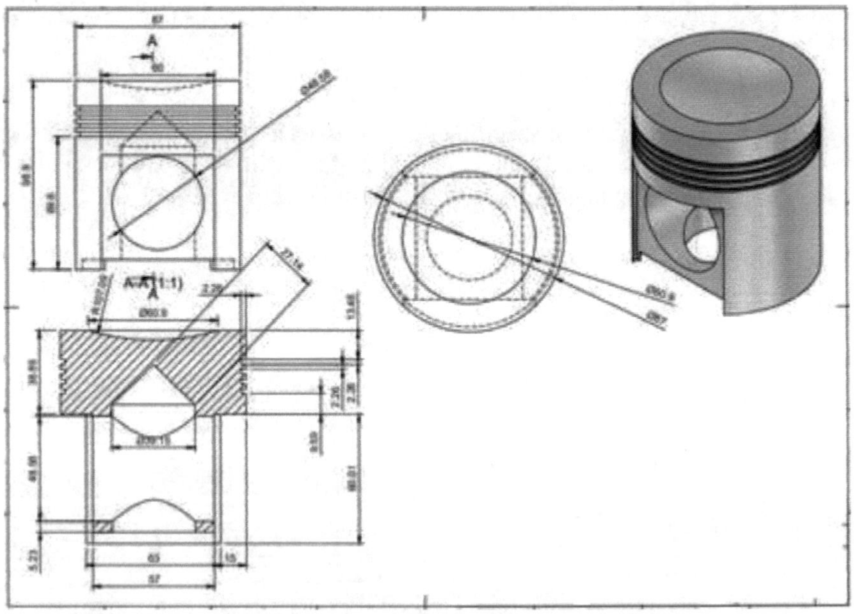

Figure 1. CAD drawing &model of piston head (all dimensions are in mm)

3.2. *Analysis*

The analysis is done by applying the set case loadings on a piston and procuring multiple GD designs. To accomplish this, Autodesk Fusion 360 software is used. The pressure is applied on piston head to investigate the stress-strains in the Generative Design based piston and conventional piston [16].The pressure applied on the piston head, the temperature effects, stresses, strains, and various other parameters are taken into consideration [9]. The density of material is taken as 2.67E-06 kg/mm^3. The yield strength (σ_y) is taken 240 MPa under high temperature which is essential for a material to be used in engine pistons which is an area susceptible to this kind of temperature. The Young's modulus (E) value is determined as 71000 MPa & Poisson's ratio as 0.33. The load applied on the face or top of the piston (called case loading) is 5MPa.

4. Results & Discussion

The generative design software involves the use of mathematical algorithms in order to optimize and get the most efficient results according the potential points of stress that might arise in the future in the component's applications. We use this algorithm to incorporate the design in our piston thereby reducing the amount of material from spots of weak failure and in turn strengthening the actually weak points in the design where failure might occur that it may curb the usage of the component.

The obstacle geometries in which the external components like piston ribs and wrist pin are to be placed have to be specified before the simulation run. The design values are as mentioned in Table 2, which throws light on the comparative results of the studies conducted where we can see that the though the mass is considerably reduced along with the surface area the density remains the same, hence the lattice structure possesses the rigidity just as the basic design which was operated on initially.

4.1. Generative Design Outcome

The mathematically obtained design is shown in figure 2(a), while the design generated by AI is shown in figure 2(b). Table 2 shows the comparison of properties for basic design and generative design. From the Table 2, it can be seen that the mass of generative design based piston is reduced by 23 percent.

Table 2. Comparison of properties (Basic design and Generative design)

Properties	Basic Design	Generative design
Mass (gms)	911.353g	696.274
Volume (mm^3)	3.413E+05	2.608E+05
Density($g/mm^{3)}$	0.003	0.003
Area (mm^2)	48101.40	39544.40

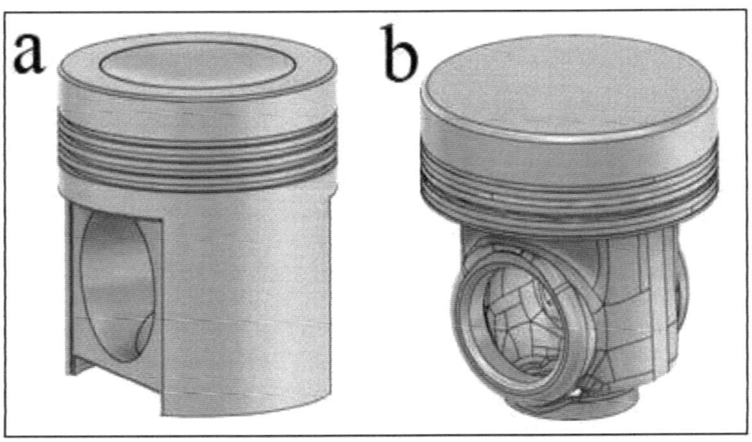

Figure 2. Piston designs: (a) Basic design & (b) Generative design

4.2. Static Stress Analysis: Basic CAD Design

As the objective is to determine the critical stresses the part undergoes as a component and to overcome any failures that may arise during operation of the part as part of the system since this is a functional part. The stress distribution on a piston depends upon the deformation, hence in order to minimize stress concentration, the piston crown should possess sufficient rigidity to reduce the deformation. The stress analysis parameters was taken to obtain values at which the piston would be damaged based on numerous research papers standard pressure force applied on the piston was 5MPa greater than 5Mpa would cause piston damage and deformation. The forces on the top face of the piston are taken into account while calculating the stresses involved through the operation. A stress analysis was carried out based on the Von Mises Criteria, and the induced stress-strains were investigated in the piston.

The tessellated model of the CAD design includes the properties of the model which denotes the maximum and minimum values of displacement, stress & strain (see figure 3). The coordinates are shown alongside the models on the right side. From figure 3(a), the displacement of the component to the maximum extent comes up to be 0.00533 mm and falls to 0 mm in the minimum side. As shown in figure 3(b), the case loadings are taken as 5 MPa and the maximum design stress is obtained as 117.4 MPa (as this can be seen in Table 3 also). The figure 3(c) depicts the original design strain ranging from 2.847E-08 to 0.00139. Table 3 depicts the outcomes of the analysis performed on the basic piston design, which was designed in order to simulate the current design characteristics of the piston.

Table 3. Stress analysis for basic design based piston

Name	Minimum	Maximum
Safety Factor		
Safety Factor (Per Body)	2.044	3
Stress		
Von Mises	0.002925 MPa	117.4 MPa
1st Principal	-49.8 MPa	97.86 MPa
3rd Principal	-174.7 MPa	13.84 MPa
Displacement		
Total	0 mm	0.00533 mm
X	-0.001356 mm	0.001352 mm
Y	-0.001392 mm	0.0014 mm
Z	-0.005159 mm	1.128E-05 mm
Reaction Force		
Total	0 N	177.9 N
X	-37.72 N	42.61 N
Y	-44.85 N	42.65 N
Z	-11.98 N	177.7 N
Strain		
Equivalent	5.224E-08	0.002493
1st Principal	2.304E-08	0.001664
3rd Principal	-0.002753	3.627E-07

IOP Conf. Series: Materials Science and Engineering 1228 (2022) 012014 doi:10.1088/1757-899X/1228/1/012014

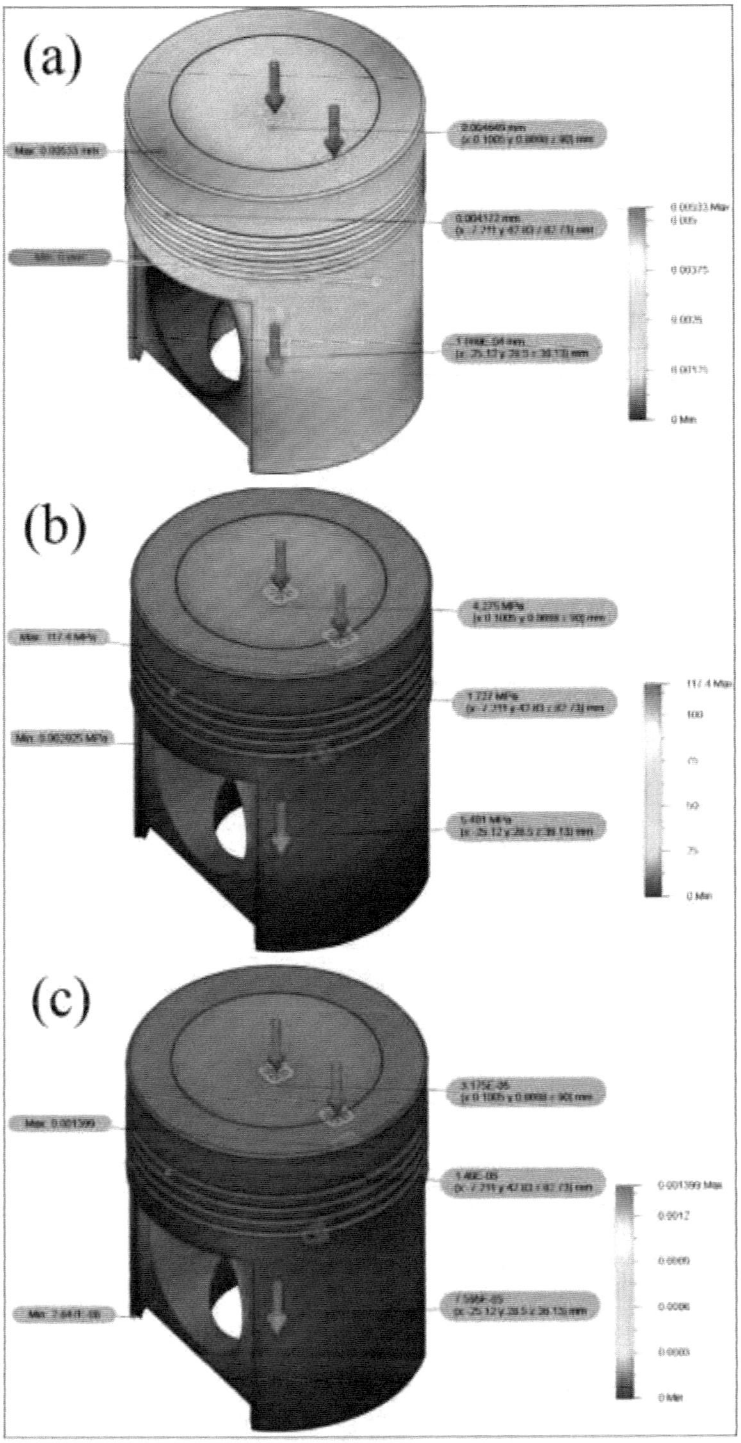

Figure 3. Stress analysis for basic design: (a) Displacement (b) Stress & (c) Strain

4.3. Static Stress Analysis: Generative Design

Similarly the stress analysis was done for generative design based piston. The displacement of the component to the maximum extent is 0.01011 mm and falls to 0 mm as shown in Figure 4(a). The loading are applies as 5 MPa and the maximum design stress is 116.9 MPa as shown in Figure 4(b). Figure 4(c) depicts the original design strain ranging from 2.245E-06 to 0.002323. Based on stress analysis, the maximum pressure was approximately 117MPa. The results observed shows to be valid, as the stress analysis parameters were taken same as the standard parameters of the piston.

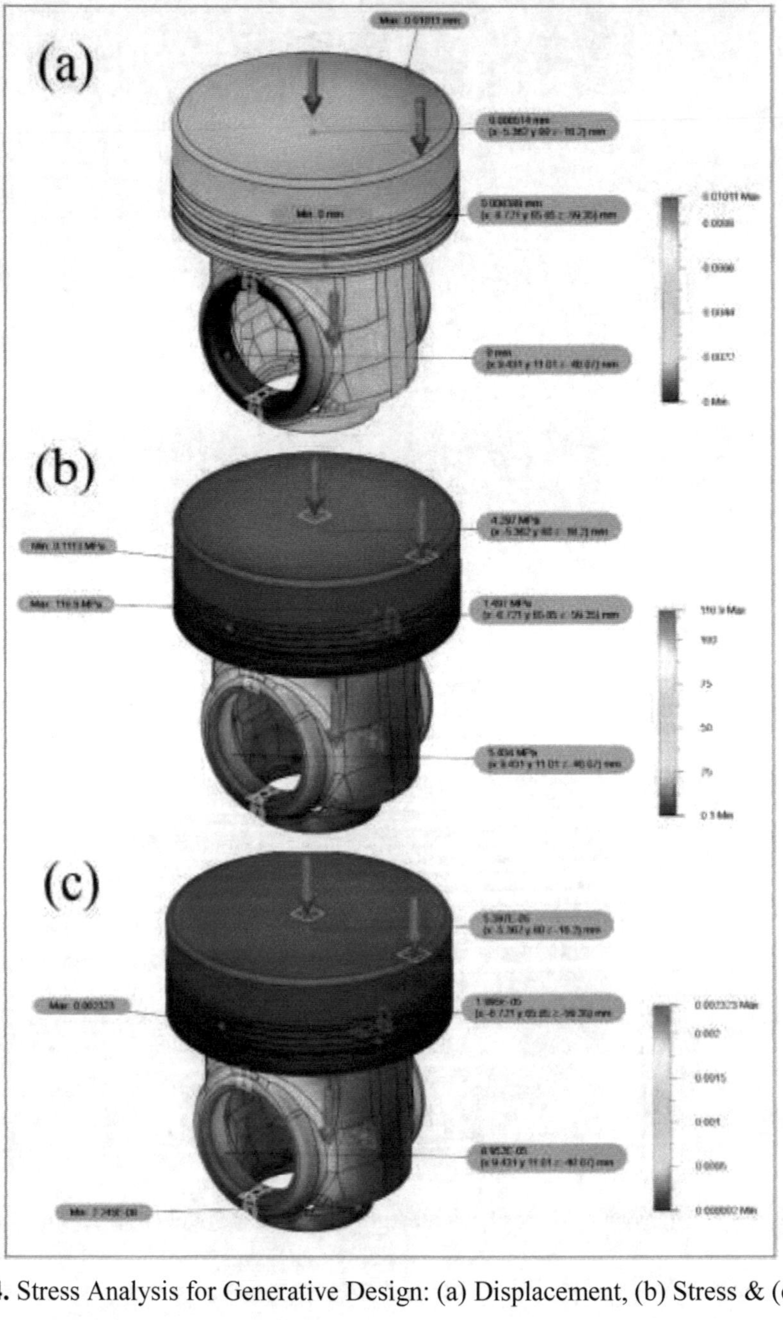

Figure 4. Stress Analysis for Generative Design: (a) Displacement, (b) Stress & (c) Strain

Table 4. Stress Analysis for generative design based piston

Name	Minimum	Maximum
Safety Factor		
Safety Factor (Per Body)	2.052	15
Stress		
Von Mises	0.1113 MPa	116.9 MPa
1st Principal	-33.71 MPa	72.37 MPa
3rd Principal	-157 MPa	9.162 MPa
Displacement		
Total	0 mm	0.01011 mm
X	-0.002955 mm	0.003032 mm
Y	-0.01006 mm	2.198E-05 mm
Z	-0.002307 mm	0.00225 mm
Reaction Force		
Total	0 N	134.7 N
X	-40.6 N	35.34 N
Y	-6.586 N	134.1 N
Z	-43.24 N	42.43 N
Strain		
Equivalent	2.245E-06	0.002323
1st Principal	5.527E-07	0.002027
3rd Principal	-0.002558	-1.008E-06

5. Conclusion

In this paper, the feasibility of Generative Design has been studied on an engineering component. The piston design has been obtained by implementing generative design approach. The GD based piston results in the less weight approximately 23 percent as compared to the basic conventional design. Furthermore the stress analysis was also done to ensure, that the reduction in mass does not results in increase in stresses. The static stress analysis was executed using Fusion360, which resulted the stresses in new GD based design are lower than the stresses induced in conventional design. Although difference is not too much but it proves that the original piston can be substituted by the Generative Design, as the stresses remains the same. Though the design process for generative design has progressed the manufacturing process is still in the development stages. A generative designs main advantage is as it eliminates the human error which in turn saves the cost in material reduction and manufacturing errors.

Appendix

A.1 Thickness of Piston Head

$$t_h = D\sqrt{\frac{3}{16}\frac{P_{max}}{\sigma_b}}$$

where D = bore dia (in mm), for Al max $\sigma_b = 90\ N/mm^2$; $P_{max} = 8$ MPa

$$t_h = 11.23\ mm$$

IOP Conf. Series: Materials Science and Engineering 1228 (2022) 012014 doi:10.1088/1757-899X/1228/1/012014

A.2 Piston Ribs Thickness

$$t_r = \frac{t_h}{3} \; to \; \frac{t_h}{2}$$

$$t_r = 3.74 \; mm \; to \; 5.615 \; mm$$

Since, ratio of stroke length to bore $\frac{1}{D} = \frac{84.01}{87} = 1.036$,
∴ D < 1.5, a cup is required.

Radius of Piston Cup = 0.7*D = 60.9 mm

A.3 Piston Rings

p_w= 0.025 MPa, radial pressure on cylinder wall

= 110 N/mm^2, tensile stress for ring material

- Radial width of ring,

$$b = D\sqrt{\frac{3pw}{\sigma_t}}$$

$$87\left(\sqrt{\frac{3}{110}}\right)0.025 = 2.26 \text{ mm}$$

- Axial thickness of piston ring, h = from (0.7*b) to (b) = 1.58mm to 2.26mm

Number of piston rings in petrol engines = 4

- Limit on minimum axial thickness

$$h_{min} = \frac{D}{10*z} = \frac{87}{40} = 2.17 \text{ mm} \quad (z=4)$$

∴ Axial thickness must be b/w 2.17 mm & 2.26 mm.

- Gaps b/w free ends

Diameter of piston ring greater than that of cylinder bore. A part of the ring is slightly cut diagonally. During the assembly, the ring is compressed diagonally & passed into liner.

G = (3.5b) to (4b)
G = 7.91 mm to 9.05 mm (before assembly)

G = (0.002D) to (0.004D)
G = 0.174mm to 0.348 mm (after assembly)

- Width of top lands & ring lands

Distance from top of piston to the first ring groove (h_1) is called top land,

$h_1 = (t_h)$ to $(1.2t_h) = 11.23$ mm to 13.476 mm

Distance b/w 2 consecutive grooves (h_2),

$h_2 = (0.75h)$ to $(h) = 1.695$ mm to 2.26 mm (Taking 'h' as 2.26 mm)

A.4 Piston Barrel

Thickness of piston barrel at the top end t_3 ,

$= (0.03D) + b + 4.9 = 9.77$ mm

Thickness of piston barrel at the lower or open end is given by t_4 ,

$= (0.025t_3)$ to $(0.35t_3)$ = 2.44 mm to 3.41 mm

A.5 Piston Skirt

Maximum gas force on piston head =

$$P_{max} \left(\frac{\pi D^2}{4}\right) = 47557.42 \text{ N} = 47.6 \text{ kN}$$

Side thrust =

$$P_{max} \left(\frac{\pi D^2}{4}\right) \mu = 0.2 * 47557.42 \text{ N} = 4755.74 \text{ N} = 4.67 \text{ kN}$$

Length of skirt, $L_s = (0.65D)$ to $(0.8D)$ = 56.55 mm to 69.6 mm

$L_s = 69.6$ mm

Total length of piston = top land + length of ring section + length of skirt

= 13.476 + [(4*2.26)+(3*2.26)] + 69.6 = 98.9 mm

Length of piston pin in connecting rod bushing,

$L_1 = (0.45D) = (0.45*87) = 39.15$ mm

Piston pin diameter, d_o,

$47557.4 = (P_b*d_o*L_1)$ $[P_b = 25 \text{ MPa}]$

IOP Conf. Series: Materials Science and Engineering 1228 (2022) 012014 doi:10.1088/1757-899X/1228/1/012014

$d_o = 48.58$mm

Inner diameter = $(0.6d_o) = 29.148$ mm

References

[1] McCormack J, Dorin A and Innocent T 2004 Generative Design: A Paradigm for Design Research *Design Research Society International Conference* **2** 8

[2] Carlos G V and Santiago M L 2020 Generative Design for Social Manufacturing *Applied Deep Generative Network AGDN 2020* ed J Hayes *et al* (Spain)

[3] Chen Z 2016 Research on the Impact of 3D Printing on the International Supply Chain *Advances in Materials Science and Engineering* **2016** 4173873

[4] Sangeun O, Yongsu J, Seongsin K, Ikjin L and Namwoo Kang 2019 Deep Generative Design: Integration of Topology Optimization and Generative Models *Journal of Mechanical Design* **141(11)** 111405

[5] Williams J C, Chen J and Ma C 2017 Additive Manufacturing of High Value Ti Components: Opportunities and Challenges *International Journal of Additive and Subtractive Materials Manufacturing* **1(2)** 119-132

[6] Bagassi S, Lucchi F, DeCrescenzio F and Persiani F 2016 Generative Design: Advanced Design Optimization Processes for Aeronautical Applications *30th Congress of the International Council of the Aeronautical Sciences ICAS 2016* (Daejon Korea) 1-7

[7] Kanada Y 2016 3D Printing of Generative Art Using the Assembly and Deformation of Direction Specified Part *Rapid Prototyping Journal* **22(4)** 636-644

[8] Shahrubudina N , Leea T C and Ramlan R 2019 An Overview on 3D Printing Technology: Technological, Materials, and Applications *Procedia Manufacturing* **35** 1286-1296

[9] Jun W, Xiaoping Q and Michael Y W 2019 Advances in Generative Design *Computer Aided Design* **116** 102733

[10] Laura T 2021 Product Customization & Generative Design *Multidiszciplináris tudományok* **11** 87-95

[11] Carolina A and Fernando V R 2016 Implementation of a Simulation Model Using the Systems Dynamics: Case Study from Study of Mass to Digital Manufacturing *International Journal of Engineering Research and Applications* **6(7)** 43-41

[12] Rao L T, Satyanarayana K, Rao M S S S, Rao T V H and Rao S V U 2015 Stress Analysis of 4 Stroke Diesel Engine Piston *International Journal of Current Engineering and Scientific Research* **2(2)** 121-125

[13] Singh A R and Sharma D P K 2014 Design, Analysis and Optimization of Three Aluminium Piston Alloys Using FEA *International Journal of Engineering Research and Applications* **4(1)** 94–102

[14] https://s2000.club/OM/Honda%20F22C%20engine.pdf

[15] Singhal S K, Jain P K and Pandey P M 2008 Adaptive Slicing for SLS Prototyping *Computer Aided Design and Applications* **5(1–4)** 412–423

[16] Kumar P, Bhandari A and Chaturvedi S K 2019 Design and Thermal Analysis of IC Engine Piston Design Using Catia and Ansys software *International Journal of Mechanical Engineering* **6(10)** 12–19.

IOP Conf. Series: Materials Science and Engineering 1228 (2022) 012015 doi:10.1088/1757-899X/1228/1/012015

Review on various materials used in Additive Manufacturing

Sangeetha N[1], Monish P[2], Brathikan V M[3]

[1] Department of Mechanical Engineering, Kumaraguru College of Technology, Coimbatore
[2] Department of Mechanical Engineering, Kumaraguru College of Technology, Coimbatore
[3] Research Cell, Kumaraguru College of Technology, Coimbatore

sangeetha.n.mec@kct.ac.in

Abstract. 3D printing or Additive manufacturing or Rapid prototyping is a technology where 3D structures are designed and printed which is currently doing good for the manufacturing sector of many industries such as automotive, aerospace, medical, jewellery, constructions etc. Additive Manufacturing is a fast-emerging technology which has been exceedingly used for mass customization and fabrication of free design sourced products. Additive manufacturing is a method where the materials are put together in a desired shape via a certain process with the appropriate material type. The property of the materials used for 3D printing is highly dependent on the type and composition of the material. The various types and compositions of materials hugely impacts their implementation in potential applications which is discussed in this paper. The dominantly used materials, their composition, their properties, their applications and their future scope are discussed. This paper gives a clear overview on the material technology used in the additive manufacturing industry.

1. Introduction

Additive manufacturing (AM) or Rapid prototyping commonly addressed as 3D printing which is a method of manufacturing 3D objects by adhering the specific materials in specific form to produce the required properties in the material. According to the American Society for Testing and Materials (ASTM) standards the additive manufacturing processes can be classified into seven major categories, namely (i) binder jetting, (ii) directed energy deposition, (iii) material extrusion, (iv) material jetting, (v) powder bed fusion, (vi) sheet lamination and (vii) vat photo polymerization. (Y. Wang et al., 2017)

In recent years the requirement of different properties on different materials for potential real-world applications is very high. Properties such as mechanical strength, resolution of object, roughness or smoothness etc., are highly influential on the material type used for additive manufacturing.(Ngo et al., 2018) Each of these material types has its manufacturing method, based on which the efficiency of printing will vary.

The crucial purpose of this review is to analyze all the materials and its corresponding technologies used in additive manufacturing. The properties, the compositions and the potential application of the

Content from this work may be used under the terms of the Creative Commons Attribution 3.0 licence. Any further distribution of this work must maintain attribution to the author(s) and the title of the work, journal citation and DOI.
Published under licence by IOP Publishing Ltd

IOP Conf. Series: Materials Science and Engineering 1228 (2022) 012015 doi:10.1088/1757-899X/1228/1/012015

materials is discussed. The review comprises the research done in different material categories for the further researches in 3D Printing technologies.

2. Categorization of materials in 3D printing

The most prominently used material categories in 3D Printing are Metals, Ceramics, Polymer, Bio-materials, and Smart-materials in the recent manufacturing sector. (Ngo et al., 2018)

2.1. Metals

Metallic 3D printing plays a very significant role in the manufacturing industries. Recent technologies in additive manufacturing use a variety of energy sources and feedstock. The high diversity in the feedstock materials had led to very significant researches across the world.

Metallic 3D printing primarily has 4 methods namely, Powder Bed Fusion (PBF), Direct Energy Deposition (DED), Binder Jetting, Laminated Object Manufacturing (LOM).(Zhang et al., 2018) Each of these methods use a different form of feedstock. The Powder Bed Fusion uses feedstock in metallic powder bed format which is melted using lasers (SLM) or electron beams (EBM) to get the desired structure. (Taminger & Hafley, 2004) Pure metals such as copper, iron, gold are processed through SLM and EBM.(Das et al., 2016) The Direct Energy Deposition (DED) uses feedstock in a powder flow or metal wire. (Oh et al., 2019) . In Electron Beam Freeform Fabrication (EBF3) a consumable metal wire is used which is melted and selectively sprayed to form a layer of the material.(Taminger & Hafley, 2004) In Laser Engineered Net Shaping (LENS) method, a powder flow is used as feedstock(Atwood et al., 2018). In Binder Jetting the feedstock used will be in powder bed form which is adhered together when the liquid binder is injected.(Ziaee & Crane, 2019). In Laminated object manufacturing (LOM) metallic sheets are used as feedstock.(Feygin & Hsieh, 1991) . Ultrasonic Additive Manufacturing (UAM) is a commonly used technique which uses an ultrasonic energy source to stack a precisely cut metal sheets to form the desired 3D structure. (Hehr & Norfolk, 2020)

Table 1. Metallic material's printing methods, process and properties (Das et al., 2016)

Method	Feedstock type	Process	Material	Youngs Modulus (GPa)	Yield Strength (MPa)	Ultimate Strength (MPa)	Density (g/cm3)	Poisson's Ratio
DED	Powder Fed	LENS	Ti6Al4V	113	945	979	4.41	0.34
DED	Wire fed	EBF3	Inconel 718	202	1195	1372	8.2	0.288
LOM	Metal sheet fed	UAM	Al6061	68	217	225	2.7	0.33
LOM	Metal sheet fed	UAM	Al3003	68.9	262	266	2.73	0.33
PBF	Powder Bed Fed	EBM	Iron (Fe)	204	50	540	7.86	0.29
PBF	Powder Bed Fed	EBM	Copper (Cu)	110	33.3	210	8.96	0.343
PBF	Powder Bed Fed	SLM	Gold (Au)	79	80	900	19.3	0.415

IOP Conf. Series: Materials Science and Engineering 1228 (2022) 012015 doi:10.1088/1757-899X/1228/1/012015

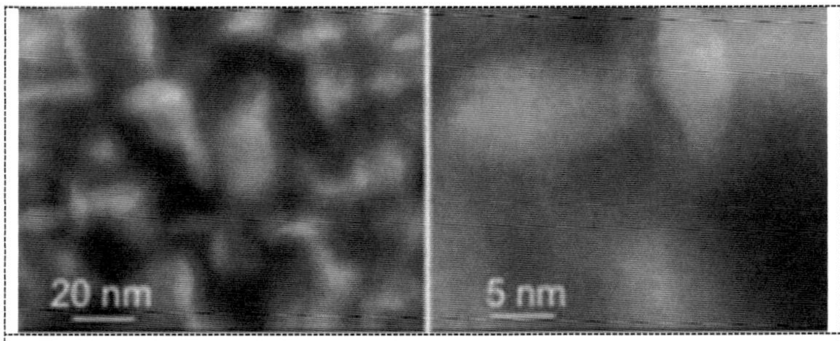

Figure 1. High-resolution image from a transmission electron microscope exhibiting the presence of γ ″ microstructure.(Pogson et al., 2003)

Binder Jetting metal printing has a time-consuming post process and the complete fabrication time is very high. They also have less geometrical accuracy and it requires more research for efficient fabrication.

2.1.1 Applications. Ti6Al4V is a widely known alloy in Metal 3D printing because of its combination of good mechanical properties with low specific weight and corrosion- resistant nature of the material. This made the material to be used in a variety of engineering environments related to aeronautics in various applications including functional prototypes, solid end-use parts, medical devices and spare parts. (Ngo et al., 2018) Inconel 718 exhibits high resistance to thermal, oxidation and corrosion which made this material an attractive choice for extreme environment applications such as turbine, engine parts. It is also suited for cryogenic components. It's a favored material in automotive and aeronautical industry. (Baicheng et al., 2017)

2.2. *Ceramics*

Manufacturing of Ceramic components via 3D printing has opened up many possibilities for potential real-world applications. Generally machining of ceramic components are hard due to their hardness and brittleness. The methods on ceramic 3D printing tries to overcome these problems through preprocessing the feed stock material before printing. This is primarily categorized into slurry-based, powder-based and bulk solid-based methods. (Chen et al., 2019)

In slurry-based technologies a semi liquid system is dispersed with ceramic particles which depending of the viscosity of the system uses a form of ink or paste as feedstock. The technology types used to print slurry based are Stereolithography (SL), Digital light processing (DLP), Two-photon polymerization (TPP), Inkjet printing (IJP) and Direct ink writing (DIW).

In powder-based technologies, loose ceramic particles are used as feed stock in powder bed format which are selectively sintered. The technology types used to print powder-based materials are Three-dimensional printing (3DP), Selective laser sintering (SLS) and Selective laser melting (SLM). (Bertrand et al., 2007)

In bulk solid-based technologies as the name mentions, either sheets of the material or filaments are used as feedstock. In case of filament, the feeding is done through extrusion or deposition. For sheets as feed material, the sheets are successively adhered together to form the output. The technologies types used to print bulk solid based 3D printing are Laminated object manufacturing (LOM) and Fused deposition modelling (FDM). (Conzelmann et al., 2020)

IOP Conf. Series: Materials Science and Engineering 1228 (2022) 012015 doi:10.1088/1757-899X/1228/1/012015

Table 2. Ceramic material's printing technology, technique and properties.

Technology used	Technique used	Feedstock form	Material reinforced with Ceramic	Properties exhibited
Slurry-based	SL	Liquid fed	Photo-polymerized silicon oxycarbide	They possess high performance and versatile functionalities which is also altered by varying compositions. (Eckel et al., 2016)
	DLP	Liquid fed	Alumina and bioactive glass	Very high relative densities and good mechanical strength (Tesavibul et al., 2012)
	TPP	Liquid fed	Zr–Si polymer ceramic scaffolds	Enables creation of structure porosity with pore dimensions (Koroleva et al., 2015)
	IJP	Liquid fed	Aqueous Si_3N_4 ink	Good mechanical properties (Cappi et al., 2008)
	DIW	Liquid fed	Robocasted Al_2O_3–ZrO_2 gel paste	Low thermal expansion coefficient and high compressive strength
Powder-based	3DP	Powder bed fed	Polymerized Silicon resin ($CaSiO_3$-based)	High porosity, biocompatible and zero cytotoxic effects
			Ti_3SiC_2	enhanced bending strength
	SLS	Powder bed fed	Hydroxyapatite–tricalcium phosphate (HA-TCP), hydroxy-apatite polycarbonate (HA-PC), hydroxyapatite polyether ether ketone (PA-PEEK) and silica–polyamide (SiO_2-PA)	Low geometrical accuracy and rough surface with macro-porous structures customizable in a controllable manner
	SLM	Powder bed fed	Al_2O_3 and SiC-based ceramics	Very fine resolutions and features were detailed; limited to fabricate small parts
Bulk solid-based	LOM	Sheet fed	Al_2O_3 and SiC	Novel light weight parts can be fabricated
	FDM	Filament fed	PCL-HA and alumina	less accuracy and rod shaped (favored for bio-medical applications)

2.2.1. Applications. Hydroxyapatite–tricalcium phosphate (HA-TCP), hydroxyapatite–polycarbonate (HA-PC), hydroxyapatite polyether ether ketone (PA-PEEK) and silica–polyamide (SiO_2-PA), these are certain major components that are fused with ceramics which provide customizable macro porous structures which are very helpful in the biomedical sector such as bone implants and tissue engineering. (G. Lee et al., 2020)

2.3. Polymers

3D printing of polymer-based materials has showed a significant impact on current manufacturing technologies. 3D printed pure polymer products present lack of strength and functionality which is overcome by polymer composites made of combined matrix and reinforcements which provides structural and functional properties.

Common 3D printing methods for polymers are Fused deposition modeling (FDM), Powder bed and inkjet head 3D printing, Stereolithography (SLA), Selective Laser Sintering (SLS), 3D plotting/direct-write. Most widely used polymer composites are Particle reinforced composites, Fiber reinforced polymer composites and Nanocomposites. (X. Wang et al., 2017)

Particle reinforcements are done to improve the desired properties of the polymer matrix. Particles can be easily mixed with the required polymers either in powder or liquid form. Printing of structural components for real-world applications is availed by the particle reinforced composites.(Singh et al., 2002) Reinforcements of fiber is done to enhance the polymer matrix properties. The feedstock used will be in pellet form. These pellets are blended with the appropriate fiber and then sent to the extruder. The extruder fabricates the filament which is further used build the 3D objects. Powder technologies are not preferred as a smooth layer powder-fiber mixture is not possible. (Herrera-Franco & Valadez-González, 2004)

Carbon nanotubes, ceramic, metal nanoparticle and graphite are common nanomaterials which exhibit unique electrical, mechanical and thermal properties. Reinforcement of these materials with polymers help us in the fabrication of high-performance functional composites.

Table 3. Polymeric material's printing technique, composite types, material and properties.

Techniq ue	Feedstock type	Composite Type	Material used	Properties enhanced
FMD	Filament fed	Particle reinforced polymer	Iron/ABS, and Copper/ABS	High storage modulus and thermal conductivity, Low coefficient of thermal expansion (Nikzad et al., 2011)
			Al and Al2O3/Nylon-6	Low frictional coefficient (Boparai et al., 2015)
			BaTiO3/ABS and CaTiO3/ Polypropylene	High dielectric permittivity and controllable resonance frequency (Isakov et al., 2016)
			Tungsten/PC	High dielectric permittivity, x-ray attenuation factor. (Shemelya et al., 2015)
			BaTiO3/ABS	High dielectric permittivity (Castles et al., 2016)
			Thermoplastic elastomer/ABS	Low anisotropy of printed components (Torrado Perez et al., 2014)
		Fiber reinforced polymer	Short glass fiber/ABS	140% of improved tensile strength (Zhong et al., 2001)
			Short carbon fiber/ABS	115% of improved tensile strength (Tekinalp et al., 2014)
		Polymer nanocomposit es	TiO2/ABS and Carbon nanofiber/ABS	High tensile modulus and strength, reduced elongation (Torrado Perez et al., 2014)
			Montmorillonite/ABS	High tensile strength and modulus, flexural strength and modulus, and thermal stability, low thermal expansion coefficient (Weng et al., 2016)

			Graphene/ABS	High electrical conductivity and thermal stability(Wei et al., 2015)
			Carbon nanofiber/Graphite/ polystyrene	Less capacitive background current (Rymansaib et al., 2016)
SLA	Liquid fed	Particle reinforced polymer	Al2O3/UV cured resin	High dielectric permittivity and reduced dielectric loss tangent (Kurimoto et al., 2015)
		Polymer nanocomposit es	CNT/epoxy	High tensile strength, reduced elongation (Sandoval & Wicker, 2006)
			Graphene oxide/photopolymer	High tensile modulus, strength and elongation (Lin et al., 2015)
			TiO2/epoxy acrylate	High tensile strength and modulus and thermal stability (Duan et al., 2011)
			BaTiO3/PEGDA	High piezoelectric coefficient [(Hwang et al., 2015)]
			CNT/acrylic ester	Improved absorption of electromagnetic energy (Guo et al., 2015)
			BST/epoxy	Ultralow thermal conductivity and high energy conversion (He et al., 2015)
DLP	Liquid fed	Particle reinforced polymer	Diamond microparticle /acrylate-based resin	High heat transfer (Kalsoom et al., 2016)
		Polymer nanocomposit es	Silver/PEGDA	High electrical conductivity (Fantino et al., 2016)
SLS	Powder fed	Particle reinforced polymer	Glass bead/Nylon-11	High tensile modulus and compressive modulus. (Chung & Das, 2006)
		Polymer nanocomposit es	Carbon black/nylon-12 Al2O3/polystyrene	Improved tensile strength and impact resistance (Zheng et al., 2006)
			TiO2/nylon-12 and graphite/ nylon-12 and Silica/Nylon-11	High tensile modulus, reduced elongation (Kim et al., 2013)
DW	Viscous liquid fed	Fiber reinforced polymer	Short carbon fiber/Silicon carbide whisker/epoxy	127% of improved tensile strength (Ning et al., 2015)

2.3.1. Applications. Polymers possess a lot of applications in biomedical, electronic and certain aerospace applications (Kumar et al., 2021). Material composites such as cell seeded multicellular spheroids/agarose are used in the construction of liver tissue.(Liao et al., 2019), Cell seeded alginate/Nano fibrillated cellulose are used for constructing of cartilage(Möller et al., 2017). Carbon black/PCL composites were used to print piezoresistive sensors and capacitive sensors. CNT/PLA composites helped in printing thin flexible electric circuits. (Manapat et al., 2017) Glass fiber and carbon fiber reinforced photopolymer composites are used in the manufacturing of airfoil and propellers used in aerospace sector(Asif et al., 2020). Carbon fiber reinforced Polyether Ether

Ketone (PEEK) composites also used for aerospace applications because of its incredibly high heat resistance and light weight. PEEK is lighter than traditional aluminium and maintain ⅔ of its strength. (A. Wang et al., 1998)

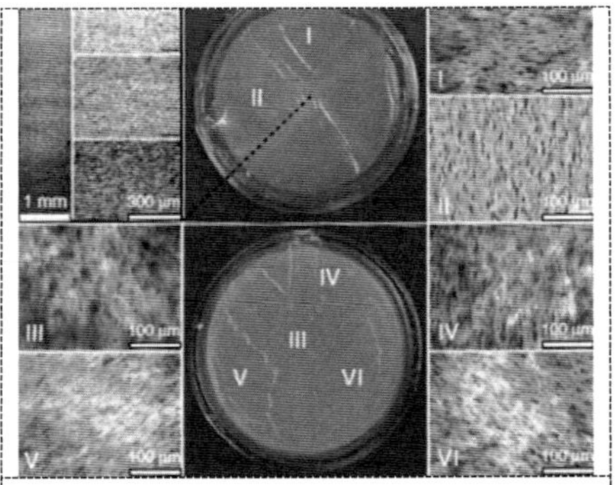

Figure 2. Actual photograph and optical microscope image of heterogeneous polymer composites structure (Wu et al., 2016)

2.4. *Biomaterials*

Additive manufacturing of Bio materials has led to many potential applications in the bio medical sector and many more. The 3D printing techniques used for bio materials are 3D plotting/direct ink writing, Laser assisted bioprinting, Selective laser sintering (SLS), Stereolithography, Fused deposition modelling (FDM) and Robotic assisted deposition. In 3D writing layer by layer deposition takes place of material takes place. In laser assisted printing, the deposition is controlled by laser pulse and its deposited on a quartz disk. In Selective laser sintering (SLS), and Stereolithography the material is selectively sintered via laser and light respectively. In FDM, the printing is done by basic extrusion and deposit method. For Robotic assisted deposition, solid loaded slurry is directly written.

Table 4 Biomaterial's printing methods, materials and properties.

3D Printing Method	Feedstock form	Material used	Property exhibited
3D plotting/direct ink writing	Ink fed	polycaprolactone (PCL), hydroxyapatite (HA), bioactive glasses, polylactic acid (PLA)/polyethylene glycol (PEG),	Easily incorporated with protein and living cells (Sobral et al., 2011)
Laser-assisted bioprinting	Liquid fed	HA, zirconia, HA/MG63	Ambience suitable for organic and inorganic cells and materials (Guillotin et al., 2010)
Selective laser sintering	Powder bed fed	PCL, HA, PLLA, tricalcium phosphate, and poly(3-hydroxybutyrate)	High biocompatibility (Kutikov et al., 2015)
Stereolithography	Liquid fed	photo-curable polymers	Enables complex design manufacturing (Rymansaib et al., 2016)

Fused deposition modelling	Filament fed	Structural and biopolymers, ceramic-polymer, or metal polymer composites	production limited to thermoplastic polymers (Kalita et al., 2003)
Robotic assisted deposition/ robocasting	Liquid fed	HA/PLA, HA/PCL, and bioactive glass (6P53B)/PCL	composited good with ceramics (Cao et al., 2003)

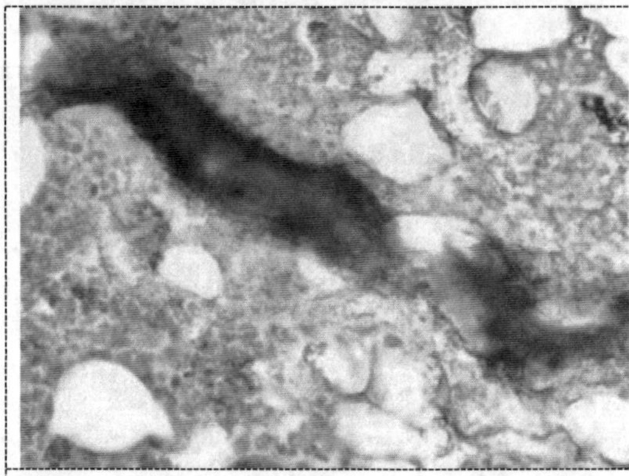

Figure 3. 3D-printed-doped tricalcium phosphate (Bandyopadhyay et al., 2015)

2.4.1Applications. Bio materials play a very crucial role in biomedical applications majorly due to their bio compatibility. Materials like polycaprolactone (PCL), hydroxyapatite (HA), bioactive glasses have the potential ability to create protein loaded bio materials which can be used to 3D print tissues that might help cure diseases. Majorly, the other common bio materials are used for tissues engineering, bone implants, etc. (Kutikov et al., 2015)

2.5. Smart Materials

The general types of 3D printable smart materials are smart Nano composites (such as piezoelectric material), Shape Memory Alloys (SMA) (which possesses super-elasticity) and Shape Memory Polymers (SMP) (which are stimuli-responsive materials) and there are certain multi materials used to print Actuators for soft robotics, Self-evolving structures and Anti-counterfeiting system. Polyjet printing is a very efficient method to print smart materials. In Polyjet printing the photo polymeric material jets dropped which solidifies on exposure to UV light. (J. Y. Lee et al., 2017)

Table 5. Smart material's type, smart composites used, printing technique and properties.

Starting material	Type of material	Smart composite used	Printing Technique	Property exhibited
Single material	Smart nanocomposite s	piezoelectric nanoparticles and photo liable polymer	Digital projection printing (DPP)	Electric charge produced when stress is applied, also works in reverse (Composite & Al, 2014)

	Shape memory alloys (SMAs)	Nickel titanium (NiTi)	Selective laser melting (SLM)	Temperature activated shape change (Griffith & Halloran, 1996)
	Shape memory polymers (SMP)	Shape memory polymer	PolyJet	Temperature activated shape change (Wu et al., 2016)
Multiple materials	Actuators for soft robotics	Dielectric elastomers and rigid material	PolyJet	Large strains are produced upon activation by electricity (Rossiter et al., 2009)
	Self-evolving structures	Hydrophilic polymer and rigid material	PolyJet	Hydrophilic polymer volume increases when exposed to water (Raviv et al., 2014)
	Anti-counterfeiting system	Mixture of quantum dots and photopolymer resin with a rigid metal	PolyJet	Quantum dots absorb UV light and emit visible lights (Ivanova et al., 2014)

2.5.1. Applications. There is a very high potential for the 3d printed smart materials in various real-world application. These are printed materials which react to their ambience and give out the response. .(Composite & Al, 2014) These materials can be favoured for biomedical implants and also for many other micro-electrochemical devices application. It can also be used in many electronics-based applications as they print materials which are stimuli responsive. (Zarek et al., 2016)

3. Summary

In this review, the material technology of additive manufacturing industry and its wide diversity in is discussed. The major material categories mentioned are metals, ceramics, polymers, biomaterials and smart-materials. Every particular material has its own type of feedstock for efficient printing. The commonly used types are filament, liquid based, powder based or in pellet form. The field of applications each material used are influenced by their properties. Metal 3d printing is on high demand in auto and aero industries, ceramics, biomaterials and smart materials have high potential applications in the biomedical industries and the polymers, which is very widely used in many sectors including auto, aero, art, medical etc. The 3D printing technology is pushing its limits with technology like 3D printing with lunar substances for the applicability for lunar colonization.

References

Asif, S., Chansoria, P., & Shirwaiker, R. (2020). Ultrasound-assisted vat photopolymerization 3D printing of preferentially organized carbon fiber reinforced polymer composites. *Journal of Manufacturing Processes*, *56*(December 2019), 1340–1343. https://doi.org/10.1016/j.jmapro.2020.04.029

Atwood, C., Griffith, M., Harwell, L., Schlienger, E., Ensz, M., Smugeresky, J., Romero, T., Greene, D., & Reckaway, D. (2018). *Laser engineered net shaping (LENS^{TM}): A tool for direct fabrication of metal parts*. *1*(1998), E1–E7. https://doi.org/10.2351/1.5059147

Baicheng, Z., Xiaohua, L., Jiaming, B., Junfeng, G., Pan, W., Chen-nan, S., Muiling, N., Guojun, Q., & Jun, W. (2017). Study of selective laser melting (SLM) Inconel 718 part surface improvement by electrochemical polishing. *Materials and Design*, *116*, 531–537. https://doi.org/10.1016/j.matdes.2016.11.103

Bandyopadhyay, A., Vahabzadeh, S., Shivaram, A., & Bose, S. (2015). Three-dimensional printing of biomaterials and soft materials. *MRS Bulletin*, *40*(12), 1162–1168. https://doi.org/10.1557/mrs.2015.274

Bertrand, P., Bayle, F., Combe, C., Goeuriot, P., & Smurov, I. (2007). Ceramic components manufacturing by selective laser sintering. *Applied Surface Science*, *254*(4), 989–992. https://doi.org/10.1016/j.apsusc.2007.08.085

Boparai, K., Singh, R., & Singh, H. (2015). Comparison of tribological behaviour for Nylon6-Al-Al2O3 and ABS parts fabricated by fused deposition modelling: This paper reports a low cost composite material that is more wear-resistant than conventional ABS. *Virtual and Physical Prototyping*, *10*(2), 59–66. https://doi.org/10.1080/17452759.2015.1037402

Cao, T., Ho, K. H., & Teoh, S. H. (2003). Scaffold design and in vitro study of osteochondral coculture in a three-dimensional porous polycaprolactone scaffold fabricated by fused deposition modeling. *Tissue Engineering*, *9*(SUPPL. 1). https://doi.org/10.1089/10763270360697012

Cappi, B., Özkol, E., Ebert, J., & Telle, R. (2008). Direct inkjet printing of Si3N4: Characterization of ink, green bodies and microstructure. *Journal of the European Ceramic Society*, *28*(13), 2625–2628. https://doi.org/10.1016/j.jeurceramsoc.2008.03.004

Castles, F., Isakov, D., Lui, A., Lei, Q., Dancer, C. E. J., Wang, Y., Janurudin, J. M., Speller, S. C., Grovenor, C. R. M., & Grant, P. S. (2016). Microwave dielectric characterisation of 3D-printed BaTiO3/ABS polymer composites. *Scientific Reports*, *6*(March), 1–8. https://doi.org/10.1038/srep22714

Chen, Z., Li, Z., Li, J., Liu, C., Lao, C., Fu, Y., Liu, C., Li, Y., Wang, P., & He, Y. (2019). 3D printing of ceramics: A review. *Journal of the European Ceramic Society*, *39*(4), 661–687. https://doi.org/10.1016/j.jeurceramsoc.2018.11.013

Chung, H., & Das, S. (2006). Processing and properties of glass bead particulate-filled functionally graded Nylon-11 composites produced by selective laser sintering. *Materials Science and Engineering A*, *437*(2), 226–234. https://doi.org/10.1016/j.msea.2006.07.112

Composite, N. À. P., & Al, K. I. M. E. T. (2014). 3D Optical Printing of Piezoelectric Materials. *Acs Nano*, *10*, 9799–9806.

Conzelmann, N. A., Gorjan, L., Sarraf, F., Poulikakos, L. D., Partl, M. N., Müller, C. R., & Clemens, F. J. (2020). Manufacturing complex Al2O3 ceramic structures using consumer-grade fused deposition modelling printers. *Rapid Prototyping Journal*, *26*(6), 1035–1048. https://doi.org/10.1108/RPJ-05-2019-0133

Das, S., Bourell, D. L., & Babu, S. S. (2016). Metallic materials for 3D printing. *MRS Bulletin*, *41*(10), 729–741. https://doi.org/10.1557/mrs.2016.217

Duan, Y., Zhou, Y., Tang, Y., & Li, D. (2011). Nano-TiODN2/DN-modified photosensitive resin for RP. *Rapid Prototyping Journal*, *17*(4), 247–252. https://doi.org/10.1108/13552541111138360

Eckel, Z. C., Zhou, C., Martin, J. H., Jacobsen, A. J., Carter, W. B., & Schaedler, T. A. (2016). Additive manufacturing of polymer-derived ceramics. *Science*, *351*(6268), 58 LP – 62. https://doi.org/10.1126/science.aad2688

Fantino, E., Chiappone, A., Calignano, F., Fontana, M., Pirri, F., & Roppolo, I. (2016). In situ thermal generation of silver nanoparticles in 3D printed polymeric structures. *Materials*, *9*(7), 21–23. https://doi.org/10.3390/ma9070589

Feygin, M., & Hsieh, B. (1991). Laminated object manufacturing: A simpler process. *Proceedings of the 2nd Solid Freeform Fabrication Symposium (SFF)*, *0*, 123–130. http://sffsymposium.engr.utexas.edu/Manuscripts/1991/1991-16-Feygin.pdf%5Cnhttp://sffsymposium.engr.utexas.edu/1991TOC

Griffith, M. L., & Halloran, J. W. (1996). Freeform fabrication of ceramics via stereolithography. In

Journal of the American Ceramic Society (Vol. 79, Issue 10, pp. 2601–2608). https://doi.org/10.1111/j.1151-2916.1996.tb09022.x

Guillotin, B., Souquet, A., Catros, S., Duocastella, M., Pippenger, B., Bellance, S., Bareille, R., Rémy, M., Bordenave, L., Amédée j, J., & Guillemot, F. (2010). Laser assisted bioprinting of engineered tissue with high cell density and microscale organization. *Biomaterials, 31*(28), 7250–7256. https://doi.org/10.1016/j.biomaterials.2010.05.055

Guo, S. Z., Yang, X., Heuzey, M. C., & Therriault, D. (2015). 3D printing of a multifunctional nanocomposite helical liquid sensor. *Nanoscale, 7*(15), 6451–6456. https://doi.org/10.1039/c5nr00278h

He, M., Zhao, Y., Wang, B., Xi, Q., Zhou, J., & Liang, Z. (2015). 3D Printing Fabrication of Amorphous Thermoelectric Materials with Ultralow Thermal Conductivity. *Small, 11*(44), 5889–5894. https://doi.org/10.1002/smll.201502153

Hehr, A., & Norfolk, M. (2020). A comprehensive review of ultrasonic additive manufacturing. *Rapid Prototyping Journal, 26*(3), 445–458. https://doi.org/10.1108/RPJ-03-2019-0056

Herrera-Franco, P. J., & Valadez-González, A. (2004). Mechanical properties of continuous natural fibre-reinforced polymer composites. *Composites Part A: Applied Science and Manufacturing, 35*(3), 339–345. https://doi.org/10.1016/j.compositesa.2003.09.012

Hwang, S., Reyes, E. I., Moon, K. sik, Rumpf, R. C., & Kim, N. S. (2015). Thermo-mechanical Characterization of Metal/Polymer Composite Filaments and Printing Parameter Study for Fused Deposition Modeling in the 3D Printing Process. *Journal of Electronic Materials, 44*(3), 771–777. https://doi.org/10.1007/s11664-014-3425-6

Isakov, D. V., Lei, Q., Castles, F., Stevens, C. J., Grovenor, C. R. M., & Grant, P. S. (2016). 3D printed anisotropic dielectric composite with meta-material features. *Materials and Design, 93*, 423–430. https://doi.org/10.1016/j.matdes.2015.12.176

Ivanova, O., Elliott, A., Campbell, T., & Williams, C. B. (2014). Unclonable security features for additive manufacturing. *Additive Manufacturing, 1*, 24–31. https://doi.org/10.1016/j.addma.2014.07.001

Kalita, S. J., Bose, S., Hosick, H. L., & Bandyopadhyay, A. (2003). Development of controlled porosity polymer-ceramic composite scaffolds via fused deposition modeling. *Materials Science and Engineering C, 23*(5), 611–620. https://doi.org/10.1016/S0928-4931(03)00052-3

Kalsoom, U., Peristyy, A., Nesterenko, P. N., & Paull, B. (2016). A 3D printable diamond polymer composite: A novel material for fabrication of low cost thermally conducting devices. *RSC Advances, 6*(44), 38140–38147. https://doi.org/10.1039/c6ra05261d

Kim, H. C., Hahn, H. T., & Yang, Y. S. (2013). Synthesis of PA12/functionalized GNP nanocomposite powders for the selective laser sintering process. *Journal of Composite Materials, 47*(4), 501–509. https://doi.org/10.1177/0021998312441812

Koroleva, A., Deiwick, A., Nguyen, A., Schlie-Wolter, S., Narayan, R., Timashev, P., Popov, V., Bagratashvili, V., & Chichkov, B. (2015). Osteogenic differentiation of human mesenchymal stem cells in 3-D Zr-Si organic-inorganic scaffolds produced by two-photon polymerization technique. *PLoS ONE, 10*(2), 1–18. https://doi.org/10.1371/journal.pone.0118164

Kumar, S., Singh, R., Singh, T. P., & Batish, A. (2021). On investigation of rheological, mechanical and morphological characteristics of waste polymer-based feedstock filament for 3D printing applications. *Journal of Thermoplastic Composite Materials, 34*(7), 902–928. https://doi.org/10.1177/0892705719856063

Kurimoto, M., Yamashita, Y., Ozaki, H., Kato, T., Funabashi, T., & Suzuoki, Y. (2015). 3D printing of conical insulating spacer using alumina /UV-cured-resin composite. *Annual Report - Conference on Electrical Insulation and Dielectric Phenomena, CEIDP, 2015-Decem*, 463–466. https://doi.org/10.1109/CEIDP.2015.7352047

Kutikov, A. B., Gurijala, A., & Song, J. (2015). Rapid prototyping amphiphilic polymer/hydroxyapatite composite scaffolds with hydration-induced self-fixation behavior. *Tissue Engineering - Part C: Methods, 21*(3), 229–241.

https://doi.org/10.1089/ten.tec.2014.0213

Lee, G., Carrillo, M., McKittrick, J., Martin, D. G., & Olevsky, E. A. (2020). Fabrication of ceramic bone scaffolds by solvent jetting 3D printing and sintering: Towards load-bearing applications. *Additive Manufacturing*, *33*(October 2019), 101107. https://doi.org/10.1016/j.addma.2020.101107

Lee, J. Y., An, J., & Chua, C. K. (2017). Fundamentals and applications of 3D printing for novel materials. *Applied Materials Today*, *7*, 120–133. https://doi.org/10.1016/j.apmt.2017.02.004

Liao, W., Wang, J., Xu, J., You, F., Pan, M., Xu, X., Weng, J., Han, X., Li, S., Li, Y., Liang, K., Peng, Q., & Gao, Y. (2019). High-throughput three-dimensional spheroid tumor model using a novel stamp-like tool. *Journal of Tissue Engineering*, *10*. https://doi.org/10.1177/2041731419889184

Lin, D., Jin, S., Zhang, F., Wang, C., Wang, Y., Zhou, C., & Cheng, G. J. (2015). 3D stereolithography printing of graphene oxide reinforced complex architectures. *Nanotechnology*, *26*(43), 434003. https://doi.org/10.1088/0957-4484/26/43/434003

Manapat, J. Z., Chen, Q., Ye, P., & Advincula, R. C. (2017). 3D Printing of Polymer Nanocomposites via Stereolithography. *Macromolecular Materials and Engineering*, *302*(9), 1–13. https://doi.org/10.1002/mame.201600553

Möller, T., Amoroso, M., Hägg, D., Brantsing, C., Rotter, N., Apelgren, P., Lindahl, A., Kölby, L., & Gatenholm, P. (2017). In Vivo Chondrogenesis in 3D Bioprinted Human Cell-laden Hydrogel Constructs. *Plastic and Reconstructive Surgery - Global Open*, *5*(2), 1–7. https://doi.org/10.1097/GOX.0000000000001227

Ngo, T. D., Kashani, A., Imbalzano, G., Nguyen, K. T. Q., & Hui, D. (2018). Additive manufacturing (3D printing): A review of materials, methods, applications and challenges. *Composites Part B: Engineering*, *143*(December 2017), 172–196. https://doi.org/10.1016/j.compositesb.2018.02.012

Nikzad, M., Masood, S. H., & Sbarski, I. (2011). Thermo-mechanical properties of a highly filled polymeric composites for Fused Deposition Modeling. *Materials and Design*, *32*(6), 3448–3456. https://doi.org/10.1016/j.matdes.2011.01.056

Ning, F., Cong, W., Qiu, J., Wei, J., & Wang, S. (2015). Additive manufacturing of carbon fiber reinforced thermoplastic composites using fused deposition modeling. *Composites Part B: Engineering*, *80*, 369–378. https://doi.org/10.1016/j.compositesb.2015.06.013

Oh, W. J., Lee, W. J., Kim, M. S., Jeon, J. B., & Shim, D. S. (2019). Repairing additive-manufactured 316L stainless steel using direct energy deposition. *Optics and Laser Technology*, *117*(December 2018), 6–17. https://doi.org/10.1016/j.optlastec.2019.04.012

Pogson, S. R., Fox, P., Sutcliffe, C. J., & O'Neill, W. (2003). The production of copper parts using DMLR. *Rapid Prototyping Journal*, *9*(5), 334–343. https://doi.org/10.1108/13552540310502239

Raviv, D., Zhao, W., McKnelly, C., Papadopoulou, A., Kadambi, A., Shi, B., Hirsch, S., Dikovsky, D., Zyracki, M., Olguin, C., Raskar, R., & Tibbits, S. (2014). Active printed materials for complex self-evolving deformations. *Scientific Reports*, *4*, 1–8. https://doi.org/10.1038/srep07422

Rossiter, J., Walters, P., & Stoimenov, B. (2009). Printing 3D dielectric elastomer actuators for soft robotics. *Electroactive Polymer Actuators and Devices (EAPAD) 2009*, *7287*, 72870H. https://doi.org/10.1117/12.815746

Rymansaib, Z., Iravani, P., Emslie, E., Medvidović-Kosanović, M., Sak-Bosnar, M., Verdejo, R., & Marken, F. (2016). All-Polystyrene 3D-Printed Electrochemical Device with Embedded Carbon Nanofiber-Graphite-Polystyrene Composite Conductor. *Electroanalysis*, *28*(7), 1517–1523. https://doi.org/10.1002/elan.201600017

Sandoval, J. H., & Wicker, R. B. (2006). Functionalizing stereolithography resins: Effects of dispersed multi-walled carbon nanotubes on physical properties. *Rapid Prototyping Journal*, *12*(5), 292–303. https://doi.org/10.1108/13552540610707059

Shemelya, C. M., Rivera, A., Perez, A. T., Rocha, C., Liang, M., Yu, X., Kief, C., Alexander, D., Stegeman, J., Xin, H., Wicker, R. B., MacDonald, E., & Roberson, D. A. (2015). Mechanical, Electromagnetic, and X-ray Shielding Characterization of a 3D Printable Tungsten–

Polycarbonate Polymer Matrix Composite for Space-Based Applications. *Journal of Electronic Materials*, *44*(8), 2598–2607. https://doi.org/10.1007/s11664-015-3687-7

Singh, R. P., Zhang, M., & Chan, D. (2002). Toughening of a brittle thermosetting polymer: Effects of reinforcement particle size and volume fraction. *Journal of Materials Science*, *37*(4), 781–788. https://doi.org/10.1023/A:1013844015493

Sobral, J. M., Caridade, S. G., Sousa, R. A., Mano, J. F., & Reis, R. L. (2011). Three-dimensional plotted scaffolds with controlled pore size gradients: Effect of scaffold geometry on mechanical performance and cell seeding efficiency. *Acta Biomaterialia*, *7*(3), 1009–1018. https://doi.org/10.1016/j.actbio.2010.11.003

Taminger, K. M., & Hafley, R. A. (2004). Electron Beam Freeform Fabrication for Cost Effective Near-Net Shape Manufacturing. *Nato Unclassified*, *March 2006*, 19.

Tekinalp, H. L., Kunc, V., Velez-Garcia, G. M., Duty, C. E., Love, L. J., Naskar, A. K., Blue, C. A., & Ozcan, S. (2014). Highly oriented carbon fiber-polymer composites via additive manufacturing. *Composites Science and Technology*, *105*, 144–150. https://doi.org/10.1016/j.compscitech.2014.10.009

Tesavibul, P., Felzmann, R., Gruber, S., Liska, R., Thompson, I., Boccaccini, A. R., & Stampfl, J. (2012). Processing of 45S5 Bioglass® by lithography-based additive manufacturing. *Materials Letters*, *74*, 81–84. https://doi.org/10.1016/j.matlet.2012.01.019

Torrado Perez, A. R., Roberson, D. A., & Wicker, R. B. (2014). Fracture surface analysis of 3D-printed tensile specimens of novel ABS-based materials. *Journal of Failure Analysis and Prevention*, *14*(3), 343–353. https://doi.org/10.1007/s11668-014-9803-9

Wang, A., Lin, R., Polineni, V. K., Essner, A., Stark, C., & Dumbleton, J. H. (1998). Carbon fiber reinforced polyether ether ketone composite as a bearing surface for total hip replacement. *Tribology International*, *31*(11), 661–667. https://doi.org/10.1016/S0301-679X(98)00088-7

Wang, X., Jiang, M., Zhou, Z., Gou, J., & Hui, D. (2017). 3D printing of polymer matrix composites: A review and prospective. *Composites Part B: Engineering*, *110*, 442–458. https://doi.org/10.1016/j.compositesb.2016.11.034

Wang, Y., Blache, R., & Xu, X. (2017). Selection of additive manufacturing processes. *Rapid Prototyping Journal*, *23*(2), 434–447. https://doi.org/10.1108/RPJ-09-2015-0123

Wei, X., Li, D., Jiang, W., Gu, Z., Wang, X., Zhang, Z., & Sun, Z. (2015). 3D Printable Graphene Composite. *Scientific Reports*, *5*, 1–7. https://doi.org/10.1038/srep11181

Weng, Z., Wang, J., Senthil, T., & Wu, L. (2016). Mechanical and thermal properties of ABS/montmorillonite nanocomposites for fused deposition modeling 3D printing. *Materials and Design*, *102*, 276–283. https://doi.org/10.1016/j.matdes.2016.04.045

Wu, J., Yuan, C., Ding, Z., Isakov, M., Mao, Y., Wang, T., Dunn, M. L., & Qi, H. J. (2016). Multi-shape active composites by 3D printing of digital shape memory polymers. *Scientific Reports*, *6*(November 2015), 1–11. https://doi.org/10.1038/srep24224

Zarek, M., Layani, M., Cooperstein, I., Sachyani, E., Cohn, D., & Magdassi, S. (2016). 3D Printing of Shape Memory Polymers for Flexible Electronic Devices. *Advanced Materials*, *28*(22), 4449–4454. https://doi.org/10.1002/adma.201503132

Zhang, Y., Wu, L., Guo, X., Kane, S., Deng, Y., Jung, Y. G., Lee, J. H., & Zhang, J. (2018). Additive Manufacturing of Metallic Materials: A Review. *Journal of Materials Engineering and Performance*, *27*(1), 1–13. https://doi.org/10.1007/s11665-017-2747-y

Zheng, H., Zhang, J., Lu, S., Wang, G., & Xu, Z. (2006). Effect of core-shell composite particles on the sintering behavior and properties of nano-Al2O3/polystyrene composite prepared by SLS. *Materials Letters*, *60*(9–10), 1219–1223. https://doi.org/10.1016/j.matlet.2005.11.003

Zhong, W., Li, F., Zhang, Z., Song, L., & Li, Z. (2001). Short fiber reinforced composites for fused deposition modeling. *Materials Science and Engineering A*, *301*(2), 125–130. https://doi.org/10.1016/S0921-5093(00)01810-4

Ziaee, M., & Crane, N. B. (2019). Binder jetting: A review of process, materials, and methods. *Additive Manufacturing*, *28*(December 2018), 781–801.

https://doi.org/10.1016/j.addma.2019.05.031

Acknowledgments

This research work is partly supported by research cell in Kumaraguru College of Technology.

Effects of stacking sequences and longitudinal parameter on the dynamic characteristics of multilayer glass epoxy composite

Saumil Desai[1], Vishal Mehta[2]*

[1,2]Department of Mechanical Engineering, School of Engineering, Institute of Technology, Nirma University, Ahmedabad, Gujarat, India - 382481
[1]saumil.desai@nirmauni.ac.in, [2]vishal.mehta@nirmauni.ac.in
*Corresponding Author

Abstract. In the current scenario, the structures made of composites are drawing attention due to their high spring rate, capacity to withstand high force, reinforcement and optimized mass. Up till now, several materials have been designed and manufactured. The components as well as structures made of these composite materials do expose to dynamic loading during their functions which leads to the failure of the respective component/structure. The dynamic characteristics depend on the material properties, physical dimension, no. of layers, orientation of the fibers and end conditions. Here an analysis is performed to determine the vibration characteristics of multilayer composite beams made of glass epoxy composite. The Cantilever boundary condition is considered. Ten layers are considered. Various stacking are considered with the different fiber orientations. The natural frequency of a multilayer composite beam is sensitive to the length and fiber orientations. As the length and number of layer increase, the natural frequency decreases. The natural frequency for fiber orientation at 90^0 is minimum as compared to other fiber orientations irrespective of the length while the natural frequencies for the 0^0 fiber orientation is more compared to other orientations for out of plane vibrations irrespective of the geometrical dimensions. The natural frequencies for other orientations are more compared to 0^0 orientation for specific modes due to high stiffness in the torsional direction.

Keywords: Glass epoxy composites, Cantilever configuration, modal characteristics, stacking sequence, fiber orientations, multi-layer

1. Introduction

The growth of composite materials with its relevant design and production techniques is considered as one of the important progress in the timeline of the materials. The composite is a material consisting of two or more phases bonded together which makes the difference between from the metal alloys in which two or more metals are mixed in one another. These are the materials for many functions with exceptional properties which can be customized to satisfy the requirements of specific purpose. These composites imbibes the qualities in terms of sustainability to wear, corrosion, and high-temperature. Such exclusive characteristics offer the engineer with breakthrough in design which is not conceivable with traditional monolithic (unreinforced) materials. Composite technology also allows a wide range of solid, ceramic materials to be used in applications Composites technology also makes possible the use of an entire class of solid materials, ceramics, in applications for which monolithic materials are inadequate due to their high strength dispersion and low resistance to mechanical and thermal shocks.

Content from this work may be used under the terms of the Creative Commons Attribution 3.0 licence. Any further distribution of this work must maintain attribution to the author(s) and the title of the work, journal citation and DOI.
Published under licence by IOP Publishing Ltd

IOP Conf. Series: Materials Science and Engineering 1228 (2022) 012016 doi:10.1088/1757-899X/1228/1/012016

Now, due to the manufacturing processes available for composites, they are very well adjusted to the construction of large, complex structures, which allows union of parts, reducing manufacturing expenses. These composites are widely used in aerospace industry. Not limited to this, they are being used for the commercial purposes in mechanical engineering ranging from I.C. engines, machine components, thermal management as well as the components used in automobile such as rakes, drive shafts, flywheels and tanks [1]. Apart from this, the composites can be used in pressure vessel equipment, sports and leisure equipments, marine structure and biomedical devices or bone and teeth because of offering the unmatched attributes in terms of chemical, mechanical, electrical, thermal and electrical properties.

Though the component made from composites conceives good static structural properties, the component has to undergo several dynamic loading during its operation. Hence it is necessary to know the dynamic behaviour of the composite material because the material is prone to rupture failure under time varying loads. While the component or structure is exposed to such time varying or dynamic nature load, the displacement of it should be taken care of to avoid the resonance phenomenon. Therefore, it is necessary to know the modal characteristics of the component or structure made of composite material. In this regard, Hemanshu et al [2] carried out an analysis to understand the dynamic behaviour of composite beams considering the different boundary configurations and stacking sequences. Another study was conducted by Tremaine et al [3] in which analytical formulation was proposed for the damped orthotropic composite material to determine the vibrational characteritics and later they compared the analytical solution with the experiment.

For this study, glass fibers and epoxy resin are combined. Fibers act as reinforcement material and epoxy resin serves as the matrix material. Fiberglass is a very popular material due to its excellent mechanical, thermal, chemical, and tribological properties. Fiberglass is an excellent competitor for product designers whose primary concern is cost. Glass and epoxy can be combined to achieve good performance characteristics in terms of good strength to stiffness ratio, excellent chemical resistance, long-term stability and fatigue resistance, and low thermal expansion coefficient.

2. Literature Review

The researchers have carried out the studies based on experimental and/or numerical method to investigate properties of composites under mechanical load as well as the dynamic characteristics considering various theoretical approaches. Venkatesan et al [4] performed the free vibration to understand the vibration characteristics of the plates which are circular and made of layers. Naymeh at al [5] performed the vibration analysis and conducted a numerical calculation to obtain the vibration characteristics simply supported multi-layered composite plates which are rectangular in shape. In the work, there is an assumption that the layers are tight bound with each other. Wu et al [6] conducted the free vibration analysis of rectangular composite plate which is loaded with mass. The free vibration analysis is performed by Ganapathi et al [7] for the composite laminates made of anisotropic material with simply supported configuration. Ferreira et al [8] performed the free vibration analysis of statically deformed plate considering various thickness to length ratio. Murat et al [9] conducted a free vibration analysis of trapezoidal laminates with first order shear deformation theory for symmetric angle and cross ply. Liz et al [10] performed the vibration of rectangular laminated plates in which symmetric fiber directions are considered at various angle. The plates were elastically supported to resist the angular and linear displacements. Topdar et al [11] suggested hybrid plate theory for the free vibration characteristics of the laminated composites which were sandwiched with the piezoelectric layer. Schwarts-Givli et al [12] proposed an analytical method for free vibration of delaminated unidirectional sandwich panels. The analytical method is based on Hamilton's principle with the high order unidirectional sandwich panel theory and modified Galerkin approach. The analytical method considered the flexibility of the core in the transverse direction. Huang et al [13] implemented the green function to investigate the free vibration characteristics of rectangular plates made of orthotropic nature with variable thicknesses. Also, they considered the aspect ratios and end conditions for the effect on free vibration characteristics. Kang et al [14] adopted Rayleih-Ritz method for free vibration of circular plates made of wood disk. Aydogdu et al [15] carried out the analytical approach for the free vibration

analysis of square plates made of plies laminated in different angular directions. They implemented Hamilton's principle for the free vibration of laminated plates with all edges considered simply supported boundary condition. Arkadiusz et al [16] carried out the vibrational analysis of wires made of shape memory alloy embedded in multilayer composite plate. In the vibrational analysis, they performed the sensitivity analysis of fundamental mode frequency considering the effects of size of the plate and properties of the material. Antunes et al [17] performed the experimental free vibration analysis of composite plates of variable stiffness and compared with the theoretical models. They considered three boundary conditions namely (1) Cantilever i.e. one edge fixed and other edges free (2) Doubly clamped i.e. two opposite edges are clamped with remaining two opposite edges free and (3) All edges free. The springs were considered along the edge in analytical model to represent a clamped condition experimentally. Mishra et al [18] performed the experimental analysis to investigate the vibrational characteristics of industrial woven fiber composite plates. The results of the experimental method were compared with finite element method. They considered the influence of number of layers, aspect ratio, orientation of the fibers and different boundary conditions on the vibrational characteristics of the composite plates. Chinta et al [19] conducted modal analysis of rectangular composite plate using finite plate with simply supported boundary condition. The symmetric orientation of the fibers is considered. They validated the results of finite element method with the analytical approach. R. Samyal [20] conducted an experimental analysis of composite plates to find the natural frequencies and mode shapes such that the optimal location for patching the piezoelectric sensor can be determined. The fibers were orientated at 0^0 and 45^0 with hand lay-up technique. Jinguang et al [21] carried out the experimental analysis of floating rafts to compare the vibrational characteristics especially damping. Two samples were considered where one sample was made of carbon fiber reinforced polymer (CFRP) and another was made of metal. They performed the experimental analysis and compared the analysis with finite element analysis, too. Despite low weight, CFRP made raft has high damping which indicated the good vibration absorption as compared to metal made raft. Medhene et al [22] conducted a comparative modal analysis among steel, E-glass epoxy and fiber reinforced plastic. They considered the cantilever configuration and found that the E-glass epoxy offers better vibration characteristics compared to the fiber reinforced particle and steel cantilevers. Tita et al [23] conducted an experimental analysis to understand the dynamic characteristics of fiber reinforced composite beams. They considered various stacking sequence for the number of beams and observed that the change in stacking sequence can make a structure made of fiber reinforced composite material more stiff for the same mass, geometry and boundary conditions. Capasso et al [24] modelled the fiber composite material with solid and shell elements respectively to investigate the structural intensity of the composite material. Gopalan et al [25] investigated effect of stacking sequences on composite numerically and experimentally and found that the natural frequency is more for 0^0 fiber orientations compared to other orientations irrespective of the boundary conditions for out of plane vibrations.
Motivated by the information, an attempt is made to conduct a numerical study to understand the effects of fibre orientations and variation in longitudinal parameter on the vibrational characteristics of multilayer composite plate.

3. Methodology

Here, multilayer i.e. 10 layers composite plates are treated as a beam with the aim of getting the undamped natural frequencies and mode shapes. This composite plate is converted into finite element model using shell181 with 4 nodes elements using ANSYS. The fibres in the laminate are oriented as shown in table 1. The cross section of the composite is considered as 5000 μm × 100 μm with longitudinal parameter varying from 5000 μm to 15000 μm. Cantilever boundary condition is applied across the width of the composite to study the vibration characteristic based on Ansys based numerical method.
The characteristic equation for the free vibration of discrete system is expressed as

$$[\mathrm{m}]\{\ddot{y}\} + [C]\dot{y} + [k]y = 0 \qquad [1]$$

In the absence of damping element, the governing equation is

$$[m]\{\ddot{y}\} + [k]y = 0 \quad [2]$$

The above elementary matrices are calculated by using

$$[m]^e = \rho_C \int_{Vol} [N]^T [N] d(vol)\, [m]^e \quad [3]$$

$$[K]^e = \int_{Vol} [B]^T [D][B] d(vol) \quad [4]$$

Where [m] = global mass matrix, [C] = global damping matrix, [K] = global stiffness matrix, $[m]^e$ = elemental mass matrix, [N] = interpolation functions matrix, $[K]^e$ = elemental stiffness matrix, [B] = displacement matrix (based on shape functions), [D] = laminate elasticity matrix (orthotropic case), $\boldsymbol{\rho_C}$ = density of composite

Table 1. Fiber orientation and its designation

Designation of fiber orientation	Fiber orientation
#1	0/0/0/0/0/0/0/0/0/0
#2	0/90/0/90/0/90/0/90/0/90
#3	90/90/90/90/90/90/90/90/90/90
#4	30/-30/30/-30/30/-30/30/-30/30/-30
#5	45/-45/45/-45/45/-45/45/-45/45/-45
#6	0/30/0/-30/0/0/-30/0/30/0
#7	0/45/0/-45/0/0/-45/0/45/0

Table 2. Material properties Glass Epoxy composite [23]

Material	Properties	Symbol	Value
Glass Fiber	Modulus of elasticity	E_{fi}	76 GPa
	Density	ρ_{fi}	2560 Kg/m^3
	Poisson's coefficient	υ_{12fi}	0.22
Epoxy Resin	Modulus of elasticity	E_m	4 GPa
	Density	ρ_m	1300 Kg/m^3
	Poisson's coefficient	υ_m	0.40
Lamina (Orthotropic)	Modulus of elasticity		
	Fiber at 0^0	E_X	44.8 GPa
	Normal to fiber	$E_Y = E_Z$	11.27 GPa
	Density of composite	ρ_C	1780 Kg/m^3
	Modulus of rigidity	$G_{12} = G_{13}$	4.86 GPa
		G_{23}	4.45 GPa
	Poisson's coefficient	υ_{12}	0.28
	Fiber volume fraction	V_{fi}	60%

The next step is the assembly of the global matrices which is done by the package using the equation. To obtain the natural frequencies of an undamped system with N degrees of freedom, the solution is obtained by solving:

$$[\boldsymbol{m}]_{NxN}\{\ddot{\boldsymbol{y}}\}_{Nx1} + [\boldsymbol{K}]_{NxN}\{\boldsymbol{y}\}_{NX1}=0 \quad (5)$$

Through this procedure, it is possible to obtain the modal frequencies of the mono layer composite material with orthotropic nature. Table 2 shows the mechanical properties of Glass Epoxy composite.

$$[D]^{-1} = \begin{bmatrix} \frac{1}{E_x} & \frac{-v_{xy}}{E_x} & \frac{-v_{xz}}{E_x} & 0 & 0 & 0 \\ \frac{-v_{yx}}{E_x} & \frac{1}{E_y} & \frac{-v_{yz}}{E_x} & 0 & 0 & 0 \\ \frac{-v_{zx}}{E_x} & \frac{-v_{zy}}{E_x} & \frac{1}{E_z} & 0 & 0 & 0 \\ 0 & 0 & 0 & \frac{1}{G_{xy}} & 0 & 0 \\ 0 & 0 & 0 & 0 & \frac{1}{G_{yz}} & 0 \\ 0 & 0 & 0 & 0 & 0 & \frac{1}{G_{xz}} \end{bmatrix} \quad (6)$$

4. Result and Discussion

Table 3. Natural for the different orientations of fibers for multilayer composite plates, 5000 μm long and 5000 μm width

Mode	#1	#2	#3	#4	#5	#6	#7	Max. frequency for each mode
1	2.9919E+04	2.4166E+04	1.5841E+04	2.4413E+04	1.9755E+04	2.8158E+04	2.7017E+04	2.9919E+04
2	4.3988E+04	4.0912E+04	3.6936E+04	5.0778E+04	5.1710E+04	4.6792E+04	4.7227E+04	5.1710E+04
3	7.0171E+04	6.6007E+04	5.3900E+04	8.1370E+04	7.1009E+04	7.8971E+04	7.9765E+04	8.1370E+04
4	1.0913E+05	1.1672E+05	8.5478E+04	1.0988E+05	9.7626E+04	1.1299E+05	1.1738E+05	1.1738E+05
5	1.3514E+05	1.3886E+05	1.1893E+05	1.2824E+05	1.3760E+05	1.3145E+05	1.2983E+05	1.3886E+05
6	1.5187E+05	1.4727E+05	1.2598E+05	1.5507E+05	1.4891E+05	1.5455E+05	1.5470E+05	1.5507E+05

Table 4. Natural for the different orientations of fibers for multilayer composite plates, 10000 μm long and 5000 μm width

Mode	#1	#2	#3	#4	#5	#6	#7	Max. frequency for each mode
1	7.96E+03	6.31E+03	4.04E+03	6.28E+03	4.95E+03	7.41E+03	7.07E+03	7.96E+03
2	1.88E+04	1.79E+04	1.68E+04	2.36E+04	2.15E+04	2.08E+04	2.12E+04	2.36E+04
3	2.79E+04	2.44E+04	1.75E+04	2.65E+04	2.46E+04	2.88E+04	2.81E+04	2.88E+04
4	4.42E+04	3.63E+04	2.43E+04	3.59E+04	2.90E+04	4.17E+04	4.01E+04	4.42E+04
5	6.28E+04	5.84E+04	5.26E+04	7.27E+04	7.36E+04	6.66E+04	6.72E+04	7.36E+04
6	9.00E+04	8.29E+04	6.30E+04	8.90E+04	7.42E+04	9.81E+04	9.85E+04	9.85E+04

Table 5. Natural for the different orientations of fibers for multilayer composite plates, 15000 μm long and 5000 μm width

Mode	#1	#2	#3	#4	#5	#6	#7	Max. frequency for each mode
1	3.58E+03	2.83E+03	1.80E+03	2.79E+03	2.18E+03	3.31E+03	3.16E+03	3.58E+03
2	1.17E+04	1.13E+04	8.39E+03	1.27E+04	1.00E+04	1.32E+04	1.35E+04	1.35E+04
3	1.47E+04	1.24E+04	1.08E+04	1.53E+04	1.32E+04	1.45E+04	1.39E+04	1.53E+04
4	2.12E+04	1.70E+04	1.11E+04	1.67E+04	1.61E+04	1.98E+04	1.89E+04	2.12E+04
5	3.80E+04	3.59E+04	3.01E+04	4.45E+04	3.60E+04	4.13E+04	4.20E+04	4.45E+04
6	5.44E+04	4.50E+04	3.33E+04	4.67E+04	4.85E+04	5.19E+04	4.98E+04	5.44E+04

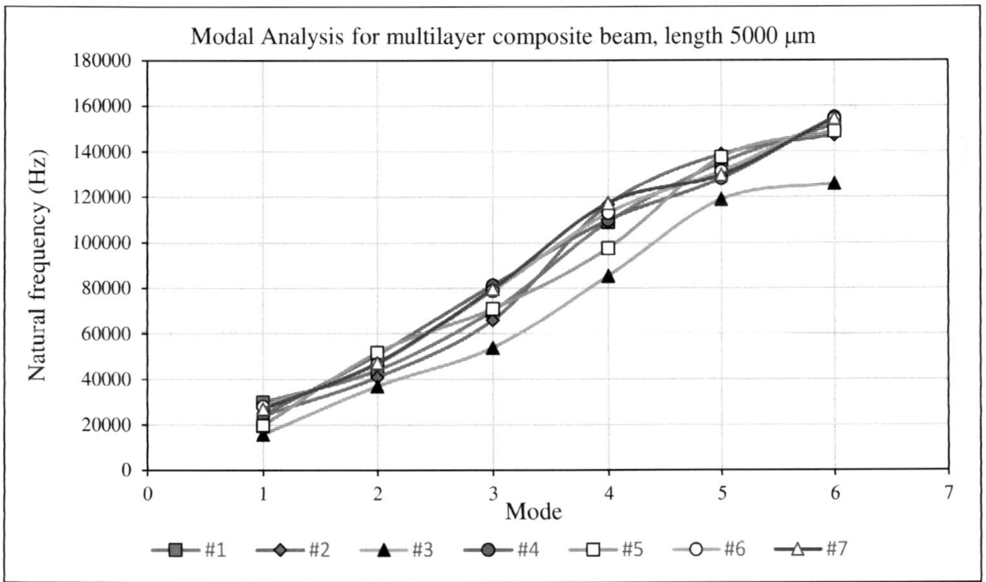

Fig. 1 Natural frequencies for the multilayer composite beam of length 5000 μm

IOP Conf. Series: Materials Science and Engineering 1228 (2022) 012016 doi:10.1088/1757-899X/1228/1/012016

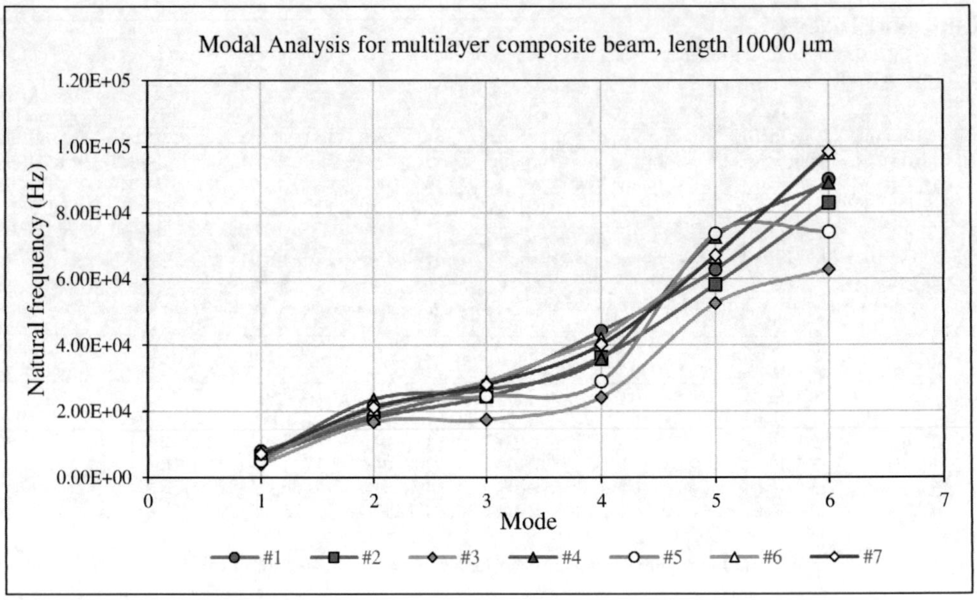

Fig. 2 Natural frequencies for the multilayer composite beam of length 10000 μm

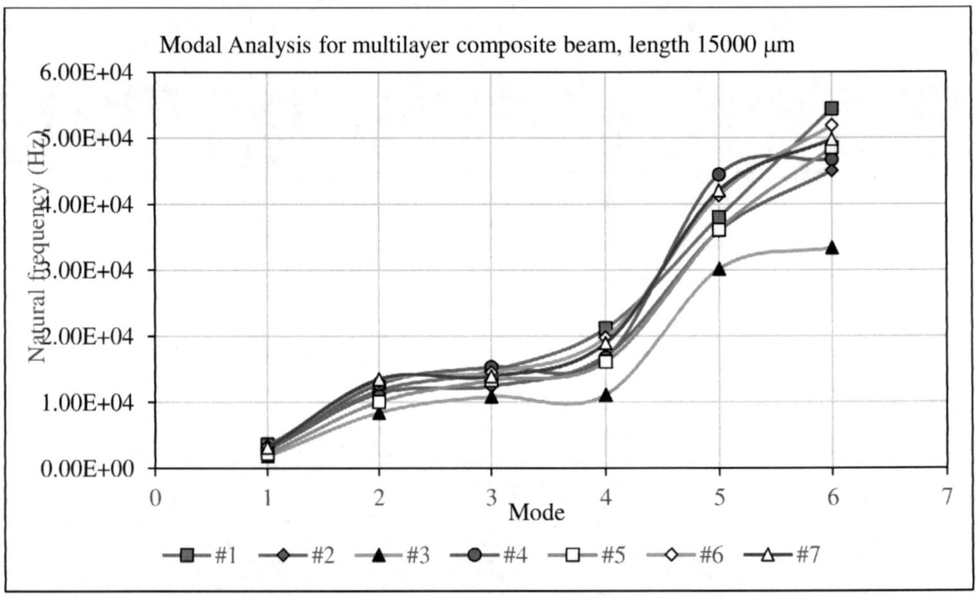

Fig. 3 Natural frequencies for the multilayer composite beam of length 15000 μm

Table 3-5 comprises of the natural frequencies of multilayer composite beams for the length 5000, 10000 and 15000 μm. The tables depict that the natural frequency decrease with increase in the length which indicate that natural frequency is inversely proportional to the longitudinal parameter of the multilayer composite beam. In the table, maximum frequency for each node is identified with which remaining frequencies have been compared for the respective node. Through this comparison, alternative fiber orientation can be used with a difference of at least 5% in comparison of maximum natural frequency for the respective node. From the table 3-5, it can also be concluded that natural frequency is depending on the stacking and orientation of the fibers along with the natural frequency. It shows that the natural frequency of fundamental mode is maximum for one orientation over another orientation of the fiber but it may not be maximum for all nodes of the same orientation of the fiber which is represented in the figure 1-3 for each length. Out of the fiber orientations considered here, the orientation #3 has least natural frequencies compared to other orientation for all lengths. The modulus of elasticity of composite

IOP Conf. Series: Materials Science and Engineering 1228 (2022) 012016 doi:10.1088/1757-899X/1228/1/012016

decreases with increase in the fiber angle. Hence, the orientation #3 has least natural frequencies for all nodes irrespective of the length of the composite. From the figure 1-3 and table 3-5, it can be considered that the fiber orientation and length of the composite may have a relation with each other. As the length increases, the natural frequency is maximum for the #1 orientation than other orientations at higher modes, too which is similar to the result obtained by [25].

Mode shapes

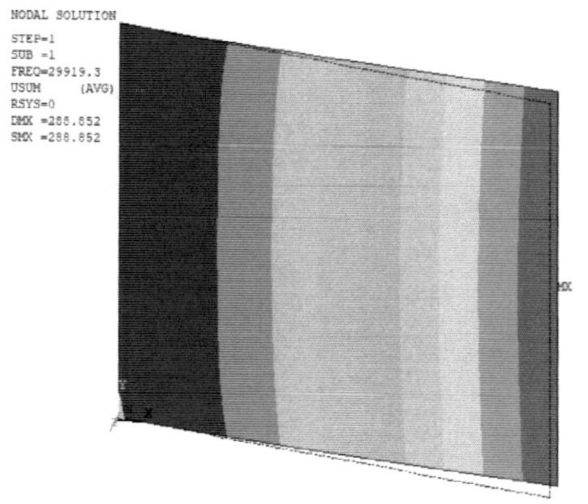

Fig.4 1st mode shape for the #1 for multilayer composite beam, length 5000 μm

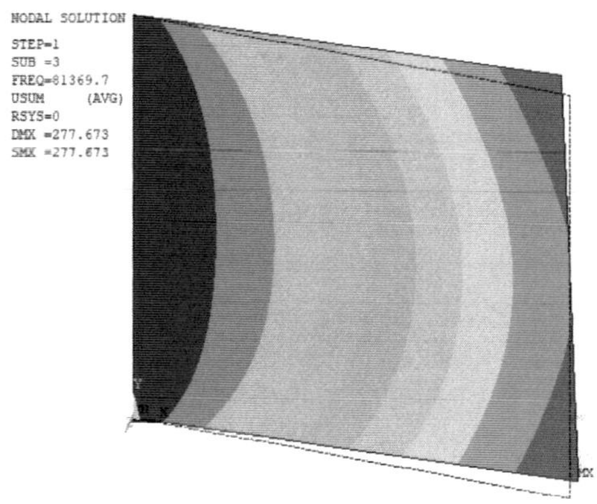

Fig.5 3rd mode shape for the #4 for multilayer composite beam, length 5000 μm

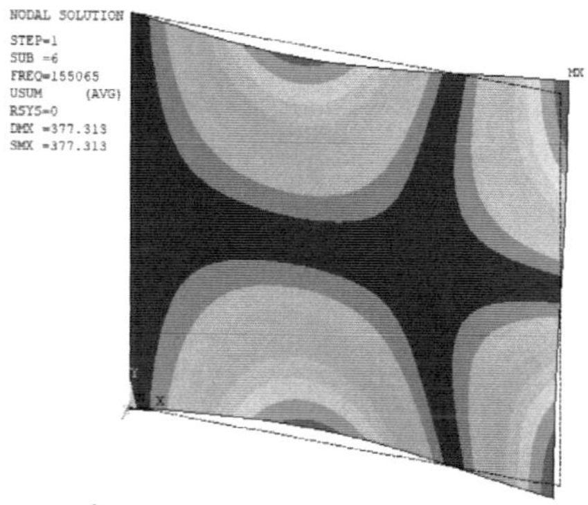

Fig.6 6th mode shape for the #4 for multilayer composite beam, length 5000 μm

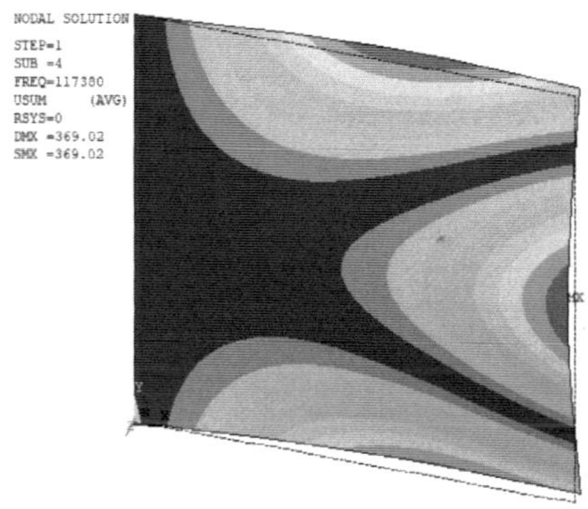

Fig.7 4th mode shape for the #7 for multilayer composite beam, length 5000 μm

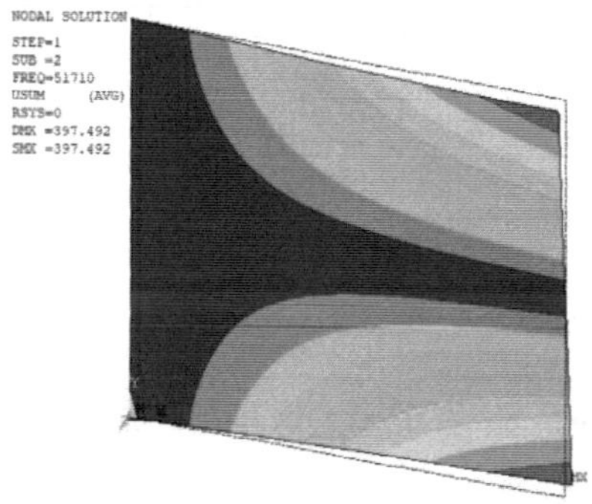

Fig.8 2nd mode shape for the #5 for multilayer composite beam, length 5000 μm

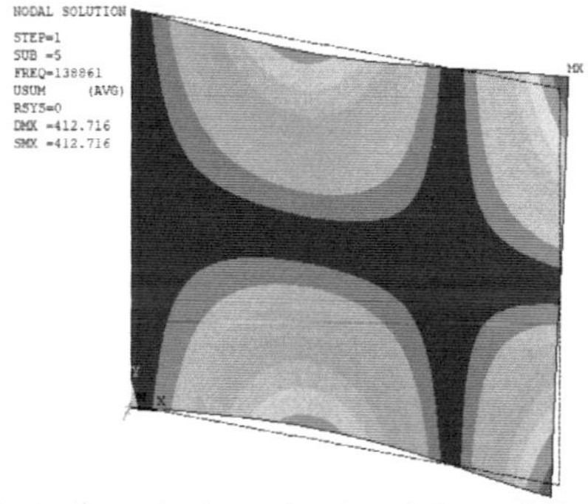

Fig.9 5th mode shape for the #2 for multilayer composite beam, length 5000 μm

IOP Conf. Series: Materials Science and Engineering 1228 (2022) 012016 doi:10.1088/1757-899X/1228/1/012016

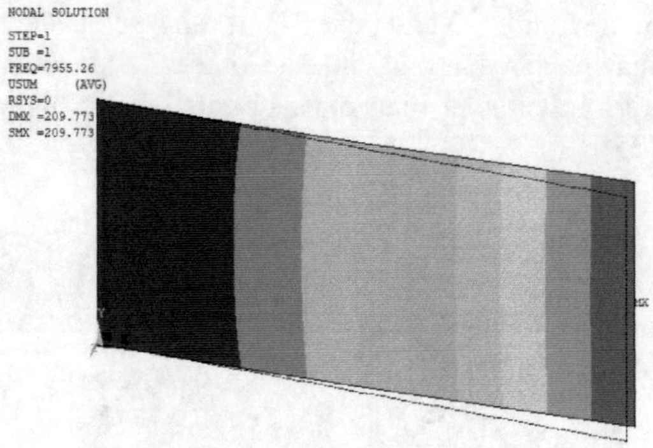

Fig.10 1st mode shape for the #1 for multilayer composite beam, length 10000 μm

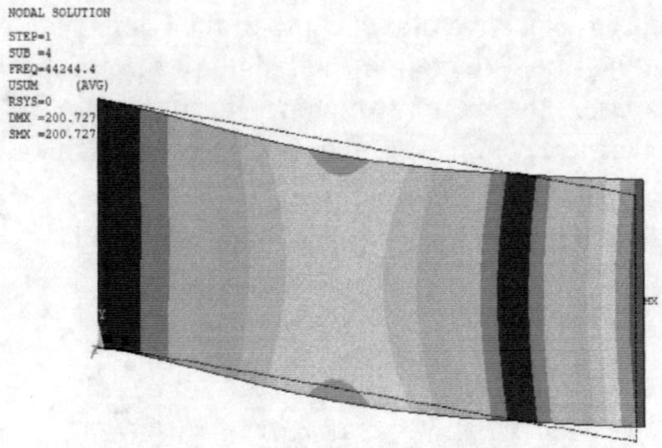

Fig.11 4th mode shape for the #1 for multilayer composite beam, length 10000 μm

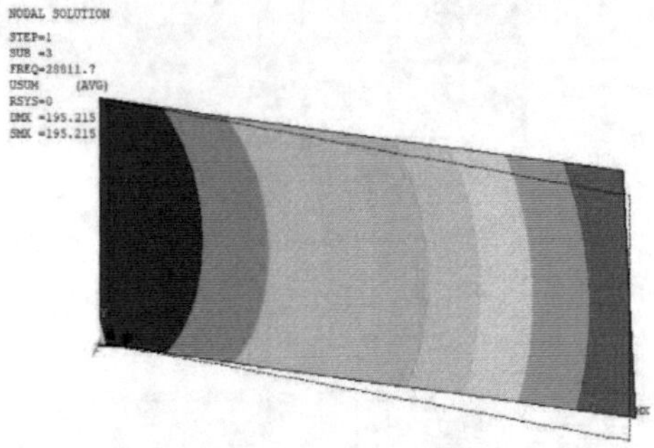

Fig.12 3rd mode shape for the #6 for multilayer composite beam, length 10000 μm

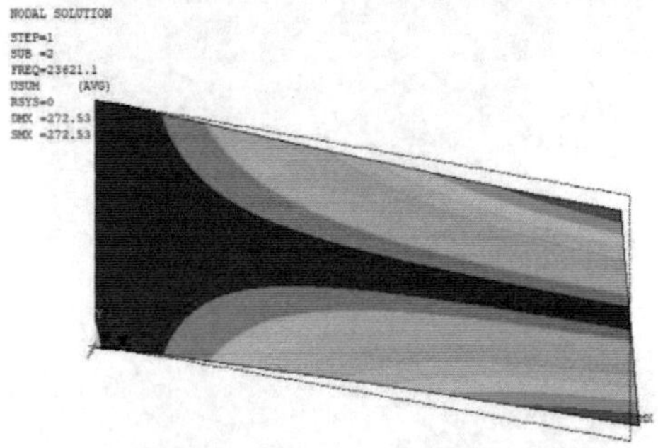

Fig.13 2nd mode shape for the #4 for multilayer composite beam, length 10000 μm

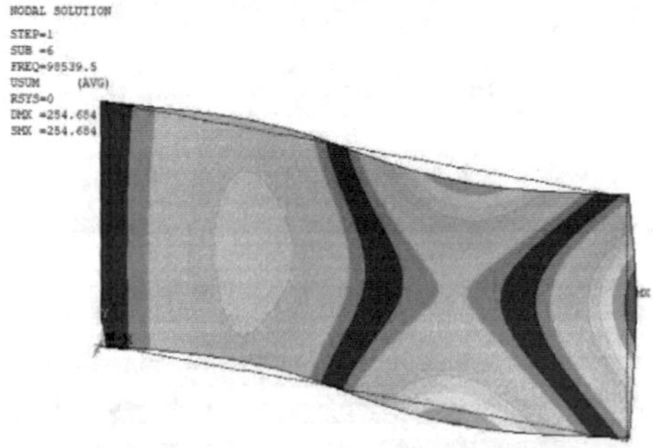

Fig.14 6th mode shape for the #7 for multilayer composite beam, length 10000 μm

Fig.15 5th mode shape for the #5 for multilayer composite beam, length 10000 μm

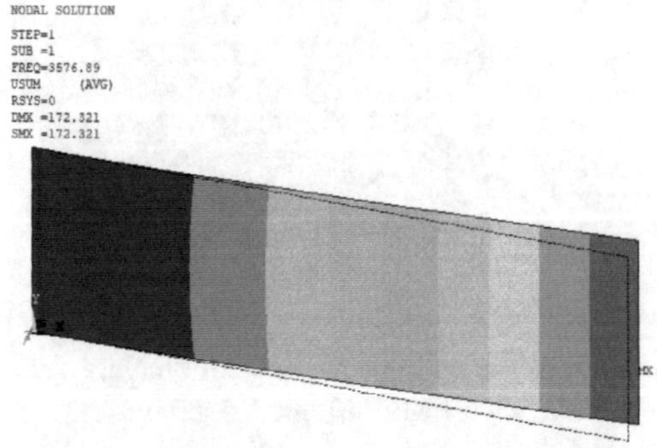

Fig.16 1st mode shape for the #1 for multilayer composite beam, length 15000 μm

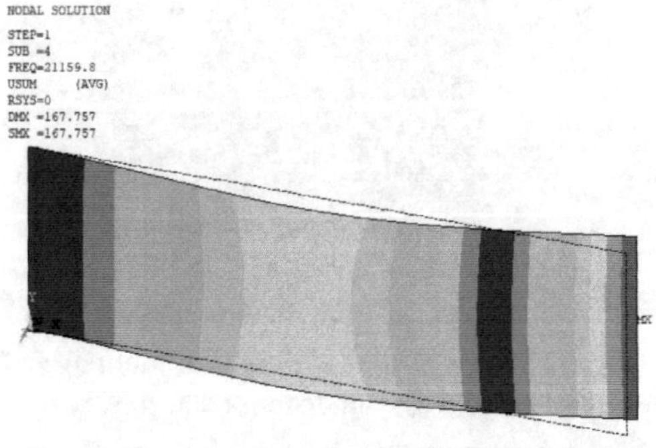

Fig.17 4th mode shape for the #1 for multilayer composite beam, length 15000 μm

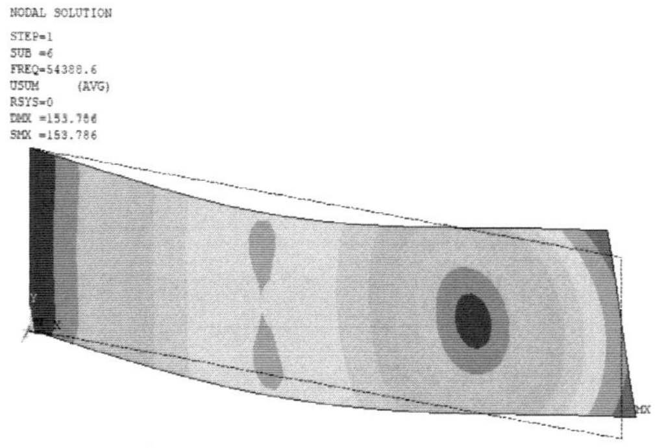

Fig.18 6th mode shape for the #1 for multilayer composite beam, length 15000 μm

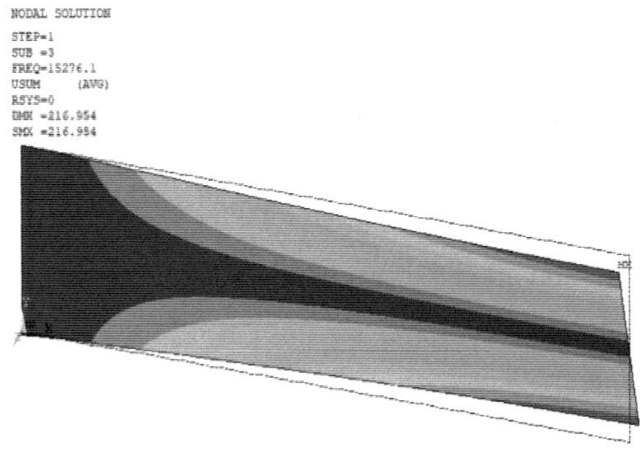

Fig.19 3rd mode shape for the #4 for multilayer composite beam, length 15000 μm

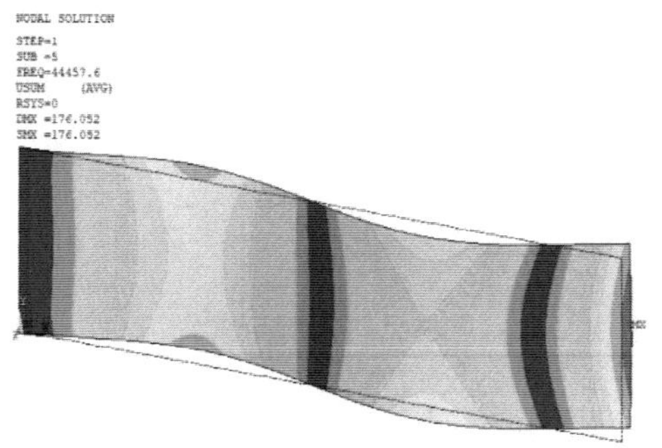

Fig.20 5th mode shape for the #4 for multilayer composite beam, length 15000 μm

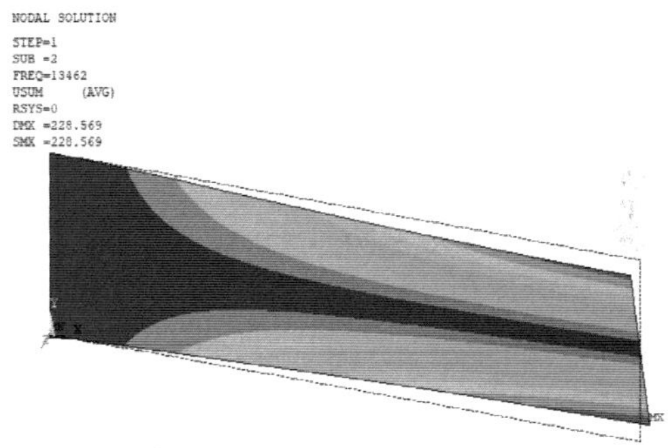

Fig.21 2nd mode shape for the #7 for multilayer composite beam, length 15000 μm

The figures 4-21 show the mode shapes of the multilayer glass epoxy composite plate. The composite plates is considered as cantilever beam configuration. These mode shapes are corresponding to the maximum frequency obtained for each fiber orientation. The figures 4-21 show that the length and number of stacking layers have the effect on the dynamic characteristics. With rise in both parameters, the natural frequency decreases. The fiber orientations also influence the natural frequencies of the composite plate for the same physical dimensions. The composite plate with longitudinal orientation possesses modulus of elasticity more than the lateral and transverse elasticity. Hence the stiffness increases in the longitudinal direction which results in the natural frequencies more compared to other orientations for higher modes. Through numerical simulation, the obtained results show that the natural frequencies for the orientations other than longitudinal direction are more than the longitudinal orientation. The other orientations have the stiffness high in the torsional direction rather than out of plane direction. For the out of plane vibrations, the fiber orientation along the length possesses high elasticity. Hence, the natural frequencies are high for the longitudinal orientation at higher modes.

Conclusion

The present analysis shows the dynamic characteristics of glass epoxy multi-layer composite considering the composite configuration. Total 10 layers are considered. The different fiber orientations are considered from #1 to #7 with the longitudinal parameter equal to 5000 μm, 10000 μm and 15000 μm. The cross section of the composite beam is 5000 μm × 100 μm. The obtained results depict that (1) Natural frequency is inversely proportional to the length of the composite. It decreases with increase in the length irrespective of the orientation of the fibers. (2) The elasticity decreases with increase in the fiber orientation from 0^0 to 90^0. Hence, the natural frequency is minimum for #3 orientation than other orientations as shown in fig. 1-3. (3) The fundamental mode frequency is more for #1 orientation from all orientations irrespective of the length but for higher nodes with the same length, the natural frequencies is more in other modes, too. It shows that the fiber orientation plays a role onto the natural frequencies. (4) From the fig. 1-3, it is observed that the natural frequency is more for #1 than other orientations for 15000 μm than 10000 μm and 5000 μm. Hence it can be concluded that the length and fiber orientations may have a relation with each other. Along with this, mode shapes for the respective maximum natural frequencies are shown in the figure 4-21 for all lengths.

References

[1]. Zweben, C. (1998). Advances in composite materials for thermal management in electronic packaging. *Jom*, *50*(6), 47-51.

[2]. Hemanshu (2013), *Vibration Analysis of Composite Beam,* Master's Thesis, National Institute of Technology, Rourkela

[3]. Kellie Tremaine (2012), MODAL ANALYSIS OF COMPOSITE STRUCTURES WITH DAMPING MATERIAL, Master's thesis, The Faculty of California Polytechnic State University

[4]. Venkatesan, S., & Kunukkasseril, V. X. (1978). Free vibration of layered circular plates. *Journal of Sound and Vibration*, *60*(4), 511-534.

[5]. Taylor, T. W., & Nayfeh, A. H. (1996). The vibration characteristics of thick rectangular multilayered plates. *Composites Part B: Engineering*, *27*(6), 623-631.

[6]. Chang, T. P., & Wu, M. H. (1997). On the use of characteristic orthogonal polynomials in the free vibration analysis of rectangular anisotropic plates with mixed boundaries and concentrated masses. *Computers & structures*, *62*(4), 699-713.

[7]. Ganapathi, M., Kalyani, A., Mondal, B., & Prakash, T. (2009). Free vibration analysis of simply supported composite laminated panels. *Composite Structures*, *90*(1), 100-103.

[8]. Ferreira, A. J. M., Castro, L. M., & Bertoluzza, S. (2009). A high order collocation method for the static and vibration analysis of composite plates using a first-order theory. *Composite structures*, *89*(3), 424-432.

[9]. Gürses, M., Civalek, Ö., Ersoy, H., & Kiracioglu, O. (2009). Analysis of shear deformable laminated composite trapezoidal plates. *Materials & Design*, *30*(8), 3030-3035.

[10].Nallim, L. G., & Grossi, R. O. (2007). Vibration of angle-ply symmetric laminated composite plates with edges elastically restrained. *Composite structures*, *81*(1), 80-83.

[11].Topdar, P., Sheikh, A. H., & Dhang, N. (2007). Vibration characteristics of composite/sandwich laminates with piezoelectric layers using a refined hybrid plate model. *International Journal of Mechanical Sciences*, *49*(11), 1193-1203.

[12].Schwarts-Givli, H., Rabinovitch, O., & Frostig, Y. (2008). Free vibration of delaminated unidirectional sandwich panels with a transversely flexible core and general boundary conditions—a high-order approach. *Journal of Sandwich Structures & Materials*, *10*(2), 99-131.

[13].Huang, M., Ma, X. Q., Sakiyama, T., Matuda, H., & Morita, C. (2005). Free vibration analysis of orthotropic rectangular plates with variable thickness and general boundary conditions. *Journal of Sound and Vibration*, *288*(4-5), 931-955.

[14].Kang, W., Lee, N. H., Pang, S., & Chung, W. Y. (2005). Approximate closed form solutions for free vibration of polar orthotropic circular plates. *Applied Acoustics*, *66*(10), 1162-1179.

[15].Aydogdu, M., & Timarci, T. (2003). Vibration analysis of cross-ply laminated square plates with general boundary conditions. *Composites Science and Technology*, *63*(7), 1061-1070.

[16].Żak, A. J., Cartmell, M. P., & Ostachowicz, W. M. (2003). A sensitivity analysis of the dynamic performance of a composite plate with shape memory alloy wires. *Composite Structures*, *60*(2), 145-157.

[17].Antunes, A. M., Ribeiro, P., Rodrigues, J. D., & Akhavan, H. (2020). Modal analysis of a variable stiffness composite laminated plate with diverse boundary conditions: experiments and modelling. *Composite Structures*, *239*, 111974.

[18].Mishra, I., & Sahu, S. K. (2015). Modal analysis of woven fiber composite plates with different boundary conditions. *International Journal of Structural Stability and Dynamics*, *15*(01), 1540001.

[19].Chinta, V. S. (2021). Modal Analysis Of Carbon/Epoxy Plate By Varying Fibre Orientation. *Turkish Journal of Computer and Mathematics Education (TURCOMAT)*, 12(10), 7580-7586.

IOP Conf. Series: Materials Science and Engineering 1228 (2022) 012016 doi:10.1088/1757-899X/1228/1/012016

[20].Samyal, R., Singh, S., & Bagha, A. K. (2019). Modal analysis of composite panel at different fiber orientations. *Materials Today: Proceedings*, *16*, 477-480.

[21].Jinguang, Z., Hairu, Y., Guozhi, C., & Zeng, Z. (2018). Structure and modal analysis of carbon fiber reinforced polymer raft frame. *Journal of Low Frequency Noise, Vibration and Active Control*, *37*(3), 577-589.

[22].Medhane, Y. U., & Talmale, P. S. (2017) Comparative Analysis of Composite Materials by Using Finite Element Method. International Journal of Engineering Research & Technology, 6(8), 315-323

[23]. Tita, V., Carvalho, J. D., & Lirani, J. (2003). Theoretical and experimental dynamic analysis of fiber reinforced composite beams. Journal of the Brazilian Society of Mechanical Sciences and Engineering, 25(3), 306-310.

[24].Capasso, P. J., Petrone, G., Kleinfeller, N., De Rosa, S., & Adams, C. (2021). Modeling of fiber composite structures for the calculation of the structural intensity. *Composite Structures*, *262*, 113631.

[25].Gopalan, V., Suthenthiraveerappa, V., Annamalai, A. R., Manivannan, S., Pragasam, V., Chinnaiyan, P., Mannayee G. & Jen, C. P. (2021). Dynamic Characteristics of Woven Flax/Epoxy Laminated Composite Plate. *Polymers*, *13*(2), 209.

IOP Conf. Series: Materials Science and Engineering 1228 (2022) 012017 doi:10.1088/1757-899X/1228/1/012017

An overview of integration of PV system into electric grid

P Rani[1], V P Arora[2] and N K Sharma[3]

[1,2]Electrical Engineering Department, MRSPTU Bathinda, INDIA

[3]Electrical Engineering Department, IKGPTU Jalandhar, INDIA

E-mail: pretty.singla15@gmail.com (Corresponding Author)

Abstract. With new developments in switching converter technology, the utilization of non-conventional energy resources as decentralized generators has gone up significantly. Integration technique has now become effective due to the globe's energy demands, which have imposed a considerable need for alternative techniques through which electricity can be generated or integrated, and also the fact that integrating renewable electricity into traditional sources is essential to minimize the utilization of fossil fuel usage. Photovoltaic (PV) systems are at the centre of this transformation, using the sun's abundant energy and converting it to alternating current. Due to present electricity consumption, degradation of fossil-fuel resources, and effect on the environment, research groups are increasingly willing to integrate renewable energy from this source into networks. PV generation integrated techniques are discussed, as well as the advantages of integration of PV system with electric grid, solar PV system features for integration, and the issues and alternatives of PV-grid integration are identified. This study would help in the installation of PV panels with grid integration in upcoming work while avoiding the apparent problems that have been experienced in previous projects, as well as provide data for academic scholars on the feasibility of integration of PV system with electric grid.

1. Introduction

Non-renewable and renewable types of energy will serve a significant part in the development since they are ecologically acceptable, simple to be used, and will eventually become more financially feasible with increased popularity [1]. Meeting growing energy needs is a difficult task for power sector authorities and utilities to tackle, as it might result in overcoming the challenges for energy supplies, quality of air, and Greenhouse gas reducing emissions [2]. The phrase "renewable energy" refers to a wide range of resources that all rely on self-renewing fuel sources such as sunshine, underground streams, geothermal heat, wind energy, and bio-energy, which includes feedstock, farming, commercial, and municipality waste [3]. Renewable sources produce few, if any, carbon dioxide, trash, or contaminants that lead to environmental degradation, urban pollution, and medical issues, and they don't cost the ecosystem to clear away. As there is a raising need for sustainable renewable power to replace carbon fuels, renewable power integration is now standard practice in many nations across the world. The International Renewable Energy Agency (IRENA) reported that nations installed a historic 260 gigawatts (GW) of new renewable power generation in 2020, up 50% from the previous year, as nations reduced their dependency on conventional energy resources [4].

Grid Connected Solar System (GCSS) is a system that enables large amounts of solar power to be integrated into the power system. This is a significant topology because integrating standardized

Content from this work may be used under the terms of the Creative Commons Attribution 3.0 licence. Any further distribution of this work must maintain attribution to the author(s) and the title of the work, journal citation and DOI.
Published under licence by IOP Publishing Ltd

GCSS into the power network optimizes the operational cost losses, and adds benefit to the utility and customer [5]. PV cells in PV panels transform incoming sun's radiation into dc power, requiring the need of a switching converter between the PV generator and the network. Although micro renewable energy plant techniques, such as generating units based on small hydroelectric or wind energy, can be linked directly to the Electrical network, initial fault conditions, electrical efficiency, and switching converter make a power electronic interface a better alternative [6]. GCSS run in parallel with the power system, therefore they don't need any backup. GCSS help to mitigate carbon dioxide emissions by replacing the required energy by the linked (localized) circuit and delivering supplementary power to the system when they produce surplus power [7].

In this article, we focus on the increasing use of GCSS and the operational problems posed by the widespread implementation of these Distributed generators on the general quality and durability of the power grid to ensure the efficiency, reliability and stability of the power system.

2. PV generation

PV modules transform energy from the sun into a DC supply, and a converter with the Maximum Power Point Tracker (MPPT) converts the force from DC to AC so it could be transferred to the electric system. When distributing power, the converter is in charge of keeping to the grid's regulatory requirements [8]. The number of PV plants is growing at a high rate, owing to lower technological costs and assistance from authorities and utility corporations for GCSS. A string is a collection of PV cells linked in series; a PV array is a collection of strings parallel-connected; each PV array is linked to one MPPT; each converter may have one or even more MPPTs. The power generated moves through a series of safety levels, including fuses, relays, and earth-leakage sensors, before arriving at the main Switchboard, which contains the energy meter and its safety systems and is the final step before power is transferred to the distribution network -to sensitive loads [9]. The aim of an MPPT, in practice, is to keep records of the peak power that may be produced from the PV array at any given moment. The MPPT is linked to the converters active system, which is in responsible of transposing DC supply into AC supply, that can then be used to power a user load or transferred to the power network as depicted in Figure 1 [10].

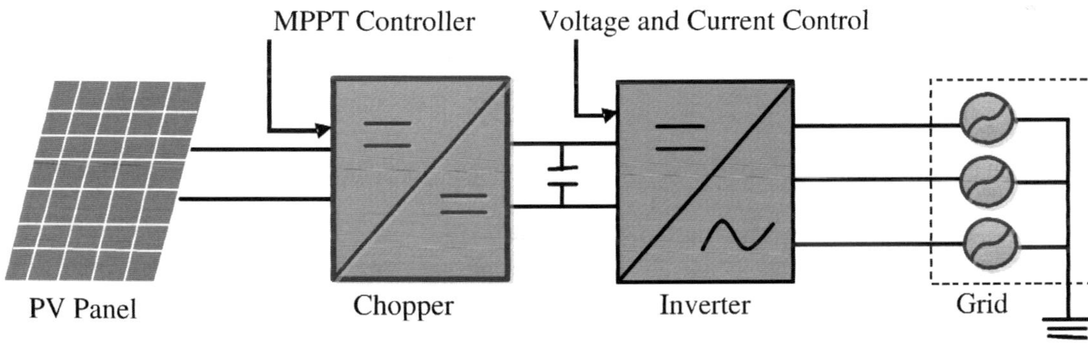

Figure 1. Main components of GCSS

2.1. *GCSS advantages [11]:*

- Installation Key
- Maximum reliability
- Maximum efficiency
- High flexibility

IOP Conf. Series: Materials Science and Engineering 1228 (2022) 012017 doi:10.1088/1757-899X/1228/1/012017

The cost of electricity is reduced with a GCSS since additional electricity may be sold to the local electric utility. GCSS is less complicated to establish since they don't require storage. Grid connectivity of PV power generating plants offers the benefit of optimizing the usage of produced electricity due to the absence of storage constraints. Throughout its lifetime, a PV plant is carbon negative, since any power generated more than that required to create the PV panel initially balances the need to replace fossil fuels. Even if the sunlight does not appear every day, any installation will result in a relatively reasonable average reduction of greenhouse gas usage [12].

3. PV-grid integration issues

The goal of this article is to outline the grid-connected challenges that are related to solar PV generating units. As a result of the enormous explosion of the PV system various difficulties and obstacles related to their insertion into electric grids at high percentages have developed. The majority of modern distribution networks are developed, managed, and regulated under the concept that every distributed line has a single source of electricity [13].

This focus of the project is, however, violated by the connecting of the small-scale GCSS power plant to the grid system. In addition, numerous breakthroughs in the creation of simplified, regulated specifications for utility connections of small-scale power plant producing facilities, notably solar photovoltaic systems, have been achieved in recent years. GCSS produces a huge amount of power at sunup, so a lot of energy is fed into the distribution network, covering a large part of the energy consumption. However, GCSS ability to contribute to load consumption in the daytime so at the end of the day (when solar radiation is relatively low) is low. This raises the possibility of increasing GCSS energy export throughout peak hours. The interconnection of a PV plant with the grid has several problems [14]. We utilize a grid-connected converter for grid connection; this converter's goal is to absorb power from the grid when PV generation is inadequate and to deliver electricity when surplus power is generated. In 100 milliseconds, the grid is integrated with the Solar panel, and the Solar panel is disconnected. Before a network is linked to the network, technological, power quality issues, insurance, measurement and tariffs, innovation and development, and regulation and guidelines concerns must be resolved between the system administrators, utilities, and the relevant regulatory authorities [15].

3.1. Technological issue

Some utilities-linked PV plants operate at a voltage level of approximately 320 dc volts before even being converted to conventional AC, relying on the network infrastructure. As it is harder to fix a DC spark than an AC spark at around the same power, the possible fire threat of DC at any of these values is higher than that of ordinary AC. Appropriate connection, on the other hand, guarantees that any dangers associated with DC electricity are greatly reduced.

A phenomenon known as islanding is amongst the most serious concerns about safety for small client PV systems. The safety problem is that, although an operator can guarantee that its generating resources are either turned off or detached from the workplace, a Solar panel-created island mode is beyond their control [16].

3.2. Power quality issue

Another technological problem for providers and consumers is Power Quality (PQ). As equipment and devices are developed to operate at or near defined voltage and power parameters and variations may cause equipment failure or destruction, the supply should be continuously delivered at a normal voltage level. A PV converter can distort the system, which might cause issues. The non-linear multilevel converters such as boost converter, inverter introduce harmonics into the grid network. Because of the variant nature, there is a power variation at the solar output, making the utility grid unreliable. As a result, a device is employed for converter design. Reactive and harmonic currents components taken from non-linear applications cannot be controlled by a GCSS [17].

IOP Conf. Series: Materials Science and Engineering 1228 (2022) 012017 doi:10.1088/1757-899X/1228/1/012017

3.3. Insurance issue

Insured is necessary to safeguard organizations and their workers in the case of any mishaps caused by the consumer's PV plant's functioning. Another liability-related problem is compensation, which refers to protection from or recompense for harm, loss, or damage. In agreements between companies and PV users, the PV user is usually required to compensate the utility for any possible losses caused as a consequence of the PV system's performance. Indemnification rules are relatively unnecessary if insurance is required [18].

3.4. Innovation and development issue

PV energy development process is a fast-paced process that necessitates creative and dynamic thinking as well as the ability to quickly adapt to shifting market needs. To establish a uniform worldwide strategy that takes into consideration the development of the many green technologies, improving the interaction between research and policy is critical. In recent years, the PV sector has made tremendous progress. For the advancement of PV, the following study approaches are essential [19].

- For demand scheduling of the unexpected production of PV plants, better quantitative and soft programming techniques should be established. Improved PQ and weather prediction techniques are needed to predict the performance of a GCSS plant in advance so that it can be appropriately controlled and linked with the other electric generation sources.
- The output of the PV system must be anticipated under various meteorological and partly shade circumstances, which necessitates modeling and analysis of the entire PV system.
- Reduced silicon and some other material usage in traditional crystalline silicon systems should be investigated to limit material and related expenses.

3.5. Regulation and guidelines

Governmental regulations and safety associations have established rules for manufacturing, maintenance, and installations to address concerns of security, protection, and voltage stability. The standards of the Bureau of Indian Standards (BIS) are based on the guidelines of the International Electro-Technical Commission (IEC) and the Institution of Electrical and Electronics Engineers (IEEE). These standards are still in the works [20].

When voltages at the point of common coupling diverge from normal base values owing to defects and other disruptions, IEEE standard 1547-2003 advises disconnecting of decentralized energy supplies. Table 1 lists the permissible variances.

Table 1. List of permissible variances in voltage level

Range of Voltage (percent of the system's rated voltage)	Setting time
<50%	0.16
50% - 88%	2.00
110% - 120%	1.00
>120%	0.16

4. Several alternatives

The growing number of Decentralized Resources (DR) necessitates innovative techniques for electrical grid planning and control to sustain or maybe even increase the power delivery dependability and efficiency. PV, wind, and other renewable technologies have hastened the move to low-carbon technologies. Renewable technologies, such as PV and wind, have expedited the shift to sustainable power sources [21]. In light of the aforementioned, a few of the major alternatives for DRs utilizations are as follows [22-23].

- Merging DR with an Energy Storage Device (ESD) can be used to achieve voltage stability by employing renewable energy sources. The advantages of an ESD are divided into three categories depending on end-users: transfer level applications, big industrial uses, and International Organization for Standardization (ISO) Market utilize.
- Switching converters has rapidly progressed in recent years has resulted in two aspects: the innovation of fast power electronic-based switches capable of switching rapidly and ability to handle high currents, and the initiation of digital computing control scheme capable of implementing sophisticated and powerful control techniques. As a result of these characteristics, processors that are both economical and power sector -friendly have been developed.
- Integration in DR power generation can be optimized by dispersing renewable energy sources of electricity over a broader geographical region in micro plants rather than big units concentrated in one location.
- The generated power of big PV plants fluctuates throughout the day, and this electricity is supplied into the grid. Constantly varying power raises concerns about the system's ability to maintain a stable power system.

5. Conclusion

The large-scale implementation of microgrids into the network has gone up significantly, according to the latest developments in power systems. Customer demands are more vulnerable to PQ. This study examines the behavior of PV panels as well as technological various GCSS concerns that are common in renewable technologies. Some technological grid integration issues, alternatives, and GCSS generation in the research have been described in this article. Switching converters are potential choices for reducing voltage instability in PV systems. The use of energy storage devices with MPPT in a PV system decreases voltage variations. The authors' major goal was to lay a foundation for readers interested in PQ concerns in PV integrated systems.

References

[1] Sterling J, Mclaren and Cory K 2014 Treatment of Solar Generation in Electric Utility Resource Planning Treatment of Solar Generation in Electric Utility Resource Planning *National Renewable Energy.*

[2] Akubude V C, Nwaigwe K N and Dintwa E 2019 Production of biodiesel from microalgae via nano-catalyzed transesterification process: a review Mater. *Sci. Energy Technol.* **2** 216–225.

[3] Parida S, Iniyan R and Goic R, 2011 A review of solar photovoltaic technologies *Renewable and Sustainable Energy Reviews* **15** 1625–1636.

[4] Nwaigwe K N, Mutabilwa P and Dintwa E 2019 An overview of solar power (PV systems) integration into electricity grids *Materials Science for Energy Technologies* **1** 629-33.

[5] Omer A M 2008 Energy, environment and sustainable development *Renew Sustain Energy Rev* 2265 – 300.

[6] Al Smadi H 2018 Overview of Grid-Connected PV Systems Challenges with Regards to Grid Stability *Solar Energy Engineering* 2018.

[7] Rani A and Sharma G 2017 A Review on Grid-Connected PV System *International Journal of Trend in Scientific Research and Development* **1** 558-563.

[8] Chatterjee S and Kumar P A techno-commercial review on grid-connected photovoltaic system. Renew. *Sustain. Energy Rev.* **81** 2371–2397.

[9] Kumar A, Gupta N and Gupta, V. 2017 A Comprehensive review on grid-tied solar photovoltaic system *J. Green Eng.* **7** 213–254.

[10] Orioli A and Gangi A D 2013 Load mismatch of grid-connected photovoltaic systems: Review of the effects and analysis in an urban context *Renewable and Sustainable Energy Reviews* **21** 13–28.

[11] Namjoshi M and Anubhaba 2013 Study of Grid tied Smart Solar Photovoltaic system: A review

International Journal of Scientific & Engineering Research **4** 811-815.

[12] Farhad M M, Ali M M, Iqbal M A, Islam N N, Ashraf N 2012 A New Approach to Design of an optimized Grid Tied Smart Solar Photovoltaic (PV) System *International Journal of Advancements in Research & Technology* **1** 107-11.

[13] Zahedi A 2011 A review of drivers, benefits, and challenges in integrating renewable energy sources into electricity grid *Renewable and Sustainable Energy Reviews* **15** 4775–4779.

[14] Kempener R, Komor P and Hoke A 2013 Smart grids and renewables–a guide for effective deployment *Int. Renew. Energy Agency*.

[15] Muni T, Vijay and Lalitha S V N L 2017 Technical Issues of Grid Connected Solar Photovoltaic Cell–A Survey *International Journal of Circuit Theory and Applications* **10** 913-920.

[16] Khyani H K and Vajpai J 2014 Integration of solar PV systems to the grid: issues and challenges *International Journal of Engineering Research & Technology* **2** 393-7.

[17] Blaabjerg F, Chen Z, and Kjaer S B 2004 Power Electronics as Efficient Interface in Dispersed Power Generation Systems *IEEE Trans. on Power Elec*tronics **19** 1184-1194.

[18] Larsen C, Brooks B and Starrs T 2000 Connecting to the Grid", A Guide to PV Interconnection Issues *Interstate Renewable Energy Council Third edition* **35.**

[19] Jamil M, Rizwan M and Kothari DP 2017 Grid Integration of Solar Photovoltaic Systems. *Crc Press*.

[20] Banerjee B, Jayaweera D and Islam S 2016 Grid Integration of Renewable Energy Systems *Smart Power Systems and Renewable Energy System Integration. Studies in Systems, Decision and Control* **57** 75-97.

[21] Anees A S 2012 Grid integration of renewable energy sources: Challenges, issues and possible solutions *IEEE International Conference on Power Electroni*cs 1- 6.

[22] Sandhu M and Thakur T 2014 Issues, Challenges, Causes, Impacts and Utilization of Renewable Energy Sources - Grid Integration *International Journal of Engineering Research and Applications* **4** 636-643.

[23] Driesen J and Belmans R 2006 Distributed generation: challenges and possible solution *IEEE Power Engineering Society General Meeting* **18** 8.

IP based Dual Data Security for Cloud Storage Against Denial-of-Service Attack

[1] Priyanka Jain, , [2] Uday Chourasia,[3] Raju Barskar

[1]Priyanka.jain0111@gmail.com

[2] udaychourasia@rgtu.net,

[3] rajubaraskar@rgtu.net

University Institute of Technology RGPV, Bhopal (M.P.), India

Abstract - The main aim of the attacker is to increase the unwanted packets flooding in the network. It is considered as a king of assaults because it floods the network with an excessive number of useless packets, causing the network capacity to be consumed and data transmission in the network to be hampered. The primary work of DDSCS is to visualize the impact of a DoS attack on a network and identify the node or nodes that are causing the network's performance to be negatively affected. The attacker ping attempts are very high in network but the data delivery attempts are almost negligible. The DDCSS security system examines the profiles of each node in the network through its IP address and the number of ping attempts. If the attacker is one of the nodes that flooded the network with superfluous packets, the DDSCS method prevented the attacker in the network. The simulation results demonstrate the performance of DoS, SPAS and proposed DDSCS but the DDSCS schemes, indicating that it is successful and demonstrating a 0% attacking rate in the presence of an attacker.

***Index Terms*— DoS, DDSCS, Cloud, Security, Flooding, Routing.**

1. Introduction

Most Information Technology (IT) organizations nowadays are confronted with problems posed by rapidly changing environments that have highly unique requirements. These circumstances exist for both current and older systems that require dependability, security, and, in certain cases, a high level of service assurance. To address these issues, several businesses began offering self-managed, well-designed online platforms, and pay-per-use services, which are referred to as cloud computing. The on-demand provision of computing resources such as servers, data storage, networking, databases, etc. The key aim of CC is to allow multiple users to access the data centers [1] [2]. Users can also view data on a remote server. The opportunity for cost savings is the key explanation for the implementation of CC by many organizations. The CC allows you the right to access resources as necessary and to pay only for what you use [3]. Cloud infrastructure has made it easier to manage IT activities as an outsourced unit without a lot of in-house personnel. Cloud Computer has evolved as a cutting-edge technology that provides developers with a variety of

Content from this work may be used under the terms of the Creative Commons Attribution 3.0 licence. Any further distribution of this work must maintain attribution to the author(s) and the title of the work, journal citation and DOI.
Published under licence by IOP Publishing Ltd

computing resources. IaaS, PaaS, SaaS and are all examples of CC resources [1].

The introduction of the Cloud represents a watershed moment in technical development for quick information processing. When a new computing system is released, academics and researchers place a high focus on its security. The success of any information processing system depends on the security of data processing across all systems. [3].

In the [4] and [5] mentioned the analysis of security issues connected with the cloud computing environment. Malware injection attacks, Spoofing attacks, flooding attacks, denial of service (DoS), distributed denial of service (DDoS), and enhanced denial of service (EDoS) are all possible in the cloud. Detecting attack packets and filtering them in order to protect cloud users is a challenging challenge to do. There are four distinct cloud models available for you to choose from, depending on your company's requirements. Cloud computing has evolved as a cutting-edge technology that provides developers with a variety of computer resources.. Cloud computing domain provides resources in the form of Infrastructure as a Service, Platform as a Service, and Software as a Service. The location information of nodes is also possible to maintain by using the location-based approach [6].

2. ATTACKS ON THE CLOUD

The attacker aim in the cloud is to drop the useful information of genuine users or consumes the available resources of cloud. Each attacker has different behavior in network [7] [8] [9] [10] [11]. The following are some examples of attacks in cloud:-

2.1 DDoS (Distributed Denial of Service) assault

DDoS attacks are often divided into three types: traffic attacks, denial of service attacks, and distributed denial of service attacks. There are two sorts of attacks: bandwidth attacks and application attacks. A distributed denial of service (DDoS) attack is a type of DoS attack that targets many machines. Database servers, web servers, cloud computing servers, and other network-connected systems are now under attack from network-based attackers. To counter a DDoS assault, a large number of internet devices must be used in tandem to overload a target online service with fictional data flow.

2.2 SQL Injection Attack

A code injection attack is one in which malicious code is introduced into a computer by introducing harmful code strings into instructions. SQL injection is an example of code injection attack. The goal of this vulnerability is to steal information from the user's computer. It is possible to modify the nature of a standard SQL query by injecting malicious code into it. As a result, hackers are able to acquire access to a database, which they can then use to perform their own malicious SQL query.

2.3 Attacks Using Phishing and Spoofing

One of the security threats that wireless networks face is spoofing attacks. When an intruder or malicious application successfully impersonates other devices on a network, the attacker is able to perform attacks near the network host or steal information. IP address spoofing, ARP spoofing, and DNS spoofing are some of the most frequent types of spoofing attacks.

2.4 The Assault on the Dictionary

In the event of a Dictionary assault, a list of terms is kept on hand. The encrypted text is subsequently sent across the internet. Using a dictionary, an intruder tries to crack encrypted text. The dictionary is searched for the encryption text, and the attack is successful if it contains one or more dictionary terms.

2.5 Denial-of-Service (DoS) attacks (DoS)

In general, denial-of-service (DoS) attackers seek to bring a server down. A vast number of clients profit from the server's services. DoS attackers behave similarly to legitimate clients; nevertheless, the attacker's purpose is to render services unreachable by submitting many pending requests and causing the service queue to overflow.

2.6 Jamming Devices Attack (G)

A jamming attack is a sort of denial-of-service attack in which the purpose is to prevent other nodes or users from getting information from the original source in a timely and suitable manner. It works by clogging up network

traffic by blocking a certain communication channel. Attacker is only active for a short length of time in this attack and exclusively targets the most vital information.

2.7 Man-in-the-Middle Attack (G.)

Data is exchanged between two parties that have agreed to share it across a network. The attacker wants to build a link between the two parties involved in this assault. Once the attackers have gotten access to the traffic, they will attempt to impersonate a sender or a receiver. As a result, they change the traffic and focus on maintaining the secrecy and integrity of the information.

2.8 Internet-based Phishing Attacks

Phishing is an type of social engineering attack that occurs online. The attackers send phishing emails to a huge number of people, including a link to a bogus website in the body of the message. Personal information like as usernames and passwords, as well as other sensitive data, are collected by deceiving users into revealing it with the attackers. Users will be sent to the bogus website after clicking on the given link, where they will be asked to input their personal information.

3 Literature Survey

In this section presents the previous work that is already done in field security in cloud. The detail description of work with drawback is mentioned in table 1. In all the previous works not a single author worked on the infection or record the attacker activity but proposed DDSCS scheme evaluated it. The drop or loss of data only evaluated by [16] and [18] but proposed is scheme showing better performance.

Table-1 Comparative Study among previous work on the basis of different parameters and Performances

S.No.	Author/s	Worked done	Overhead	Infection	Loss	Efficiency	Technique Used	Result
1	Premlata singh et al. [12]	They compare AES, RSA- and SHA security algorithms.	High	-	-	Good	The network security algorithms are used for secure communication	Only concept of security has explained
2	Iva Ranjan et al. [13],	They proposed a scheme to secure from malware injection attack.	High	-	-	Good	SQLIA for stop the injection of packets on server.	Measure the speed of server
7	M. K. Sin chana [14]	They are discuss about the factors affected the cloud performance.	Not Evaluated	-	-	Not Evaluated	Security schemes overview and flaws	Only the Description of techniques are discussed.
8	H.Dhakrey [15]	Proposed a Hybrid encryption technique for encrypted data.	Low	-	-	Good	Use the combination of AES, DES and Blowfish or	Only the encryption and decryption time comparison evaluated.
9	Huang	Proposed a	Low	-	Low	Good	Distinct	Detect the

	Shaohua et al [16]	security scheme to detect the abnormal flow of data in the cloud.					Feature Number (DFN) for Detecting traffic.	unwanted flow of packets by measuring the entropy.
10	Binanda Sengupta et al. [17]	Proposed a data storage scheme for arrange and handle data efficiently.	High	-	-	Good	cloud storage protocols for dynamic data (DSCSI and DSCSII)	Calculate the storage overhead, communica-tion cost in different file size.
11	Min Huang [18]	Proposed a scheme to calculate the credibility of loss and apply the security after the judgment of loss.	High	Based on credibility	Low	Good	Only focus on the security factors not affect on cloud or not able to provides better security.	Measure the vulnerability of confidentiality, integrity, availability, and reliability
12	G Soni et al. [19]	Proposed scheme secure the 6G network from flooding attack in WSN	Low	Low	Low	Good	A SIMMF scheme proposed for attacker detection and prevention.	Evaluate overhead, energy consumption , attacker infection and network lifetime.

4 Proposed Approach in Cloud

In cloud computing, there are several users which are connected and share information. If more users try to fetch the information then the load on servers will increase. The user in the network perceives, compute, and communicates. The genuine user communication can begin with initial deployment, after which the nodes can disperse and receive information from the connected router. When two nodes are not in range of another then the information acquired by one can be transmitted to the other to the cloud through the router. Due to the broadcast nature of the transmission channel, cloud networks are particularly vulnerable to security threats. Additional vulnerabilities exist because nodes are frequently placed in hostile or hazardous environments where they are not physically protected, resulting in an increased risk of compromise. The primary goal of the research is to protect the network from being attacked by outsiders or hackers. As a result of the massive quantity of packets delivered in the network by the DoS attacker, the network's performance is negatively damaged in this dissertation. By utilizing a profile-based security strategy, an attacker can't access the network, and an infection-free environment may be maintained.

A. Proposed Algorithm

Algorithm: Compute **DDSCS**

Input:

N_n: set of mobile nodes *// n=0,1,2,3.....n-1.*

Si: set of Sender Nodes *// i =0,1,2,3....i-1*

Ij: set of intermediate nodes *// j=0,1,2,3....j-1*

R_i: Set of Receiver Nodes *// i =0,1,2,3....i-1*

DDSCS: Detection and Prevention Scheme

C_s= Cloud storage

A_T: suspected node or Attacker

$A_T \in N_i$: set of attacker nodes whose spread infection

M_ID= Malicious ID or Attacker ID

Output:

Delay, PDR, drop percentage

S ← execute-route(S_i, R_i, I_j)

While (N in range) **do**

I_j ← receive routing packets

For each I_j in range , DDSCS watch the I and set A_T

Call_ DDSCS (I_j ,packet in C_s) // the data store in cloud and after that forward to receiver

While** $I_j \neq R_i$ **do

Calculate Ratio of I_j: ((forward/receives) ratio) // Attacker PDR always down in network

If** (I_j== R_i) **then

Send Ack to S node

Call data-pkt()

Else

R not in zone

End if

End do

End do

DDSCS (node, packet_type) *// packets are data or unwanted packets injected by small malicious programs*

If *(I_j update data of S_i node)*

Check updates data by I_j

  If *(update is True && receiver ID !=R_j)*

  I_j ϵ M_ID // it is Attacker (A_T)

  Identifies (infected data count, M_ID symptoms, Packet_loss)

  End if

End if

While *(symptoms == excessive ping_attemts)* ***Do***

  If *(Infection==high) // heavy flooding in network*

    Symptoms are not normal

    Capture node number

    Count heavy flooding status

    Analyze behavior

    Trace time of data update

  Else

    Forward packets to nest node or destination

  End if

End do

Prevention-manager *(DDSCS, capture, M_ID, abnormal-table)*

  Analyze attack type with abnormal ping_attempt

  Send normal msg to A_T

If *(Attacker (A_T) profile == normal-profile && excessive ping_attempt==true)*

  Attacker (A_T) ϵ normal profile node

  Disable attack communication from C_s

  Block the node (symptoms not favorable)

  DDSCS Broadcast attacker node info and its symptoms to all connected node

  Disable attacker existence in network

  Continuously analyze the link behavior

  Calculate performance of the network

Else

  Nodes are normal or not malicious (M)

End if

Now in case of normal routing no need to identified the attacker in network but if in actual attacker s are infected the normal activity of network then in that case it is essential to apply a effective security scheme to protect network from attacker. This flow chart represents the sequential attacker identification and protection scheme.

1. Identified attacker through proposed DDSCS scheme by counting the number of pings and entry of attacker on router in cloud.
2. Capture vulnerable node information & Rate in case of attacker that infected vulnerable nodes in cloud network
3. Identified the data type and rate to confirm the presence of DoS attacker.
4. The attacker behavior is to identifying the route record and the vulnerable receiver nodes are easily affected from it.
5. Then call DDSCS to identify the attacker in cloud.
6. DDSCS protector is checking the attacker through higher sending data rate in network i.e. beyond the capacity.
7. Send rate control message at different Inter Arrival Time to block the DoS activity through DDSCS.

5 Simulation Parameters

The number of parameters consider for simulation of previous security scenario RSA and novel Profile_IP (DDSCS) are mentioned in table1. The number of users, number of routers and length of data or size of data in the form of packets and different layer protocols like TCP, UDP etc. all are same in both the security scenarios. Simulation of all the protocols are done by using Omnet++[20]. This simulator provides the better visualization of network as compare to the other simulators. It is the collection of number of packages for cloud computing, sensor network and other streams

Table-2 Simulation Parameters

Parameters	Value
Number of Users	9
Cloud Access Router	8
Length	72Bytes
Storage	1TB
Cloud Data Center	1
PPP Frame	83Bytes
Transport Data	TCP, UDP
Attack	Denial of

IOP Conf. Series: Materials Science and Engineering 1228 (2022) 012018 doi:10.1088/1757-899X/1228/1/012018

Type	Service
Security Mechanism	RSA, Profile_IP (DDSCS)
Initial Joule	100Joule
Simulation Time	100Seconds

6 Result Analysis

The results comparison of previous and proposed scheme mentioned in this section and the performance of DDSCS scheme is better than the previous RSA scheme in Cloud.

6.1 Security Architecture of Proposed DDSCS

In the given scenario of communication, a separate route is used to connect the senders and receivers. One or more than of the senders is a malicious user or attacker in the network. In the given scenario, users send the data in the cloud and cloud transfer the data to the respective router and finally router forward the data to the receiver. The DoS attacker floods unwanted packets in the network because of that the bandwidth and router processing capability are affected. Each sender and receiver has a unique IP address. The attacker in the network is present and continuously ping for flooding unwanted information.

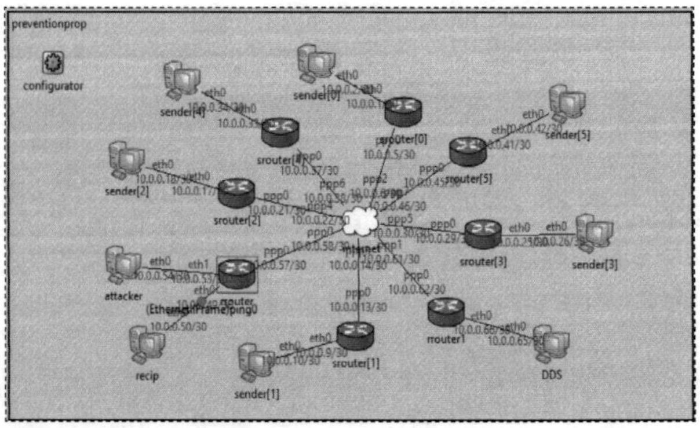

Figure- 6.1: Simulation Architecture of Proposed DDSCS

6.2 PDR Performance

The Packet Delivery Ratio (PDR) performance of DoS, previous SPAS and novel DDSCS approach are mentioned in the given graph. The number of senders sends the data to the receiver and how much amount of data successfully received at the destination is measure through PDR In the graph, up to the end of the simulation time, about 95% of data successfully received at the destination. Same as in the case of a DoS attacker the performance of receiving the data gradually decreases in the network. The performance of SPAS is also better but DDSCS is more secure for communication in the cloud.

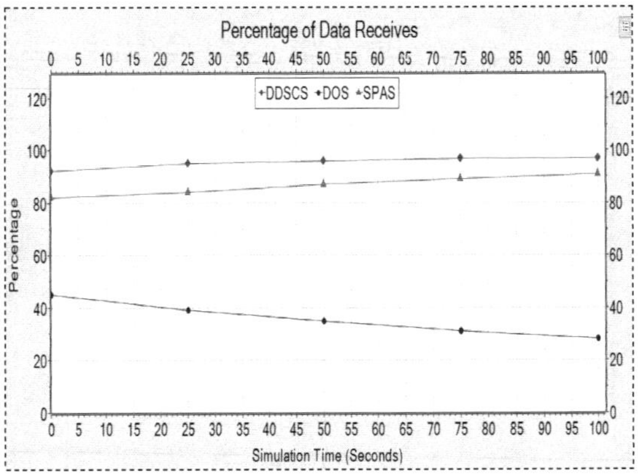

Figure-6.2 PDR Performance Analysis

6.3 Secured Cloud Access Percentage

Security is one of the major issue in cloud communication. The cloud provides fast access to the users. If the data is available on the cloud then no need to forward the request of the user on the cloud. Cloud computing provides on-demand services to users and users fulfill the task by using the services as well as hardware resources virtually. The secure cloud access percentage of DDSCS is about 90% in network but the accessibility of the old SPAS scheme is only 80%. The proposed scheme shows a 10% improvement in performance. More cloud access means more utilization of resources.

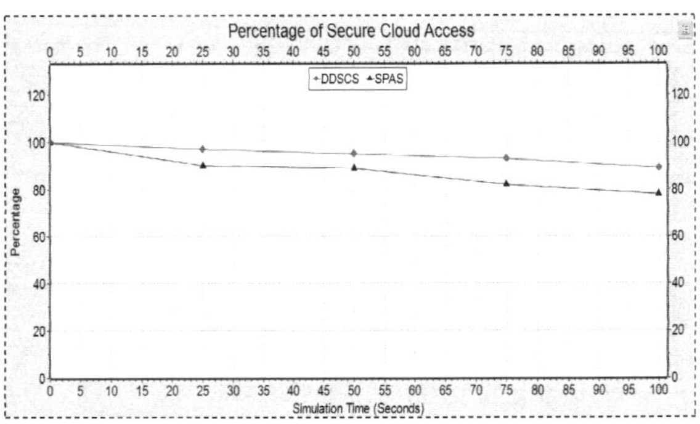

Figure-6.3 Access Percentage of user in Cloud

6.4 Energy Utilization Analysis

The number of senders and receivers consumes some amount of energy for communication. If the energy consumption is high and performance is poor it means the poor performance of the cloud and its resources and if the energy consumption is high and performance is better shows the utilization of energy. In the given graph the energy consumption in the DoS attack is high that's why the utilization is poor. The proposed DDSCS scheme consumes less energy, which means energy utilized in information forwarding and receiving. The SPAS approach also consumes less energy but not less than the novel DDSCS approach.

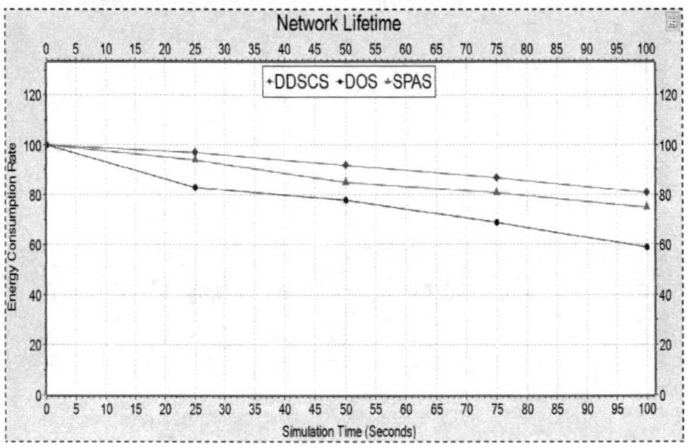

Figure-6.4 Energy Utilization Performance Analysis

6.5 Analysis of Attacker Ping in Cloud

The behavior of DoS attacker is to continue forwarding the unwanted information in network and the router or other connected devices are routed or forwarding the information to the other devices in the network. The number of ping percentage analyses of DoS attackers is mentioned in the given graph. The more flooding of packets means the attacker ping many times in network for forwarding the useless packets. The attacker is 10.0.0.53/30 ping in the network through this IP and the IP information is also forward to the router and the router is directly connected to the cloud. The ping percentage is 100% up to the end of simulation time. The number of ping attempts is very high.

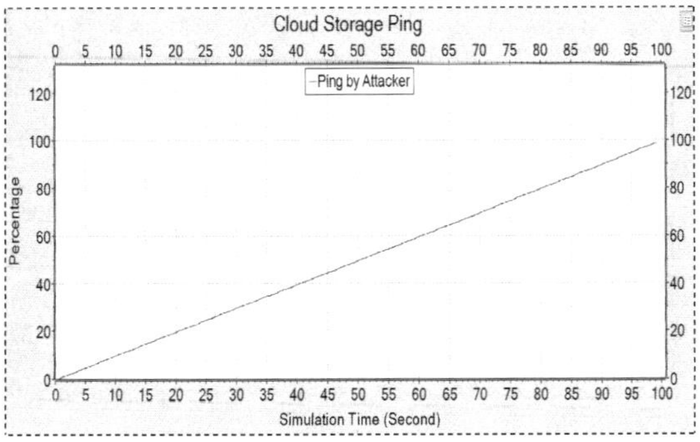

Figure-6.5 Attacker Ping Analysis

6.6 Number of Frame Access per Seconds

The sender, receiver, router, and switch are the major components used for forwarding the data from a network to the cloud and cloud to the receiver. Without the internet, it is not possible to access the cloud and its services. The number of frames access per second performance is measured through the given graph and the performance of the attacker is equivalent to the router and receiver because the attacker generates a huge amount of unwanted packets in the network. The routers access the highest quantity of frames attacker access the lowest but the difference is very minute. The given analysis shows the attacker's bad effect on the network.

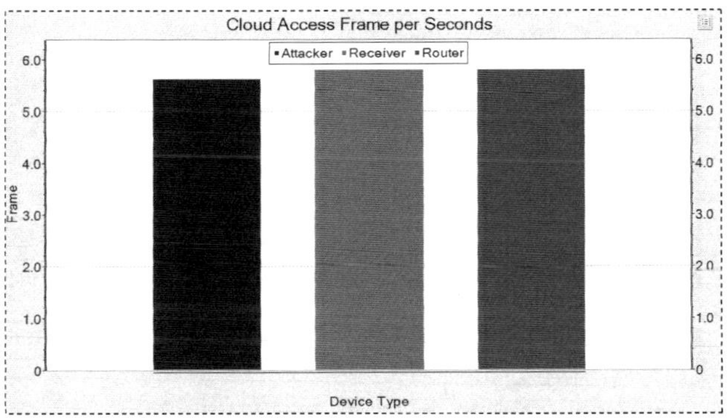

Figure-6.6 Number of Frames Access in unit Second

7. Conclusion and Future Scope

In cloud communication, the nodes are constantly exchanging information with one another in the network. The attacker ping attempts are very high in network. The DDSCS novel approach first checks the number of ping attempts of users. If the number of ping attempts is more than point out the IP address of the sender. The attacker is the node that does not receive the content from any other node in the network. The DDSCS suggested technique does away with the requirement for a centralized authority, which is impractical in such networks. The findings indicate that the existence of a DOS significantly increases the amount of packet loss and routing burden experienced by the network. Protecting the network through a self-organized, completely dispersed, and locally-based method is the goal of the proposed DDSCS mechanism. The energy consumption in presence of the DoS attacker is more and the percentage of data received is 28% up to the end of simulation time. The DDSCS scheme showing 98% of data packets receiving and improves energy utilization.

The DoS attacker flooding slowdown the communication of the sender and response of the server. The other attackers are also very harmful and their aim is to degrade the network performance. The Sybil attacker is one of the attackers that communicate with others through multiple identities. In the future try to propose the scheme to prevent the cloud from Sybil attacker.

.

References

[1] Weiwei Lin & Gaofeng Peng et.al. "Scheduling Algorithms for Heterogeneous Cloud Environment: Main Resource Load Balancing Algorithm and Time Balancing Algorithm" *Journal of Grid Computing; Dordrecht* **Vol. 17 (4),** 2019.

[2] Zhao G, Liu, Tang Y.Sun,W.Zhang F, Ye X,Tang N, "Cloud Computing : A statistics aspect of users," First International Conference on Cloud Computing (CloudCom), Beijing, *Springer,Berlin, Hiedelberg,* pp 347 – 358, 2009.

[3] Harsh Mishra, Jatin Makad, Nikhil Pathak, Dr. Gaurav Soni, "Cloud Based Career Guidance System," International Journal of Scientific Research in Computer Science", *Engineering and Information Technology (IJSRCSEIT)* ISSN No.2456-330, April 2021.

[4] S. Yu, S. Guo, and I. Stojmenovic, ''Can We Beat Legitimate Cyber Behavior Mimicking Attacks from Botnets?'' in *Proceeding of INFOCOM*, pp.2851-2855, 2012.

[5] J. Francois, I. Aib, and R. Boutaba, ''Firecol, a Collaborative Protection Network for the Detection of Flooding DDoS Attacks,'' *IEEE/ACM Transaction* Network**, Vol. 20**, No. 6, pp. 1828-1841, Dec. 2012.

[6] Gaurav Soni, Kamlesh Chandravanshi, Mahendra Ku. Jhariya, Deepak Singh Tomar "A Multipath Location based Hybrid DMR Protocol in MANET", *IEEE International Conference on Emerging Technologies in Computer Engineering (ICETCE),* Feb 2020.

[7] Bikram Khadka, Chandana Withana, Abeer Alsadoon, Amr Elchouemi, "Distributed Denial of Service attack on Cloud: Detection and

Prevention," *IEEE International Conference and Workshop on Computing and Communication*, 2015.

[8] Li Qian, Zhenyuan Zhu, lun Hu, Shuying Liu, "Research of SQL Injection Attack and Prevention Technology". *IEEE, International Conference on Estimation, Detection and Information Fusion*, 2015.

[9] Gaurav Soni, S Sudhakar, "A Flooding based IDS Security against Jamming Attack in MANET," *EasyChair Preprint*, 2020.

[10]Oliver Eigner, Philipp Kreimel, Paul Tavolato," Detection of Man-in the-Middle Attacks on Industrial Control Networks," *IEEE International Conference on Software Security and Assurance*, 2016.

[11]Jakov Andri, Dijana Oreski, Tonimir Kisasondi," Analysis of Phishing Attacks against Students". MIPRO, May 30 - June 3, 2016.

[12]Premlata Singh, Sushil Kr. Saroj, "A Secure Data Dynamics and Public Auditing Scheme for Cloud Storage," IEEE 6th International Conference on Advanced Computing & Communication Systems (ICACCS), 2020.

[13]Iva Ranjan, Ram Bhushan Agnihotri, "Ambiguity in Cloud Security with Malware-Injection Attack," *Proceedings of the Third International Conference on Electronics Communication and Aerospace Technology (ICECA)*, 2019.

[14]M. K. Sinchana, R. M. Savithramma, David Q. Liu, "Survey on Cloud Computing Security," *Innovations in Computer Science and Engineering* pp 1-6, 2020.

[15]Hrithik Dhakrey , "Secure File Storage in Cloud Computing," *School of Computer Science Engineering, Galgotias University*, 2018.

[16]Huang Shaohua, Xiao Nanfeng, "Abnormal Traffic Monitoring Methods Based on a Cloud Computing Platform," *IEEE 5th International Conference on Cloud Computing and Big Data Analytics*, 2020.

[17]Binanda Sengupta, Akanksha Dixit, and Sushmita Ruj, "Secure Cloud Storage with Data Dynamics Using Secure Network Coding Techniques," *IEEE Transactions on Cloud Computing*, 2020.

[18]Min Huang, "Design of basic Process of Information Security Risk Assessment in Cloud Computing Environment," *IEEE International Conference on Consumer Electronics and Computer Engineering (ICCECE 2021)*, 2021.

[19]Gaurav Soni, Kamlesh Chandravanshi, "Security Scheme to Identify Malicious Maneuver of Flooding Attack for WSN in 6G," *8th International Conference on Signal Processing and Integrated Networks*, 26-27 Aug. 2021.

[20]OMNeT++ Home Page. http://www.omnetpp.org [access from July, 2021]

A review on various types of densification/briquetting technologies of biomass residues

Sunny Vaish[1], Naveen Kumar Sharma[2], Gagandeep Kaur[3]

Department of Electrical Engineering, Main Campus, IKG PTU, Kapurthala-Jalandhar, Punjab[123].

Abstract. The growing development, industrialization and population is the main reason of the increment in the global power demand around the world. The conventional power resources are limited and not eco-friendly as well as diminished in the future. Now, every country is move towards the more utilization of renewable energy resources. The biomass renewable energy source is one of them. The biomass residues material can be utilized for the production of power and also overcome the problem of decomposition of agricultural wastes and other wastes products by conversion into solid biomass fuels. The conversion of scattered agricultural residues/wastes into the useful product (biomass fuel) through the technology known as densification. The densification process supports the more utilization of renewable source and clean energy demands. The environmental challenges related with burning of biomass residues in the open field can be minimized by the densification of biomass residues. The densification is the process of compaction or densifying the biomass residues or waste through the application of various pressure levels. The challenges associated with handling, storage and transportation of biomass residues can also be minimize by densification technology. Briquettes and pellets are the end product of densification applications. The various types of densification pressure application and comparison have been discussed in the paper.

Keywords. Densification, Biomass residues, briquettes, Pellets

1. Introduction

The necessity of power is increased by increasing the advancements and developments in the world. The conventional energy resources and fossil fuels are the major share to generate power which is responsible for the global warming and environmental changes. The contribution of product of fossil fuels like petroleum, coal etc. are the key energy sources which produce the near about 80% of the total world's energy requirements [1]. The non-renewable source of energy such as fossil fuels, oil and natural gas are having the possibility of getting diminished in the future. On the contrary, renewable source of energy such as solar, wind, biomass, geothermal and tidal are available abundantly in the nature and can be renewed by itself. Among the all-renewable resource of energy, biomass is the third largest source of energy in the entire world. Biomass can produce matter biologically from the materials and lining matters [2]. Dry grass, wood, straws, solid waste of municipality and kitchen, waste of poultry and forest are the type of Biomass energy source. The wood is the vital usable biomass source for domestic (cooking etc.) activities, space heating and also for industrial purposes [3, 4]. The utilization of biomass in the power generation sector is better than the usage of fossil fuels because it reduces the greenhouse gases (carbon dioxide) which makes it an appropriate renewable fuel for power generation [5]. The usage of biomass directly for cooking and heating purposes is responsible for air pollution and producing bad impact on living organisms and environment. To avoid this problem, agricultural waste or residues can be converted into some solid biomass fuels by using the biomass conversion techniques

Content from this work may be used under the terms of the Creative Commons Attribution 3.0 licence. Any further distribution of this work must maintain attribution to the author(s) and the title of the work, journal citation and DOI.
Published under licence by IOP Publishing Ltd

such as briquettes and pellets [6, 7]. The basic biomass waste used for making briquettes or pellets are not only wood and its residues (sawdust, chips and shavings of wood), but also agricultural waste such as straws, waste materials of the food industries and fuel crops [7]. The large residual of wood collected from the forest and wood companies. This collection of wood residues could be used as the fuel or source of energy for the production of briquettes. The briquetting technology ensuring the total utilization of trees and decreases the wastage of remaining wood materials coming from industries. This will help in control the pollution from the direct usage of burning of firewood and charcoal [8]. The resultant products from biomass conversion techniques emitting less carbon dioxide after combustion process and produce low ash content.

The problem arises through a usage of biomass residues (both agricultural and agro-industrial wastes) have been studied and give appropriate solution of associated problems in [9]. Although, it is not economical to utilized biomass residues directly or in traditional method. The collection of various types of indigenous biomass residues is very difficult to tackle and transportation [10]. Also, the storage and usage (after storage) in its original form is main problem because the property of biomass residues (Moisture content, shape and size etc.) depends on time and environmental behavior [11]. The process for compressing to form the briquettes in various shapes and sizes is known as densification. Due to conversion of raw biomass residue in the dense form (briquettes) in various shapes and sizes, the cost of storage, handling and transportation of briquettes can be reduced [12-14].

This paper consists of five sections. The first section presents the introduction of the paper. The second section depicts the densification of biomass residues/raw materials and third section depicts technologies of densification or briquetting. Section fourth shows pre-treatment methods of solid biomass fuels and fifth section depicts the conclusion of the paper.

2. Densification of biomass residues/raw material

Agricultural solid biomass residues are spread over bulky area. The availability of these types of residues dependent on seasonal variations and causing a problem of storage and handling [15]. The biomass residues having low density associated with challenging in managing, transporting, storing and combustion process. The densification/briquetting technology plays a key role to minimizing problem of high storage capacity and easy transportation. Briquettes/pellets is the best solution for utilizing the raw agriculture waste as a solid biomass fuel [16]. The various steps are involved for preparation of densified biomass fuel (briquettes & pellets) are

2.1 Accumulation of scattered Biomass residues

The collection of material, remaining after the harvesting (agricultural wastes) which is ready to used directly as a fuel for domestic (cocking or heating) purposes in the inconvenient size and shape.

2.2 Preparation of biomass residues

The various process to preparing the biomass residue for compaction. Drying, decrease in size, mixing of different biomass residues in suitable ratios and mixing of binder with biomass residues etc. are some preparing process. There are various types of binders used to improve the quality of briquettes. Biosolids, Cassava, corn, gelatin, microalgae, Molasses and starch are some examples of binders [17, 18]. The binders are used for increased the compressive strength and mechanical durability of biomass briquettes [16].

2.3 Compaction of biomass residues

The process in which raw biomass residues converted into dense form (briquettes & pellets) known as densification or compaction. Figure 1 shows Biomass solid fuels (Briquettes/pellets) manufacturing process. Generally, biomass pellets/briquettes machine are required for compaction process. The process depends on the type of densification technology is employed. Three different compacting pressure are considered for briquettes compaction process and the higher compaction pressure (150 kPa) is better for to improve the quality of briquettes by using cassava as a binder [17].

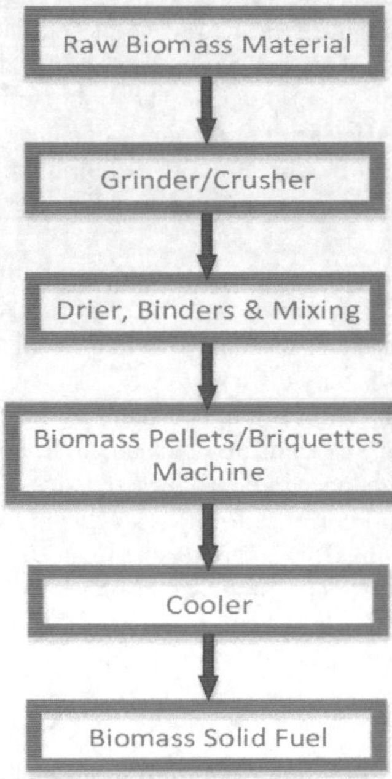

Figure 1 – Biomass solid fuels (Briquettes/pellets) manufacturing process [17] [19]

3. Technologies of densification/briquetting

The compaction of biomass/crops refer to as Densification/Briquettes technology. It is also defined as the technique of compaction of particles (biomass residues) with the help of application of different level of pressure to form a solid biomass fuel. Extrusion, hydraulic piston press, screw press machine, piston type machine, roller press machine and pallet press (ring & flat die) machines are used to withdraw a briquettes/pellet by applying a suitable pressure. The quality of briquettes/pellets are depended upon the strength, mechanical durability, moisture content, calorific values, density etc. which plays the important role while handling, storage and transportation of the solid biomass fuels [20, 21]. Binder are the material used to strengthening or solidification of the crushed biomass residues during densification. Molasses, Cassava, starch, corn, biosolids,. gelatin, microalgae etc. are fall in the category of binders [22]. During densification of biomass residues, the moisture in the raw material or residues materials produced steam under high pressure, hydrolyzing, lignin and hemicellulose into lower molecule carbohydrates, sugar polymers, lignin products and many more derivatives. These types of products (under heat and pressure in the die of machines), act as the glues/gums which helps the particles to bind together tightly [23]. The major types of densification technique are described below that generally used to manufacture the solid biomass fuels (briquettes or pellets). Table 1 shows the different types of densification/briquetting technologies and its comparison.

3.1 Screw Compaction/Extruder Machine

Screw extruder has a rotating screw which exerts a force on biomass residues coming from the feedstock for compression. A separate mechanical drive can be used to operate a screw inside the machine. A significant amount of energy is needed to reduce the particle friction because biomass residues are moderately compressed and crammed. At high temperature (200^{0}C -250^{0}C) and initial temperature, particles of biomass squander their elasticity properties and become softer [23]. It causes interlinking of particles and local bonds formation. A particle having size range between 2mm to 4mm, are ideal for extrusion period because they are responsible for optimum binding of biomass residues material. When biomass enters in tapered die section, a remaining moisture is again evaporated because of high temperature of range 280^{0}C. This process is helps to intensify the compression on biomass residues materials. The high temperature and heavy compression make the biomass material more appropriate for burning and co-firing process [23, 26, 29]. A well labelled diagram of biomass screw type extruder is depicted in Figure 2.

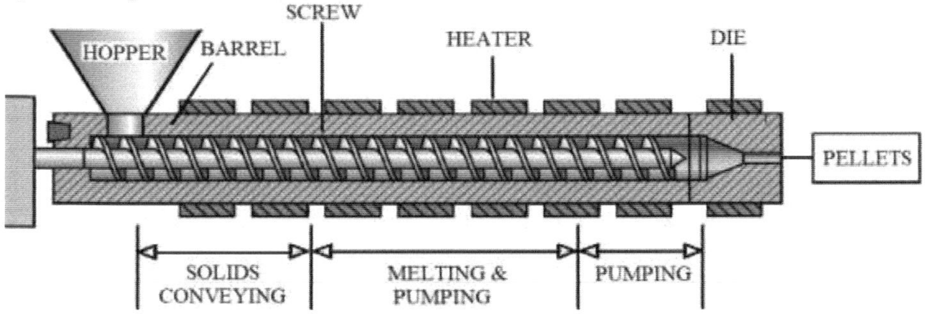

Figure 2 – Biomass screw type extruder [26]

3.2 Piston Press/Pump Machine

Piston press used piston which is driven by flywheel or some hydraulic mechanism for the production of densified biomass fuel (briquettes/pellets). A biomass residues material coming from feedstock unit into the cylindrical shaped piston drive which can be driven mechanically/hydraulically. For heating of biomass residues material, temperature range lies between 1500^{0}C to 3000^{0}C during compression process. The incoming biomass residues material is compressed towards the conical shaped die by the piston and the briquettes is extrude from the opening face of the die [29, 30]. A biomass piston press/pump type machine depicted in Figure 3.

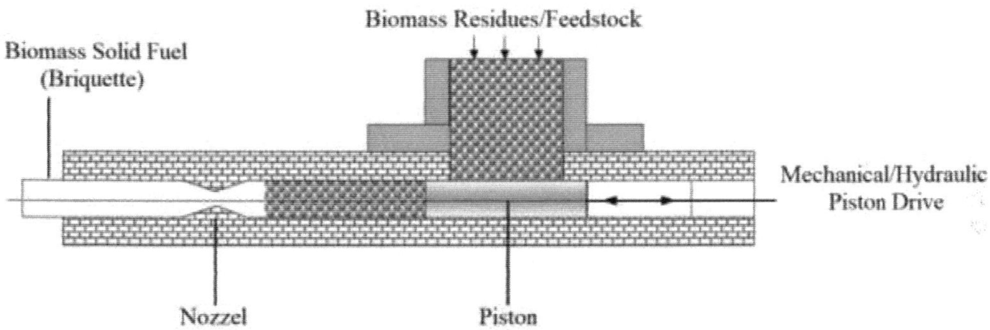

Figure 3 – Biomass piston press/pump type machine [29]

3.3 Roller Press/Mill Machine

A biomass roller press/mill type machine depicted in Figure 4. The main unit of roller mill is two rollers that both rotates in the opposite direction with each other. The biomass residues material is fed through the hopper. The flow of biomass residues material is controlled by control mechanism [23]. The forces exerted on the predetermined quantity of pre-compressed biomass residues materials by the two rollers

Table 1- Different types of densification/briquetting technologies and its comparison

Comparing Parameters	Screw Extruder /Press	Piston Pump/ Press	Roller Mill/ Press	Pellet Mill/ Press
References	[23-27, 29]	[23-27, 29]	[26, 28, 29]	[26, 28, 29]
Invention	Japan (in 1945)			
Requirement of binders	Not required	Not required	Required	Not required

Shape	Cylindrical	Cylindrical	Generally Elliptical (relies on shape of die)	Cylindrical
Particle size (in mm)	2 to 6	6-12	Less than 4	Less than 3
Optimum moisture content of the raw material (in %)	8-9	10-15	10-15	10-15
Wear of contact parts	High	Low	High	High
Output from machines	Continuous	In Strokes	Continuous	Continuous
Power consumption (kWh/ton)	36-150	37-77	29-83	16.4-74.5
Density of Biomass solid fuel (briquettes/pellets) (in g/cm^3)	1 to 1.4	1-1.2	0.6-0.7	0.7-0.8
Maintenance	Low	High	Low	Low
Combustion performance of Biomass solid fuel (briquettes/pellets)	Very Good	Moderate	Moderate	Very good
Suitability in (a) Gasifiers (b) Co-firing (c) Biochemical conversion	 Suitable Suitable Suitable	 Suitable Suitable Suitable	 Suitable Suitable Suitable	 Suitable Suitable Suitable
Homogeneity	Homogeneous	Not Homogeneous	Not Homogeneous	Homogeneous
Pressure type	High-Pressure	High-Pressure	Low-Pressure	

which continuously rotates in opposite direction [29]. The biomass residues material can be drag out from one side and densified product is extrude from other side. The gap between the two rollers is relies on type of biomass residues material used, content of moisture, size of particle, ratio of binders and several other factors [30, 31]. Table 2 presents the Features, merits and demerits of densification/briquettes technologies.

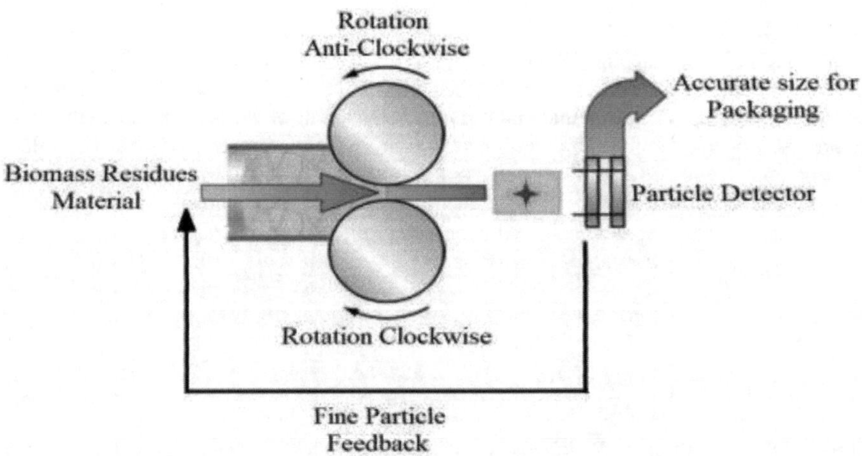

Figure 4 – Biomass roller press/mill type machine [31]

Table 2- Features, merits and demerits of densification/briquettes technologies [21, 23, 32, 42]

	Screw Press	Piston Press	Roller Press	Pellet Mill
Features	Extrusion process done by special type of screw, continuous output from the machine, High pressure type, Produced product is completely solid and Homogeneous, Binder is not needed	Extrusion of biomass products done by piston, High pressure type, Binder is not needed	Elliptical shape (relies upon the die shape), Briquettes suitable for gasifiers and cofiring, Binder is needed	Cylindrical shape, Binder is not needed Suitable for gasifier and cofiring, Homogenous
Merits	Produce less noise, Generate carbonized briquettes, Low maintenance, Very good combustion performance	Less wear and tear, Less power consumption, Less noise, Moderate combustion performance	Moderate combustion performance, Low maintenance	Continuous output from the machine, Low maintenance Very good combustion performance
Demerits	High tear and wear, High power consumption	Do not used for making carbonized briquettes, Non-homogenous	High wear of contact parts, Carbonization of coal is not possible	Carbonization of coal is not possible

3.4 Pellet Mill Machine

A biomass pellet press type machine depicted in Figure 5. A pellet mill machine used for generating pellets from biomass residues material of size whose length is 13-19 mm and diameter is 6.3-6.4 and having cylindrical shape and density range between 1125-1190 kg/m^3 [24]. This type of pellet press machine has a roller and circular matrix. The biomass residues material which is to be pelletized is compressed between the circular matrix and roller mechanism. The circular matrix contains a large no of orifice. The size of orifice is pre-determined. The pelletized biomass residues materials are extruded automatically from these orifices of circular matrix. A cutter (knife) is installed at the unit which cuts the pellets in the desired length and the final product is collected [29, 30].

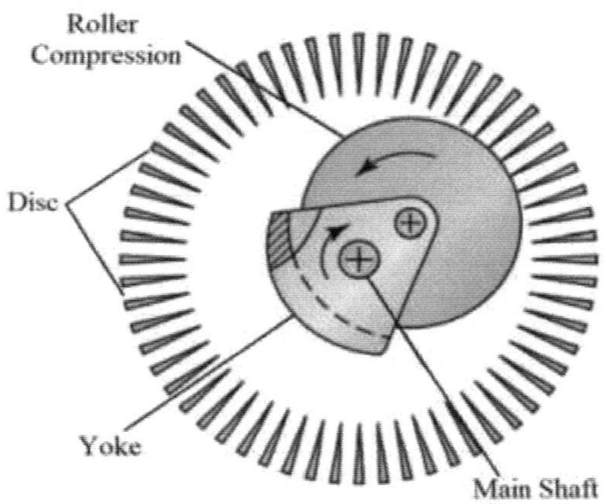

Figure 5 – Biomass pellet press type machine [26]

4. Pre-treatment methods of solid biomass fuels

The process of pretreatment is extensively used before the briquetting or densification processes because it improves the quality of the product. Table 3 depicts the methods of pre-treatment of biomass residues materials. The biomass raw material and other waste material collected from the open fields

Table 3 – Methods of pre-treatment of biomass residues materials [34-38]

Pre-Treatment Methods	Procedures	Effects	Benefits
Steam Explosion [41]	It is method in which high pressure steam is applied for a short time and then released.	Due to this, there is a physical, chemical and structural changes in the biomass residues materials.	Help in binding during palletization and briquetting.
Torrefaction [38, 39]	It is a process of slow heating of biomass residues materials to a temperature of near about 300^0C.	Due to this, volatile materials and smoke evolving products are reduces to make final product better.	Help in binding during palletization and briquetting.
Ammonia fiber expansion [40]	It is a process in which ammonia aqueous is used at high pressure and temperature to obtain more hydrolysis yields for several gramineous feedstocks.	Due to this process, recrystallizing of cellulose in the biomass residues material reduces hemicellulose and lignin.	Help in binding during palletization and briquetting.
Grinding [39]	It is a process of grinding of the biomass raw/residues material at specific size depends upon the densification technology selected.	Due to this, conversion of the biomass raw or residues material in appropriate size and breakdown the lignin present in it.	This increases the surface area of the biomass raw or residues material which improves the binding process.
Pre-heating	It is a process of heating of biomass raw or residues material before densification or briquetting process to enhance the quality of the product.	Due to this, energy consumption of manufacturing the briquettes or pellets can be decreased.	Generate more lignin which help in binding during densification

(agricultural) , industries, municipal corporation etc. have very low heating value. It is very important to enhance the characteristics of raw biomass material to increase the efficiency of usage of the biomass [32]. To improves the quality, make more densified and stable product, many manufacturers of solid biomass fuel (briquettes/pellets) use pretreatment procedures [33].

5. Conclusion

The problem of decomposing of large agricultural residues and waste material obtained from rural and urban areas can be rectifying by densification technology for biomass residues. There are various types of densification technology which offers the different types of solid biomass fuels. Briquettes or pellets can be use directly for domestic purposes and also for power generation. The densification of biomass residue application has different ratings and produce different shape and size of briquettes or pellets. This study described the various densification technologies and its comparison. On basis of this, an effective densification technique can be used for a specific biomass residue material. Pretreatment methods are used to improve the biomass quality before densification process which enhances the overall quality of briquettes or pellets.

References

[1] S.K. Sansaniwal, K. Pal, M.A. Rosen and S.K. Tyagi, "Recent advances in the development of biomass gasification technology: A comprehensive review", Renewable and Sustainable Energy Reviews, vol. 72, pp. 363-384, 2017.

[2] M.S. Javed, R. Raza, I. Hassan, R. Saeed, N. Shaheen, J. Iqbal and S.F. Shaukat, "The energy crisis in Pakistan – A possible solution via biomass-based waste", Journal of Renewable and

Sustainable Energy, vol. 8, pp. 1-20, 2016.
[3] C. Antwi-Boasiako and B.B. Acheampong, "Strength properties and calorific values of sawdust-briquettes as wood-residue energy generation source from tropical hardwoods of different densities", Biomass and Bioenergy, vol. 85, pp. 144-152, 2016.
[4] J. Nagarajan and L. Prakash, "Preparation and characteristics of biomass briquettes using sugarcane bagasse, corncob and rice husk", Materials Today- Proceedings, 2021.
[5] L. Junfeng and H. Runqing, "Sustainable biomass production for energy in China", Biomass and Bioenergy, vol. 25 (5), pp. 483-499, 2003.
[6] L. Chen, Li. Xing and L. Han, "Renewable energy from agro-residues in China: Solid biofuels and biomass briquetting technology", Renewable and Sustainable Energy Reviews, vol. 13 (9), pp. 2689-2695, 2009.
[7] L. Jiang, X. Yuan, Z. Xiao, J. Liang, H. Li, L. Cao, H. Wang, X. Chen and G. Zeng, "A comparative study of biomass pellet and biomass-sludge mixed pellet: Energy input and pellet properties", Energy conversion and management, vol. 126, pp. 509-515, 2016.
[8] P. Purohit, A.K. Tripathi and T.C. Kandpal, "Energetics of coal substitution by briquettes of agricultural residues", Energy, vol. 31, pp. 1321-1331, 2006.
[9] E.R.K. Fernandes, C. Marangoni, O. Souza and N. Sellin, "Thermochemical characterization of banana leaves as a potential energy source", Energy Conversion and Management, vol. 75, pp. 603-608, 2013.
[10] J.L. Bowyer and V.E. Stockmann, "Agricultural residues-An exciting Bio-based raw material for the global panels industry", Forest Products Journal, vol. 51 (1), 2001.
[11] N. Manicham, D. Ravindran and P. Subramanian, "Biomass densification methods and mechanism", Cogeneration and Distributed Generation Journal, vol. 21 (4), pp. 33-45, 2006.
[12] C. Karunanithy, Y. Wang, K. Muthukumarappan and S. Pugalendhi, "Physiochemical characterization of briquettes made from different feedstocks" Biotechnology Research International, vol. 2012, pp. 1-12, 2012.
[13] N. Kaliyan and R.V. Morey, "Factors affecting strength and durability of densified products", ASAE Annual International Meeting, American Society of Agricultural and Biological Engineers, 2006.
[14] J. Singh, B.S Panesar and S.K. Sharma, "A mathematical model for transporting the biomass to biomass based power plant", Biomass and Bioenergy, vol. 34, pp. 483-488, 2010.
[15] V.M. Victor, S.V. Jogdand and A.K. Chandraker, "Biomass densification technologies to obtain briquettes for energy application – A review", International Journal of Engineering Research and Technology, vol. 3 (20), pp. 1-4, 2015.
[16] W. Stelte, J.K. Holm, A.R. Sanadi, S. Barsberg, J. Ahrenfeldt and U.B. Henriksen, "A study of bonding and failure mechanisms in fuel pellets from different biomass resources", Biomass and Bioenergy, vol. 35 (2), pp. 910-918, 2011.
[17] E.F. Aransiola, T.F. Oyewusi, J.A. Osunbitan and L.A.O. Ogunjimi, "Effect of binder type, binder concentration and compacting pressure on some physical properties of carbonized corncob briquette", Energy Reports, vol. 5, pp. 909-918, 2019.
[18] T. Olugbade, O. Ojo and T. Mohammed, "Influence of binders on combustion properties of biomass briquettes: Arecent review", BioEnergy Research, pp. 1-19, 2019.
[19] S. Clarke, P. Eng. And F. Preto., "Biomass densification for energy production", Factsheet-Ministry of agriculture, Food and Rural Affairs, Ontario, pp. 1-6, 2011.
[20] T.I. Mohammed and T.O. Olugbade, "Characterisation of briquettes from rice bran and palm kernel shell", International Journal of material Science Innovations, vol. 3 (2), pp. 60-67, 2015.
[21] L.G. Tabil, S. Sokhansanj and R.T. Tyler, "Performance of different binders during alfalfa pelleting", Canadian Agricultural Engineering, vol. 39 (1), pp. 17-27, 2017.
[22] T. Olugbade, O. Ojo and T. Mohammed, "Influence on binders on combustion properties of biomass briquettes- A recent review", Bioenergy Research, 2019.
[23] P.D. Grover and S.K. Mishra, "Biomass Briquetting- Technology and practices", Food and Agriculture Organisation of the United Nations, Bangkok, 1996.
[24] J.S. Tumuluru, C.T. Wright, K.L. Kenny and J.R. Hess, "A review on biomass densification technologies for energy application", Idaho National laboratory, Biofuels and Renewable

Energies Technologies Department, 2010.

[25] S.E. Ibitoye, T.C. Jen, R.M. Mahamood and E.T. Akinlabi, "Densification of agro-residues for sustainable energy generation: an overview", Bioresources and Bioprocessing, vol. 8 (75), PP. 1-19, 2021.

[26] J.S. Tumuluru, C.T. Wright, K.L. Kenny and J.R. Hess, "A review of biomass densification systems to develop uniform feedstock commodities for bioenergy application", Biofuels, Bioproducts and Biorefining, vol. 5, pp. 683-707, 2011.

[27] R. Sanap, M. Nalawade, J. Shende and P. Patil, "Automatic screw press briquette making machine", International Journal of Novel Research in Electrical and Mechnaical Engineering, vol. 3 (1), pp. 19-23, 2016.

[28] S.Y. Kpalo, M.F. Zainuddin, L.A. Manaf and A.M. Roslan, "A review of technical and economic aspects of biomass briquetting", Sustainability, vol. 12, pp. 1-30, 2020.

[29] N. Manickam, D. Ravindran and P. Subramanian, "Biomass densification methods and mechanism", Cogeneration and distributed generation journal, vol. 21 (4), pp. 33-45, 2009.

[30] S.C. Bhattacharya, S. Sett and R.M. Shrestha, "State of art for biomass densification", Energy Sources, vol. 11 (3), pp. 161-182, 2007.

[31] K.A. Yehia, "Estimation of roll press design parameters based on the assessment of a particular nip region", Powder Technology, vol. 177 (3), pp. 148-153, 2007.

[32] A. Anukam, S. Mamphweli, P. Reddy, E. Meyer and O. Okoh, "Pre-processing of sugarcane bagasse for gasification in a downdraft biomass gasifier system- A comprehensive review", Renewable and Sustainable Energy Reviews, vol. 66, pp. 775-801, 2016.

[33] S.C. Bhattacharya, "State of the art of utilizing residues and others types of energy source", RERIC International Energy Journal, vol. 15 (1), 1993.

[34] V.B. Agbor, N. Cicek, R. Sparling, A. Berlin and D.B. Levin, "Biomass pre-treatment- Fundamentals toward application", Biotechnology Advances, vol. 29, pp. 675-685, 2011.

[35] M. Galbe and O. Wallberg, "Pretreatment for biorefineries- a review of common methods for efficient utilization of lignocellulosic materials", Biotechnology for Biofuels, vol. 12, 2019.

[36] P. kumar, D.M. Barrett, M.J. Delwiche and p. Stroeve, "Methods for pretreatment of lignocellulosic biomass for efficient hydrolysis and biofuels production", Industrial and Engineering Chemistry Research, vol. 48 (8), pp. 3713-3729, 2009.

[37] G. Brodeur, E. Yau, K. Badal, J. Collier, K.B. Ramachandran and S. Ramakrishnan, "Chemical and physicochemical pretreatment of lignocellulosic biomass- A review", Enzyme Research, vol. 2011, pp. 1-17, 2011.

[38] G. Wang, Y. Luo, J. Deng, J.H. Kuang and Y.L. Zhang, "Pretreatment of biomass by torrefaction", Chinese Science Bulletin, vol. 56, pp. 1442-1448, 2011.

[39] S.B. saleh, B.B. Hansen, P.A. Jensen and K.D. Johansen, "Efficiency fuel pretreatment- Simultaneous torrefaction and grinding of biomass", Energy fuels, vol. 27, pp. 7531-7540, 2013.

[40] B. Bals, C. Wedding, V. Balan, E. Sendich and B. Dale, "Evaluating the impact of ammonia fiber expansion pretreatment conditions on the cost of ethanol production", vol. 102 (2), pp. 1277-1283, 2011.

[41] W. Stelte, "Steam explosion for biomass pre-treatment", Energy and Climate, Centre for renewable energy and transport, pp. 1-15, 2013.

[42] A. Kaur, M. Roy and K. Kundu, "Densification of biomass by briquetting- A review", International Journal of Recent Scientific Research, vol. 8 (10), pp. 20561-20568, 2017.

IOP Conf. Series: Materials Science and Engineering 1228 (2022) 012020 doi:10.1088/1757-899X/1228/1/012020

Modelling and Analysis of Magnesium-Fiber Composite Material for Aircraft Applications

Nathi Ram Chauhan[1], Arhantika Khera[2], Shubhi Agarwal[3], Mehak Bagri[4], Aditi Aggarwal[5]
[1-5]Department of Mechanical and Automation Engineering, IGDTUW

E-mail: mehakbagri1999@gmail.com, akhera27@gmail.com

Abstract. Weight to strength ratio of materials used in the aerospace industry is one of their most important characteristics. Thus, the low density of Magnesium alloys is making them surge as the most sought-after material in the field of aeronautics. Actually, the density of Magnesium alloys is only two third than that of Aluminium alloys. This asset creates an enormous opportunity in making aircraft structures of less weight. In addition to this, good machinability and castability is displayed by Magnesium alloys. Fiber Metal Laminates (FML), which are bound together through epoxy, are being used extensively inside the aircraft enterprise. In this illustration, two varieties of FML (fiber material between 2 layers of magnesium alloys of 2 mm thickness each) are studied. FML-1, which has carbon fiber sandwiched between magnesium alloy, and FML-2 which has Jute fiber sandwiched between magnesium alloy layers. The laminates are then analyzed in the software ANSYS for tensile and bending load systems. The FMLs are compared to basic model of Magnesium alloy having 6 mm thickness and the results are compared for three parameters: Total Deformation, Eq von mises strain, and Eq von mises stress. The investigation concluded that carbon FML performs best under Tensile load system, while FML made of jute fiber is best suited under Bending load system.

1. INTRODUCTION

Composites are made up of several different materials. Mechanical, electrical, and chemical qualities are all unique to each substance. They combine to form a reinforced material with increased stiffness, strength, and other desirable characteristics. Composites are extensively being used for numerous applications in the real world; for instance, their utility inside the oil and gas refineries and on construction sites is pervasive. The increased use of these composites can be attributed to their obvious blessings of improved properties and low weight. As our planet is becoming mindful of environmental needs, there is a continuous urge in industries for better suited environment friendly materials. Here, natural fiber composites can be an appropriate choice.

1.1 Fiber Metal Laminates

Fibre metal laminates are made with combination of metal layers strengthened with layers of fiber [1]. Fibre metal laminates offer high modification capabilities with regard to material properties. Owing to this extensive customizability, FMLs are becoming popular particularly in airplane ribs [2]. As such, the vital need for them is to be light-weight and showcase excessive flexural and tensile strengths. Presently, fibre-glass is being examined for this purpose whereas a new trend inside. This enterprise indicates an urge for more reasonably-priced, light weight, and stronger substances for building the structure [4]. To work on the economical aspect, jute fibre has been selected as an opportunity as it is a natural material which can obtain easily. Fibre metal laminates provide us with a particular advantage as metals provide strength and fibers can help in reducing the weight of the final substance. Through cautiously choosing the additives, the material specific requirements can be fulfilled. Because they possess adaptable properties, they find their use case in aircraft enterprise. ARALL (Aramid Reinforced Aluminium Laminate) is manufactured from Aluminium sheets which possess excessive tensile power and are combined with aramid fiber sheets

Content from this work may be used under the terms of the Creative Commons Attribution 3.0 licence. Any further distribution of this work must maintain attribution to the author(s) and the title of the work, journal citation and DOI.
Published under licence by IOP Publishing Ltd

bolstered with epoxy(resin). GLARE (Bolstered glass) also evolved at the same time and composed of numerous metal layers brought together with layers of excessive strong fiberglass bolstered with epoxy(resin). GLARE comprises of fiber-glass in comparison to Aramid which has higher adhesion and therefore utilised for a much broader variety of industrial use cases [3-7].

1.2 Ribs of Aircraft

Curved profiles, C-shaped profiles and thin-walled tubes are the fundamental geometries used in airplane rib structure. They are very rigid and have excessive strength over weight ratio. According to recent studies, it can be concluded that Fibre metal laminates are optimal for forming the above-mentioned shapes and provide the suitable properties as well. A growing need in the aircraft enterprise suggests an urge for low price, light-weight and material with more strength for structure construction [8]. Most vital traits which need attention are component's tensile, flexural and compression strength along with buckling and wear resistance [9]. Earlier, research on resin was carried out to increase the damage tolerance of the body but that proved to be a costly affair. The airplane enterprise wanted low cost, strong and rigid systems that were capable of resisting the flight turmoil with minimum damage to the body. In order to solve this hassle, studies were carried out to find a solution to capitalize the capability of fabric bolstered composite to supply airplane bodies with a cheap and strong structure.

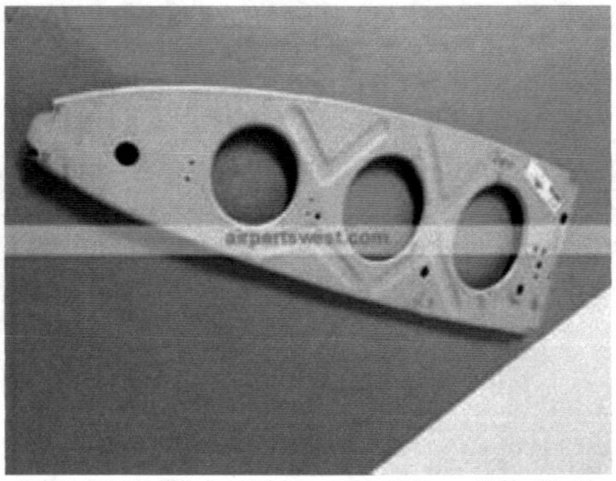

Figure 1: Piper (Aircraft) Wing Rib [10]

1.3 Magnesium Alloys

Due to their great recycling potential and low density, alloys of magnesium are proving to be useful for a lot of demands in the industry for good machining abilities. Their light weight makes them especially suited to the aircraft industry to save fuel and makes air travel economical. The magnesium alloy composite material is replacing the traditional material to bring down the cost and enhance mechanical properties. The crucial dimensions like material's tensile, flexural and compression strength and buckling, wear resistance require consideration and are to be looked at [5].

1.4 Carbon Fiber

One majorly used substance to include widespread forms of composite is carbon fiber. In accordance with its strength to weight ratio, they have an edge over other materials as they can alter the mechanical and physical properties by changing shape, orientation, and content material [15]. Using this in composite materials, current studies suggest that it has potential to show incredible strength as well as being extraordinarily less in weight. By this we checked if the fibre metal laminate combined with carbon-fiber is feasible for airplane rib application.

1.5 Jute Fiber

Up till now, researches in aerospace industry on Fibre metal laminates include ARALL, GLARE, and additionally on artificial fibers bolstered with different metals, such as titanium or magnesium [11-12].

IOP Conf. Series: Materials Science and Engineering 1228 (2022) 012020 doi:10.1088/1757-899X/1228/1/012020

Natural fibres like Jute which is a cheap and environment friendly alternative is introduced with an attempt to decrease the price of carbon-fibre and glass-fibre and to minimize the pollution footprint.

2. RESEARCH BACKGROUND

Numbers of studies have been researched on magnesium (Mg) composite and fiber metal laminate (FML). particularly, "handbooks of composites" edited by Lubin.[1], Richard et.al.[2] said that a many layers of metal, strengthened with many layers of fiber, forms a metal/fibre laminate composite fabric, Asundiet.al [3] added fiber metal laminates (FML) as a pertinent material for upcoming airplanes. Vlot et.al [4] researched the damage resistance of numerous FML. Felix [6] have completed an evaluation of the information on strength under tensile load of aramid bolstered aluminium laminate (ARALL). Mark et.al [7] researched a number of Magnesium (Mg) Alloy relevance in automotive systems. Suresh et.al [9] has done an analysis on flexural and tension responses of Angle Ply Carbon-Jute Aluminium FML that could be used as wing ribs in airplane. Xun[11] studied wooden-fiber combined polypropylene composite assessment of life cycle. Cortes [12] referred to titanium (Ti) based thermoplastic Fibre metal laminate structure. Truong et al. [15] studied the multiply and multi axial combined performance of composites of carbon and its conclusions in destruction of sheared bolstered composite materials and its mechanical qualities. Edwardson [19] studied the manufacturing method laser forming of Fibre metal laminate. Vlot [20] introduced most of the material's applicable aspects, ranging from mechanical capabilities statistically, effect of fatigue and impact on aircraft body layout, manufacturing and maintenance.

On enormous assessment of literature, it became recognized that the usage of FMLs require a time-ingesting and steeply-priced certification method to shield the public from poorly designed and dangerous aircrafts and the crucial deficiencies of aramid reinforced aluminium laminate (the most used FML) which are still being considered [13]. It was diagnosed that despite magnesium having considerably more strength over weight ratio as compared to aluminium (Al), the cost of extraction for magnesium (Mg) alloys is a lot more than alloys of aluminium(Al) [14]. In addition, magnesium can offer light-weight transportation and subsequent reduction of fuel. Commonly occurring fiber like, jute-fiber are not expensive and can assist in reducing the price if used as an option compared to traditional FMLs that include glass or carbon.

Hence, magnesium (Mg) composite material is suggested for replacing aluminium (Al) metal. The advantages of magnesium (Mg) metal alloys over aluminium metal (Al) alloys are because of its increased potential for electromagnetic protection, lower density, outstanding corrosion resistance and widespread weight saving. [12].

3. METHODOLOGY

3.1 Selection of material

Earlier, because of high reactivity with oxygen, magnesium wasn't considered for aircraft enterprise. Although Magnesium is more expensive than aluminium, their weight is lighter and they possess greater stiffness. Thus, they are being accepted as a safer alternative. AZ31(Magnesium alloy) consisting of Mg, Al and Zn has thc following composition of 2.83%wt of Aluminium, 0.80%wt of Zinc, Copper and Manganese.

After comparing numerous materials, 2 arrangements were finalized; Magnesium with Carbon Fiber (FML-1) and Magnesium with Jute (FML-2). Material properties are showed in Table 1.

Although carbon (synthetic fiber) gives greater strength, they are injurious for health. Thus, comparison is done with jute fibers (natural fiber) as it is a light-weight option and economical with being environment-friendly [16].

Table 1: Properties of materials used

Properties	Density(g/cm3)	Tensile Strength (MPa)	Elastic Modulus (GPa)
Mg	1.780	250-290	45
Carbon fiber	1.75-2.00	3000-7000	200-500
Jute fiber	1.50	393-773	10-30

3.2 Analysis

FML-1 and FML-2 are designed on ANSYS-R1 2020 software. Load is simulated to observe and compare their performances. The method of analysis is iterated in Figure 2.

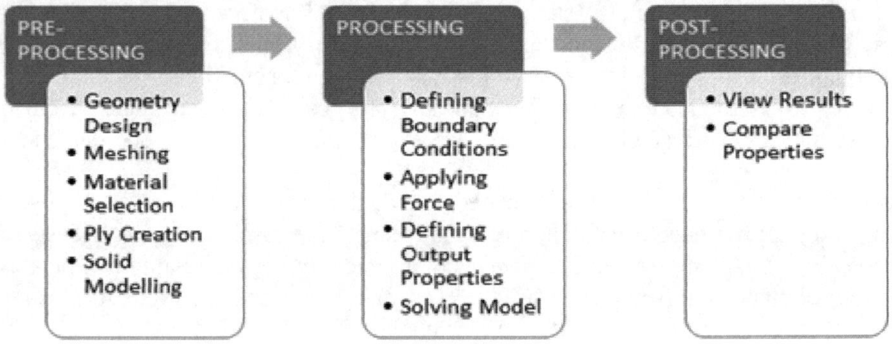

Figure 2: Methods employed

a. Pre-processing

ANSYS-R1 was used to model the designs, Space-Claim was used to make the required shape and ANSYS-mechanical was used for meshing. Triple layered composite geometries with two mm thickness of material were used. ACP (ANSYS Composite PrepPost) was used for layering. The dimensions are provided in Table 2.

Table 2: Dimensions of geometry

Area (Cross Section)	Thickness of layers	Complete thickness
*100*100 mm*	*2 mm*	*6 mm*

b. Processing

Here we fixed the geometry faces and subjected them to UDL (uniformly distributed load). Force direction was assigned with respect to analysis that is being done. Three parameters are being checked namely: Total Deformation, Eq. (Von Mises) Strain and Eq. (Von Mises) Stress. ANSYS-Mechanical is used after defining all the properties and results are checked accordingly.

c. Post processing

Comparison of properties was done in this stage with the basic geometry of alloy of magnesium.

IOP Conf. Series: Materials Science and Engineering 1228 (2022) 012020 doi:10.1088/1757-899X/1228/1/012020

4. RESULTS AND DISCUSSION

The experiment started with the geometry construction followed by meshing which is the breaking down of geometry into small shapes and the conclusion is obtained for different values of applied force. Space-Claim was used for creating the model which was used. From Space-claim the geometry was brought into ANSYS-WORKBENCH-R1. Force was given to all four models in 10 steps from 1000-10,000N.

The force is given in 2 different ways namely: *Bending Analysis* and *Tensile Analysis*

To get variation we used different values of Force:

a. *Total Deformation*- Total Deformation expressed in *mm* is the measurement of dimensional change that occurs when the geometry is given an external load.
b. *Eq. Von Mises Stress*- Eq. (Von Mises) Stress predicts the fracture point or yielding of a substance under applied load. It is put on to ductile material like metals. It's unit is *MPa.*
c. *Eq. Von Mises Strain* - It is defined as the amount of deformation experienced by the geometry the direction of force divided by the original dimensions of the body. Eq. (Von Mises) Strain is a scalar and dimensionless quantity.

4.1 Boundary conditions

a. *Tensile analysis*: Face with dimensions of [100 mm * 6 mm] is fixed and UDL which is uniformly distributed load is put on the directly opposite face. Due to this force is evenly distributed over a particular face.

b. *Bending analysis*- Two directly opposite faces with dimensions [100 mm * 6 mm] are fixed and UDL is put on the side which is perpendicular to both fixed supports.

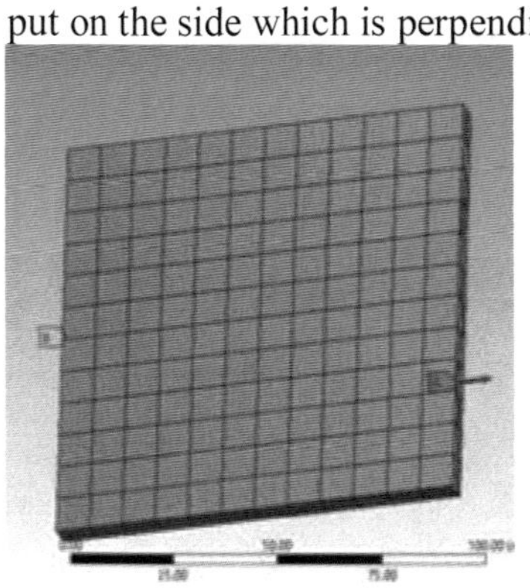

Figure 3: Tensile Analysis

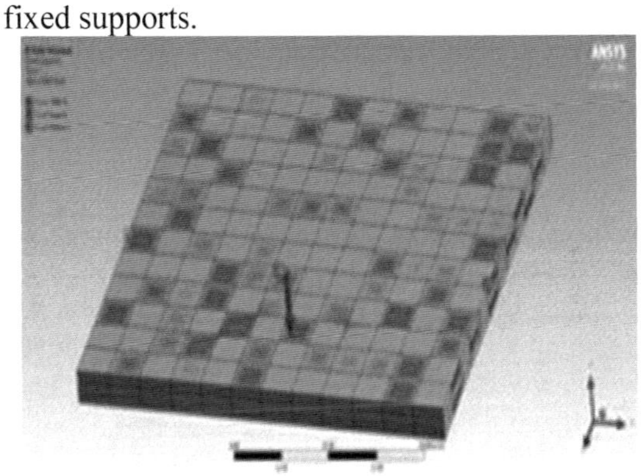

Figure 4: Bending analysis

4.2 Analysis under tensile load

Tensile analysis comprises of a force (pull type) which is put on a body and observe the response. By this analysis we can calculate how much the body can stretch before it breaks and the strength of material being used. UTM (Universal testing machines) are generally used for this analysis and it is nature wise destructive. For this experiment ANSYS-R1 was used for simulations.

Table 3: Results when 10kN tensile load was applied on laminates

Material/ Test	Total Deformation (mm)	Eq. Von Mises Stress (MPa)	Eq. Von Mises Strain
Solid Mg Alloy	3.72E- 02	27.411	6.09E-04
Magnesium and Carbon Fiber FML	2.89E-04	24.767	2.89E-04
Magnesium and Jute Fiber FML	4.30E-02	25.627	5.72E-04

IOP Conf. Series: Materials Science and Engineering 1228 (2022) 012020 doi:10.1088/1757-899X/1228/1/012020

4.3 Bending analysis

Bending analysis (also called Flexural analysis) measures the reaction of a body which is subjected to simple beam loading. Generally, it is performed on polymers and composites which possess flexibility.

Table 4: Results when 10kN Bending load was applied on laminates

Material/ Test	Total Deformation (mm)	Eq. Von Mises Stress (MPa)	Eq. Von Mises Strain
Solid Mg Alloy	0.32753	107.53	2.41E-03
Magnesium and Carbon Fiber FML	0.32666	94.064	2.11E-03
Magnesium and Jute Fiber FML	0.32651	90.2	2.02E-03

4.4 Performance Comparison

Results received through the above analysis is now being compared for- Magnesium (Mg) Alloy Solid, Magnesium-Carbon (Mg-C) Fiber Composite and Magnesium-Jute(Mg-jute) Composite.

4.4.1. Performance comparison in case of Tensile analysis

1. Maximum deformation can be observed by magnesium-jute which is followed by magnesium (Mg) alloy and the least deformation can be observed for magnesium-carbon fiber laminates for above taken steps varying from force of 1kN-10kN.
2. Maximum Eq. stress can be observed with magnesium-alloy laminate when taken in comparison with different laminates. Likewise, similar force is applied step-by-step to other laminates till 10kN and a recurring pattern is observed in all 10 steps. Moreover, Magnesium-alloy laminate showcased the maximum stress and the Magnesium-Carbon Fiber laminate showcased the minimum stress.
3. When a force of 1000N is applied, magnesium-alloy laminate showcased the maximum equivalent strain and Magnesium-Carbon Fiber laminate showcased the minimum strain.

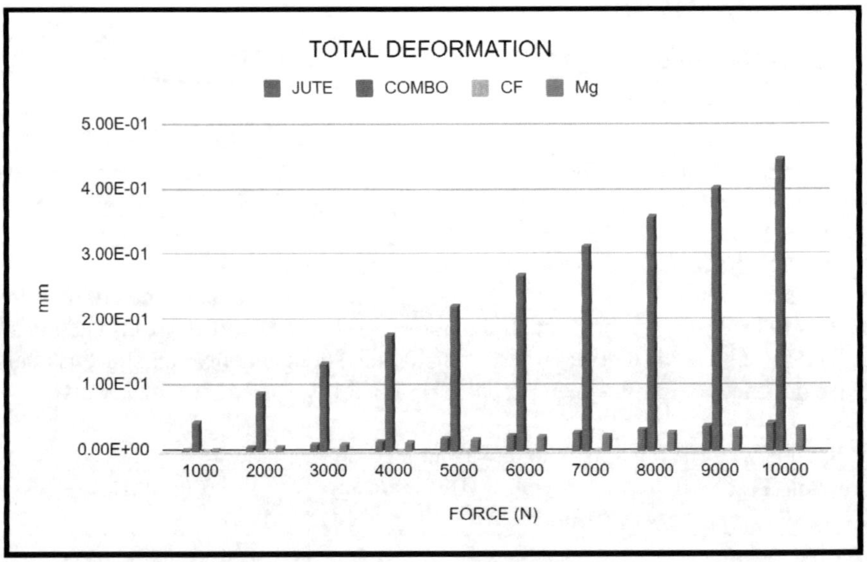

Figure 5: Total deformation under tensile loading

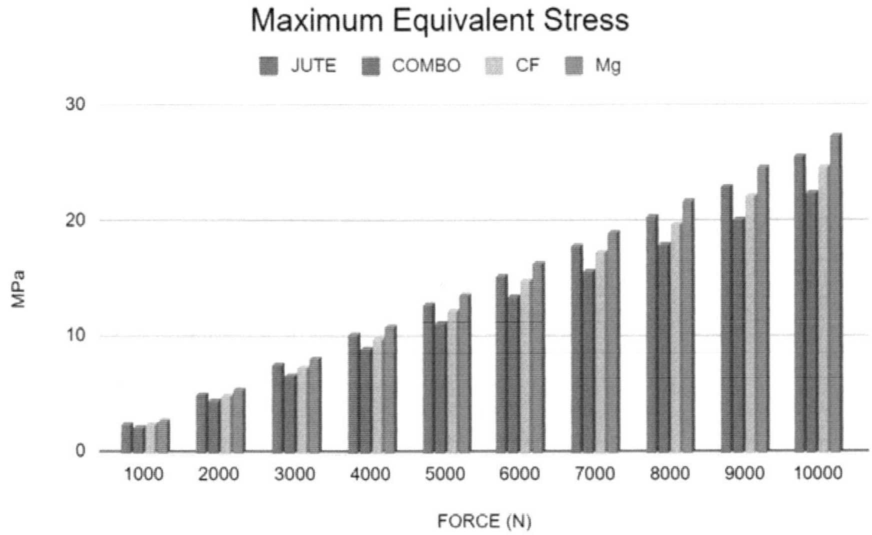

Figure 6: Maximum Stress under tensile loading

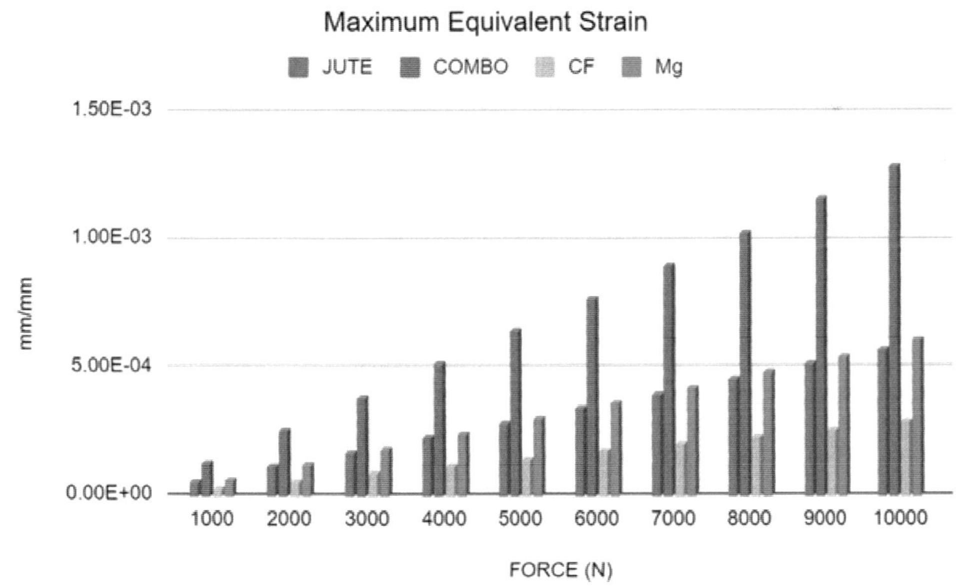

Figure 7 : Maximum Strain under tensile loading

4.4.2 *Performance comparison in case of Bending analysis*

1. Magnesium-jute provides the maximum deformation followed by magnesium-alloy and minimum deformation can be observed in magnesium-carbon fiber laminates.

2. Similar force can be applied step-by-step throughout the laminates till 10kN and similar pattern can be observed until last step. Moreover, Magnesium-alloy laminate showcased the maximum stress and the Magnesium-Jute Fiber laminate showcased the minimum stress.
3. Magnesium alloy laminate showcased the maximum strain and Magnesium-Jute laminate showcased the minimum strain.

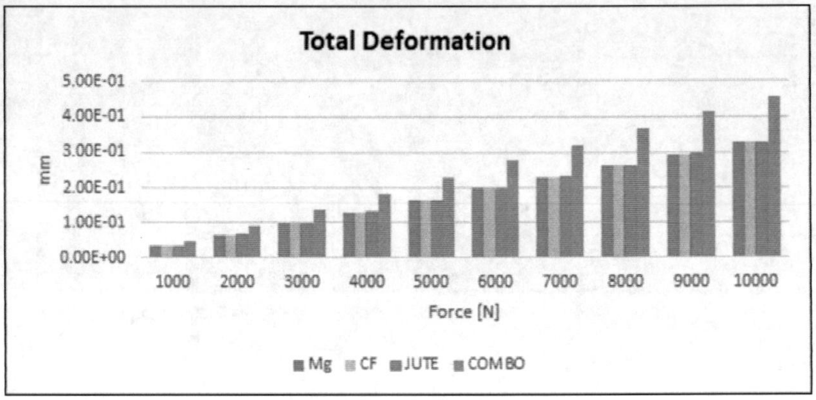

Figure 8: Total deformation under Bending load

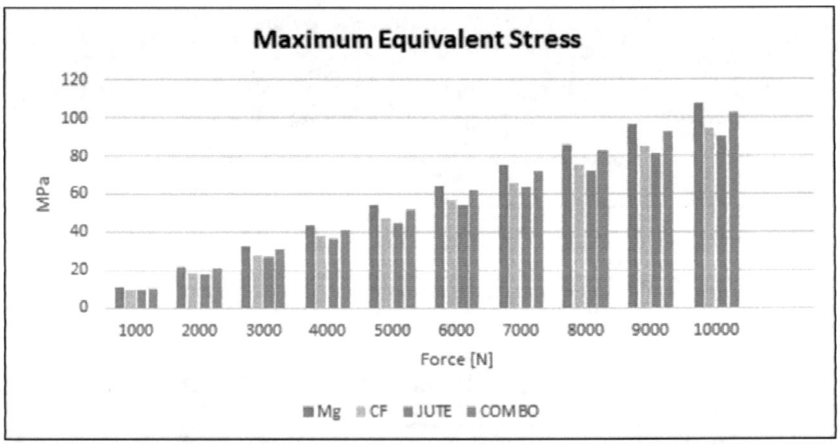

Figure 9: Maximum Stress under bending load

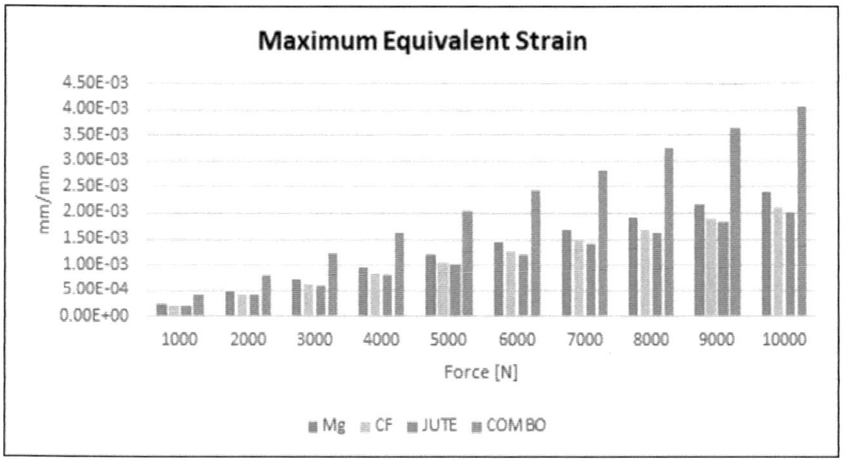

Figure 10 : Maximum Strain under bending load

5. CONCLUSIONS

By observing the analysis under bending and tensile loads, the following results can be concluded:

I. The minimum equivalent strain and minimum total deformation under tensile loading was observed in sample FML 1 which is made by adding carbon-fiber in between the two Magnesium (Mg) alloy layers. The prime reason for this can be because of the high stiffness of carbon fiber which strengthens the composite and reduces deformation.

II. On the other hand, for bending analysis, the minimum equivalent strain, stress and total deformation is observed for sample FML2 which is made by adding Jute in between two Magnesium (Mg) alloy layers when compared with different combinations at the provided conditions.

That makes us conclude that jute can be a feasible strengthening material in order to design parts that are prone to loads at various angles, such as wing ribs in airplanes.

References

[1] Lubin, George. “Handbook of composites”. Springer Science & Business Media, 2013.
[2] Pettit, Richard G. "Fiber/metal laminate." U.S. Patent No. 5,227,216. 13 Jul. 1993.
[3] A. Asundi, Alta Y.N. Choi. "Fiber metal laminates: An advanced material for future aircraft", Journal of Materials Processing Technology, 1997
[4] A. Vlot, M. Krull. "Impact Damage Resistance of Various Fiber Metal Laminates", Le Journal de Physique IV, 1997
[5] J. F. Laliberte, C. Poon, P. V. Straznicky, and A. Fah, "Applications of Fiber-Metal Laminates", Polymer Composites, April 2014.
[6] H. FelixWu, "Statistical Analysis of Tensile Strength of ARALL Laminates", Sage journal, October 1989.
[7] Mark Easton, Aiden Beer, Matthew Barnett, Chris Davies, Gordon Dunlop, Yvonne Durandet, Stuart Blacket, Tim Hilditch, and Peter Beggs, "Magnesium Alloy Applications in Automotive Structures.", JOM, vol 60, November 2008, pp.57-62.
[8] Land, J. E., “HUMS. The Benefits: Past, Present and Future,” Proceedings of the IEEE Aerospace Conference, Vol. 6, Inst. of Electrical and Electronics Engineers, New York, 2001, pp. 3083–3094.
[9] Suresh, Sanoj P., and V. Sankar. "Numerical Analysis on Tension and Flexural Behaviour of Angle-Ply Carbon Jute Aluminium Fiber Metal Laminates.” IJMTES vol. 4, June, 2017
[10] Image of Piper Wing Aircraft Wing sourced from website airpartswest.com
[11] Effective Approach for the Manufacture of Fiber Metal Laminates (F - Proceedings of the International Conference on Aerospace Science and Technology, pp. 26-28.
[12] Xun Xu, Krishnan Jayaraman, Caroline Morin, Nicolas Pecqueux, 2008. Life cycle assessment of wood-fiber-reinforced polypropylene composite, Journal of Materials Processing Technology 198, p. 168.
[13] Cortes, P., Cantwell, W.J., 2006. Structure-properties relations in titanium-based thermoplastic fiber-metal laminates, Polymer Composites, p. 264.
[14] Cortes, P., Cantwell, W.J., 2006. The prediction of tensile failure in titanium-based thermoplastic fiber-metal laminates, Composite Science Technology 66, p. 2306.
[15] Cortes, P., Cantwell, W.J., 2004. Fracture properties of a fiber-metal laminates based on magnesium alloy, Journal of Material Science 39, p. 1081
[16] T.C. Truong, M. Vettori, S. Lomov, and I. Verpoest, Compos. Part A, 36, 1207 (2005).
[17] M.Vasumathi, Vela Murali “Effect of alternate metals for use in natural fiber reinforced fiber metal laminates under bending, impact and axial loadings” International Conference on design and manufacturing, icondm 2013
[18] Polmear, I. J. "Magnesium alloys and applications." Materials science and technology 10.1, 1994
[19] Avedesian, Michael M., and Hugh Baker, eds. “ASM specialty handbook: magnesium and magnesium alloys”, ASM international, 1999.
[20] Edwardson S.P., French P., Dearden G., Watkins K.G., Cantwell W.J., laser forming of fiber metal laminates, Lasers in Eng., 2005.
[21] Vlot A., Gunnik J.W., Fiber Metal Laminates an Introduction, Kluwer Academic Publishers, 20

IOP Conf. Series: Materials Science and Engineering 1228 (2022) 012021 doi:10.1088/1757-899X/1228/1/012021

Performance Analysis of Fully Depleted SOI (FD-SOI) MOSFET Incorporating Dielectric Engineering

[1]Prashant Kumar, [2]Munish Vashishath, [3]Neeraj Gupta, [4]Rashmi Gupta

Assistant Professor, J.C. Bose University of Science & Technology, Faridabad, India
Professor, J.C. Bose University of Science & Technology, Faridabad, India
Assistant professor, ECE Department , Amity University Haryana, India
Assistant professor, CSE Department , Amity University Haryana, India

Neerajsingla007@gmail.com

Abstract. This paper presents an efficacy of an FD SOI MOSFET structure in suppressing short channel effects is investigated through 2-Dimenional simulation on ATLAS Silvaco. The electrical comparison between Bulk MOSFET, FD-SOI MOSFET and FD-SOI MOSFET using high-k gate dielectric at 45nm technology has been carried out. The structures have been designed and study of threshold voltage, sub threshold slope, drain current (Ion), leakage current (Ioff), Ion/Ioff, transconductance has been done. The features offered by the new device structure are found to be better than the conventional bulk MOSFET. Aluminum Oxide (Al_2O_3) has been compared to SiO_2 as dielectric material. As the dielectric constant of the materials increased, there is a reduction in the leakage current. It provides higher transconductance and better subthreshold slope, thus resulting in improvement of the device. A comparative analysis has been carried out among bulk MOSFET and SOI MOSFET using aluminum oxide, silicon oxide and silicon nitride as gate dielectric materials. Aluminum Oxide also provides higher drain current as compared to other dielectric materials presented in the paper. Thus, Aluminum Oxide is a good candidate as a dielectric for the future devices as compared to traditional oxides. The proposed device can be useful for low power applications.

Keywords: MOSFET, Sub threshold slope, trans conductance, High-k, leakage current.

1. Introduction

The Increased Performance, reduction in Power and minimized Area has a great demand today. 45nm and 32nm nodes are used to design chips today. Silicon on Insulator (SOI) is considered as a substitute over conventional MOSFET as it is not providing improvement in the performance. Silicon-On-Insulator thus provides high performance and consumes less power. The year 1994 saw the development of a "fully" functional power PC using CMOS SOI technology as an outcome of those research efforts [1]. After this development, the study of current-voltage characteristics of the SOI which is modeled analytically has been done.

MOSFET characteristics degrade due to continuous scaling, and some effects known as short channel effects appear, worsening the correct operation of CMOS devices [2]. The importance of these effects and the complexity of the methods used to counter them led to the development of alternative transistor structures and new materials. These new architectures using SOI MOSFET can be the possible alternative

Content from this work may be used under the terms of the Creative Commons Attribution 3.0 licence. Any further distribution of this work must maintain attribution to the author(s) and the title of the work, journal citation and DOI.
Published under licence by IOP Publishing Ltd

to the bulk traditional MOSFETs. The SOI MOSFETs have the intrinsic advantage of diminished area of source and drain regions and absence of depletion region beneath the source and drain which reduces leakage current and junction capacitance making these devices an attractive option for reduced power reduced-voltage applications [3-5].

SOI has several advantages over other technological solution like better sub threshold slope, no latch up, lower threshold voltage [6-7]. SOI with high-k gate dielectric increase capacitance allowing high drain current and lower leakage current.

Even though SOI shows superior performance over conventional bulk MOSFET, there are some unavoidable issues like drive current overshoot, heating effect which rises with SOI devices in nano-regime. To overcome these issues different SOI structures have been proposed during recent time. The International Technology Roadmap for Semiconductor (ITRS) report 2009 predicts that development and research on SOI MOSFET structure will be continued till 2024 to achieve devices of dimension less than 16nm with modified SOI structure. Some of the different improved SOI structures are thin body FD SOI, FD SOI with high-k gate dielectric [8-9].

The leakage current of conventional bulk MOSFET exponentially increases with temperature and causes high power dissipation. Leakage current and sub threshold slope are increasing day by day because of the rise in the scaling. These problems of bulk MOSFET can be controlled by the adoption of advanced structures. Silicon-on-Insulator (SOI) technology can be used instead of bulk MOSFET. SOI has been invented with a buried oxide layer (BOX) in the silicon substrate. Due to the isolation created by this BOX layer, various short channel effects have been reduced [10].

It is observed that as the high dielectric material is used instead of SOI with SiO_2, the performance is increased, reduction in OFF state current. Also, the DIBL and the sub threshold slope is reduced. The high-k dielectric material provides higher Transconductance than conventional SiO_2. It provides higher voltage gain; hence SOI MOSFET device with high-k dielectric can be used for amplification purpose. Thus, it is better option for upcoming SOI MOSFET devices [11-13].

2. Device Structure

To eliminate the problems associated with SCEs in deep sub-micron SOI MOSFETs, several improved device structures have been proposed. One such device structure is fully depleted (FD)-SOI MOSFET with high-k dielectrics. The dielectric material should be such that it is thermally stable thus increases on state driving current in comparison to FD-SOI MOSFET, reduces the leakage current at off state. A schematic view of a FD-SOI MOSFET with Silicon nitride (Si_3N_4) at 45nm Technology using ATLAS is shown in Figure 1. In this structure, instead of gate dielectric silicon dioxide (SiO_2) k=3.9, high-k gate dielectric Silicon nitride (Si_3N_4) k=7 is used.

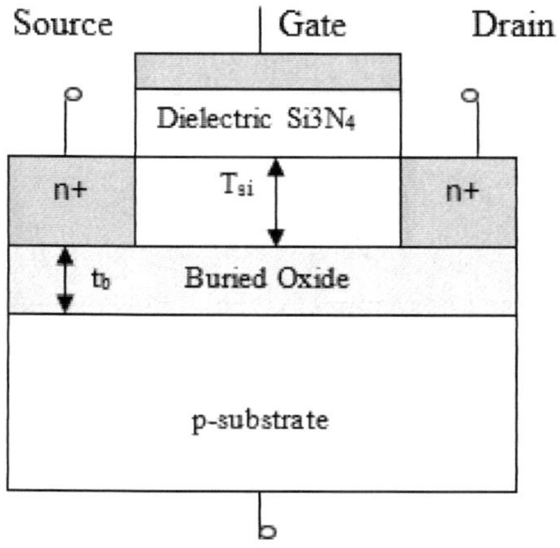

Figure 1: Schematic view of FD-SOI MOSFET with Si_3N_4 as gate dielectric

One of the interested candidates for such novel insulator is Aluminum Oxide (Al_2O_3) k=9. It has relatively high thermal conductivity compared to silicon dioxide (30W/m-K versus 1.4 W/m-K), high compressive strength. It has excellent stability and excellent dielectric properties. Thus, the use of Al_2O_3 as gate dielectric may mitigate the short channel effects like subthreshold slope and DIBL and hence reduces leakage current which further enhances the device performance by reducing power dissipation.

3.Results & Discussion

Numerical simulation studies using SILVACO TCAD TOOL are used to find out the unique features of a FD-SOI MOSFET with different dielectrics and compare with that of a conventional Bulk MOSFET [14].

3.1 Bulk MOSFET

Figure 2(a) illustrates the simulated device structure of Bulk MOSFET. Figure 2(b) illustrates the doping profile of simulated Bulk MOSFET.

Figure 2(a) indicates the simulated structure of Bulk MOSFET where the oxide layer present at the top insulates the gate from the channel is 2nm. The thickness of silicon is 4nm.

Figure 2(b) indicates the doping profile of simulated structure where p-type doping of 6e17cm^{-3} is present in the entire silicon region, and the areas on the both the side of the channel has a n-type doping of 6e20cm^{-3}.

IOP Conf. Series: Materials Science and Engineering 1228 (2022) 012021 doi:10.1088/1757-899X/1228/1/012021

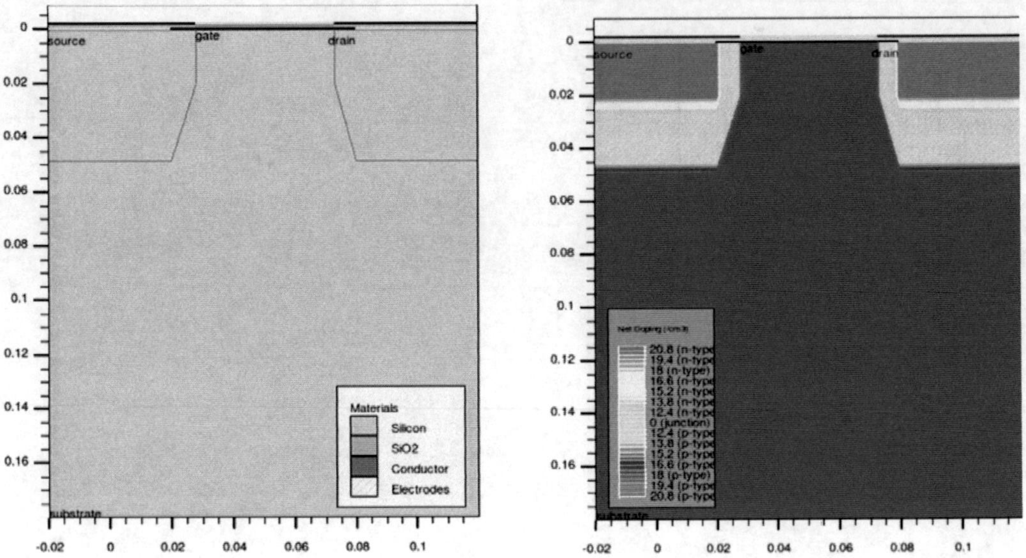

Figure2 (a) Simulated structure of Bulk MOSFET (b) Doping profile of Simulated Bulk MOSFET

3.2 FD-SOI MOSFET

Table 1 depicts the parameters consider for simulation of FD-SOI MOSFET. FD-SOI MOSFET is simulated using SILVACO TCAD Tool. Figure 3(a) illustrates the simulated device structure of FD-SOI MOSFET. Figure 3(b) illustrates the doping profile of simulated FD-SOI MOSFET.

Table 1 Device parameters used for simulation of FD-SOI MOSFET using Si_3N_4 as gate dielectric

Device parameters	**SOI n-MOSFET**
Gate oxide thickness, t_{ox}	2nm
Channel Length, L	45nm
Body doping, N_A	6 x $10^{17}cm^{-3}$
Source/Drain doping	6 x $10^{20}cm^{-3}$
Thickness of thin-film, T_{Si}	4nm
Thickness of Buried Oxide, t_b	6nm
Gate Oxide Material	Silicon nitride

IOP Conf. Series: Materials Science and Engineering 1228 (2022) 012021 doi:10.1088/1757-899X/1228/1/012021

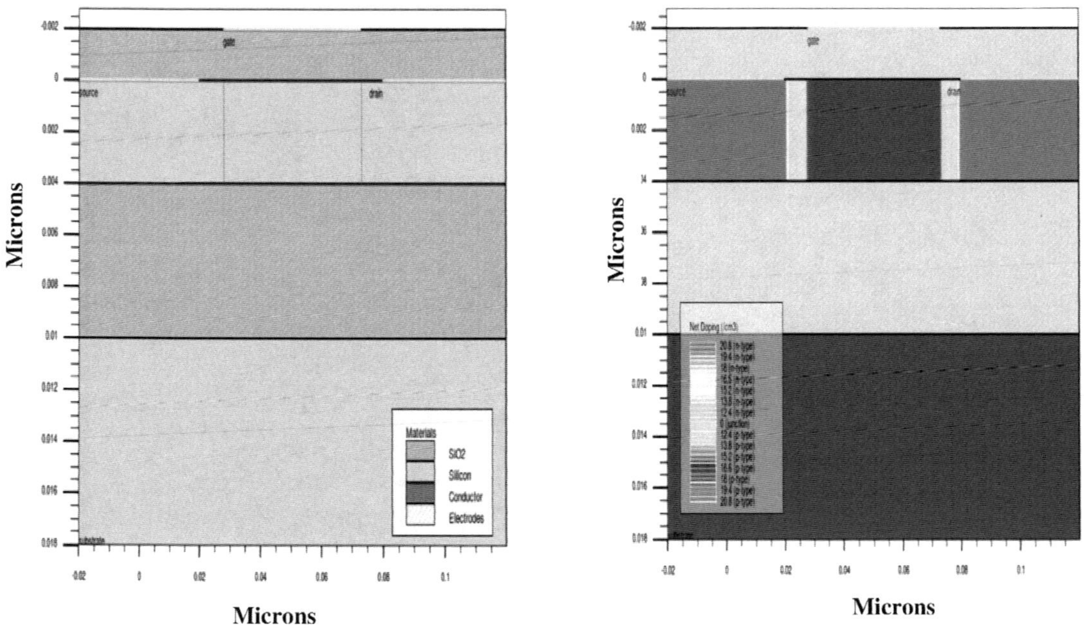

Figure 3 (a) Simulated structure of FD-SOI MOSFET (b) Doping profile of simulated FD-SOI MOSFET.

3.3 FD-SOI MOSFET using Silicon nitride (Si_3N_4)

FD-SOI MOSFET using Si_3N_4 as gate dielectric is fabricated and simulated using SILVACO TCAD Tool. Figure 4(a) depicts the simulated structure of FD-SOI MOSFET using Silicon nitride (Si_3N_4). Figure 4(b) illustrates the doping profile of simulated FD-SOI MOSFET using Silicon nitride (Si_3N_4).

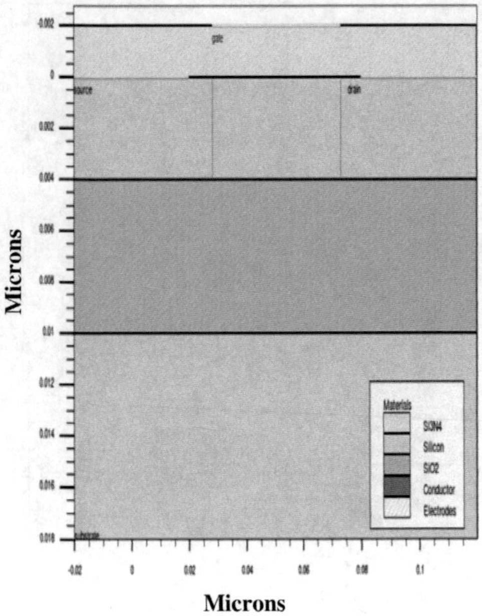

Figure 4 (a) Simulated structure of FD-SOI MOSFET using Silicon nitride (Si_3N_4)

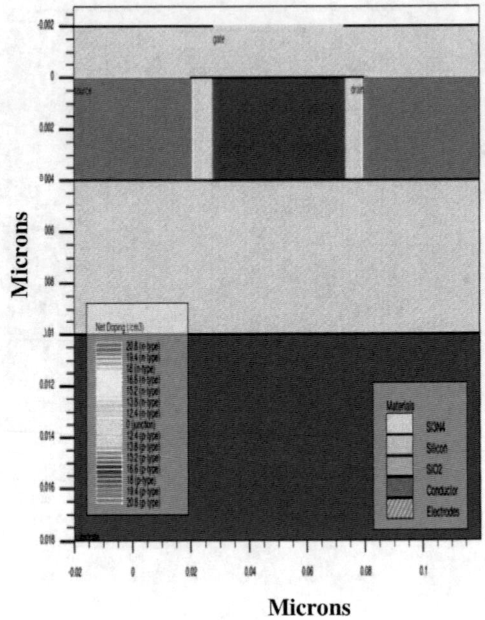

(b) Doping profile of simulated FD SOI MOSFET using Silicon nitride (Si_3N_4)

3.4 FD-SOI MOSFET using Aluminum oxide

FD-SOI MOSFET with Al_2O_3 as gate dielectric is simulated using SILVACO TCAD Tool.

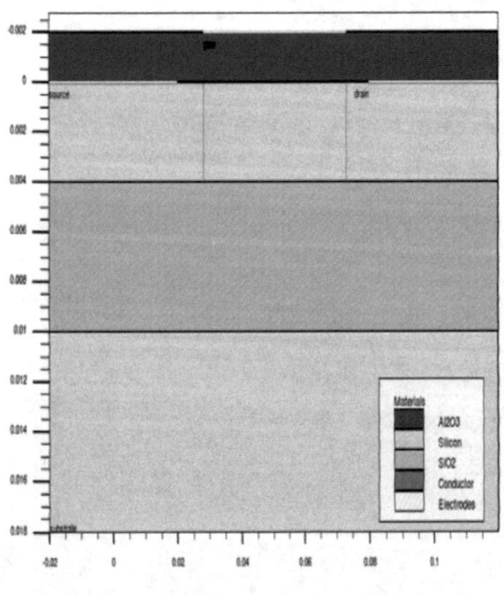

Figure 5(a) Simulated structure of FD-SOI MOSFET Aluminum oxide (Al_2O_3)

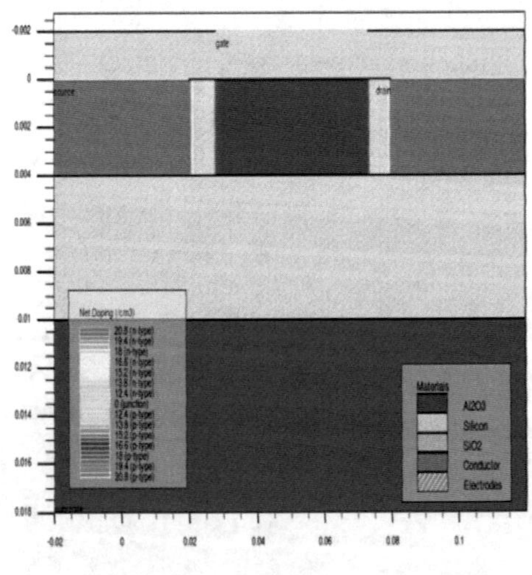

(b) Doping profile of simulated FD-SOI MOSFET using using Aluminum oxide (Al_2O_3)

IOP Conf. Series: Materials Science and Engineering 1228 (2022) 012021 doi:10.1088/1757-899X/1228/1/012021

Figure 5(a) illustrates the simulated device structure of FD-SOI MOSFET using Aluminum oxide (Al_2O_3). Figure 5(b) illustrates the doping profile of simulated FD-SOI MOSFET using Aluminum oxide (Al_2O_3).

4. Comparative Analysis

Table 2 highlights the comparison of output parameters of various simulated device.

Table 2 output parameters of various simulated devices

Device (45nm technology)	**Maximum Drain Current (A)**	**Transconductance G_m(A/V)**
Bulk NMOSFET	5.60×10^{-4}	0.0013
FD-SOI MOSFET	5.74×10^{-4}	0.0015
FD-SOI MOSFET using Silicon nitride (Si_3N_4)	5.97×10^{-4}	0.0018
FD-SOI MOSFET using Aluminum oxide (Al_2O_3)	7.47×10^{-4}	0.0022

The transconductance has a key role as it determines the speed of a circuit at which it switches and gain of amplifiers. The device with more value of transconductance will provide more speed.

Figure 6 highlights the disparity of transconductance vs gate bias for various simulated devices.

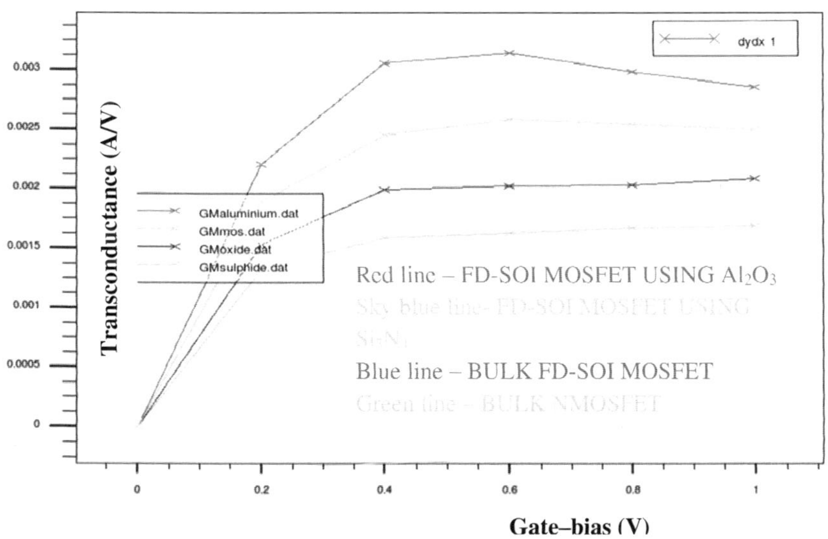

Figure 6: Variation of transconductance with gate bias

IOP Conf. Series: Materials Science and Engineering 1228 (2022) 012021 doi:10.1088/1757-899X/1228/1/012021

Figure 6 depicts that FD-SOI MOSFET using Aluminum Oxide (Al_2O_3) is a highest transconductance device and provide high speed for operation. This device also possesses a large voltage gain. It is concluded that FD-SOI MOSFET using Aluminum Oxide (Al_2O_3) has the highest value of drain current as there is no heating effect and short channel effects, which are responsible for degradation of current. Table 3 reveals the short channel key parameters for different simulated devices. Threshold voltage is less for FD-SOI MOSFET using Al_2O_3 as compared to other devices.

Table 3 Short channel parameters of different simulated devices

Device (45nm technology)	**Subthreshold Leakage Ioff (A)**	**Subthrehold Slope St (V/ decade)**	**I_{on}/I_{off}**	**Threshold Voltage (V)**
Bulk NMOSFET	$2.28x10^{-7}$	0.102	$2.5x10^3$	0.172
FD-SOI MOSFET	$1.81x10^{-7}$	0.095	$3.17x10^3$	0.15
FD-SOI MOSFET using Silicon nitride (Si_3N_4)	$6.63x10^{-8}$	0.081	$1.08x10^4$	0.14
FD-SOI MOSFET using Aluminum oxide (Al_2O_3)	$5.51x10^{-8}$	0.078	$1.36x10^4$	0.13

St specifies how efficiently a device can restrict the flow of drain current when $V_{gs} < V_{th}$. A device having a steeper subthreshold slope has a faster conversion from off to on current states.

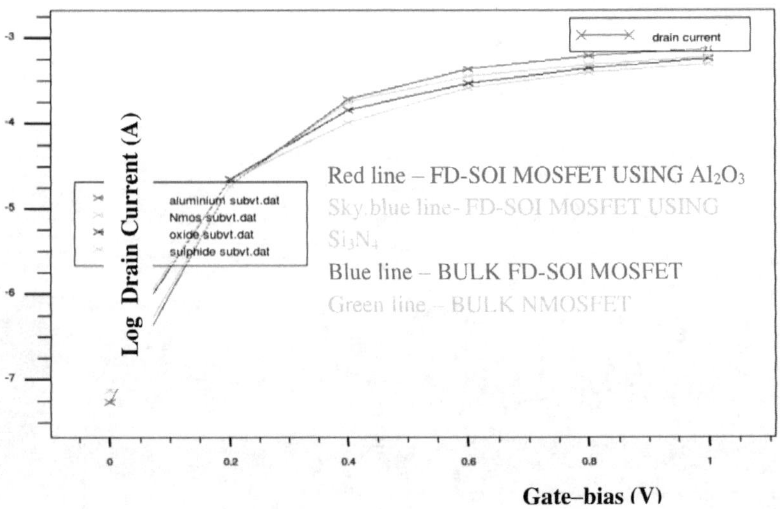

Figure 7 Change in Log drain current with gate bias

Figure 7 shows variation of log drain current with gate voltage. From the figure it is concluded that FD-SOI MOSFET using Aluminum Oxide (Al_2O_3) has the steeper subthreshold slope and therefore exhibits fast transition between off and on state.

Subthreshold leakage current is the current when the transistor is partially conducting in sub threshold regime. From the Table 3 it is concluded that FD-SOI MOSFET using Aluminum Oxide (Al_2O_3) possess lowest value of subthreshold leakage. From the Table 3 it is concluded that as we move from Bulk MOSFET to FD-SOI MOSFET to FD-SOI MOSFET using Aluminum Oxide (Al_2O_3), on state current increases and off state current decreases. So, I_{on}/I_{off} increases, and it is maximum for FD-SOI MOSFET using Aluminum Oxide (Al_2O_3).

5. Conclusion

The developed 2-D simulation analyses the effectiveness of FD-SOI MOSFET using Aluminum Oxide (Al_2O_3) to suppress the hot carrier effect. The suppression of hot carrier effect is attributed to high thermal conductivity of Aluminum Oxide. Analysis of the impact of different device parameters on the threshold voltage, maximum drive current, subthreshold leakage and subthreshold slope has been carried out. From simulated results it can be concluded that there is improvement in performance of FD-SOI MOSFET using Aluminum Oxide (Al_2O_3) as compared to SiO_2 dielectric material. As the dielectric constant of the materials increased, there is a reduction in OFF state leakage current and improvement in Ion/Ioff ratio. It provides higher transconductance and better subthreshold slope, thus resulting in improvement of the device. So, results indicate that Aluminum Oxide is indeed a suitable alternative to silicon dioxide as gate dielectric in SOI. This makes the proposed device is suitable for ultra-low power applications.

References

[1] P. Kumar, D. Joy and B.K. Jeblin, "Nanoscale tri-gate MOSFET for Ultra low power applications using high-k dielectrics," Nanoelectronics, 2013, [INEC 2013, 5th IEEE International Conference on, 2-4, pp.12-19, Jan. 2013

[2] K. Suzuki, Y. Tosaka, and T. Sugii, "Analytical threshold voltage model for short channel double-gate SOI MOSFETs", IEEE Trans. Electron Devices, vol. 43, no. 7, pp. 1166– 1168, Jul. 1996.

[3] Y. Taur, "Analytic solutions of charge and capacitance in symmetric and asymmetric double-gate MOSFETs," IEEE Trans. Electron Devices, vol. 48, no. 12, pp. 2861–2869, Dec. 2001.

[4] Neeraj Gupta, Janak B. Patel and A.K.Raghav, "A Study of Conventional and Junctionless MOSFET Using TCAD Simulations", IEEE Conference on Advanced Computing & Communication Technologies, 2015, Haryana, pp. 53-56.

[5] Sonam Rewari, Subhasis Haldar, Vandana Nath, S.S. Deswal, and R.S. Gupta, "Numerical Modeling of Subthreshold Region of Junctionless Double Surrounding Gate MOSFET (JDSG) ," Superlattices and Microstructures Journal, vol. 90, pp 8-19, Feb 2016.

[6] Neeraj Gupta, Janak B. Patel and A.K.Raghav, "Performance and a new 2-D analytical modeling of a Dual-Halo Dual-Dielectric Triple-Material Surrounding-Gate-All-Around (DH-DD-TM-SGAA) MOSFET" ,Journal of Engineering Science and Technology, Vol. 13, No. 11, Nov., 2018

[7] Neeraj Gupta, Janak B. Patel and A.K.Raghav, "An accurate 2D Analytical Model for Transconductance-to-Drain Current ratio (g_m/I_d) for a Dual-Halo Dual-Dielectric Triple-Material Cylindrical-Gate–All-Around (DH-DD-TM-CGAA) MOSFETs", International Journal of Engineering, Vol. 31 No. 07,July 2018

[8] Neeraj Gupta, Prashant Kumar, Nitin Sachdeva, Tarun Sachdeva, Munish Vashistha, "Performance investigation of Dual-Halo Dual-Dielectric Triple Material Surrounding Gate MOSFET with High-K dielectrics for Low Power Applications", Journal of Semiconductor Technology and Science, Vol 20, No. 3, pp.1-8,June 2020.

[9] Neeraj Gupta, Janak B. Patel and A.K.Raghav, "A Comparative study of conventional and Double Gate SOI MOSFET using TCAD Simulations" proceedings of 3rd International conference on computing sciences, April 2016

[10] N. B. Balamurugan, K. Sankaranarayanan, P. Amutha, and M. Fathima John, "An Analytical Modeling of Threshold and Subthreshold Swing on Dual Material Surrounding Gate Nanoscale MOSFETs for high speed Wireless Communication", Journal of Semiconductor Technology and Science, vol. 8, no. 3, pp. 221-226, 2008.

[11] Jay hind kumar verma, Subhash Haldar, R. S. Gupta and Mridula Gupta," Modeling and simulation of subthreshold behavior of cylindrical surrounding double gate MOSFET for enhanced electrostatic integrity", Superlattices and microstructures, vol. 88, pp. 354-364, 2015.

[12] Prashant Kumar, Munish Vashisht, Neeraj Gupta, Rashmi Gupta, "Subthreshold Current Modeling of Stacked Dielectric Triple Material Cylindrical Gate All Around (SD-TM-CGAA) Junctionless MOSFET for Low Power Applications", Silicon 13(9):1-9, 2021.

[13] Neeraj Gupta, Prashant Kumar, "Elicitation of Scattering Parameters of Dual-halo Dual-dielectric Triple-material Surrounding Gate (DH-DD-TM-SG) MOSFET for Microwave Frequency Applications", Advances in Electrical and Electronics Engineering, Vol. 19, No. 1, pp. 66-73, April 2021.

[14] ATLAS: 3D Device Simulator, SILVACO International, 2015.

IOP Conf. Series: Materials Science and Engineering 1228 (2022) 012022 doi:10.1088/1757-899X/1228/1/012022

Artificial electric field algorithm based superconducting magnetic energy storage system for damping inter area oscillations

Aaqib Manzoor, Mairaj ud Din Mufti and Mohammad Younis Nabi

Department of Electrical Engineering National Institute of Technology Srinagar, India

Abstract: This paper presents an artificial electric field algorithm (AEFA) based optimal control of superconducting magnetic energy storage (SMES) system for damping inter area oscillations. The artificial electric field algorithm is based upon Coulomb's law of electrostatics and Newton's law of motion. The proposed technique is used to optimize the SMES controller parameters for effective damping of inter area oscillations. Because of the potential for system damage and the subsequent constraints on power transmission along select lines, low frequency inter-area oscillations have been identified as a serious problem in utility systems. With the advancement in power electronics technologies, a new option for damping of power system oscillations has been created by integrating energy storage devices to it. The proposed scheme is tested on a two area interconnected power system which is suitably modelled and it's modal analysis is done. The participation of each area towards the inter area mode is analyzed by using participation matrix and the suitable placement of SMES units is decided accordingly. MATLAB Simulations are carried out to prove the efficacy of proposed scheme. The simulation results show that proposed SMES controller along with the incorporation of SMES units to each generator bus effectively damps out the inter area oscillations.

1. Introduction

In practical operation of power systems the electromechanical or low frequency oscillations arise due to chain structure of transmission systems, weak connection between different areas, imbalance in power, shortage of reserve capacity, etc. The LFOs are divided into two forms: local area oscillation mode and inter-area oscillation mode. In local area mode, the generators of a single area or plant swing against each other and in inter-area mode the generators of different areas swing against each other. The frequency of inter-area oscillations are in the range of 0.2Hz-2Hz. These oscillations may persist for longer periods and may lead to full or partial interruption of power if not controlled or damped properly as mentioned by *Cui et al.* in [2].

The behavior of interactions in the inter-area modes makes it challenging to design the damping control. The design technique therefore, should aim at enhancing the damping performance of one mode while having least effect on other modes. Further the controller should assure stable operation over full range of operating conditions. The ability of ESSs to enhance the transient behavior of power systems and the increase of integration of renewable energies in recent years has led to massive expenses in prototyping, research and integration of storage devices and evolution of huge variety of energy storage technologies. The most promising technologies that come into analysis are Battery Energy Storage system (BESS), Superconducting Magnetic Energy Storage Systems (SMES), Compressed Air Energy Storage System (CAESS), Flywheel Energy Storage System (FESS) and Electrochemical Capacitor

Content from this work may be used under the terms of the Creative Commons Attribution 3.0 licence. Any further distribution of this work must maintain attribution to the author(s) and the title of the work, journal citation and DOI.
Published under licence by IOP Publishing Ltd

Storage System (ECESS). A generalized mode of storage devices of energy for voltage and angle stability analysis is presented by *Zhu et al.* in [3] which is useful to synthesize and differentiate different control techniques. An analytic eigenvalue solution based algorithm for a multi-machine system with BESSs is presented by *Ortega et al.* in [4] to improve the damping ratio for a target mode oscillation.

Advantages which SMES possess over other technologies are extremely fast response time, long life of storage system, high efficiency and environmental friendly behavior of system as mentioned by *Liu et al.* in[5]. The option of real power modulation via the SMES is utilized effectively to improve stability by damping out inter area oscillations.

This paper aims at designing an optimal controller for SMES in a two-area four machine power system which should be robust in its performance for changing operating conditions and parameters of the system. In this paper an artificial electric field algorithm (AEFA) based optimal control of superconducting magnetic energy storage (SMES) system for damping inter area oscillations is presented. The proposed algorithm belongs to a metaheuristic algorithm which is more popular in recent years. The metaheuristic algorithms are problem agnostic, simple to implement, and have a high computational capacity. Metaheuristic algorithms prevent local solution stagnation by thoroughly searching the whole search space as mentioned by *Anita et al.* in [6]. The participation of each area towards the inter area mode is analyzed by using participation matrix and the suitable placement of SMES units is decided accordingly. Simulations show that a controlled sequence of charging and discharging of the SMES coil successfully damps out these oscillations.

2. Modelling of two area system

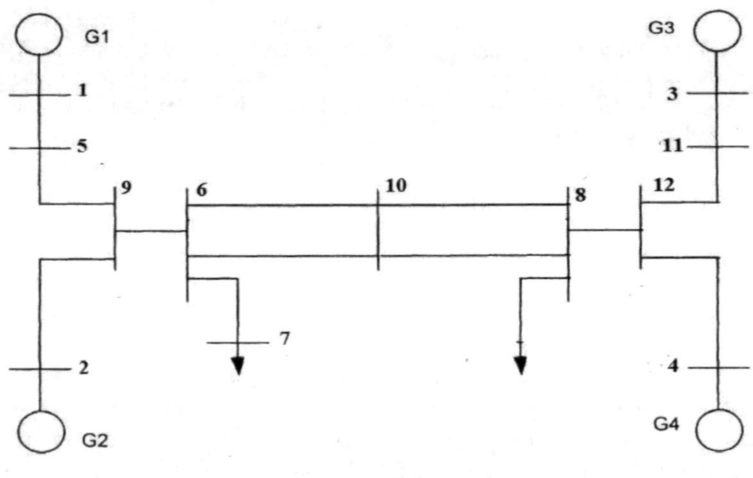

Figure 1. Single line diagram of two area system

The studied system is a two area four machine system taken from book by *Rogers* [7]. The data of given system is taken from power system toolbox. In classical modelling of generators each generator model has two state variables or dynamic variables:

Change in rotor angle ($\Delta\delta$) and Generator speed deviation from synchronous speed ($\Delta\omega$).

So the states of two area four machine system are:

$$x' = [\ \Delta\delta_1\ \Delta\omega_1\ \Delta\delta_2\ \Delta\omega_2\ \Delta\delta_3\ \Delta\omega_3\ \Delta\delta_4\ \Delta\omega_4]$$

The swing equation that represents the dynamics of rotor is

$$\frac{H}{\pi f} \times \frac{d^2\delta}{dt^2} = P_m - P_e \tag{1}$$

Steady state representation

$$x_1 = \delta$$
$$x_2 = \dot{\delta}$$
$$\dot{x_1} = x_2$$
$$\frac{H}{\pi f} \times \dot{x_2} = P_m - P_e$$
$$\dot{x_2} = \frac{\pi f}{H}(P_m - P_e)$$

Where $x_1 = \delta$ in electrical radians
$x_2 = \omega_r - \omega_s$ is relative speed between rotor and synchronous speed in rad/sec

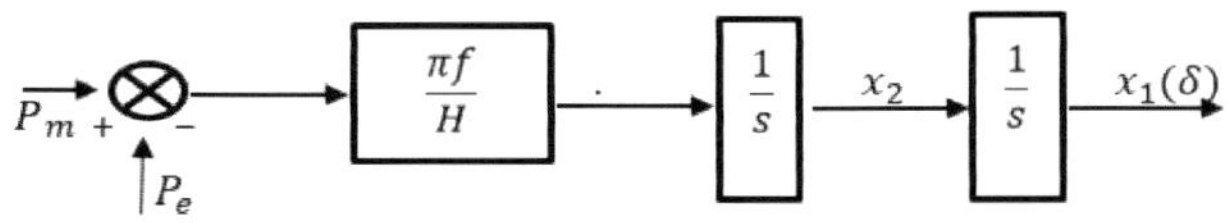

Figure 2. Block diagram of swing equation

First the Y matrix of the system is calculated and by eliminating all the nodes excluding generator nodes we calculated the Ybus matrix of the reduced network. The reduction is done by taking in to consideration that all the nodes have zero input injecting currents except the generator internal nodes. The nodal equation of power system with p number of generators is given by

$$\begin{bmatrix} I_p \\ 0 \end{bmatrix} = \begin{bmatrix} Y_{pp} & Y_{pq} \\ Y_{qp} & Y_{qq} \end{bmatrix} \begin{bmatrix} V_p \\ V_q \end{bmatrix}$$

p = number of generator nodes
q = number of remaining nodes
By eliminating V_q , we find

$$I_p = [Y_{pp} - Y_{pq} {Y_{qq}}^{-1} Y_{qp}]\, V_p$$

Desired reduced matrix can be written as

$$Y_R = [Y_{pp} - Y_{pq} {Y_{qq}}^{-1} Y_{qp}]$$

The dimensions of Y_R is pxp, where p is the quantity of generators. This network reduction technique called Kron reduction, is used only when loads are assumed as constant impedances. The reduced Y matrix of our system is given in appendix A.

The power into the system at node m, is the electrical output power of machine m and is given by

$$P_{e\,m} = {E_m}^2 G_{mm} + \sum_{\substack{n=1 \\ n \neq m}}^{p} E_m E_n Y_{mn} \cos(\theta_{mn} - \delta_m + \delta_n) \qquad m=1,2,\ldots\ldots..p \quad (2)$$

Where $\boldsymbol{Y}_{mn} = Y_{mn} < \theta_{mn} = G_{mn} + jB_{mn}$ is the negative of transfer admittance between m and n node and $\boldsymbol{Y}_{mm} = Y_{mm} < \theta_{mm} = G_{mm} + jB_{mm}$ is driving point admittance at node m.

The swing equation can be now written as

$$\frac{H}{\pi f} \times \frac{d^2\delta}{dt^2} = P_{mm} - [{E_m}^2 G_{mm} + \sum_{\substack{n=1 \\ n \neq m}}^{p} E_m E_n Y_{mn} \cos(\theta_{mn} - \delta_m + \delta_n)] \qquad (3)$$

Prior to disturbance at t=0

$$P_{mm0} = P_{em0} = E_m{}^2 G_{mm0} + \sum_{\substack{n=1 \\ n\neq m}}^{p} E_m E_n Y_{mn0} \cos(\theta_{mn0} - \delta_{m0} + \delta_{n0})$$

3. Modelling of SMES

SMES devices store energy in the form of magnetic field created by the flow of DC current in coil operated at cryogenic temperature so that resistive losses are negligible. The SMES units are charged from the AC supply by using a converter, which converts the Ac into Dc. The energy stored can be utilized by converting it back into ac using converters. On account of this the SMES units can be used to damp the oscillations of the system by absorbing or supplying the active power from or to the system. The main parts of SMES devices are ferromagnetic core, coolant, superconducting coil for storing and driving circuit for inserting or withdrawal of power from SMES.

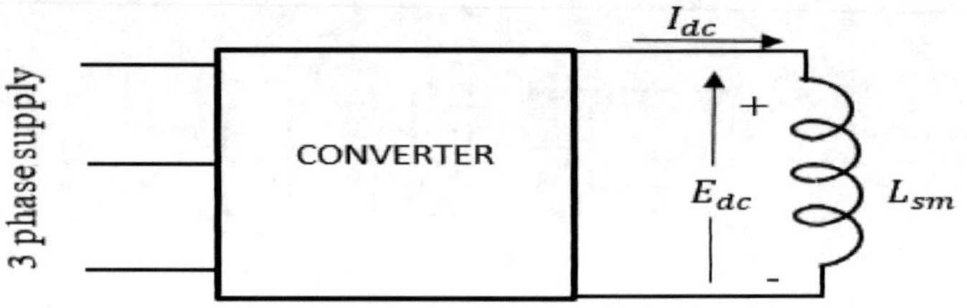

Figure 3. SMES system

The active power supplied by SMES unit is

$$P_{sm} = -E_{dc} I_{dc} \tag{4}$$

Where E_{dc} and I_{dc} are the coil voltage (volts) and coil current (amperes) respectively.
In floating operating state

$$E_{dc} = 0$$

And $I_{dc} = I_{dc0}$ is the nominal current.
Linearizing equation (4) about the operating state

$$\Delta P_{sm} = \left.\frac{\partial P_{sm}}{\partial E_{dc}}\right|_0 \Delta E_{dc} + \left.\frac{\partial P_{sm}}{\partial I_{dc}}\right|_0 \Delta I_{dc}$$

$$\Delta P_{sm} = -I_{dc0} \Delta E_{dc}$$

$$\Delta E_{dc} = -\frac{\Delta P_{sm}}{I_{dc0}} \quad \text{where } \Delta P_{sm} \text{ is in pu.} \tag{5}$$

Also change in SMES current is given by

$$\Delta I_{dc}(s) = \frac{\Delta E_{dc}}{s L_{sm}} \tag{6}$$

Thus considering the converter dynamics and SMES current derivation feedback we have the following block diagram.

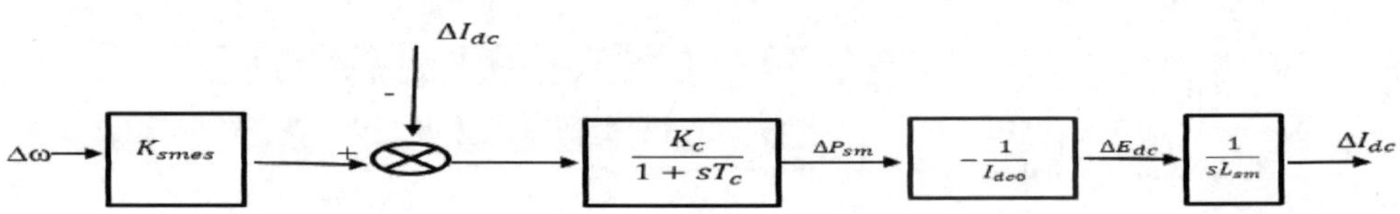

Figure 4. Block diagram of SMES system

4. Objective function

The main purpose of this paper is to design an artificial electric field algorithm (AEFA) based controller for considered test system. Adaptation of a proper objective function as a performance index is used to construct any controller. In this paper, integral of square error (ISE) is used to assess the fitness of the provided AEFA algorithm. The speed deviation is taken as input to controller, that error or deviation is minimized using proposed AEFA over the ISE subjected to constraints. The proportional gain of best fitness value for four proportional controllers are taken as optimum parameters.

$$ISE = \int_0^{T_{sim}} [(\Delta\omega_1)^2 + (\Delta\omega_2)^2 + (\Delta\omega_3)^2 + (\Delta\omega_4)^2]\, dt\ , \ 0 \leq K_{SMES} \geq 10 \qquad (7)$$

5. Artificial electric field algorithm

AEFA can be considered as an isolated system of charges which follows Coulomb's law of electrostatic force and the law of motion. Agents are treated as charge particles in the proposed algorithm, and their strength is decided by their charges. An electrostatic force attracts or repels these particles, and objects advance in the search space as a result of this force. As a result, charges are used as a direct method of communication via electrostatic force, and the position of the charge corresponds to a problem solution. The charges are based on the proposed solution's and population's fitness. We only consider the electrostatic attraction force in the suggested algorithm so that a charge particle with the highest charge ("best") attracts all other particles with lesser charges and moves slowly across the search space[6]

5.1. Flow chart of AEFA

The flow chart of the procedure of artificial electric field algorithm is given in figure 5.

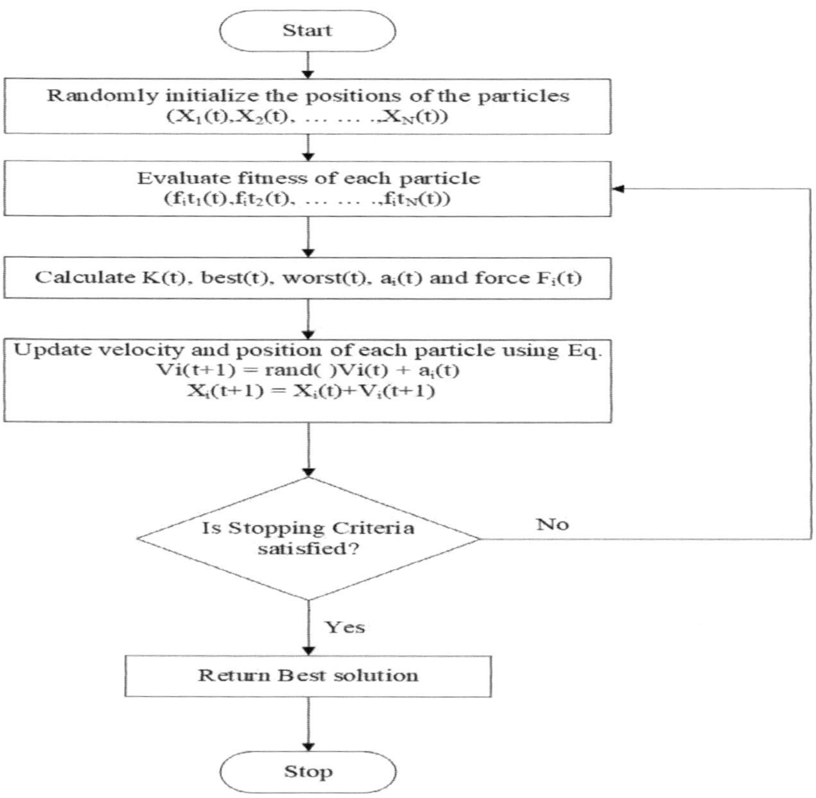

Figure 5. Flow chart of AEFA

IOP Conf. Series: Materials Science and Engineering 1228 (2022) 012022 doi:10.1088/1757-899X/1228/1/012022

6. Simulation results and discussion

The mathematical model described in section 2 is simulated in MATLAB. First the load flow analysis is done in order to find the operating equilibrium conditions given in appendix A. The participation factor corresponding to eigenvalue related to inter-area mode is given in table 1. From the table it can be observed that the participation of state variables of all the four generators to inter-area mode is almost equal. So, SMES units are placed at each generator bus. The SMES control unit is designed, by using the AEFA, to optimize the ISE performance index given in equation (7). The optimized gain of different controllers is given in table 2. SMES current and voltage limiters as mentioned in appendix A are used because of constraint of hardware implementation.

Table 1. Participation factor matrix

	Participation matrix
$\Delta\delta_1$	0.046
$\Delta\omega_1$	0.046
$\Delta\delta_2$	0.049
$\Delta\omega_2$	0.049
$\Delta\delta_3$	0.055
$\Delta\omega_3$	0.055
$\Delta\delta_4$	0.048
$\Delta\omega_4$	0.048

We excite the inter-mode by increasing the input mechanical power of generator 2 in area 1 by 0.01 pu and decrease the input mechanical power of generator 4 in area 2 by 0.01 pu. The eigenvalues and damping ratio corresponding to this mode is given in table 3. It can be observed that the damping ratio is zero, so oscillations are not damped out. After the incorporation of proposed SMES controller the eigenvalue and the corresponding damping ratio is shown in table 3. Here we can observe that the damping ratio for inter-area mode gets improved after the application of SMES units.

The damping effect of the proposed controller can also be analyzed from the results shown below. Figure 6 shows the speed of four generators without SMES and figure 7 shows the effect of proposed control scheme on it. From the results of SMES voltage, current and power, shown below, we can observe that there is proper charging and discharging of SMES units. Since, the mechanical power of area1 is increased, the SMES units in area1 will absorb excess power (charge) so the current is increasing from initial current with positive voltage. Also when the mechanical power of area 2 is decreased, the SMES units in area 2 will supply the power (discharge), so the current is decreasing from initial current with negative voltage. It can also be seen from the results when oscillations are damped out the voltage across the SMES units is zero so the energy is stored in these units in the form of magnetic field.

Table 2 Optimized gains of SMES controllers

K_{SMES1}	K_{SMES2}	K_{SMES3}	K_{SMES4}
10	9.9653	8.5705	10

Table 3 Eigenvalues of inter area mode

	Eigenvalues	Damping ratio
Without SMES	±4.75i	0
With SMES	-24.37 ± 54.77i	0.47

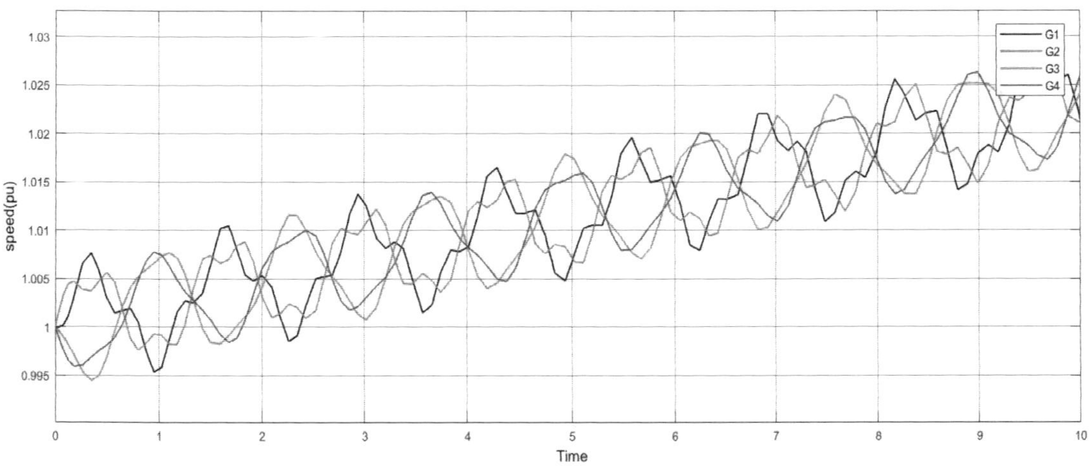

Figure 6. Speed of generators before SMES control

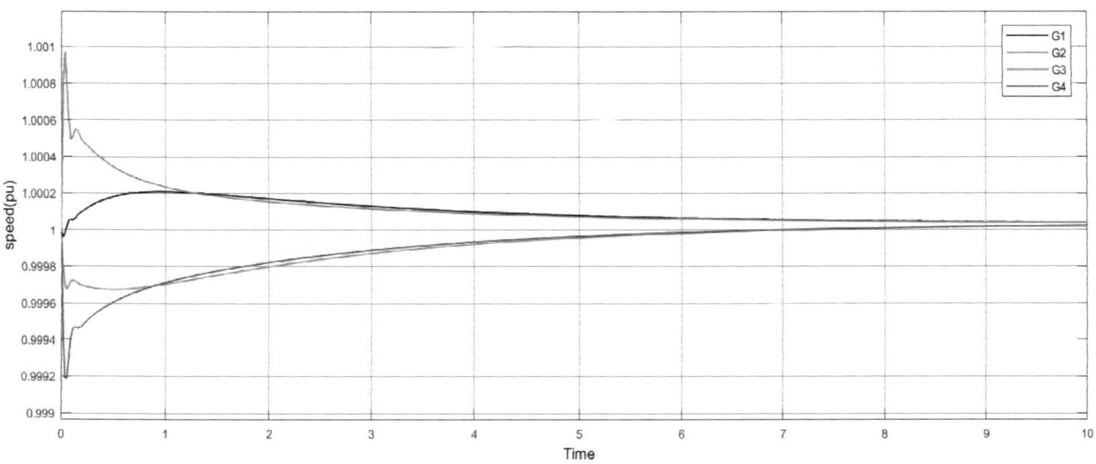

Figure 7. Speed of generators after SMES control

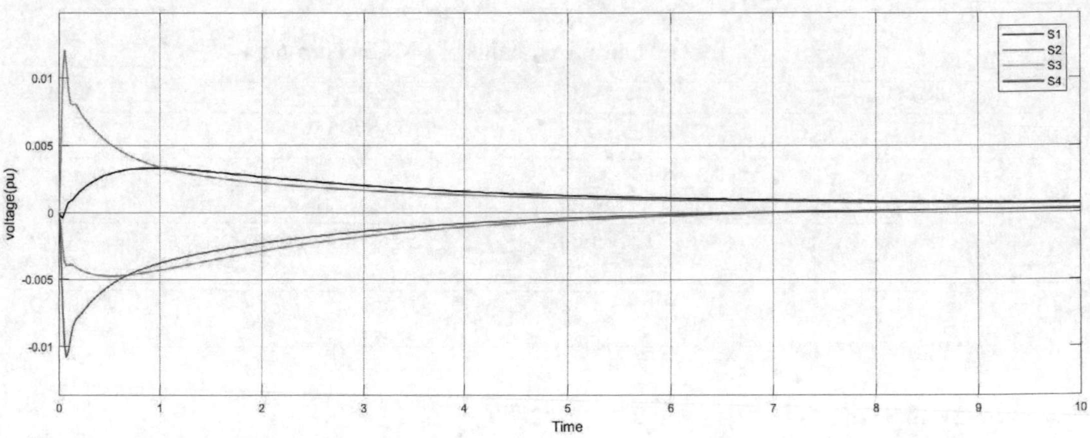

Figure 8. SMES voltage

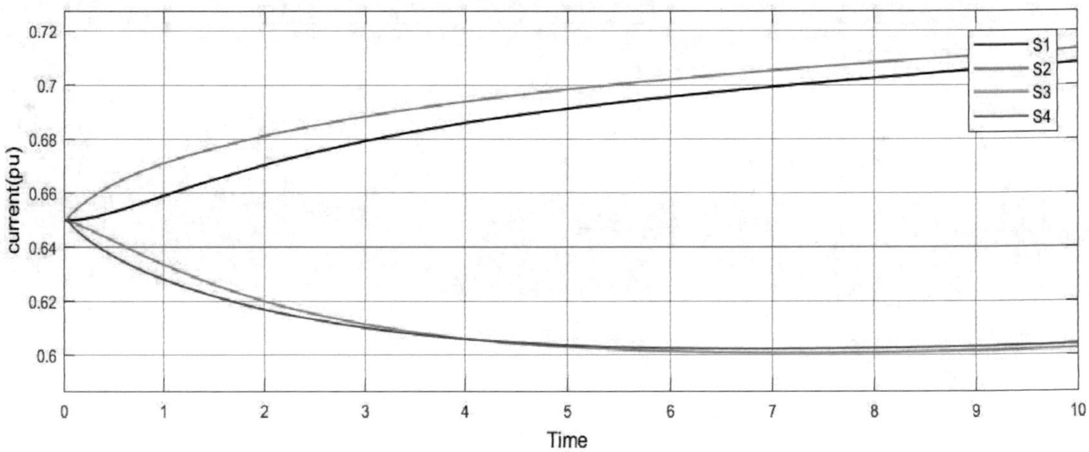

Figure 9. SMES current

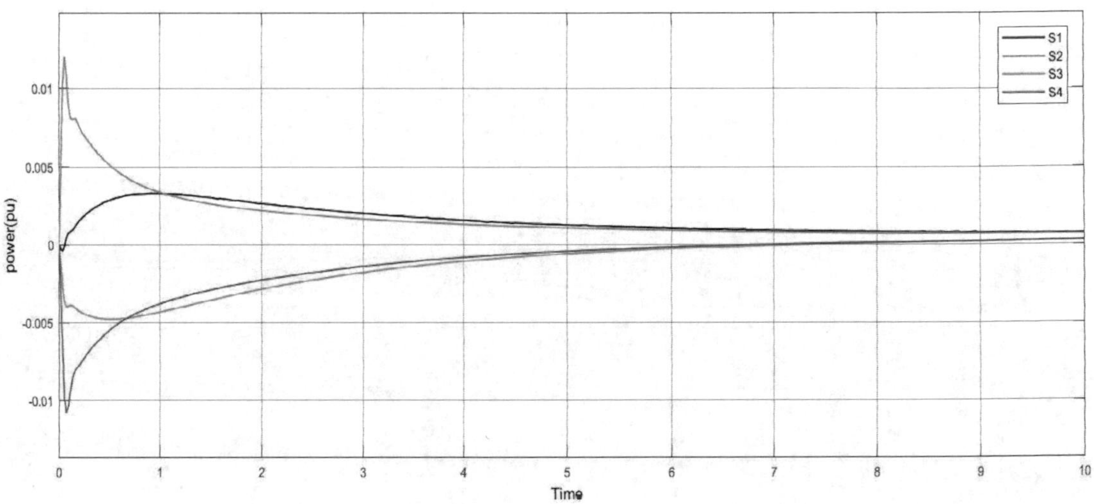

Figure 10. SMES power

7. Conclusion

An artificial electrical field algorithm based optimal control of SMES is applied successfully in this work to damp out the inter area oscillations in an interconnected two area power system. The simulation results show that by using the proposed scheme the power flow from or to SMES can be controlled in such a way that the inter area oscillations are damped out effectively.

Appendix A

System data

Generator power ratings = 900 MW

Inertia constants $H_1 = 3.5s$, $H_2 = 4.5s$, $H_3 = 5.5s$, $H_4 = 6.5s$

$E_1 = 1.10641, E_2 = 1.11641, E_3 = 1.11051, E_4 = 1.12231$

$\delta_1 = 21.5270,\ \delta_2 = 11.7, \delta_3 = -4.51$, δ_4 =-14.81

Transient reactance of each generator X'_d=0.25 pu

SMES data:

$I_{sm0} = 0.65\ pu, V_{sm0} = 0, L_{sm} = 0.264\ H, T_{dc} = 0.026\ s,\ \ K_c = 1$

$I_{sm\ max} = 1.38 I_{sm0}, I_{sm\ min} = .3 I_{sm0}, V_{sm\ max} = 0.5\ pu, V_{sm\ min} = -0.5\ pu$

Load flow results

BUS	VOLTAGE (PU)	ANGLE	P_g (PU)	Q_g (PU)	P_l (PU)	Q_l (PU)
1	1.0300	19.8154	7.0000	2.2332	0	0
2	1.0100	9.9483	7.0000	3.2636	0	0
3	1.0300	-6.2990	7.1600	2.3622	0	0
4	1.0100	-16.5895	7.0000	3.4806	0	0
5	1.0003	13.3002	0	0	0	0
6	0.9340	-5.6694	0	0	0	0
7	0.9664	-8.6445	0	0	9.7600	1.0000
8	0.9283	-32.4057	0	0	17.6700	1.0000
9	0.9630	3.0453	0	0	0	0
10	1.0000	-19.3000	0.1657	3.0202	0	0
11	0.9985	-12.9758	0	0	0	0
12	0.9595	-23.5182	0	0	0	0

Reduced Y bus of the system

1.4228 -17.7521i	0.8923 +15.3164i	0.1636 + 0.4270i	0.4724 + 1.0416i
0.8923 +15.3164i	4.5201 -21.7739i	0.4724 + 1.0416i	1.3352 + 2.5299i
0.1636 + 0.4270i	0.4724 + 1.0416i	1.8886 -17.9603i	2.0241 +14.7268i
0.4724 + 1.0416i	1.3352 + 2.5299i	2.0241 +14.7268i	7.2577 -23.4152i

References

[1] Li Y, Yang D, Liu F, Cao Y and Rehtanz C 2016 *Interconnected power systems* (Springer)

[2] Cui L, Li D, Wen J, Jiang Z and Cheng S 2008 Application of superconducting magnetic energy storage unit to damp power system low frequency oscillation *IEEE conference* Doi: 10.1109/DRPT.2008.4523785

[3] Zhu Y, Liu C, Wang B and Sun K 2018 Damping control for a target oscillation mode using battery energy storage *J. Mod. Power Syst. Clean Energy* **6**, 4 pp 833–845

[4] Ortega A and Milano F 2015 Generalized Model of VSC-based energy storage systems for transient stability analysis *IEEE Transactions on power system*

[5] Liu J, Zhang H and Zhang Y 2015 Coordinated Control Strategy of Scalable Superconducting Magnetic Energy Storage *IEEE Transactions on Smart Grid*

[6] Anita and Yadav A 2019 Artificial electric field algorithm for global optimization, Swarm and Evolutionary Computation base data, **48**, pp 93-108

[7] Rogers G 2000 *Power System Oscillations* (Norwell MA Kluwer)

[8] Mufti M D, Balasubramanian R and Tripathy S C 1998 Simultaneous frequency and voltage control of wind-diesel power systems using energy storage, *Int. J. Energy Res.* **22**, pp 221-235

[9] Tan Y L and Wang Y 1998 Augmentation of transient stability using a superconducting coil and adaptive nonlinear control, *IEEE Trans. Power Systems*, **13**, pp 361-365

[10] Quintero J and Venkatasubramanian V M 2005 A real-time wide-area control framework for mitigating small-signal instability in large electric power systems, Proceedings of the *38th Hawaii International Conference on System Sciences*, pp 66–75

[11] Neely J C, Ryan R H, Elliott T, Silva-Monroy C A, Schoenwald D A and Trudnowski D J July 2013 Damping of inter-area oscillations using energy storage *IEEE Power & Energy Society General Meeting*

[12] Anderson P M and Fouad A A 2003 *Power System Control and Stability*, 2nd Edition, (Wiley-IEEE Press)

[13] Klein M, Rogers G, and Kundur P 1991 A fundamental study of interarea oscillation in power systems, *IEEE Trans. Power System*, **6** pp 914-921

[14] Dechanupaprittha S et-al 2007 Stablization of tie-line power flow by robust SMES controller for interconnected power system with wind farms, *IEEE transactions on Applied Superconductivity*, **17**, 2

[15] Ortega A and Milano F 2016 Modeling, simulation and comparison of control techniques for energy storage systems, *IEEE transactions on Power Systems*

Comparison of the emissions of different biofuels with fossil fuel through the utilization of GREET Model

Jaspreet Hira[1], Nalin Chaudhary[2], Bharat Ankur Dogra[3], Rohit Sharma[4], Vikas Kumar[5] and Kavita Rani[6]

[1] Department of Mechanical Engineering, Raffles University, India
[2] Department of Computer Science and Engineering, Raffles University, India
[3] University of Petroleum and Energy Studies, Dehradun, India
[4]Department of Mechanical Engineering, Amity University Uttar Pradesh Noida, India
[5] Department of Automotive Technology, Mechanical Department division, Ethiopian Technical University, Addis Ababa, Ethiopia
[6] Department of Computer Science and Engineering, Lloyd institute of Engineering and technology, Greater Noida

Corresponding author's e-mail address: jas.iitdelhi@gmail.com

Abstract. Biofuels are the alternate solutions in the fossil fuel family and has been mainly utilized for the reduction in the emissions that are been generated through I.C. engine combustion. This comparison study basically works on the emission characteristics of various types of biofuels with the existing fossil fuel utilized in an internal combustion engine. Through the research work it has been concluded that the biofuels or contributing to the lower emission levels as compared to fossil fuels. According to the Greet model study, methanol produced by gasification is the best recommended fuel for combustion with the lowest emission levels. This study shown that using methanol fuel produced by gasification results in the lowest GHG emission value of 11.44 gm using the GREET model.
Keywords. Greet Model, Biofuels, Fossil fuels, Internal Combustion Engine

1. Introduction

Fossil fuel consumption has been increased tremendously, and, with an increase in trend, the global emission level has also increased. GHG emissions have raised the global temperature, so it's very important to find alternative renewable energy resources. The DSSAT and GREET models have been connected using the service-oriented architecture using Python programming and connected to integrated systems, and the authors concluded that fertiliser use, and plant population are the two major factors affecting GHG emissions [1]. Compared to the life cycle analysis for the first generation and second generation, the GHG emissions are higher for the first generation, i.e., 218 kg CO_2/GJ for the corn-to-ethanol type [2]. The calculated emissions for the three feedstock types (sugarcane, corn, and wheat ethanol) range from the higher emission value, 68 g CO_2 eq MJ-1, respectively [3]. Biofuels can provide significant GHG emission reductions by converting to 85–100% biofuels without requiring major engine modifications [4]. The adaptation of existing biofuels into new systems, such as fuel cell hybrids, reduced production costs and GHG emissions significantly, but biofuels continue to be an adapting technology for a sustainable future [5]. Ethanol replacement for fossil fuels saves about 33% of the social cost of gasoline and is cost effective up to E5-E10 [6]. Renewable fuels emit

Content from this work may be used under the terms of the Creative Commons Attribution 3.0 licence. Any further distribution of this work must maintain attribution to the author(s) and the title of the work, journal citation and DOI.
Published under licence by IOP Publishing Ltd

fewer greenhouse gases than low-carbon electricity, and current energy trends show a new modification policy being implemented in which various new sectors are explored [7]. Acetone, butanol, and ethanol achieve 80% fossil fuel savings and deliver fewer GHG emissions compared to gasoline and diesel fuels [8]. The third generation of the feedstock is algae biofuels, where climatic conditions are required for the growth of these biofuels, and the advantages are that they can be produced in an artificial environment [9]. The comparison results showed that the switchgrass ethanol combined with a limited portion of gasoline showed a significant reduction in GHG emissions [10].

2. Methodology

In the research work, various types of fuels have been used to find out the emissions levels. The most recent Greet model software was used, and the value obtained was converted to Indian standards. Various types of fuels are used, including ethanol, methanol, gasoline E10, and gasoline E85.

3. Results and Discussion

Table 1 shows the various fuels and their emissions per MJ of the fuel produced through various routes.

Table 1. Emissions through various types of fuels

Emissions	Fuels			
	Ethanol (sugarcane) Per MJ	Methanol (Biomass Gasification) Per MJ	Gasoline (Reformulated Gasoline E 10) Per MJ	Gasoline (Reformulated Gasoline E 85) Per MJ
CO_2	0.1786 Kg	10.8661 g	17.3107 g	29.7677 g
VOC	56.2824 mg	14.516 mg	27.8317 mg	49.5201 mg
PM 2.5	63.5435 mg	1.565 mg	1.8103 mg	4.7927 mg
CH4	68.1524 mg	16.0755 mg	0.1086 g	99.0055 mg
GHG- 100	23.9322 g	11.4489 g	21.8706 g	46.1346 g

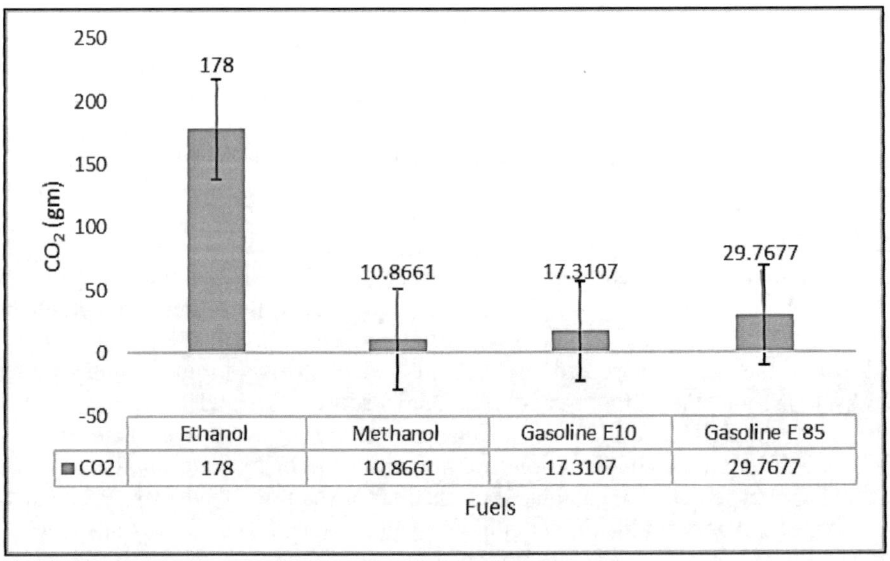

Figure 1. CO_2 (gm) for various fuels

Figure 1 shows the level of CO_2 emission in gm and, through the greet model, the maximum CO_2 emitted into the environment is 178 gm CO_2, which is the highest amongst the various fuels. It could be due to the ethanol produced through the sugarcane route. While the methanol produced through the gasification produces a minimum amount of CO_2 with a value of 10.8661 gm CO_2.

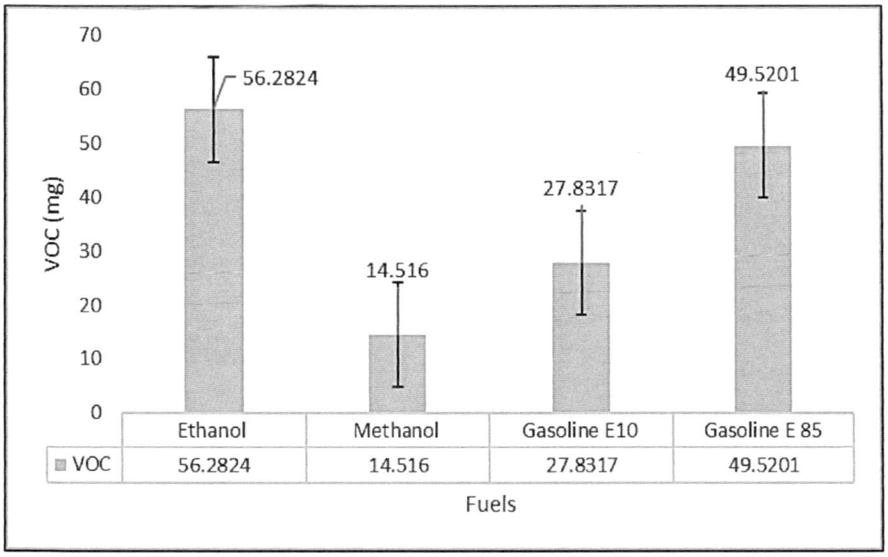

Figure 2. VOC (mg) for various fuels

Figure 2 shows the volatile organic compounds (VOC) emitted by the various fuels, and it has been concluded through the GREET model that ethanol and gasoline E85 are the maximum contributors of VOC amongst these fuels. Ethanol produces 56.284 gm of VOC and gasoline E85 49.5201 gm of VOC. The handling of these fuels plays a major role in the emission level of these compounds.

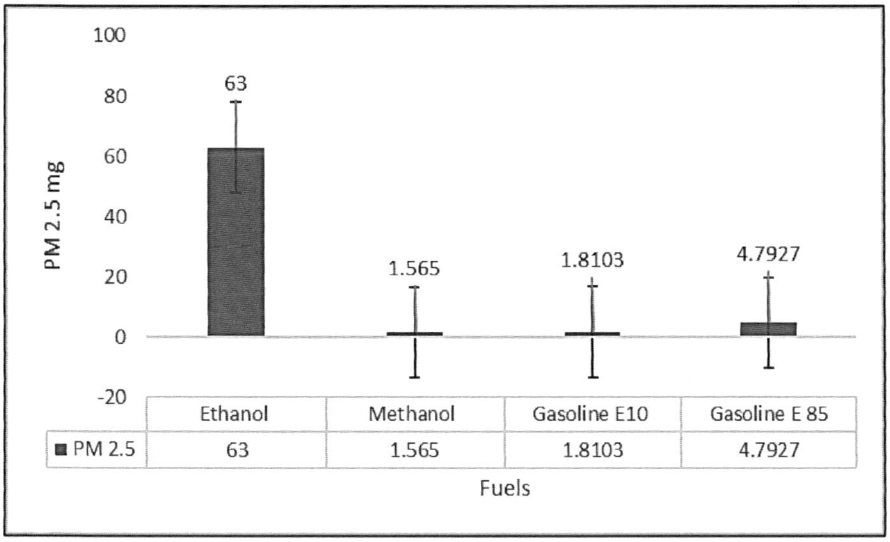

Figure 3. PM 2.5 for various fuels

Figure 3 shows that through the GREET model analysis, ethanol produces the maximum value of the PM (2.5) with a value of 63 mg. Methanol produces a minimum value of PM (2.5), i.e., 1.565 mg. The level of PM (2.5) has been the major concern globally, and therefore, special attention should be given to the level of these PM (2.5) during the combustion of internal combustion engines.

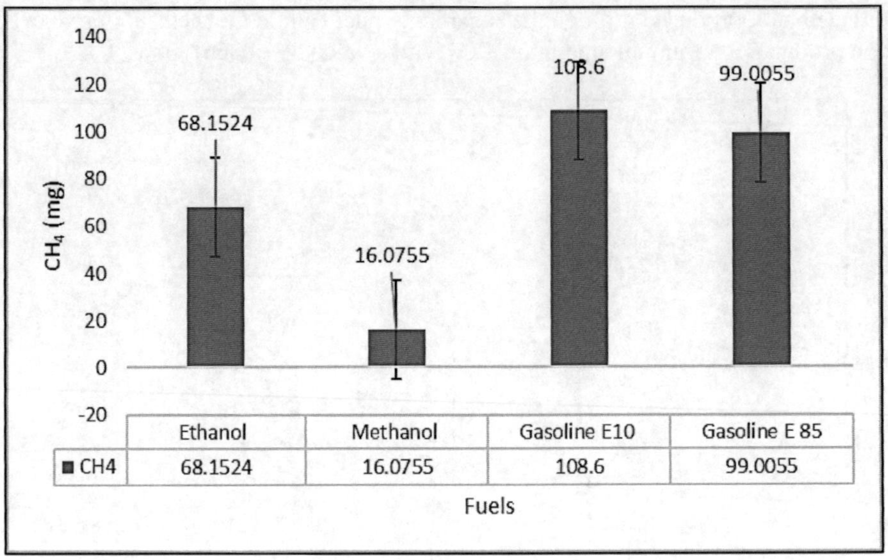

Figure 4. CH_4 for various fuels

Figure 4 shows the value of CH_4 emitted from various fuels. It can be seen through the GREET model analysis that the maximum CH_4 emission is from E10, followed by E85. Therefore, through Figure 4, the lowest value obtained is for the methanol fuel, with a value of 16.0755 mg per MJ of the fuel.

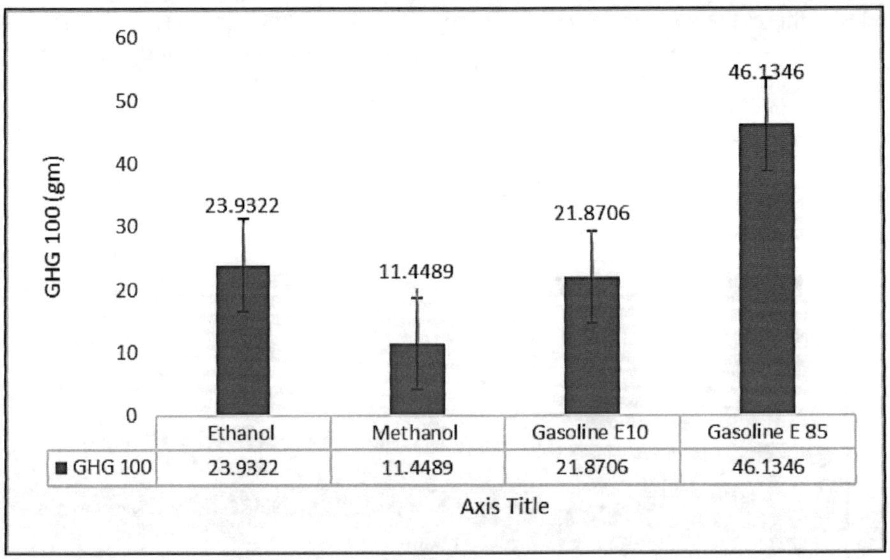

Figure 5. GHG emissions for various fuels

Figure 5 shows the total GHG emissions of various fuels, and it has been found through the GREET model that the maximum GHG emission is through the E85 fuel, and its value is 46.1346 gm per MJ of fuel produced through the reformulated gasoline route. Through the research work analysis through the GREET model, the selection of fuel for combustion and for production should be the main criteria, and the emission level produced through significant routes should also be countered.

4. Conclusions

- Using the Greet model, a few predictions on fuel utilisation for power generation and GHG emissions in the environment can be made.
- This research work has proven to utilise the methanol fuel produced through the gasification to be the lowest emitter of GHG emission value, i.e., 11.44 gm through the GREET model.
- The values developed through the Greet model have been calculated per MJ of the fuel.
- Through the Greet model analysis, the methanol produced through the gasification route is the best selected fuel for combustion with minimum emission levels.
- The Greet model is very helpful in the selection of the right fuel for environmental considerations.

References

[1] Anderson, R., Keshwani, D., Guru, A., Yang, H., Irmak, S. and Subbiah, J., 2018. An integrated modeling framework for crop and biofuel systems using the DSSAT and GREET models. *Environmental Modelling & Software*, *108*, pp.40-50.

[2] Liu, W., Xu, J., Xie, X., Yan, Y., Zhou, X. and Peng, C., 2020. A new integrated framework to estimate the climate change impacts of biomass utilization for biofuel in life cycle assessment. *Journal of Cleaner Production*, *267*, p.122061.

[3] Pereira, L.G., Cavalett, O., Bonomi, A., Zhang, Y., Warner, E. and Chum, H.L., 2019. Comparison of biofuel life-cycle GHG emissions assessment tools: The case studies of ethanol produced from sugarcane, corn, and wheat. *Renewable and Sustainable Energy Reviews*, *110*, pp.1-12.

[4] Ternel, C., Bouter, A. and Melgar, J., 2021. Life cycle assessment of mid-range passenger cars powered by liquid and gaseous biofuels: Comparison with greenhouse gas emissions of electric vehicles and forecast to 2030. *Transportation Research Part D: Transport and Environment*, *97*, p.102897.

[5] Baral, N.R., Asher, Z.D., Trinko, D., Sproul, E., Quiroz-Arita, C., Quinn, J.C. and Bradley, T.H., 2021. Biomass feedstock transport using fuel cell and battery electric trucks improves lifecycle metrics of biofuel sustainability and economy. *Journal of Cleaner Production*, *279*, p.123593.

[6] Le, L.T., van Ierland, E.C., Zhu, X., Wesseler, J. and Ngo, G., 2013. Comparing the social costs of biofuels and fossil fuels: a case study of Vietnam. *biomass and bioenergy*, *54*, pp.227-238.

[7] Andersson, Ö. and Börjesson, P., 2021. The greenhouse gas emissions of an electrified vehicle combined with renewable fuels: Life cycle assessment and policy implications. *Applied Energy*, *289*, p.116621.

[8] Desta, M., Lee, T. and Wu, H., 2021. Well-to-wheel analysis of energy use and greenhouse gas emissions of acetone-butanol-ethanol from corn and corn stover. *Renewable Energy*, *170*, pp.72-80.

[9] Bradley, T., Ling-Chin, J., Maga, D., Speranza, L.G. and Roskilly, A.P., 2021. Life Cycle Assessment (LCA) of Algae Biofuels.

[10] Chang, W.R., Hwang, J.J. and Wu, W., 2017. Environmental impact and sustainability study on biofuels for transportation applications. *Renewable and Sustainable Energy Reviews*, *67*, pp.277-288.

[11] Hira, J., Sharma, R., Kamboj, K., Kumar, V. and Sharma, P., 2019. NOx Emission Reduction in Diesel Engine Through Developed Cooled EGR Setup. In *Advances in Interdisciplinary Engineering* (pp. 1-13). Springer, Singapore.

[12] Hira, J., Sikarwar, B.S., Sharma, R., Kumar, V. and Sharma, P., 2020. Design and development of surge tank for NOx emission reduction in B20 biofueled diesel engine. *International Journal of Vehicle Structures & Systems*, *12*(5), pp.494-497.

[13] Sharma, P., Hira, J. and Anand, P., 2021. Development of a Surge Tank Set-up and Its Utilisation in the Diesel Engine for NO X Emission Reduction. In *Emerging Trends in*

Mechanical Engineering (pp. 187-197). Springer, Singapore.

IOP Conf. Series: Materials Science and Engineering 1228 (2022) 012024 doi:10.1088/1757-899X/1228/1/012024

Exploitation of polypyrrole nanostructured thin films for detecting poisonous lead metal ions using surface plasmon resonance technique

A Paliwal[1*], U P Tyagi[1*]
[1]Department of Physics, Deshbandhu College, University of Delhi, Delhi, India
***Corresponding author e-mails: 87.ayushi@gmail.com, uptyagi@yahoo.co.in**

Abstract

Optical refractive index sensor based on highly sensitive and reliable technique i.e. Surface Plasmon Resonance (SPR) uses the principle of refractive index change at the interface of metal and dielectric for sensitive detection of analyte. In SPR configuration, Surface Plasmon (SP) mode is excited at the interface. Polymer film of polypyrrole nanostructured thin film was prepared using chemical route and spin coating method. Lead metal ions solution was prepared of different concentration. The resonance parameters of the prepared sensor structure showed a systematic variation with continuous rise in the concentration of lead ions. The prepared sensor structure i.e. prism/Au/PPy is found to be highly sensitive towards detecting lead ions at room temperature. The sensor was tested for a concentration range varying from 0.5 to 50 ppm and found to exhibit fast response and recovery time. Sensitivity ~ 0.767 °/ppm was achieved for the presently prepared sensor for detecting low concentration of 0.5 to 10 ppm.

Keywords: Surface Plasmons, Reflectance, Refractive index

Content from this work may be used under the terms of the Creative Commons Attribution 3.0 licence. Any further distribution of this work must maintain attribution to the author(s) and the title of the work, journal citation and DOI.
Published under licence by IOP Publishing Ltd

1. Introduction

The sensors based on highly sensitive and renowned technique i.e. Surface Plasmon Resonance (SPR) are highly sensitive optical sensors. SPR occurs when SP modes (propagating quasi-free electron wave) are excited by the incident Electromagnetic radiation. The SPR sensors senses the change in effective refractive index of the dielectric medium adjacent to the plasmonic metal surface layer. The most commonly exploited configuration to excite the SPs is the Attenuated Total Reflection (ATR) arrangement. In ATR structure, an incident light gets total internally reflected when traveling from higher to lower refractive index medium using a coupler (i.e. optical fiber, prism or waveguide). Some of the light is partially reflected whereas some is transmitted and transmitted light travels through the lower refractive index medium in the form of Transverse Magnetic (TM) electromagnetic field i.e. evanescent wave. This evanescent wave having similar properties as that of Surface Plasmon Wave (SPW) is further used to excite the SPs. The excitation of SPs happens by transferring energy from the evanescent field to SPs and resonance condition is achieved when the propagation constants of the two matches identified by minimum reflected intensity. It is already discussed above that whatever changes in refractive index occurs near the metal surface are reflected as the variation in the resonance parameters like the resonant wavelength of the incident light [1] or resonance angle [2]. There are two well-known ATR arrangements based on prism coupler i.e. Kretschmann configuration and Otto configuration. Out of the two, Kretschamnn is the most exploited structure due to easy sample preparation, as this structure requires one face of the prism coated with the plasmonic metal layer. This leads to the generation of evanescent wave which further excites the SPs at the metal thin film-semi-infinite medium interface.

There is a category of metals like heavy metals which are natural components of the earth's crust. They are basically produced by human waiste industrial and mining activities [3]. This has led to the increase in the levels of these heavy metals which has highly toxic effects to human [4]. Mercury and lead ions are the most harmful heavy metals environment pollutant [5]. Lead metal ions have enormous deleterious effects on human like badly affects the nervous system, highly toxic to children and unborn babies [6, 7]. Lead also causes the inhibition of enzymes by binding with -SH group of enzymes and can also affect bones adversely [7]. Due to its increased discharge and dangerous toxic effect to human, methods for sensing these metal ions are very important. Deleterious effects of lead metal ions on human health urges their detection and technique of detection based on optical sensing is highly effective and sensitive [8]. And it has been already discussed that SPR is potential technique for the sensitive detection of lead metal ions.

Conducting polymers are the new class of materials which have strong potential to sense the heavy metal i.e. lead ions and polypyrrole is good conducting polymer having all the required properties to detect lead metal ions effectively [9]. Not only for detecting lead metal ions, Polypyrrole (PPy) is being exploited for the development of biosensors, composite materials, microelectronic devices etc. [10]. Tabassum and Gupta have exploited polypyrrole composite thin film for detection of manganese ions based on SPR optical fiber technology [11].

In the present work, SPR based lead ions sensor has been developed using polypyrrole nanostructured thin film. Polymer film of polypyrrole nanostructured thin film was prepared using chemical route and spin coating method. Lead metal ions solution was prepared of different concentration and the response of the resonance parameters with increase in the concentration of lead ions was recorded. The experimental results of sensing response of the prepared sensor structure i.e. prism/Au/PPy clearly indicates that the sensor is found to be highly sensitive towards detecting lead ions. The sensor is tested over a wide concentration range from 0.5 to 50 ppm at room temperature and is found to give linear response at lower concentration range from 0.5 to 10 ppm with sensitivity ~ 0.767 °/ppm.

2. Experimental details

2.1 Preparation of Polypyrrole (PPy) thin films

Monomer pyrrole was used for the synthesis of Polypyrrole (PPy) thin films utilizing chemical oxidative polymerization technique [12]. The oxidizing agent used for the reaction is ammonium per-sulphate (APS) of analytical grade. 0.1 M aqueous solution of pyrrole and 0.1 M solution of APS were mixed in equal volume in a beaker. This is the polymerization process which was carried out for 3 h

IOP Conf. Series: Materials Science and Engineering 1228 (2022) 012024 doi:10.1088/1757-899X/1228/1/012024

until a precipitate was obtained which was further filtered [13]. In order to transfer the prepared filtrate in thin film form, PPy powder were dissolved in m-cresol and casting solution was obtained by stirring it for 11 hours. PPy thin films were spin coated on a glass substrate at 3000 rpm for 40 s. the prepared films were further dried on a hot plate for 10 min.

2.2 Fabrication of prism/Polypyrrole/Au structure for sensing purpose

The experimental setup for detecting highly poisonous lead metal ions using prism/Au/PPy structure is shown in figure 1. A liquid sample cell specially made from the same material as that of the prism (i.e. BK-7 prism) is attached to the prism/Au/PPy structure such that target analyte containing varying concentration of lead metal ions is directly exposed to sensing structure.

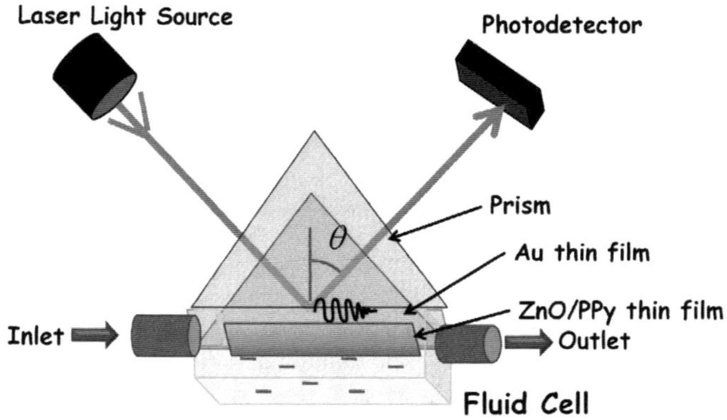

Figure 1. Schematic of prism/Polypyrrole/Au structure based highly poisonous lead metal ions sensor.

The BK-7 prism having refractive index of 1.517 was used to deposit PPy thin film for fabricating sensor structure. The prisms were thoroughly cleaned by sequentially washing them with Tri Choloro Ethylene (TCE), acetone and Iso Propyl Alcohol (IPA) chemicals. The plasmonic metal used over is gold (Au) and hypotenuse face of the prism is coated with Au thin film using well known metallization technique i.e. thermal evaporation. Subsequently, the PPy nanostructured film was deposited of optimized thickness of 200 nm.

Detection of lead ions in the present work involves sensing the shift in resonance parameters Surface Plasmon (SP) mode in prism/Au/PPy system. 1000 ppm concentration of lead metal ions was prepared by dissolving 1 g of lead sulphate ($PbSO_4$) in 1 l of distilled water [14]. Then, other lower concentration (0.5, 1, 5, 10, 30, 50 ppm) were prepared by diluting 1000 ppm solution [14]. A complete SPR reflectance curve is recorded for prism/Au/PPy system in a fixed concentration of lead ions and the corresponding shift in the resonance parameters as determined from SPR reflectance curve is monitored with increase in the concentration of lead ions in the liquid sample cell attached to it using the SPR developed explained in other reports [15, 16].

3. Results and discussions

Firstly, the experimental SPR reflectance curve for a single layer system is recorded utilizing the excitation wavelength of 633 nm in Kretschmann configuration which is used as the reference as shown in figure 2. The sharp SPR curve having minimum absorption losses is attained for prism/air/Au system with resonance angle (θ_{SPR}) at 42.6° which shift to higher angle i.e. 50.1° for prism/Au/PPy structure. Also, FWHM increases with further increase in minimum reflectance which clearly indicates the extra absorption losses induced by the addition of another layer i.e. PPy. Since, all the measurements are to be done in aqueous solution, thus SPR reflectance curve has also been recorded for prism/Au/PPy structure in aqueous solution. The SPR reflectance curve for prism/Au/PPy/aqueous solution structure is shown in figure 2 representing the further increase in FWHM and θ_{SPR}. The major reason for this rise in both the parameters is due to enhancement in the refractive index of semi-infinite medium in contact with metal layer i.e. air medium is changed to aqueous solution medium.

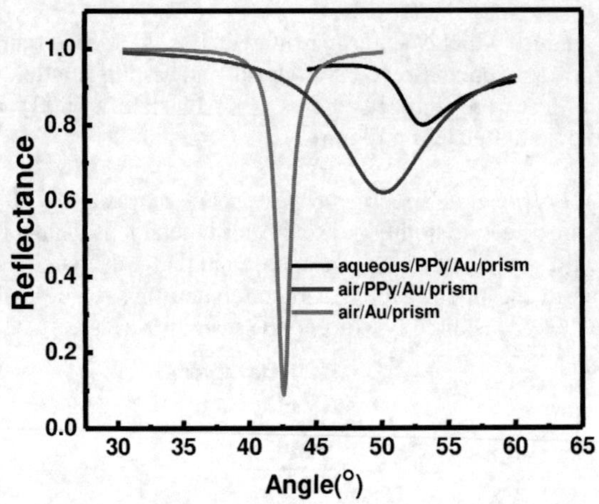

Figure 2. SPR reflectance curve for prism/Au/air, prism/Au/PPy and prism/Au/PPy/aqueous solution structures

Figure 3 shows the experimental SPR reflectance curves obtained for prism/Au/PPy/air system by varying thickness of polymer (PPy) nanostructured thin films. As evident from figure 3, the resonance parameters like R_{min} , θ_{SPR} and FWHM shows the inverse response with increase in thickness of PPy thin films from 50 nm to 200 nm. FWHM is observed to be minimum for 200 nm thick PPy film. The obtained results show that 200 nm thin PPy exhibits the sharp SPR reflectance curve (figure 3) and hence, low losses. Thus, PPy thin film of 200 nm thickness seems to be suitable for realization of SPR based lead ions sensor.

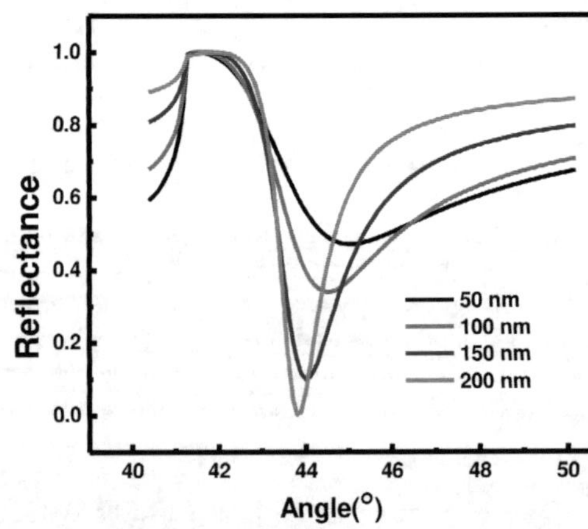

Figure 3. Experimentally recorded SPR reflectance curve for prism/Au/PPy/air structure where the thickness of PPy thin film is varied

3.1 Refractive index sensor: prism/LiF/Au sensor

Figure 4(a) shows the sensing results of the detection of lead ions in aqueous solution using the prepared sensor. The sensing results clearly indicates that SPR reflectance curve shifts in terms variation in θ_{SPR} by varying the concentration from 0.5 ppm to 50 ppm of lead ions in aqueous solution. The SPR measurements have been performed in continuous rotation mode using a photodetector interfaced with the power meter (model : make Newport) for a fixed concentration of lead ions. Now, increasing the concentration of lead ions in aqueous solution further increases the refractive index of semi-infinite bulk medium in contact with the sensing surface of Au/PPy interface. This is directly related to the variation in resonance parameters of the SPR reflectance curves. The graph of variation of θ_{SPR} with increase in the concentration of lead ions solution is shown in Figure 4 (b) (calibration curve).

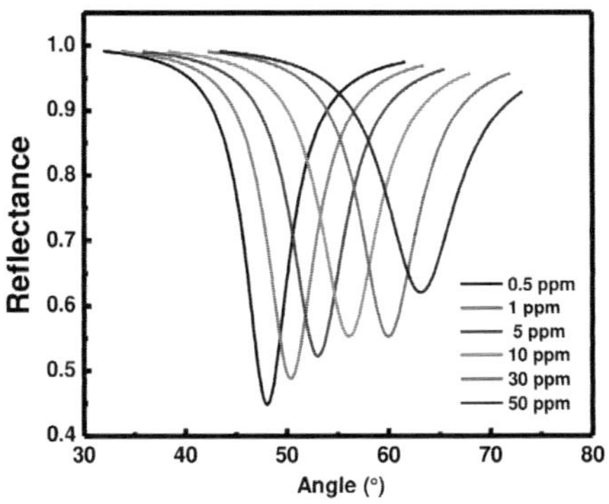

(a)

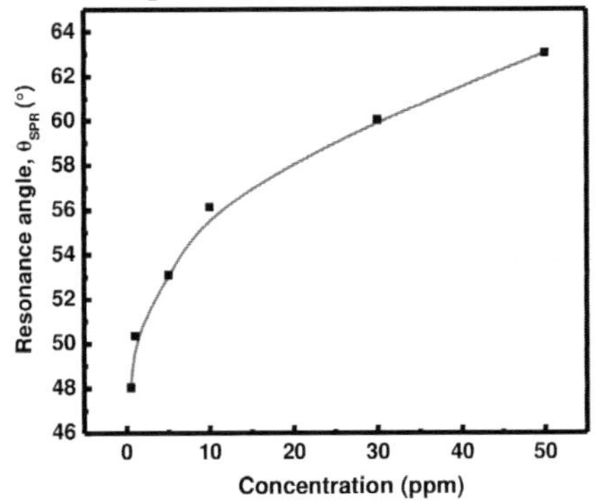

(b)

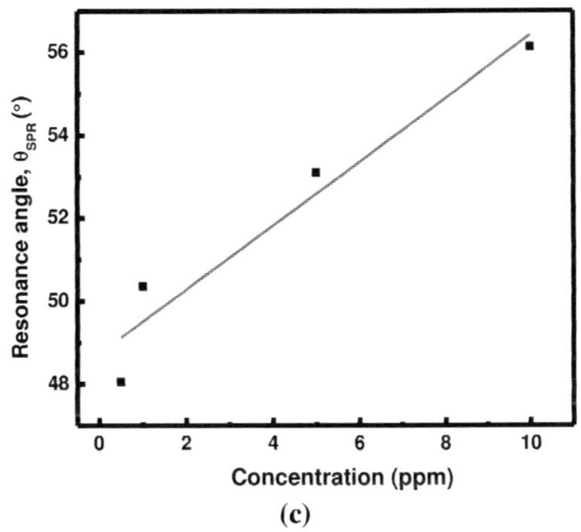

(c)

Figure 4. Sensing results of the prepared prism/Au/PPy/analyte sensor showing (a) the shift in SPR reflectance curve, (b) calibration curve s and (c) Calibration curve showing linear response in the concentration range from 0 ppm to 10 ppm.

This is found to be linear for smaller concentration of lead ions whereas with further increase in the concentration, the variation of θ_{SPR} saturates. This indicates that the sensor is linear upto the 10 ppm concentration range (figure 4(c)) and response of the sensor saturates with further increase in the concentration to 50 ppm. The significance of the calibration curve is to estimate the sensor's sensitivity which was found to be about 0.767°/ppm (from linear graph figure 4(c)).

3.2 Transient Response:

The variation of the response with time i.e. transient SPR response which is measured at fixed resonance position for increasing concentrations of lead ions (i.e. from 0.5 to 50 ppm) is shown in figure 5 (a). The laser beam is incident at a particular angle (i.e. SPR dip angle, $\theta = \theta_{SPR}$) and the variation in the response with increasing concentration of lead ions is recorded using a CCD camera. In recording the transient response, a specific concentration of target analyte solution of lead ions is poured in the liquid sample cell and it is observed that the reflectance shown by CCD increases to a stable saturated high value (figure 5 (a)). While recovering the response, the response regains its original value by flushing out the target analyte of fixed concentration. The formula governing the calculation of the sensing response for prism/Au/PPy sensor structure was calculated using the equation [17]

$$\Delta R = (R_{max} - R_{min})\, X\, 100\,\% \quad (1)$$

where R_{max} and R_{min} are the maximum and minimum values of reflectance in the presence and absence of target analyte respectively.

As evident from figure 5 (a), as the concentration of target lead ions increases from 0.5 to 50 ppm, the sensing response (ΔR) from equation (1) increases systematically. The prepared sensor's significant achievement is that the time for response and recovery are discovered to be less than 2 seconds. In addition, there is no drift in the base line (R_{min}) of prism/Au/PPy sensor structure after a series of cycles of ON and OFF in contact with the target analyte solution, showing that the constructed SPR is stable sensor. Now, the permissible value of lead ions in water is around 0.5 ppm as recommended by WHO, hence accurate detection of lead ions in aqueous solution at very low concentration is highly crucial. In the present work, lower concentration of target analyte which was detected is 0.5 ppm. Figure 5 (b) shows the dynamic mode measurements calibration curve where the response was obtained at a fixed incidence angle $\theta = 53.2^{o}$. The sensor's response indicates a perfect linear growth (ΔR) as the quantity of lead ions rises (figure 5 (b)). The findings show that the prism/Au/PPy based SPR lead ion sensor is extremely sensitive to lead ions over a wide concentration range (0.5 to 50 ppm) at room temperature with fast response and recovery time.

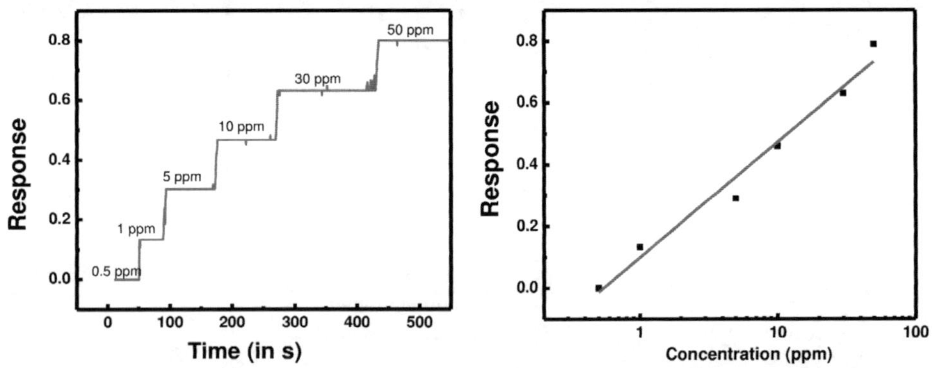

(a) (b)

Figure 5. (a) Prism /Au/PPy structure transient response after repeated exposure to lead ions solution of increasing concentration from 0.5 ppm to 50 ppm, (b) The sensor's dynamic mode calibration curve shows a linear rise in response with rise in concentration of analyte

4. Conclusion

The SP mode has been excited in Au metal layer/PPy dielectric interface. SPR reflectance curve has been recorded by first depositing Au thin film on prism i.e. for prism/Au/ air system and then coating PPy thin film on prism/Au system to form prism/Au/PPy/ air structure. SPR curve has also been recorded for prism/Au/PPy/aqueous solution structure as all the measurements are to be in aqueous environment. The thickness of the prepared PPy nanostructured film is being optimized at 200 nm thick. The optimized configuration has been exploited successfully for sensing lead ions. Lead ions have been detected over a wide concentration range (from 0.5 to 50 ppm) using prism/Au/PPy/aqueous solution structure with high sensitivity (0.767°/ppm), good linearity (in lower concentration range) and high stability due to the choice stable sensing material i.e. PPy.

References

1. Homola J, Schwotzer G, Willsch R, Ecke W and Bartelt H 1995 *Proc. SPIE* **2508** 324.
2. Liedberg B, Lundstorm I and Stenberg E 1993 *Sens. Actuators B: Chem.* **11** 63-72.
3. Fen Y W and Yunus W M M 2013 *Sensor Review* **33/4** 305.
4. Fen Y W and Yunus W M M 2011 *Optics and Photonics Journal* **1** 116-123
5. Boruah B S, Biswas R 2018 *Optical fiber technology* **46** 152.
6. Sathawara N G, Parikh D J, Agarwal Y K 2004 *Bulletin of Environmental Contamination and Toxicology* **73** 756.
7. Pearce J M 2007 European neurology **57(2)** 118.
8. Fen Y W, Yunus W M M, Talib Z A 2013 *Optik* **124** 126.
9. Migahed M D, Fahmy T, Ishra M, Barakat A 2004 *J Polym* **23** 361.
10. Sadrolhosseini A R, Noor A S M, Bahrami A, Lim H N, Talib Z A, Mahdi M A 2014 *Plos One* **9** 93962.
11. Tabassum R, Gupta B D 2015 *Sens Act. B* **220** 903.
12. Navale S T, Mane A T, Chougule M A, Sakhare R D, Nalage S R, Patil V B 2014 *Synthetic Metals*, **189** 94.
13. Chougule M A, Pawar S G, Patil S L, Raut B T, Godse P R, Sen S, and Patil V B 2011 *IEEE Sensors Journal* **11**.
14. Sadrolhosseini A R, Naseri M, Rashid S A 2017 *Optics and laser technology* **93** 216.
15. Paliwal A, Sharma A, Tomar M, Gupta V 2014 *J Appl. Phys.* **115** 043104.
16. Paliwal A, Tomar M, Gupta V 2014 *J. Appl. Phys.* **116** 023109.
17. Raether H 1986 *Surface Plasmons on Smooth and Rough surfaces and on Gratings* (Springer, Verlag, Berlin).

IOP Conf. Series: Materials Science and Engineering 1228 (2022) 012025 doi:10.1088/1757-899X/1228/1/012025

Investigation of Stability Analysis and Control of synchronous generator connected infinite bus system under fault condition

A RUPESH KUMAR
EED, National Institute of Technology
Hamirpur, India
rupesh047@gmail.com

Dr. Zakir Hussain
EED, National Institute of Technology

Abstract. This paper focuses on the Investigation of stability analysis and control of synchronous generator integrated to infinite bus system . The stability of power system is analysis of taking different configurations for suppress the oscillations of various dynamic results by using proper facts devices and manages the gains of synchronous machine regulators for better results. However, increasing the stability of system when synchronous generator integrated to infinite bus system under fault conditions and Different oscillatory modes are obtained with their eigenvalues, and the results from eigenvalue analysis are presented to demonstrate the various configurations of system.

List of Acronyms

AVR Automatic Voltage Regulator

PSS Power System Stabilizer

FACTS Flexible AC Transmission system

PSAT Power System analysis tool

Rlocus Root locus

1.Introduction

The existing system should remain in stable condition even though certain disturbance that taking place in the existing system. The power system having stability criteria provided that system should have synchronous machine because most of the generating stations are having synchronous generators so that there is a stability concept has a more significant in the power system analysis.

Content from this work may be used under the terms of the Creative Commons Attribution 3.0 licence. Any further distribution of this work must maintain attribution to the author(s) and the title of the work, journal citation and DOI.
Published under licence by IOP Publishing Ltd

2.CAUSES FOR INSTABILITY:

Why the stability may be lost means the disturbance is common at load side whenever there is a change of load at load side simultaneously there is not change of mechanical input hence the input and output not equal for long duration that will leads to resulting power not zero and it results to acceleration or deceleration. The synchronous machine is working with oscillations when the oscillations very high the stability of synchronous machine may be lose. There are two types of disturbances 1) small and gradual variation of load 2) sudden and large variation of load . The first type disturbance will not sever on system and it will not effect the generator but the second disturbance will effect the generator .A severe disturbance such as a L-L-L (LINE-LINE-LINE) fault Could cause to a voltage dip and if the generators are unable to remain connected It could leads to an excessive loss of generation. This could cause stability problems and in the long run lead to cascade tripping of other generators. whenever the system draws the reactive power the Voltage of the system's varies. Mainly the variation of the voltage depends on the load power factor. The change in voltage required the change in Excitation of synchronous generator has to be change . if it is fails to Change the Excitation then it will causes the lost of synchronism in the synchronous generator. So understood that the lack of voltage stability causes system unstable and it is only because of drawing more reactive power reasons to entire system has drags into stable region to unstable region and it leads to grid failure so in order to avoid this problem we need a rigorous study and research.

3.Methods

3.1Stability enhancement of single machine infinite bus system

The purpose of this paper is to basic understanding of the dynamic impact[2][5] on the stability of a Grid connected synchronous generator under fault condition and investigate the possibility of improving the stabilization and damping of the grid by AVR and PSS controllers[7][8] connected to the synchronous generators[1]. To investigate the effect of a disturbance on the power system stability has discussed in two cases and it has been simulated in power system analysis tool (PSAT) software. I have proposed controller that can be withstand and stable even after the sever disturbance occurs on the system, A severe disturbance such as a L-L-L fault can lead to a voltage dip that causes drawing more reactive power and if the generators are unable to remain connected it could lead to an excessive loss of generation. This could cause stability problems and may ultimately lead to cascaded tripping of other generators .

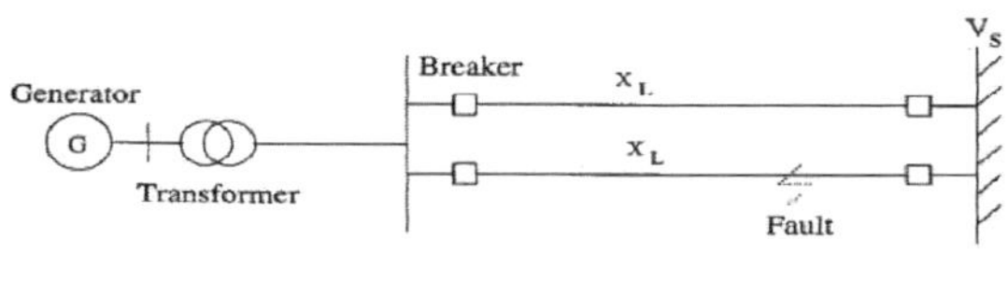

Figure-1.1

As shown in figure 1.1 The Synchronous Generator (SG) integrated to the infinite bus system. Figure 1.1 the fundamental single line diagram for power system under transient stability [6] In The diagram using analysis for the grid connected system under the consideration three phase symmetrical fault .It has taken two different conditions such as The infinite bus bar connected with single synchronous generator with and without controller like AVR and PSS[5] and Various result has been analysed the above mentioned cases by using PSAT tool box

IOP Conf. Series: Materials Science and Engineering 1228 (2022) 012025 doi:10.1088/1757-899X/1228/1/012025

4.Results and discussion :

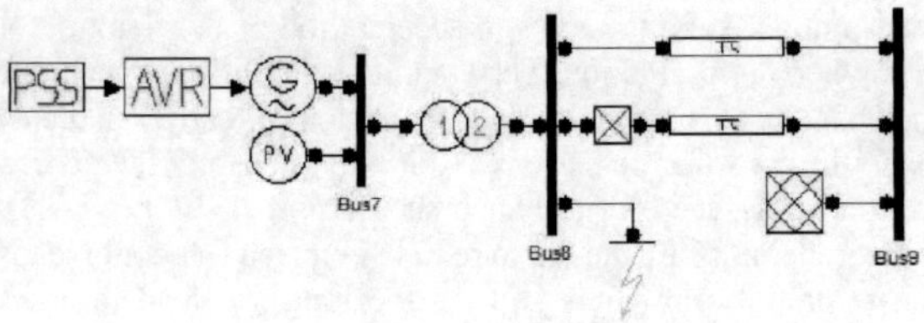

PSAT model Figure 4.1

The results for the above model Figure2.1, 2220MVA,400kv Synchronous Generator connected model obtained different configuration under several network circumstances such as before fault occurs on system , during disturbance or fault and after the fault are analysed and the different changes are taking place in synchronous generator that have been plotted below for before-fault (both lines are in service) up to time t = 1.0 sec. after one second the fault occurs at point F in line 2 that is nearer to bus8 that should consider the fault is occurs on bus8 only and the after fault (line2 is out of service). The moment fault occurs the power will not transfer to infinite bus. The fault current diverted through faulty line. Hence, as we know that under fault condition the synchronous generator acts like pure lagging pf so it is require more reactive power for that excitation need to be increase and the active power flows in the machine become zero. On that situation the maximum power transfer occurs at δ = 90 degree called as steady state stability limit and more than 90 degree the system become unstable.

The below plot clears that system having oscillation under the absence of synchronous regulator such as AVR and PSS[1][5] and rotor oscillate around the synchronous speed and shuttle down to constant due the presence of synchronous regulators.

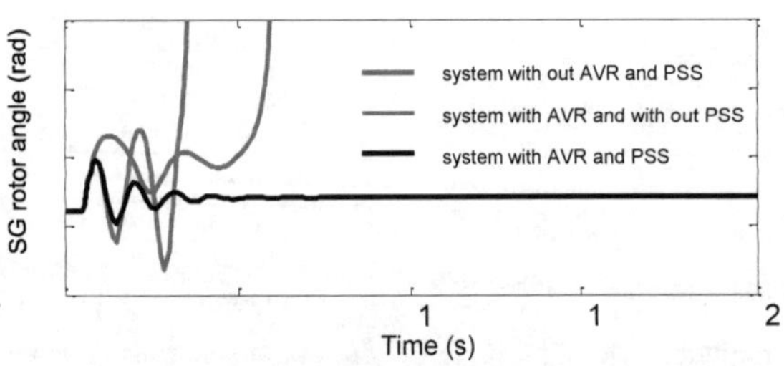

Figure 4.2 synchronous generator load angle vs time

The below plot shows when the of synchronous controller [1] AVR and PSS remain not included then here exists a continual oscillation [1] of speed which is describes that speed of generator synchronous speed it is oscillate around the synchronous speed. Hence, there exists oscillation instability

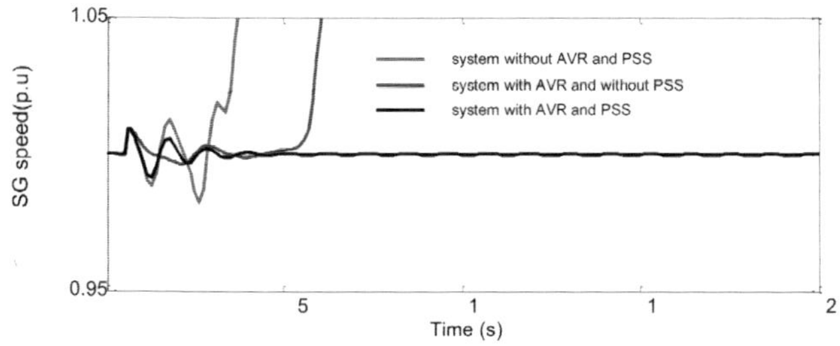

Figure 4.3 synchronous generator speed vs time

The below plot shows when the of synchronous controller [5] of excitation like AVR and PSS are not included, there exists a continual voltage fluctuations which is describes below Hence, there exists oscillation instability.

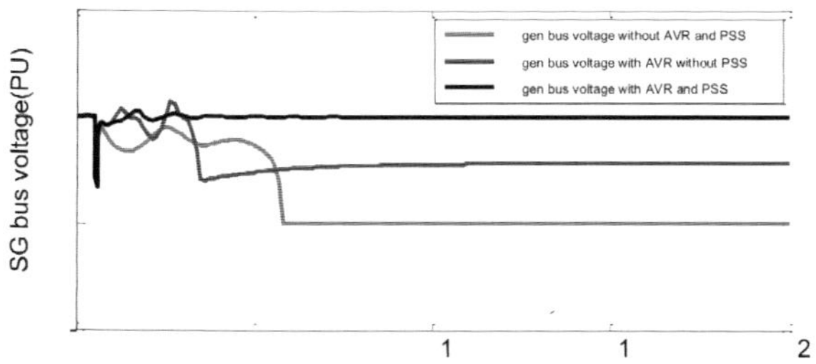

Figure 4.4 synchronous generator Voltage vs time

The below figure indicates the AVR and PSS are not included, there exists a continual reactive power a variation which is describes below .Hence, there exists oscillation instability.

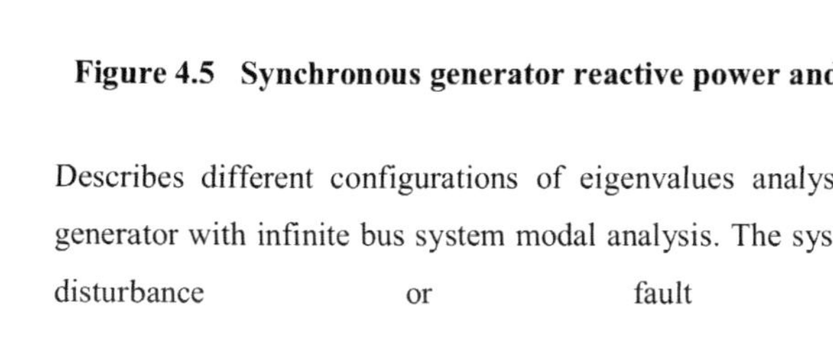

Figure 4.5 Synchronous generator reactive power and time

Describes different configurations of eigenvalues analysis. Provides the understanding of single synchronous generator with infinite bus system modal analysis. The system stability analysis for the dynamical variations under disturbance or fault causes oscillate around

The below plot shows that when the effect of synchronous generator excitation controller like AVR and controller PSS are not included, there exists a active power oscillations which is describes below. Hence, there occurs oscillation instability.

IOP Conf. Series: Materials Science and Engineering 1228 (2022) 012025 doi:10.1088/1757-899X/1228/1/012025

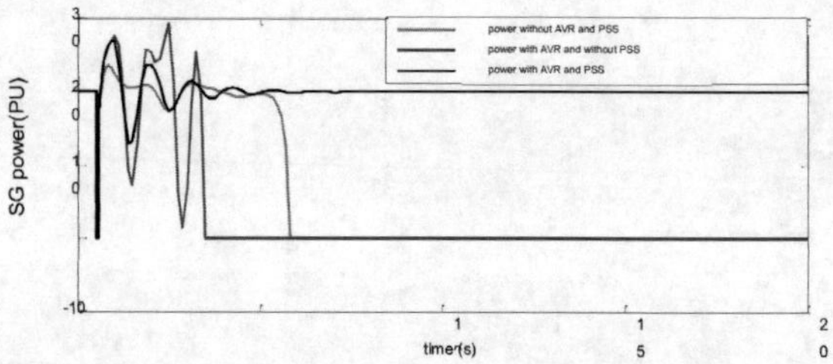

Figure 4.6 synchronous generator active power vs time

The below plot shows that when the effect of synchronous controller of excitation like including diverging points normally shown by power systems different controllers connected to the system .

5.System eigen value analysis.

5.1 SYSTEM EIGEN VALUE ANALYSIS WITH OUT AVR and PSS

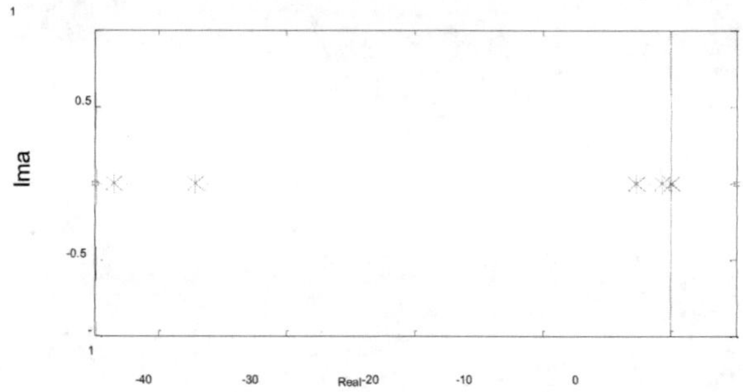

Figure 5.1

5.2 SYSTEM EIGEN VALUE ANALYSIS WITH AVR & WITH OUT PSS

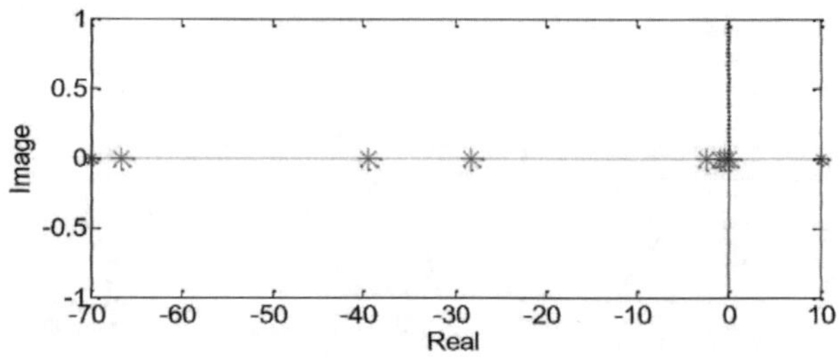

Figure 5.2

5.3 SYSTEM EIGEN VALUE ANALYSIS WITH AVR & PSS

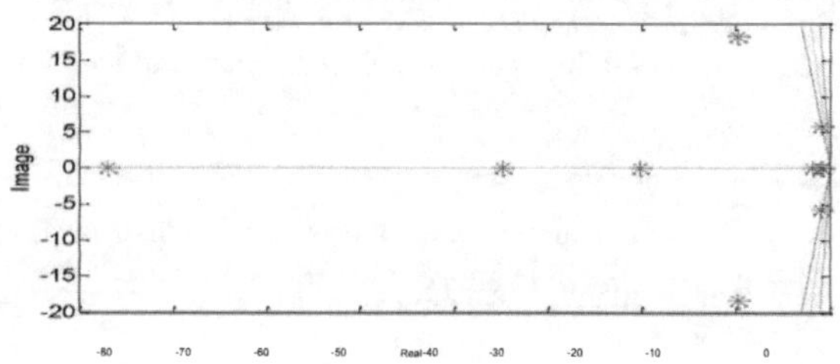

Figure 5.3

IOP Conf. Series: Materials Science and Engineering 1228 (2022) 012025 doi:10.1088/1757-899X/1228/1/012025

The above Eigen value plot explains that as we know when the poles are left side of the imaginary axis have more stable than the poles on imaginary right side of the axis. So from above analysis we seen that if the system contains with AVR and PSS have more stable than without AVR and PSS controllers.

6.1 TABLE OF EIGEN VALUES OF DIFFERENT MODEL

Without AVR and PSS	With AVR and without PSS	With AVR and PSS
-36.4944+j0	-66.6667+j0	9999.9999+j0
-27.6378+j0	-39.4346+j0	-76.9931+j0
0.25606+j6.633	-28.2212+j0	-34.8562+j0
0.25606-j6.633	0.00995+j0	-9.7511+18.2801
-1.9568+j0	-0.00995+j0	-9.7511-j5.7846
-0.08604+j0	0.46203+j0	-20.1523+j0

Table 6.1

7. Dynamic system Root locus plot

The root locus syntax wrote in the MATLAB MathWorks to obtain the plot.

7.1 Description

Root locus calculates locus plot of a single input and single output system, open -loop model. The root locus gives the relative stability of closed pole system the closed loop pole paths as a function of the feedback gain and analysed system to design controller with the defined range of gain k. the range gain k is varied between zero to infinite. with this information the locations of poles can give the information about time and frequency responses.

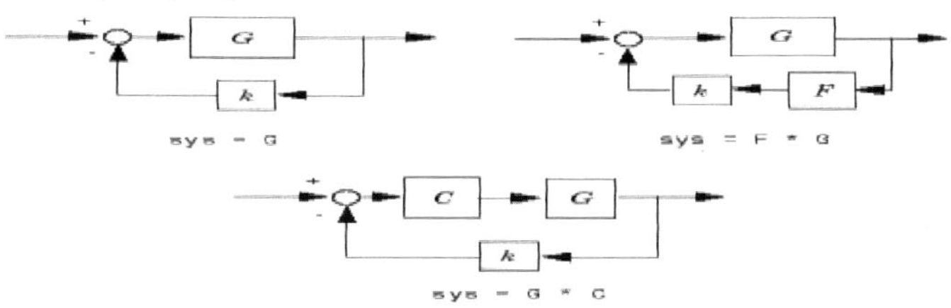

Figure 7.1

If system has transfer function

$$H(s) = \frac{C(S)}{R(S)}$$

the closed-loop poles are the roots of

$R(s)+kC(s)=0$ rlocus select the optimize positive gain k to produce a flat plot. rlocus(sys,k) uses the default value of k of gains to plot the root locus as of user defined. rlocus(sys1, sys2,.) the syntax which have to write in PSAT command to generate the root locus which will provide the plot of particular system follows below,

IOP Conf. Series: Materials Science and Engineering 1228 (2022) 012025 doi:10.1088/1757-899X/1228/1/012025

Below syntax for obtain rlocus above model figure

2.1

```
>> fm_abcd

Computation of input/output matrices A,B,C,D:

[Delta dx/dt] = [A] [Delta x] + [B] [Delta u]
    [Delta y] = [C] [Delta x] + [D] [Delta u]

These matrices are stored in structure LA.
* * * No turbine governor found
* * * No SVC device found
* * * No TCSC device found
* * * No Statcom device found
* * * No SSSC device found
* * * No UPFC device found
* * * No HVDC device found
Computation of input/output matrices completed.
>>
```

8. Without changing gain (gain=200) of AVR and PSS controller root locus From the written syntax will extract the root locus . After the obtain root locus in MATLAB property editor can customize the loci by clicking right button on that particular point it will gives the magnitude and gain, maximum overshoot of system at that particular point and we can find the optimize gain of the system by moving the through locus path left of the s plane which actually drag system for oscillations mitigation and take that particular gain which is give more stable point that gain value keep it Further analysis.

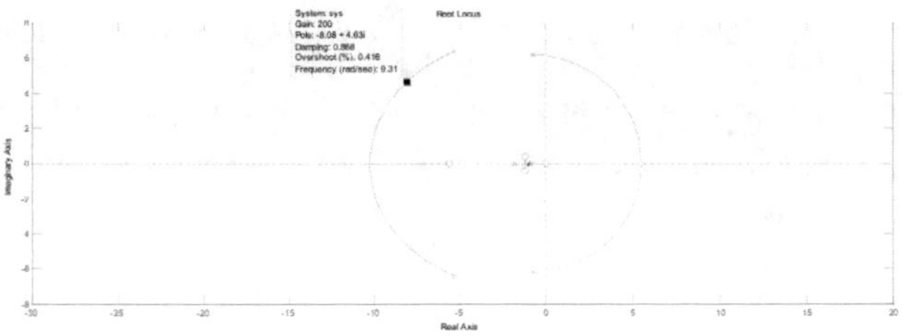

Figure 8.1

The stability of any system depends only on the location of poles but not on the locations of zeros. If the poles are located left side of plane then system is stable. When gain parameter (K) is varied from zero to infinite , if the poles are positioned on left side of (S) Plane it is always stable or absolutely stable only . If the system is stable only for limited values of parameters (K) then it is called conditionally stable. As of this discussion if move on curve and taken a point from far away to the origin and at that exact point take the gain value and expected that value use for controller AVR and PSS[5] design so that system drag from unstable to stable region or making more stable Therefore changed the gain parameter from 200 to 100 in the design of controller AVR and PSS which has been showing more stable the system as before gain reduction ,the results showing below .

Figure 8.2 After changing the gain to 100 the obtained root locus plot

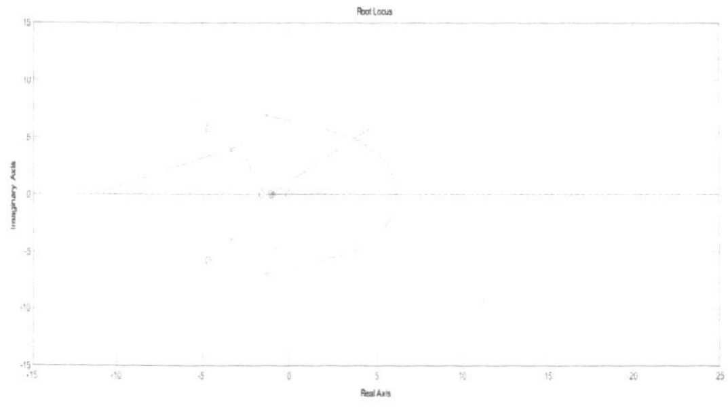

Figure 8.2

After getting root locus in figure 8.2 we dragged the system unstable to stable region by moving the optimum gain point on root locus then obtained value taken as the AVR new gain.

9.Then after obtained new results for system figure Which has showing below

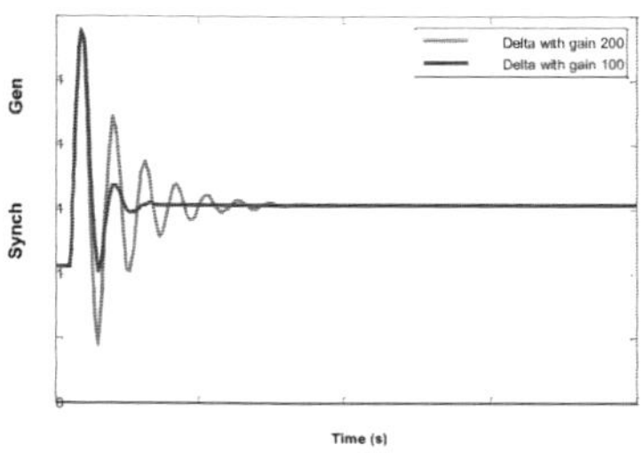

Figure 9.1

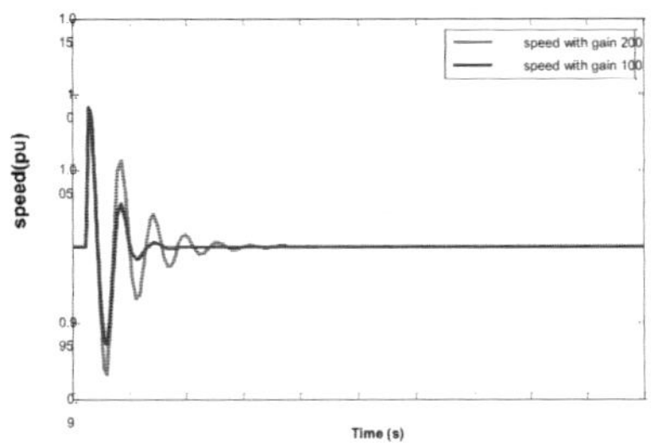

Figure 9.2

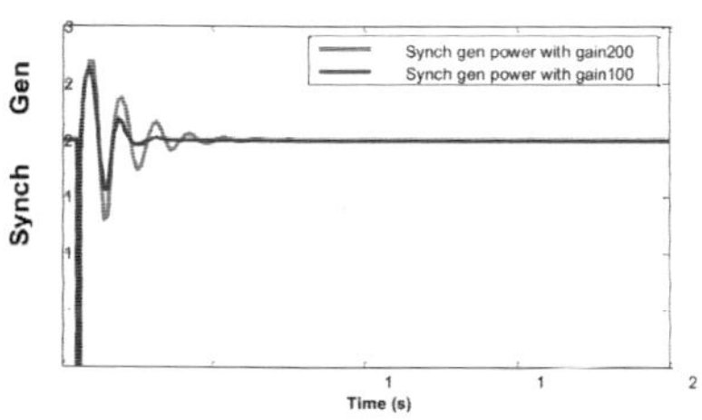

Figure 9.3

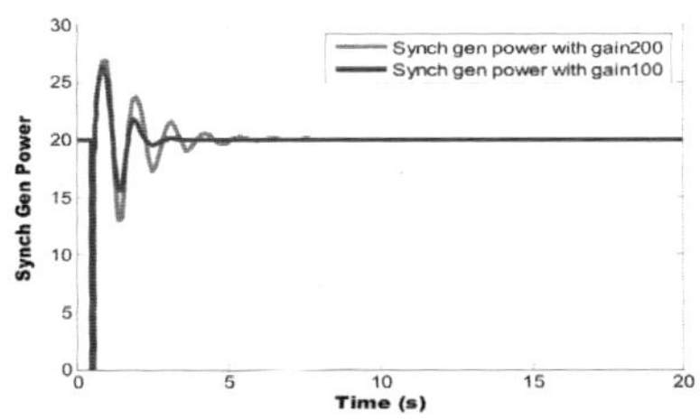

Figure 9.4

10. Summary

The stability system such as voltage stability and transient stability studies has carried out For the single synchronous generator connected to the system and various cases have studied by connecting the synchronous voltage controller such as AVR and PSS. The system dragged into unstable to stable region by considering optimum point of AVR gain even under the severe fault occurs on the system and damped out oscillations of various electrical quantities such voltage, active power, load angle and reactive power of synchronous generator to near to constant.

IOP Conf. Series: Materials Science and Engineering 1228 (2022) 012025 doi:10.1088/1757-899X/1228/1/012025

11.References

[1] Kundur P. Power System Stability and Control. New York: McGraw-Hill. 1993

[2] Reza Yousefian -Transient Stability Enhancement of Power Grid With Integrated Wide Area Control of Wind Farms and Synchronous Generators

[3] Chanbin hu,khayu chen ,li ding -Small signal modeling and stability analysis of virtual synchronous generators

[4] Sujith Mohandas, Ashwani Kumar Chandel “transient stability enhancement of the power system with wind generation” TELKOMNIKA, ,

[5] Md.Lutfur Rehaman, Md.Abu hena shatil, Design and Implementation of a Synchronous Generator with Rotor Angle Stability Control for Damping Interarea Oscillations of Interconnected Power Systems via PSS

[6] M.R.Aghamohammadi , M.Saeedi,M.zare, Impact of small synchronous generator at medium voltage on transient stability of bulk power system.

[7] A.I .Konara and U.D.Annakkage "Robust Power System Stabilizer Design Using Eigenstructure Assignment," IEEE Transactions on Power Systems.

[8] A. Dysko, W. E. Leithead, and J. O'Reilly, "Enhanced Power System Stability by Coordinated PSS Design," Power Systems, IEEE Transactions

Topological and Reliable based selection of Inverter module for PV application under standard parameters of Cost and life-Span

Shivam Sharma[1], Gagandeep Kaur[2], Naveen K. Sharma[3]

[1] Research Scholar, Department of Electrical Engineering, IKGPTU, Kapurthala, Punjab, India
[2] Associate professor, Department of Electrical Engineering, IKGPTU, Kapurthala, Punjab, India
[3] Assistant professor, Department of Electrical Engineering, IKGPTU, Kapurthala, Punjab, India

Abstract: With the advancement in manufacturing and fabrication process, utilization of PV energy systems is increased throughout the globe which leads to the increase of research over lifespan and reliability index on an entire PV system. This increment results in low average manufacturing as well running costs of the entire PV system. The most important portion of the PV system apart from the PV module is the inverter unit which provides conversion and synchronization between the PV system and OLGS. With vast research carried over design as well controlling the architecture of Inverter, it becomes a tedious task to choose the optimal design based on the application area and capital cost. In this paper, we discuss issues faced by the inverter units on the platform of reliability and life span along with the suitable survey. Further to conclude the work result of Inverter unit connected to selected PV module when connected to OLGS.

Keywords: Photo-Voltaic (PV), On-Line Grid System (OLGS), Renewable Energy Source (RES), Solid State Semi-Conductor Device (SSSD), Total harmonic Distortion (THD), Wide Band Gap (WBG), Sinusoidal Pulse Width modulation (SPWM), Space-Vector Pulse Width Modulation (SVPWM).

1. Introduction

Rising population growth and increasing dependency over automation results in increase demand of electric power which is either from fulfilled by conventional means of generation or from renewable energy sources. With change of processing techniques and high utilization of non-linear sources, buffer between grid and services is required to first provide power supply and also eliminate harmonics transferred to grid due to non-linear nature of load. All the Renewable based energy sources required buffer system which changes nature of electric power generated by RES and change it to Alternating nature for synchronization with grid. Due to no rotating part and wantage losses present in Solar power extraction system, PV panels are more suitable entangled with Inverter for Distribution generation. For better working of PV based Distribution system, PV penetration level in distribution network must be increased which is only possible by reducing average cost of entire PV system long with storage unit and grid buffer system (Inverter block). It is highly recommended that average cost of Inverter block must be reduced in order to raise power production of PV system [1]. As per recent studies cost of PV system will be reduced from 0.18 USD/KWh in 2016 to 0.05 USD/KWh in 2030 by only reducing cost of silicon extraction process and fabrication process [2].

1.1 Literature Survey:

For clubbing of PV system with local grid, inverter unit most important block of present solar power research. In late 1930's concept of inverter was introduced which was way earlier that introduction of PV module. Due to long research period lots of inverter designs and controlling topologies are presented and selecting suitable design for PV application is current gravity of power market. Normally voltage rating of PV module is low in the range of 30 to 60V which cannot meet the demand voltage of service mains which is 311 (Peak) [3-4]. Series connection of PV modules raises voltage with contrary relation of less efficiency and reliability in comparison to parallel connection [5]. With parallel connection of PV module for better efficiency and reliability, High-Frequency Transformer is required which further raises issues of less efficiency and high average costing [6-7]. Hence transformer-less inverter designs are gaining more popularity in-comparison to is insister design [3]. Two types of transformer designs are more common and popular which is (a) Multi-Stage Power Conversion (MSPC) and (b) Single-

Content from this work may be used under the terms of the Creative Commons Attribution 3.0 licence. Any further distribution of this work must maintain attribution to the author(s) and the title of the work, journal citation and DOI.
Published under licence by IOP Publishing Ltd

Stage Power Conversion (SSPC) [8-9]. For better efficiency and low average cost single stage power conversion unit in service mains is more prominent in-comparison to MSPC unit [10].

1.2 Proposed Work:
In case of Solar PV module inverter unit is the main link which connect Solar electric power to Online-grid system. Introduction to PV inverter along with contribution and foray of literature is discussed in section 1. Further to this inverter architecture under varies irradiance profile along with selective and highly used control strategy is presented in section 2. In section 3, impact of PV system on the reliability and lifetime of inverter is discussed. Present utilization of PV inverter for providing reactive power at the time of low or no irradiance is discussed in section 4. At last simulated results of three-phase PV-Inverter supplying 100KW powered PV system to Indian service mains is shown is section 5 continued with conclusion in section 6.

2. Inverter architecture under varies PV system and irradiance profile

Inverter unit is one of the main components of PV system which provides power form PV module to on-line grid system (OLGS). Although inverter unit in PV energy system acquire 50% of the initial investment [88]. In that share 20% of the cost is related to switching technology used and supporting driver circuit for SSSD [11]. Most of the research done in the section of inverter unit is either providing comparison study of different inverter control topology along with inverter design factor or cover singe Inverter unit with application area [12-13]. Recent times focus of inverter is shifted toward the cost analysis of any inverter unit under specified topology along with application area [14]. Although very few studies have been mentioned which selects the converter topology along with converter design for application are where cost act as one of the objectives for optimal selection. Inverter topologies for PV based applications are discussed below:

2.1 String Inverter
String inverters are commonly used topology for 'n' number of PV modules connected in series and provides better control of AC voltage provided to medium voltage AC-grid [15]. Due to series connection of PV modules, losses are more prominent as all the module do not exhibit same hereditary properties [16]. Fig. 1 shows the concept of string inverter along with boost converter which is further connected to OLGS. One of the studies on string inverter is done under MATLAB/Simulink platform by solar technology AG and provide us energy curve shown in fig. 2.

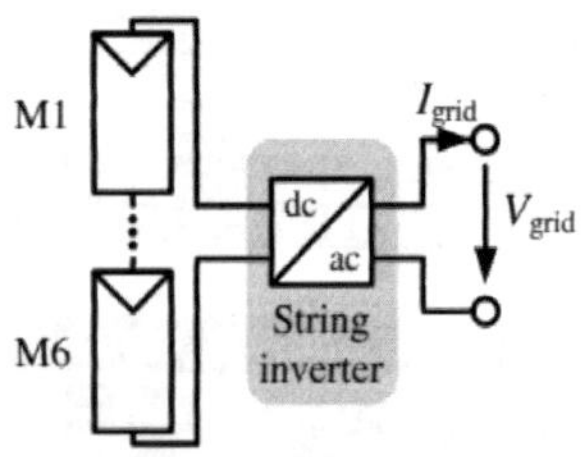

Fig.1: Architecture of String Inverter

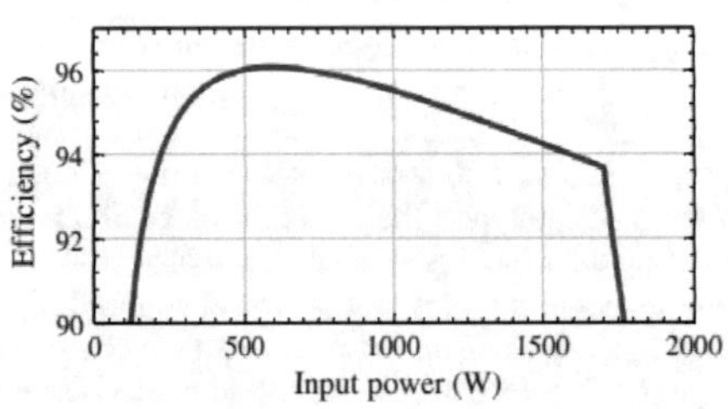

Fig.2: Efficiency curve of string inverter with normalized parameters

IOP Conf. Series: Materials Science and Engineering 1228 (2022) 012026 doi:10.1088/1757-899X/1228/1/012026

2.2 Power Optimizer

For better power utilization either all the PV modules are connected in matrix and changes their tilt angle for maximum power output or providing DC-DC micro-boost converter with any maximum power point technique and further connecting them in series to feed next to Inverter unit. Due to partial shading influence the irradiance, optimization technique was built to make sure power always come in maximum way [17]. Fig. 3 shows the connection of power optimizer converter with individual PV module. In practical simulation environment power optimizer is designed for increasing power from individual PV module [18].

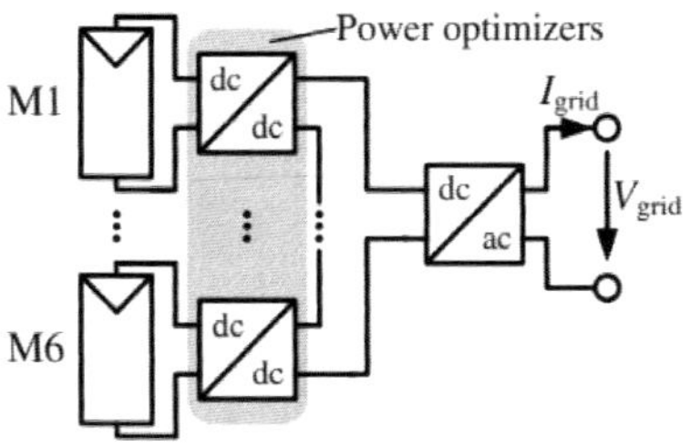

Fig.3: Architecture of Power Optimizer Topology

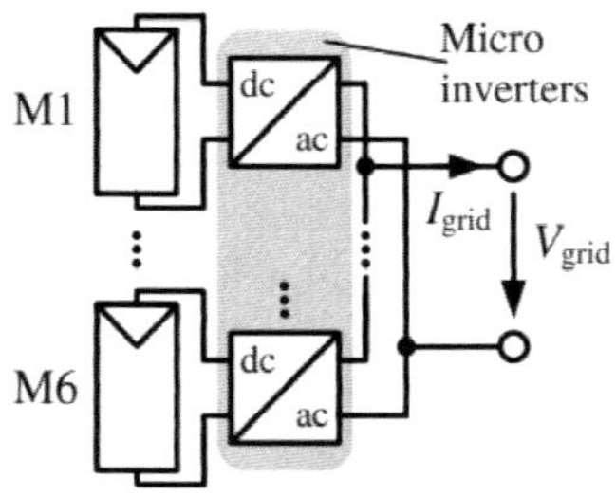

Fig.4: Architecture of shunting Micro-Inverter

2.3 Micro Inverter

Just like power optimizer acquire micro-scale boost converter unit, in micro-Inviter architecture, small sized inverter units are connected with each PV module [19]. Further to this Each sub-section of AC unit is cascaded to form perfect scale of output AC-voltage for OLGS as shown in fig. 4. Many research had been done on simulation level to justify theory under efficiency platform, very few work is been confirmed on cost prospective and power to cost platform to understand how much cost can be recovered. One of the researches had been on experimental platform along with comparison between string-based architecture and MI-based architecture which shows micro-inverter harvest 5% more energy than string Inverter [20]. On the grounds of cost and life-span, due to large number of small components involved MI architecture is costlier than SI architecture.

From all the above architecture of PV inverter, on the bases of selecting parameters either cost, efficiency or power extraction every technique comes their own perks.

3. Control Scheme of PV inverter

Most reliable and highly used control technique for inverter unit in PV system is shown in fig.5 [21-22]. In normal working condition, selection of voltage and current provided by PV module is analysed for utilizing maximum power. This strategy required DV-DC converter unit for constant voltage supply and control technique for maximum power transfer. Further constant DC voltage source with variable current parameter is delivered to inverter unit for changing nature from DC-to-AC. This conversion requires DC-AC module cascaded with filter unit for shaping AC power provided by inverter unit to

IOP Conf. Series: Materials Science and Engineering 1228 (2022) 012026 doi:10.1088/1757-899X/1228/1/012026

almost sinusoidal. Control scheme required running values of current and voltage coming out of filter unit for shaping of both AC current and voltage provided by inverter unit. The entire closed loop control technique provides additional phase angle regulation between AC voltage and current to minimize flow of reactive power [23]. THD of provided current from inverter unit maintain its value in-between 1-2 depending on THD of grid voltage in grid-tied low voltage AC-microgrid system.

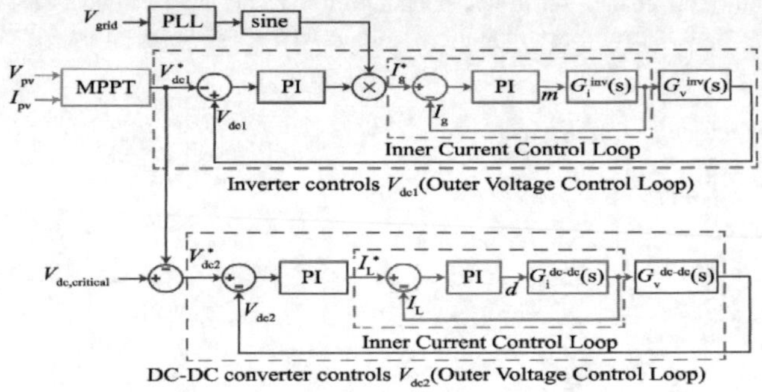

Fig.5: control scheme for PV based renewable source with cascaded inverter unit for synchronization with Low voltage AC-microgrid.

4. Impact of PV system on the reliability and lifetime of inverter

In [24], life span of crystalline silicon based 90 modules is evaluated for almost 11 years in full exposure of solar irradiance. This evaluation comes with linear degradation model suited for any PV module and provides lifespan of solar PV panel. This degradation of PV module impacts the running and recovery cost of PV system as PV module itself consumes 50% of the installed Solar PV grid cost [25]. Testing of PV modules for estimated lifespan in manufacturer unit is more than the actual lifespan of PV module when used commercially [26-28]. This is due to the fact that manufacturer testing comes with linear calculation whereas in actual working state lifespan of PV module bends toward slight non-linearity due to non-constant irradiance throughout the PV working life. Due to on-sight working of solar PV many times failure occurs even at peak points of high input power [29] and more often by changing position of PV module with different angle provides accurate mathematical model of power production which in-further impact the lifespan /degradation of SSSD used in inverter module. For the prediction of lifespan of SSSD many test cases both under theoretical as well practical environment are done by using thermal properties and cycling as main objective to justify the degradation [30-31]. Life time model for any SSSD under accelerating power cycle is derived in (1) along with all notable parameters came from detailed thermal study of SSSD,

$$N_f = A\,\Delta T_j^{\alpha}\, exp\left(\frac{E_\alpha}{k_b\, T_{jm}}\right)\, \alpha r^{\beta_1 \Delta T_j + \beta_0} \left(\frac{C+t_{on}^{\gamma}}{C+1}\right) f_D \qquad (1)$$

Where N_f is the failure cycle number, T_{jm} average junction temperature, ΔT_j and t_{on} are the cycle amplitude and period. All the other parameters are presented in table1.

Table 1. Lifetime model parameters shown in equation (1)

Parameters	*Unit*	*Value*	*Practical environment*
A	[1]	3.4368×10^{14}	
α	[1]	-4.923	$64\ K \le \Delta T_j \le 113\ K$
β_1	[1/K]	-9.012×10^{-3}	
β_0	[1]	1.942	$0.149 \le ar \le 0.42$
C	[1]	1.434	
γ	[1]	-1.208	$0.07\ s \le t_{ON} \le 63\ s$

f_D	[1]	0.6204 for diode 1 for IGBT	32.5 °C ≤ T_J ≤ 122 °C
E_a	[eV]	0.06606	
K_b	[eV/K]	8.6173324x10-5	

The life span of reactive filter components is calculated using one of the miner's rules in (2).

$$LC = \sum_i \frac{n_i}{N_{fi}} \tag{2}$$

Where n_i is the number of cycles LC circuit gone through and N_{fi} is the number of cycles of failure. Value of LC reaches unity makes obvious that component reaches its end life [32].

4.1 PV Size Impact on reliability and lifespan of inverter

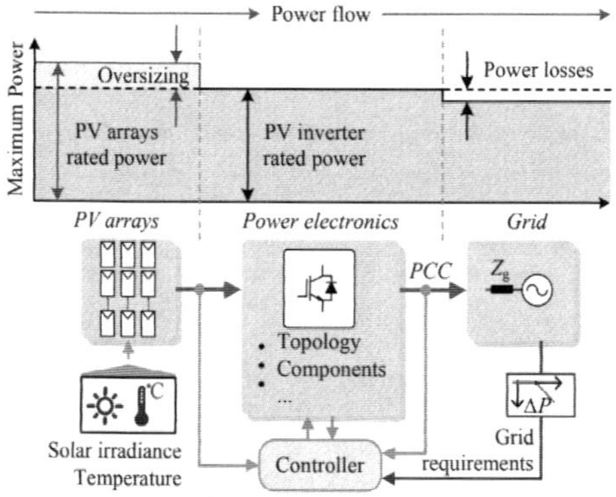

Fig. 6: Power delivering with maximum range under oversizing condition of PV system while maintaining inverter size constant.

Although PV energy system is most reliable and highly recommend in utilization of RES throughout the globe. In-comparison to last decade average cost of solar panel itself is reduced due to increase in the manufacturer units, but still for large number of populations in third world nation installing PV system is expensive. One of the efficient ways of utilizing maximum energy from selective PV system is by increasing power production capacity of PV modules after increasing its size and maintaining rating of inverter unit constant. This process will reduce average cost of PV system by 3 times in the near future. This method of over sizing PV panel along with same rating inverter will reduce overall cost of PV system as shown in fig 6 [33-34]. Although increasing of PV panel will increase the value of input current to inverter unit which either raises heat generation across SSSD or reduces the life cycle of reactive element due to increase in level of ripple current. With every year passing costing of PV panel is decreasing by the rate of 6%, moreover increasing size is recent advancement [35]. Extraction of PV power during daytime with oversized panel is shown in fig. 7 in which it clearly understood that even under lower irradiance level production is high.

Getting more energy from previous sized PV inverter reduces the average cost and increases the percentage of investment recovery under any type of inverter control topology [36]. Optimum designing of PV inverter under this technique is discussed in [37] along with cost comparison. With reducing cost of PV inverter design for large sized PV module impact the reliability and lifespan of reactive elements and SSSD attached in chosen architecture of Inverter design. The lifespan evaluation of all topologies of inverter under this condition of PV system is simulated in [38]. This makes very easy to choose inverter architecture while analyzing the initial cost spend on inverter along with maximum lifespan and cost recovery percentage.

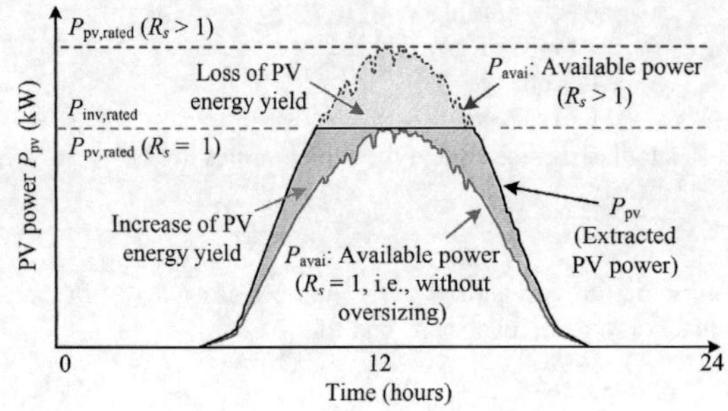

Fig. 7: Impact of Oversizing PV array on power extraction from PV system under complete irradiance chart.

Where, R_S is the sizing ratio ($\frac{P_{Pv,rated}}{P_{inv,rated}}$); P_{PV} is power extracted from module; P_{inv} is rated power provided by inverter.

4.2 DC-Link Voltage Impact on Reliability and lifespan of inverter

In recent year power production and industrial manufacturer growth of PV system had increased exponentially in-comparison to other renewable based energy source [39]. In PV system power electronics components are the weakest portion with less reliability and acquires more complex control system [40]. According to [41] and [42], high switching provides better power controlling and, on the contrary, increases the probability of failure in reactive elements along with few numbers of SSSD. For the improvement of converter reliability along with all filter units (prior or after the converter) several strategies are proposed in respective section [43] show in fig. 8. For better lifespan SSSD with wide band gap (WBG) are better suited in increasing overall lifespan of PV system [44]. The only drawback of this strategy is the increase in average costing of inverter unit which itself consumes 50 % of the cost of entire PV system [45].

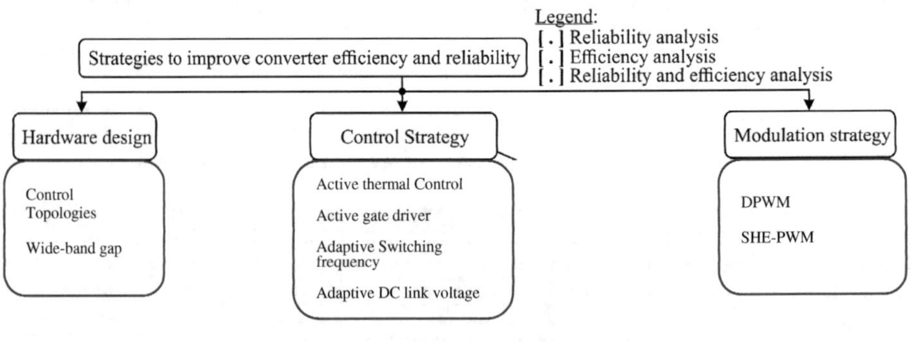

Fig. 8: Overview of strategies explored under development of PV inverter.

The other way of increasing lifespan of reactive unit in PV inverter is by reducing the benchmark of ripple current and make sure that charging-discharging of capacitive unit remains natural even in high-frequency working situation which can be achieved either by shifting toward discontinuous Pulse width modulation (PWM) [46] or by switching toward selective harmonic PWM [47]. With low DC voltage, less ripple factor viewed from capacitor side which increases the lifespan of capacitor. As per the standard of IEC 61727, voltage of grid musty be in the limit of 0.85 to 1.1p.u while PV inverter is connected across [48]. With constant grid voltage, DC-bus voltage across the PV inverter remains constant which results in less ripple current across capacitor. In actual practice, DC voltage is maintained to be low for increasing the lifespan of capacitor on the other hand variation of voltage index between the PV inverter input voltage and output voltage is varied to make sure the power provide by PV system must shadowed in-between grid synchronism.

4.3 Impact of Electromagnetic interference on Reliability and lifespan of PV inverter

High switching frequency converters provide excellent result for constant parameter variation in PV inverter module with constant DC input voltage. However, high frequency operation under any switching technique (SPWM or SVPWM) with constant frequency number leads to more electromagnetic interference [49-51]. Providing another parameter of electromagnetic interference in control technique of inverter along with other loop parameters control the switching frequency of PV inverter at every instance for better resistance to electromagnetic interference [52]. This type of control structure for PV inverter is considered to be chaotic SPWM (CSPWM) and presently valuable research work on inverter module controlling involves CSPWM [53-54]. Based on wavelet transformation, spectrum characteristics of PV inverter with CSPWM control technique is discussed in [55], whereas power loss calculation and working THD provided by inverter module under different load situation is discussed in [56]. Mathematical formulation of CSPWM along with common mode Electromagnetic interference is also expressed and evaluated in [57].

4.4 Reactive Power Control and supply provided by PV inverter

Reactive power is one of the important parameters of power required these day and PV system based Reactive power support system is receiving more interest in-comparison to other reactive compensation units. Mostly work has been done on analyzing thermal profile of PV inverter which provides reactive power both at inside or outside feed-in hours in main utility grid. Techniques like Palmgren minor rule is used to analyze lifespan of PV inverter due to switching losses generated across SSSD used in inverter. In [58], cost comparison is done along with economic impact of reactive power feeding by PV inverter on PV module running and future economics. Solar based electrical energy is more viable in-comparison to other RES which provides less running cost and having huge abundance across the globe [59]. With increase in the advancement of SSSD, extraction of electrical power from solar become easy and more economical both in efficiency as well initial investment [60-61]. Due to Increase of maturity in PV technology and reductio in average pricing of PV modules government bodies are investing hugely and increasing penetration of PV based Electrical generation system in main grid unit [62-63]. Future of PV entanglement with main grid system requires a lot of multi-functional activities just like any other Conventional generating unit which involves almost all type of ancillary services like reactive power support, load leveling, load factor averaging, Low voltage ride through etc [64-67]. All these ancillary services can be achieved by means of utilizing SSSD along with adaptive and less complex control techniques [68-69]. In-comparison to other generating units PV based energy system is more economical and fusible in feeding reactive power for both static as well dynamic loads in utility grid [70]. For providing reactive power PV unit can work in night which impact the life span of inverter unit and may impact the running cost or intermediate cost of PV system [71]. Providing reactive power flow from PV system is less expensive in comparison to other compensating systems which further compensate the intermediate cost required by PV system [72-73].

4.5 Optimization technique for selection of inverter unit for active and reactive power control

For making voltage range within acceptable limits while providing entanglement of PV based inverter-battery system, of the coordinated technique is illustrated in [74-75] along with optimization of sizing and location discussed in [76]. Based on optimization technique particle swarm optimization is used along with PV generation for IEEE-bus system discussed in [77] which is extended along with additional parameter of reactive and active power control under same module of bus system is illustrated in [78]. With utilizing battery energy storage system and dispatching inverter unit with selective power control under optimization and selective bus environment is discussed [79]. Multi-objective stochastic mixed linear programming (S-MILP) is used for optimizing the use of PV inverter in 119-bus system for better utilization in providing both active and reactive power [80]. Use of similar MILP technique to check the optimal location of PV-inverter and BESS in 69-bus system is along with additional objective of economic consideration is illustrated for showing better utilization of Binary intelligence [81-82]. Encompassing of DGs for any type of bus system requires additional information like power requirement, congestion management and economic barriers which can only be clubbed though the environment of optimization under load flow studies [83-84].

IOP Conf. Series: Materials Science and Engineering 1228 (2022) 012026 doi:10.1088/1757-899X/1228/1/012026

In [85], a new optimization technique known as epsilon multi-objective genetic algorithm (E-MOGA) is discussed which decides the size and selection of RES for reactive power generation. Environmental technical and economic aspects are the objective function for the selection of RES in 162-bus system.

5. Simulation and Results

Table 2. Schematics for test module as PV system to feed power to Indian service mains

PV Module	1Soltech 1STH-215-P
Total PV connected (Parallel Series)	470 (47 10)
PV-Grid Voltage	320 V
DC-bus Voltage	800V
Capacitor across PV-section, Converter Section	1000µf, 3227µf
Inverter Module	3-phase six switches
Inverter control Unit	Clark/perk current control
Inverter-side filter (LCL)	500µH, 100µf, 500µH
Voltage at PCC	415V (Phase-to-Phase)
Grid parameters (R_G, L_G)	100µΩ, 1µH
Grid Frequency	50Hz

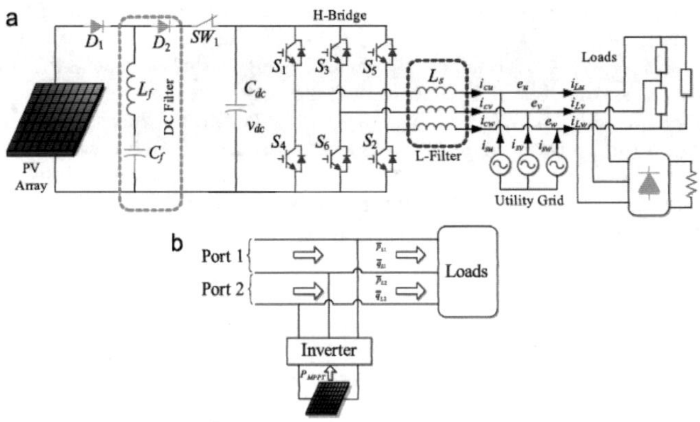

Fig.9: Schematics of PV module used for Simulation along with step-up converter and PV Inverter

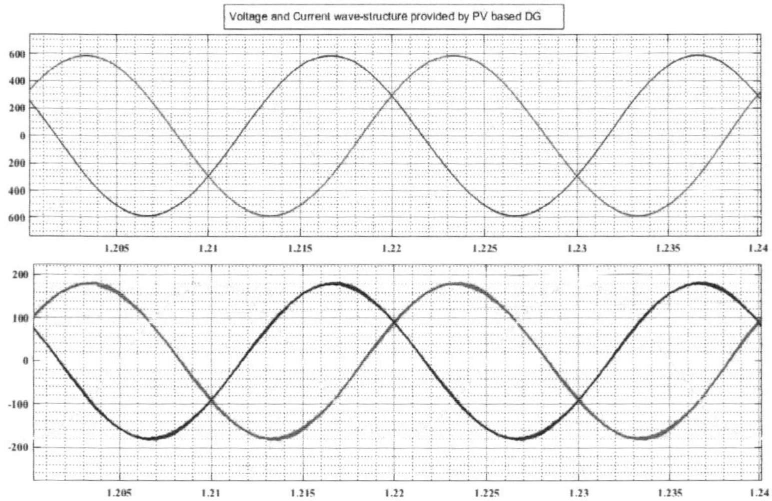

Fig. 10: Voltage and current profile of Inverter output connected to service main.

6. Conclusion

With advancement in the field of integration between renewable based inverter module and on-line grid system, liquidity of power becomes easier. With choosing right inverter for respective application system reliability and life span of converter unit increases. All the issues raised due internal parameters of inverter and their impact on life span of PV module is discussed above. Further to this result of three phase inverter connected to OLGS is presented which defines THD of supplied current in-between the range of 1 -2. Providing transformer-less inverter design with droop controlling strategy reduces the efficiency under high source current although comes with less capital cost and greater reliability. In the field of irregular source current inverter, multi-stage inverter unit is required to meet the THD demand of AC current coming out of Inverter unit. Architecture with this strategy uses less complex control system and requires higher capital cost. Therefore, it is mandatory to first analyse the requirement and then select the inverter topology for higher investment returns.

References

[1]. A. Hensel, C. Gasser, C. Schner, and A. H. Wienhausen, "Integrated 3-phase PV SiC inverter to increase self-consumption with innovative topology, flexible connection options and high efficiency," in Proc. Energy Convers. Congr. Expo., 2015, pp. 1–8.

[2]. National Renewable Energy Laboratory, "On the path to sunshot: The role of advancements in solar photovoltaic efficiency, reliability, and costs," Tech. Rep. No. NREL/TP-6A20-65872, 2016.

[3]. Kouro, S., Leon, J., Vinnikov, D., Franquelo, L.G., et al., 2015. Grid-connected photovoltaic systems: an overview of recent research and emerging PV converter technology. Indust. Electron. Mag. IEEE 9 (1), 47–61.

[4]. Monirul Islam, Saad Mekhilef, Mahamudul Hasan, Single phase transformerless inverter topologies for grid-tied photovoltaic system: A review,Renewable and Sustainable Energy Reviews, Volume 45, 2015, Pages 69-86.

[5]. Iván Patrao, Emilio Figueres, Fran González-Espín, Gabriel Garcerá, Transformerless topologies for grid-connected single-phase photovoltaic inverters, Renewable and Sustainable Energy Reviews, Volume 15, Issue 7, 2011, Pages 3423-3431.

[6]. Joydip Jana, Hiranmay Saha, Konika Das Bhattacharya, A review of inverter topologies for single-phase grid-connected photovoltaic systems, Renewable and Sustainable Energy Reviews, Volume 72, 2017, Pages 1256-1270.

[7]. B. K. Bose, "Global Energy Scenario and Impact of Power Electronics in 21st Century," in IEEE Transactions on Industrial Electronics, vol. 60, no. 7, pp. 2638-2651, July 2013, doi: 10.1109/TIE.2012.2203771.

[8]. S. Jain and V. Agarwal, "A Single-Stage Grid Connected Inverter Topology for Solar PV Systems With Maximum Power Point Tracking," in IEEE Transactions on Power Electronics, vol. 22, no. 5, pp. 1928-1940, Sept. 2007, doi: 10.1109/TPEL.2007.904202.

[9]. Wu, T.F., Chang, C.H., Lin, L.C., Kuo, C.L., 2011. Power loss comparison of single- and two-stage grid-connected photovoltaic systems. IEEE Trans. Energy Convers. 26 (2), 707–715.

[10]. Kasa, N., Ogawa, H., Iida, T., Iwamoto, H., 1999. A transformer-less inverter using buckboost type chopper circuit for photovoltaic power system. In: Proceedings of the IEEE 1999 International Conference on Power Electronics and Drive Systems, 1999. PEDS'99, vol. 2, pp. 653–658.

[11]. Y. Xue et al., "Towards next generation photovoltaic inverters," in Proc. Energy Convers. Congr. Expo. (ECCE), 2011, pp. 2467–2474.

[12]. S. B. Kjaer et al., "A review of single-phase grid-connected inverters for photovoltaic modules," IEEE Trans. Ind. Appl., vol. 41, no. 5, pp. 1292–1306, Sep./Oct. 2005.

[13]. G. Notton et al., "Optimal sizing of a grid-connected PV system for various PV module technologies and inclinations, inverter efficiency characteristics and locations," Renew. Energy, vol. 35, no. 2, pp. 541–554, Feb. 2010.

[14]. G. Bettenwort et al., "Benchmarking the different PV system concepts focussing on their total cost of ownership," in Proc. 25th European Photovoltaic Solar Energy Conf. Exhib./5th World Conf. Photovoltaic Energy Convers., Valencia, Spain, Sep. 2010, pp. 6–10.

[15]. J. M. A. Myrzik and M. Calais, "String and module integrated inverters for single-phase grid connected photovoltaic systems—A review," in Proc. IEEE Power Tech Conf., vol. 2. Bologna, Italy, 2003.

[16]. A. Mäki and S. Valkealahti, "Power losses in long string and parallelconnected short strings of series-connected silicon-based photovoltaic modules due to partial shading conditions," IEEE Trans. Energy Convers., vol. 27, no. 1, pp. 173–183, Mar. 2012.

[17]. A. Sanz et al., "Centralized vs distributed (power optimizer) PV system architecturefieldtest results undermismatched operating conditions," inProc. 37th IEEE Photovoltaic Specialists Conf. (PVSC), 2011, pp. 2435–2440.

[18]. Enecsys Micro Inverters SMI-S240W-60 Datasheet, Ver. 1.5, Enecsys Limited, Cambridge, U.K., 2012.

[19]. Enecsys Ltd., "Enecsys micro inverter SMI-240W-60," SMI-240W-60 datasheet, 2012.

[20]. Y. S. Kim and R.Winston, "Power conversion in concentrating photovoltaic systems: Central, string, and micro-inverters," Prog. Photovoltaic: Res. Appl., 2013, doi: 10.1002/pip.2317.

[21]. Venkatesh Boddapati, T. Sathesh Kumar, N. Prakash, B. Gunapriya, Current droop control of parallel inverters in an autonomous microgrid, Materials Today: Proceedings, Volume 45, Part 2, 2021, Pages 2034-2039, ISSN 2214-7853, https://doi.org/10.1016/j.matpr.2020.09.496.

[22]. Amir Khaledian, Masoud Aliakbar Golkar, Analysis of droop control method in an autonomous microgrid, Journal of Applied Research and Technology, Volume 15, Issue 4, 2017, Pages 371-377, ISSN 1665-6423, https://doi.org/10.1016/j.jart.2017.03.004.

[23]. Ogbonnaya Bassey, Chen Chen, Karen L. Butler-Purry, Linear power flow formulations and optimal operation of three-phase autonomous droop-controlled Microgrids, Electric Power Systems Research, Volume 196, 2021, 107231, ISSN 0378-7796, https://doi.org/10.1016/j.epsr.2021.107231.

[24]. A. H. Arab, Y. Bakelli, S. Semaoui, F. Bandou and I. H. Mahammed, "Long term performance of crystalline silicon photovoltaic modules," 2015 3rd International Renewable and Sustainable Energy Conference (IRSEC), Marrakech, 2015, pp. 1-5.

[25]. K. Agroui, "Indoor and Outdoor Characterizations of Photovoltaic Module Based on Mulicrystalline Solar Cells", Energy Procedia, vol. 18, pp. 857-866, 2012.

[26]. V. Sharma and S. Chandel, "Performance and degradation analysis for long term reliability of solar photovoltaic systems: A review", Renewable and Sustainable Energy Reviews, vol. 27, pp. 753-767, Nov 2013.

[27]. A. Skoczek, T. Sample and E. Dunlop, "The results of performance measurements of field-aged crystalline silicon photovoltaic modules", Progress in Photovoltaics: Research and Applications, vol. 17, no. 4, pp. 227-240, Dec 2009.

[28]. O. Sastry et al., "Performance analysis of field exposed single crystalline silicon modules", Solar Energy Materials and Solar Cells, vol. 94, no. 9, pp. 1463-1468, Sep 2010.

[29]. F. Blaabjerg, K. Ma and Y. Yang, "Power electronics - The key technology for Renewable Energy Systems," 2014 Ninth International Conference on Ecological Vehicles and Renewable Energies (EVER), Monte-Carlo, 2014, pp. 1-11.

[30]. U. Scheuermann and R. Schmidt, "Impact of load pulse duration on power cycling lifetime of Al wire bonds", Microelectronics Reliability, vol. 53, no. 9-11, pp. 1687-1691, Sep. 2013.

[31]. U. Scheuermann, R. Schmidt and P. Newman, "Power cycling testing with different load pulse durations," 7th IET International Conference on Power Electronics, Machines and Drives (PEMD 2014), Manchester, 2014, pp. 1-6.

[32]. H. Huang and P. A. Mawby, "A Lifetime Estimation Technique for Voltage Source Inverters," IEEE Trans. Power Electron., vol. 28, no. 8, pp. 4113-4119, Aug. 2013.

[33]. J. Fiorelli and M. Zuercher-Martinson, "How oversizing your arrayto-inverter ratio can improve solar-power system performance," Solar Power World, vol. 7, pp. 42–48, 2013.

[34]. M. Hussin, A. Omar, S. Shaari, and N. M. Sin, "Review of stateof-the-art: Inverter-to-array power ratio for thin-film sizing technique," Renewable and Sustainable Energy Reviews, vol. 74, pp. 265–277, 2017.

[35]. SMA, "7 reasons why you should oversize your PV array," Dec. 2015. [Online]. Available: http://en.sma-sunny.com/en/7-reasons-whyyou-should-oversize-your-pv-array-2/.

[36]. F. He, Z. Zhao, and L. Yuan, "Impact of inverter configuration on energy cost of grid-connected photovoltaic systems," Renewable Energy, vol. 41, pp. 328–335, 2012.

[37]. R. S. Faranda, H. Hafezi, S. Leva, M. Mussetta, and E. Ogliari, "The optimum PV plant for a given solar dc/ac converter," Energies, vol. 8, no. 6, pp. 4853–4870, May 2015.

[38]. A. Sangwongwanich, Y. Yang, D. Sera, F. Blaabjerg and D. Zhou, "On the Impacts of PV Array Sizing on the Inverter Reliability and Lifetime," in IEEE Transactions on Industry Applications, vol. 54, no. 4, pp. 3656-3667, July-Aug. 2018, doi: 10.1109/TIA.2018.2825955.

[39]. Renewables 2019: Global status report. REN 21, May. 2019. [Online]. Available: https://www.ren21.net/wp-content/uploads/2019/05/ gsr_2019_full_report_e n.pdf

[40]. R. Kaplar et al., "PV inverter performance and reliability: What is the role of the IGBT?," in Proc. 37th IEEE Photovolt. Specialists Conf., 2011, pp. 001842–001847.

[41]. H. Wang, M. Liserre, and F. Blaabjerg, "Toward reliable power electronics: Challenges, design tools, and opportunities," IEEE Ind. Electron. Mag., vol. 7, no. 2, pp. 17–26, Jun. 2013.

[42]. S. Yang, A. Bryant, P. Mawby, D. Xiang, L. Ran, and P. Tavner, "An industry-based survey of reliability in power electronic converters," IEEE Trans. Ind. Appl., vol. 47, no. 3, pp. 1441–1451, May/Jun. 2011.

[43]. J. M. S. Callegari, A. F. Cupertino, V. d. N. Ferreira and H. A. Pereira, "Minimum DC-Link Voltage Control for Efficiency and Reliability Improvement in PV Inverters," in IEEE Transactions on Power Electronics, vol. 36, no. 5, pp. 5512-5520, May 2021, doi: 10.1109/TPEL.2020.3032040.

[44]. M. Khodabandeh, E. Afshari, and M. Amirabadi, "A single-stage softswitching high-frequency ac-link PV inverter: Design, analysis, and evaluation of Si-based and SiC-based prototypes," IEEE Trans. Power Electron., vol. 34, no. 3, pp. 2312–2326, Mar. 2019.

[45]. A. Hensel, C. Gasser, C. Schner, and A. H. Wienhausen, "Integrated 3-phase PV SiC inverter to increase self-consumption with innovative topology, flexible connection options and high efficiency," in Proc. Energy Convers. Congr. Expo., 2015, pp. 1–8.

[46]. A. Tcai, I. M. Alsofyani, and K. Lee, "DC-link ripple reduction in a DPWM-based two-level VSC," in Proc. IEEE Energy Convers. Congr. Expo., 2018, pp. 2558–2565.

[47]. A. Moeini, H. Iman-Eini, and M. Bakhshizadeh, "Selective harmonic mitigation-pulse-width modulation technique with variable dc-link voltages in single and three-phase cascaded h-bridge inverters," IET Power Electron., vol. 7, no. 4, pp. 924–932, Apr. 2014.

[48]. Photovoltaic (PV) Systems—Characteristics of the Utility Interface, IEC 61727 IEC Std, Dec. 2004.

[49]. X. Pei, J. Xiong, Y. Kang and J. Chen, "Analysis and suppression of conducted EMI emission in PWM inverter", Proc. IEEE Int. Electr. Mach. Drives Conf., vol. 3, pp. 1787-1792, Jun. 2003.

[50]. A. Nejadpak, A. Sarikhani and O. A. Mohammed, "Analysis of radiated EMI and noise propagation in three-phase inverter system operating under different switching patterns", IEEE Trans. Magn., vol. 49, no. 5, pp. 2213-2216, May 2013.

[51]. K. Mainali and R. Oruganti, "Conducted EMI mitigation techniques for switch-mode power converters: A survey", IEEE Trans. Power Electron., vol. 25, no. 9, pp. 2344-2356, Sep. 2010.

[52]. L. Premalatha and T. A. Raghavendiran, "A simple approach of generating CPWM for reducing conducted emission in switched mode power supplies", IEEE Electrmagn. Compat. Mag., vol. 2, no. 4, pp. 67-71, 2013.

[53]. Z. Zhang, K. T. Chau, Z. Wang and W. Li, "Improvement of electromagnetic compatibility of motor drives using hybrid chaotic pulse width modulation", IEEE Trans. Magn., vol. 47, no. 10, pp. 4018-4021, Oct. 2011.

[54]. Z. Wang, K. T. Chau and C. Liu, "Improvement of electromagnetic compatibility of motor drives using chaotic PWM", IEEE Trans. Magn., vol. 43, no. 6, pp. 2612-2614, Jun. 2007.

[55]. R. Yang, B. Zhang, D. Qiu and Z. Liu, "Time–frequency and wavelet transforms of EMI dynamic spectrum in chaotic converter", IEEE Trans. Power Electron., vol. 24, no. 4, pp. 1083-1092, Apr. 2009.

[56]. B. Y. Wang, "Research on the loss calculation method and temperature testing of the IGBTs in AC-DC converters with chaotic SPWM control", 2014.

[57]. H. Li, Y. Liu, J. Lü, T. Zheng and X. Yu, "Suppressing EMI in power converters via chaotic SPWM control based on spectrum analysis approach", IEEE Trans. Ind. Electron., vol. 61, no. 11, pp. 6128-6137, Nov. 2014.

[58]. A. Anurag, Y. Yang and F. Blaabjerg, "Thermal Performance and Reliability Analysis of Single-Phase PV Inverters With Reactive Power Injection Outside Feed-In Operating Hours," in IEEE Journal of Emerging and Selected Topics in Power Electronics, vol. 3, no. 4, pp. 870-880, Dec. 2015, doi: 10.1109/JESTPE.2015.2428432.

[59]. M. A. G. de Brito, L. P. Sampaio, L. G. Junior and C. A. Canesin, "Evaluation of MPPT techniques for photovoltaic applications", Proc. IEEE ISIE, pp. 1039-1044, Jun. 2011.

[60]. F. Blaabjerg, R. Teodorescu, M. Liserre and A. V. Timbus, "Overview of control and grid synchronization for distributed power generation systems", IEEE Trans. Ind. Electron., vol. 53, no. 5, pp. 1398-1409, Oct. 2006.

[61]. K. Strunz, E. Abbasi and D. N. Huu, "DC microgrid for wind and solar power integration", IEEE J. Emerg. Sel. Topics Power Electron., vol. 2, no. 1, pp. 115-126, Mar. 2014.

[62]. S. M. Sreechithra, P. Jirutitijaroen and A. K. Rathore, "Impacts of reactive power injections on thermal performances of PV inverters", Proc. 39th Annu. IECON, pp. 7175-7180, Nov. 2013.

[63]. K. Turitsyn, P. Sulc, S. Backhaus and M. Chertkov, "Local control of reactive power by distributed photovoltaic generators", Proc. 1st IEEE SmartGridComm, pp. 79-84, Oct. 2010.

[64]. N. P. Papanikolaou, "Low-voltage ride-through concept in flyback inverter-based alternating current–photovoltaic modules", IET Power Electron., vol. 6, no. 7, pp. 1436-1448, Aug. 2013.

[65]. A. Cagnano, E. De Tuglie, M. Liserre and R. A. Mastromauro, "Online optimal reactive power control strategy of PV inverters", IEEE Trans. Ind. Electron., vol. 58, no. 10, pp. 4549-4558, Oct. 2011.

[66]. M. Morjaria, D. Anichkov, V. Chadliev and S. Soni, "A grid-friendly plant: The role of utility-scale photovoltaic plants in grid stability and reliability", IEEE Power Energy Mag., vol. 12, no. 3, pp. 87-95, May/Jun. 2014.

[67]. L. Liu, H. Li, Y. Xue and W. Liu, "Reactive power compensation and optimization strategy for grid-interactive cascaded photovoltaic systems", IEEE Trans. Power Electron., vol. 30, no. 1, pp. 188-202, Jan. 2015.

[68]. Y. Yang, H. Wang and F. Blaabjerg, "Reactive power injection strategies for single-phase photovoltaic systems considering grid requirements", IEEE Trans. Ind. Appl., vol. 50, no. 6, pp. 4065-4076, Nov./Dec. 2014.
[69]. M. Braun et al., "Is the distribution grid ready to accept large-scale photovoltaic deployment? State of the art progress and future prospects", Progr. Photovolt. Res. Appl., vol. 20, no. 6, pp. 681-697, Sep. 2012.
[70]. Profitable Night Shift, Dec. 2014, [online] Available: http://www.sma.de/.
[71]. D. P. Bagarty and D. Kastha, "Reactive power allocation for loss minimisation in a stand alone variable speed constant frequency double output induction generator", IET Power Electron., vol. 6, no. 5, pp. 851-861, May 2013.
[72]. M. Braun, "Reactive power supplied by inverters: Cost benefit analysis", Proc. EU PVSEC, pp. 2940-2946, Sep. 2007.
[73]. E. Koutroulis and F. Blaabjerg, "Design optimization of transformerless grid-connected PV inverters including reliability", IEEE Trans. Power Electron., vol. 28, no. 1, pp. 325-335, Jan. 2013.
[74]. M. N. Kabir, Y. Mishra, G. Ledwich, Z. Y. Dong, and K. P. Wong, “Coordinated control of grid-connected photovoltaic reactive power and battery energy storage systems to improve the voltage profile of a residential distribution feeder,” IEEE Trans. Ind. Inform., vol. 10, no. 2, pp. 967–977, May 2014
[75]. S. Jin, W. He, J. Zhang, X. Zhu, and H. Liu, “Optimal strategy for distribution system with PV based on voltage violation regulation,” J. Eng., vol. 2017, no. 13, pp. 1124–1128, Jan. 2017.
[76]. K. Turitsyn, P. Šulc, S. Backhaus, and M. Chertkov, “Distributed control of reactive power flow in a radial distribution circuit with high photovoltaic penetration,” in Proc. IEEE PES Gen. Meeting, 2010, pp. 1–6.
[77]. H. X. Nguyen and T. Tsuji, “Optimal PV inverter dispatch with reactive power constraint in medium voltage distribution system,” in Proc. IEEE Power Energy Soc. Gen. Meeting, 2017, pp. 1–5.
[78]. H. Dharmawardena, A. Arzani, and G. K. Venayagamoorthy, “Distributed voltage control for distribution feeder with photovoltaic systems,” in Proc. IEEE Power Energy Soc. Innov. Smart Grid Technol. Conf., 2018, pp. 1–8.
[79]. M. A. Darfoun and M. E. El-Hawary, “Multi-objective optimization approach for optimal distributed generation sizing and placement,” Elect. Power Compon. Syst., vol. 43, no. 7, pp. 828–836, Jul. 2015.
[80]. S. Santos, D. Fitiwi, M. Cruz, C. Cabrita, and J. Catalão, “Impacts of optimal energy storage deployment and network reconfiguration on renewable integration level in distribution systems,” Appl. Energy, vol. 185, pp. 44–55, 2017.
[81]. P. Meneses and J. Contreras, “Optimal placement of energy storage and wind power under uncertainty,” Energies, vol. 9, no. 7, 2016, Art. no. 528.
[82]. A. Narimani, G. Nourbakhsh, G. F. Ledwich, and G. R. Walker, “Storage optimum placement in distribution system including renewable energy resources,” in Proc. Australas. Univ. Power Eng. Conf., 2016, pp. 3–5.
[83]. A. Rabiee and S. M. Mohseni-Bonab, “Maximizing hosting capacity of renewable energy sources in distribution networks: A multi-objective and scenario-based approach,” Energy, vol. 120, pp. 417–430, 2017.
[84]. M. Ahmadi, M. E. Lotfy, R. Shigenobu, A. M. Howlader and T. Senjyu, "Optimal Sizing of Multiple Renewable Energy Resources and PV Inverter Reactive Power Control Encompassing Environmental, Technical, and Economic Issues," in IEEE Systems Journal, vol. 13, no. 3, pp. 3026-3037, Sept. 2019, doi: 10.1109/JSYST.2019.2918185.
[85]. S. Reichelstein and M. Yorston, “The prospects for cost competitive solar PV power special section: Long run transitions to sustainable economic structures in the European Union and beyond,” Energy Policy, vol. 55, pp. 117–127, 2013.

Performance Assessment of IEEE/IEC Method and Duval Triangle technique for Transformer Incipient Fault Diagnosis

Manisha, Kulraj Kaur, Naveen Kumar Sharma, Jashandeep Singh, Deepika Bhalla
I.K.G Punjab Technical University, Kapurthala
kulraj22@gmail.com

ABSTRACT

Power transformers are among the most important equipment in a power system. The assessment of the transformer's condition which is connected to a system is needed to avoid disastrous failures and the subsequent power outages. Although transformers are highly dependable and efficient, but are always under the influence of thermal, electrical stresses which are caused due to the deterioration of transformer insulation. Internal faults in a transformer can be classified as incipient faults and internal short circuit faults. The former type of faults develops over an extended period of time and lead to the degradation of insulation. Incipient faults are generally the result of deterioration of the dielectric strength of transformer mineral oil due to hydrolysis, pyrolysis, and oxidation, which result in the formation of gases that are collected in the gap between the oil and tank and partially get dissolved in it. If incipient conditions are not corrected timely, they may finally lead to collapse of the system where such a transformer is installed. Dissolved Gas Analysis (DGA) methods are extensively accepted by utilities for detection of developing faults. In this work, the brief study of two of the prevalent DGA methods for identification of incipient faults has been carried out. The methods that have been studied are IEEE/IEC Ratio Method and Duval's Triangle Method. The published DGA data where the actual fault is known is used for identification of the incipient faults and the results have been compared.

Keywords: DGA, Duval triangle, Electrical fault, ratio method

1. Introduction

In the transmission system a power transformer is a significant part, it transfers the power from one circuit to another circuit at different voltage levels while maintain the same frequency with a corresponding change in the current. So, transformer reliability significantly contributes to the reliability of the system it is installed in. Mineral oil provides insulation in the transformer, for protection against the live parts coming in contact with each other and along with it takes care of the cooling needed. The presence of explicit gasses in the transformer oil helps in identifying the faults. As a transformer operates and ages, the solid insulation in it decomposes and release hydrocarbon gases, also present along with it are atmospheric gases all these get dissolved in the insulating oil and also accumulate in the conservator tank. Natural ageing also generates gases and Dissolved Gas Analysis (DGA) is a strategy to detect incipient faults in a transformer [1]. Incipient faults occur over a period and can be said to be self-clearing faults and due to this no protective device operates. Such

Content from this work may be used under the terms of the Creative Commons Attribution 3.0 licence. Any further distribution of this work must maintain attribution to the author(s) and the title of the work, journal citation and DOI.
Published under licence by IOP Publishing Ltd

IOP Conf. Series: Materials Science and Engineering 1228 (2022) 012027 doi:10.1088/1757-899X/1228/1/012027

faults are detected by the analysis of the gases dissolved in the transformer oil. DGA is a globally accepted technique for detecting faults in a transformer without taking it out of service. DGA is used for the supervising the condition a transformer. It is similar to how the testing of blood helps in assessing the health of a person, similarly the analysis of the amount of gases in the oil is used for assessing the health of a transformer [2].

DGA includes taking a sample of the mineral oil from the transformer to be tested and subject the oil sample to gas chromatography to identify the gases present in the oil and also their respective concentration. The purpose of DGA is to identify the gas and fix of the level of each gas using appropriate methods like thermal conduction detector, flame ionization detector, and examination for issues by utilizing reasonable indicative techniques to discover the reason for the gas concentration/levels [3].

The main objective of this paper is to apply the IEC/IEEE Method and the Duvals Triangle (DT) method to IEC TC 10 Database and develop a MATLAB Programme for computation of Duval's Triangle method for analyzing the performance of these two DGA techniques. There are standards that different countries follow for analyzing the dissolved gases in a transformer oil and identifying the fault developing/ fault that has already occurred without opening the transformer. Five prevalent DGA methods are briefly studied and for two of these methods; one ratio method and the other graphical method DGA data is used to check their accuracy. The figure 1 shows the flow diagram for the proposed work using DGA Methods for identification of fault in a transformer.

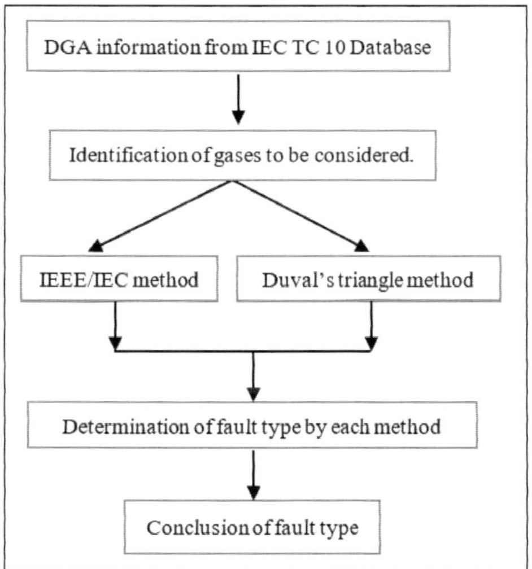

Figure1: Flow Diagram for the proposed work using DGA Methods for identification of fault in transformer

IOP Conf. Series: Materials Science and Engineering 1228 (2022) 012027 doi:10.1088/1757-899X/1228/1/012027

MS excel has been used for the computation of the results of IEEE/IEC method and a MATLAB programme has been developed for fault identification for the graphical method-DT method. The results of these two methods have been compared. In this work, the electrical faults considered are: low energy discharge and high energy discharge, the partial discharge type faults are not considered. The thermal faults considered are categorized as: faults which have temperature less than 700° C (low temperature thermal) and those which have temperature more than 700° C (high temperature thermal faults). DGA samples from IEC TC 10 database are used. The medium temperature thermal faults data has been clubbed with low temperature thermal faults.

2. Identification of Faults by Physical Inspection

On physical inspection of a transformer, faults can be identified and classified as electrical faults and thermal faults [5].

2.1 Electrical faults: Electrical faults can be further divided as high energy discharge and low energy discharge. The high energy discharges can be identified by: arcing between bushing to tank, between windings and core, between windings and porcelain, and arcing in windings, oil, and magnetic circuit (ground), bushing swollen, flashover from LV to ground, short circuit between conductors, from LV to connector and in windings. Partial discharge is also a type of electrical fault.The low energy discharges can be identified by loosening of the bushing rings, marks of tracking and sparking, low energy arcing on bushing, sparking between isolated copper tube and HV braided connection, sparking froma bushing to tank wall, and tracking in bushing[9].

*2. 2. Thermal faults:*Thermal faults can be classified as low temperature and high temperature thermal faults which have temperature < 700° C and > 700° C respectively. The low temperature thermal faults can be identified by: circulating currents in core, LV winding, magnetic tank shunt during heat run tests, carbonized windings during heat run tests, high circulating currents between conductors with winding damage, overheating of a conductor, carbonization of paper strips, and vertical bar buses burnt out. The high temperature thermal faults which have temperature can be identified by: burnt lamination during heat run test, bad contact in windings, circulating currents in clamping bolt, high contact resistance of winding terminal, circulating currents between yoke clamps and connecting bolts, ground wiring burnt, ruptured in ground wire due to circulating current, hot-spot at connectionand steel lamination eroded.

3. Identification of Faults using Dissolved Gas Analysis

Dissolved Gas Analysis (DGA) is a well-accepted technique for condition evaluation of oil filled power equipment. DGA uses the data of the gases present in the mineral oil due to oxidation and hydrolysis. This is achieved by finding the content of each type of gas in ppm (particle per million) by utilizing Gas Chromatography. The atmosphere contributes to the presence of N_2 and O_2, Partial discharge generates H_2, overheated cellulose contributes to the CO_2 presence, CO presence is also due to heated cellulose, C_2H_2 due to arcing in oil, while overheated oil is accountable for the presence of CH_4, C_2H_6, and C_2H_4. The faults are classified as thermal faults and discharge faults. Electrical faults are of two-type corona or partial discharge (PD), and discharge or arcing, while thermal heating creates faults that are classified according to the temperature range. The most severe intensity of energy dissipation occurs with heating, the least with PD followed by arcing. Five traditional DGA methods to recognize faults in a transformer are: Key Gas technique, Rogers Ratio method, Doernenburg Ratio Method, Duval's Triangle method, and IEEE/IEC Method [10].

3.1 IEEE/IEC Method

IEEE/IEC Method is recommended by both Institute of Electrical and Electronics Engineers (IEEE) and International Electrotechnical Commission (IEC) [7]. Its origin is based on the Rogers' ratio method, however in this method ratio C_2H_6/CH_4 is not considered. The reason for not considering it attributed to the fact that it only indicates a small temperature range. Absence of any fault, electrical faults of various degrees of severity, thermal faults of various degrees of severity, partial discharge of low and high energy density, are the four situations that are measurable. Both electrical faults and thermal faults are classified into more accurate subtypes. The table 1 gives the IEEE/IEC method ratios and the ratio codes. The table 2 gives the diagnostic table for the IEEE/IEC method ratio codes [2].

Table 1: IEEE/IEC method range for ratios and the respective codes [1,2]

Range	**Ratio C_2H_4/C_2H_6**	**Ratio C_2H_2/C_2H_4**	**Ratio CH_4/H_2**
<0.1	0	0	1
0.1-1.0	0	1	0
1.0-3.0	1	1	2
>3.0	2	2	2

Table 2: IEEE/IEC method incipient fault diagnostic table

Type of fault	**Fault code**	**Ratio C_2H_4/C_2H_6**	**Ratio C_2H_2/C_2H_4**	**Ratio CH_4/H_2**
No Fault	F_0	0	0	0
Partial discharge of low energy	F_1	0	Not significant	1
Partial discharge of high energy	F_2	0	1	1
Discharge of low energy	F_3	1-2	1-2	0
Discharge of high energy	F_4	2	1	0
Thermal fault < 150° C	F_5	1	0	0
Thermal fault 150°-300°C	F_6	0	0	2
Thermal fault 300°-	F_7	1	0	2

IOP Conf. Series: Materials Science and Engineering 1228 (2022) 012027 doi:10.1088/1757-899X/1228/1/012027

700°C				
Thermal fault > 700°C	F_8	2	0	2

3.2 Duval's Triangle Technique

Duval Triangle (DT) Technique was developed by Michel Duval in early 1970s. This method considers only three gases Methane (CH_4), Ethylene (C_2H_4) and acetylene (C_2H_2) and their percentage concentration is plotted in a triangular graph. The 3 vertices and 3 edges form a regular polygon. The resultant regular polygon/triangle is subdivided into fault zones which are polygons of different dimensions. The intersection point is calculated, and fault zone is identified [5]. The triangular coordinate system specifies each point uniquely in a plane by a set of numerical coordinates. Given in the figure triangle ABC having vertices A (a_1, a_2), B (b_1, b_2), C (c_1, c_2). The coordinates of the vertex A, B, C has been taken as [50 50], [0 0] and [100 0] respectively. Figure 2 shows the Duval triangle and table 3 gives the fault type identified in each zone. The DT technique divided the triangle into polygons which are separate for partial discharge (PD), low energy discharge (D1) and high energy discharge (D2), thermal faults, thermal Fault<300^{0}C (T1), Thermal fault300^{0}C-700^{0}C (T2) and a combination of thermal and electrical fault (DT).

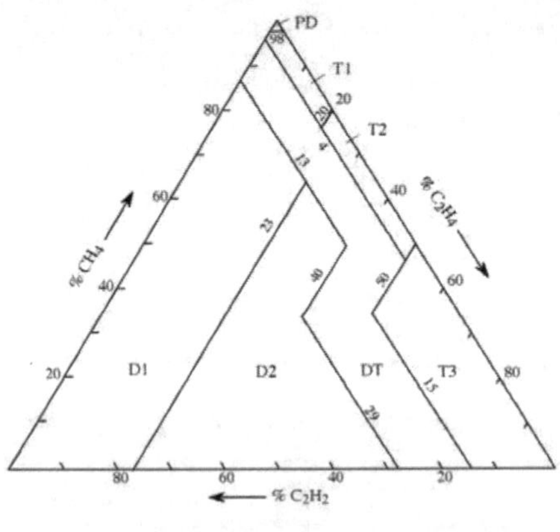

Figure 2: Duval Triangle Representation

Table 3: Fault type and symbols used in Duval's triangle method			
Sr No.	**Fault Type**	**Faults include**	**Symbol**
1.	Low energy Discharge	PD of sparking type like carbonized punctures, pinholes.	D1
2.	High energy Discharge	Discharges in oil or paper	D2
4.	Thermal fault of temperature <300^{0}C	Paper turns brown	T1
5.	Thermal fault of	Carbon particles	T2

	temperature 300⁰C-700⁰C	formation,carbonization of paper	
6.	Combined of thermal and electrical	Discharges of low energy, pinhole	DT
7.	Partial discharge	Cold plasma discharges (corona), voids(or)gas bubbles	PD

The normal ageing is not detected in this method so it could give a false indication, so before application if this technique, it must be assessed for the permissible limit of gasses dissolved for transformers under normal conditions [3].

4. Database

The database used in this work is of DGA gases ppm concentration of transformers that were physically inspected after the occurrence of fault; the actual fault type is known. The four types of fault considered in the work are: Discharge of low energy (DLE), Discharge of high energy (DLE), thermal fault of less than 700°C (TF <), and thermal fault of more than 700°C (TF >). The table 4 gives dissolved gas database from transformer for which the actual fault type is known from physical inspection [8].

Table 4:Dissolved Gas Database from transformer for which the actual fault type is known from physical inspection [8]

S.no.	**H_2**	CH_4	C_2H_2	C_2H_4	C_2H_6	**Actual fault type**
1	8800	64064	-	95650	72128	TF >
2	78	20	28	13	11	DLE
3	305	100	541	161	33	DLE
4	1270	3450	8	1390	520	TF <
5	3420	7870	33	6990	1500	TF <
6	6709	10500	750	17700	1400	TF >
7	1230	163	692	233	27	DLE
8	645	86	317	110	13	DLE
9	95	10	39	11	-	DLE
10	545	130	239	153	16	DHE
11	1570	1110	1830	1780	175	DHE
12	1820	405	634	365	35	DHE
13	3090	5020	2540	3800	323	DHE
14	7150	1440	1760	1210	97	DHE
15	755	229	460	404	32	DHE
16	440	89	757	304	19	DHE
17	290	966	57	1810	299	TF >
18	2500	10500	6	13500	4790	TF >
19	48	610	-	10	29	TF <
20	400	940	24	820	210	TF >
21	6	2990	67	26076	29990	TF >
22	12	18	-	4	4	TF <
23	66	60	-	7	2	TF <

IOP Conf. Series: Materials Science and Engineering 1228 (2022) 012027 doi:10.1088/1757-899X/1228/1/012027

24	595	80	244	89	9	DLE
25	75	15	26	14	7	DHE
26	260	215	277	334	35	DHE
27	60	5	21	21	2	DHE
28	2770	660	763	712	54	DHE
29	1500	395	323	395	28	DHE
30	20000	13000	57000	29000	1850	DHE
31	1170	255	325	312	18	DHE
32	3700	1690	3270	2810	128	DHE
33	290	1260	8	820	231	TF >
34	107	143	2	222	34	TF >
35	14	44	1	7	124	TF <

4.1 Computation for IEEE/IEC method

The calculation of the ratios for the IEEE/IEC method has been carried out in MS excel.

4.2 Modelling of Duval's Triangle method

The DT technique triangle is coded in MATLAB using the following steps:

Step 1: Each of the zone of the regular triangle is a polygon. Calculate boundaries of these zones using respective equations.

Step 2: Feed the percentage values of CH_4, C_2H_4, C_2H_2. The intersection points are found successively and are located within one of the zones.

Step 3: Check the location of the point if it lies within a zone and hence display that zone name.

The pseudo code developed in MATLAB for Duval's Triangle method of DGA for detection of faults in transformer is given below:

```
%Plot the triangle in cartesian coordinates, where A is vertex1, B is vertex2, and C is vertex 3%
Function (vertex1)
  (vertex2)
    (vertex3)
%Plot each side using the vertex %
Function{(vertex1, vertex2)(vertex2, vertex3)(vertex1, vertex3)}
Create zones Z1, Z2, Z3, Z4, Z5, Z6
{zone Z1 using linewidth}
 {zone Z2 using linewidth}
{zone Z3 using linewidth}
   {zone Z4 using linewidth}
        {zone Z5 using linewidth}
              {zone Z6 using linewidth}
%Enter the values%
Input "enter percentage of CH4, C2H2, C2H4"
```

```
{
If "1"
Print response
"point lies in zone 1"
End
If "2"
Print response
"point lies in zone 2"
If "3"
Print response
"point lies in zone 3"
If "4"
Print response
"point lies in zone 4"
If "5"
Print response
"point lies in zone 5"
If "6"
Print response
"point lies in zone 6"
Else
Print response
"point is not lying in the given zone"
}
```

The MATLAB based result of sample where the ppm values are: CH_4 = 130, C_2H_2= 0, C_2H_4 = 96 and thus the percentage values are: % CH_4 = 57.5, % C_2H_4 = 42.5, % C_2H_2 = 0. Using the developed code, the fault detected lies in T2 region is displayed in figure 3 and fault type is thermal faults > 700 °C. The MATLAB code is applied to the database given in table 4 for identification of fault type and the results are compared with actual fault.

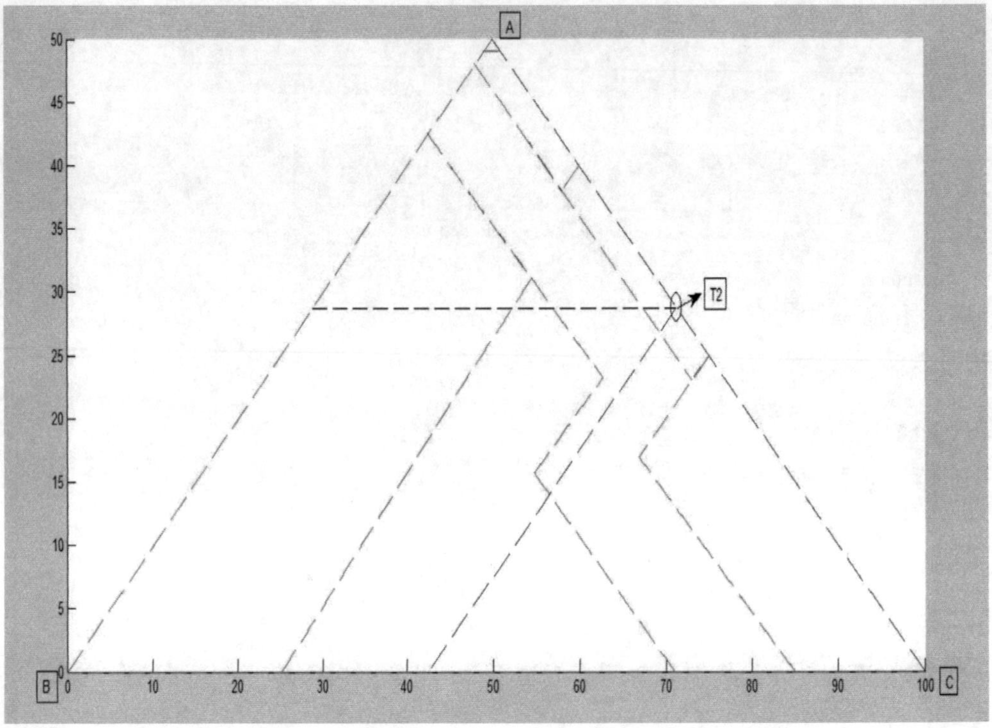

Figure 3: Result of Duval triangle for % CH_4 = 57.5, % C_2H_4 = 42.5, % C_2H_2 = 0 sample using MATLAB.

5. Results

Results of IEEE/IEC Method: The table 5 gives values of the three ratios: C_2H_2/C_2H_4, CH_4/H_2, C_2H_4/C_2H_6 of the IEEE/IEC Method their ratio codes, and the fault detected by it. As per the code generated by the ratios F0 is for no fault condition, F1 and F2 are for low and high energy partial discharge, F3 and F4 are for low and high energy discharge, F5 and F6 are for Low temperature thermal fault, F7 is for medium temperature thermal fault and F8 is for high temperature thermal fault.

Table 5: Ratio codes of samples as per IEEE/IEC Method and comparison with actual fault.

Sample no	Ratio C_2H_4/C_2H_6	Ratio C_2H_2/C_2H_4	Ratio CH_4/H_2	Fault by IEEE/IEC Method	Result Comparison
1	1	0	2	F7	X
2	1	1	0	F3	√
3	2	2	0	F3	√
4	1	0	2	F7	X
5	2	0	2	F8	X
6	2	0	2	F8	√
7	2	1	0	F4	X
8	2	1	0	F4	X
9	0	2	0	ND	--

10	2	1	0	F4	√
11	2	1	0	F4	√
12	2	1	0	F4	√
13	2	1	2	ND	--
14	2	1	0	F4	√
15	2	1	0	F4	√
16	2	1	0	F4	√
17	2	0	2	F8	√
18	1	0	2	**F7**	X
19	0	0	2	F6	√
20	2	0	2	F8	√
21	0	0	2	F6	X
22	0	0	2	F6	√
23	2	0	0	ND	--
24	2	1	0	F4	X
25	1	1	0	F3	X
26	2	1	0	F4	√
27	2	1	1	ND	--
28	2	1	0	F4	√
29	2	1	0	F4	√
30	2	1	0	F4	√
31	2	1	0	F4	√
32	2	1	0	F4	√
33	2	0	2	F8	√
34	2	0	2	F8	√
35	0	1	2	ND	--
ND: Not defined, √: correct, X: incorrect					

Results of DT Technique: The table 6 gives the results of the plots of the DT technique of DGA that are obtained using the MATLAB code and comparison of these results with the actual fault type.

Table 6: Results of DT Technique and the analysis of results compared with the actual fault type.

Sample no	**Fault by DT Method**	**Result Comparison**	**Sample no**	**Fault by DT Method**	**Result Comparison**
1	T3	√	19	PD	X
2	D1	√	20	T2	X
3	D1	√	21	T3	√
4	T2	√	22	T1	√
5	T2	√	23	T1	√
6	T3	√	24	D1	√
7	D1	√	25	D2	√
8	D1	√	26	D2	√
9	D1	√	27	D2	√
10	D2	√	28	D2	√
11	D2	√	29	D2	√
12	D2	√	30	D2	√
13	D2	√	31	D2	√
14	D2	√	32	D2	√

15	D2	√	33	T2	X
16	D2	√	34	T3	X
17	T3	√	35	T1	√
18	T3	√			

5.1 Analysis of the results

Of the 35 samples taken from IEC TC 10 Database, the samples are of only thermal faults and electrical discharges. There are no samples of partial discharge and combined electrical and thermal faults. There are 15 samples of discharge of high energy, 6 samples of discharge of low energy, 6 samples of thermal fault < 700°C and 8 samples of thermal fault > 700°C. The table 7 draws a comparison between the results of IEEE/IEC method and the DT technique.

Table 7: Comparison between the results of IEEE/IEC method and the Duval's triangle method

	IEEE/IEC Method	**DT Technique**
Not Defined	5	0
Incorrect results	9	4
Correct Results	21	31

The IEEE/IEC method could make 60% correct predictions while the DT technique made 88.6% correct predictions. IEEE/IEC method could not make any type of predictions for 14.28% of the cases due to absence of ratio codes or due to absence of a gas, while the DT technique made all predictions as a point could be placed in one of the polygons within the triangle. The performance of the DT technique is found to be better than the IEEE/IEC method.

6. Conclusion

The IEEE/IEC method being a ratio method gives satisfactory results on account of all the ratio codes set not being complete and also due to no interpretation in case C_2H_2, H_2 and C_2H_6 gas not present/or detected. The results of Duval triangle technique are good for the given set of DGA data, but it could give out a false alarm of presence of a fault under normal operating condition as no zone in the triangle is designated to normal operation.

Acknowledgement:

M Duval for sharing the MS excel sheet.

REFERENCES

[1] DeepikaBhalla, Raj Kumar Bansal, Hari Om Gupta, *"Application of Artificial Intelligence Techniques for Dissolved Gas Analysis of Transformers,"* International Journal of Electrical, Computer, Energetic, Electronic and Communication Engineering, Vol:4, No:2, 2010, pp 261-269, ISSN:2070-3724.

[2] DeepikaBhalla, Raj Kumar Bansal, Hari Om Gupta, *"Function Analysis Based Rule Extraction from Artificial Neural Networks for Transformer Incipient Fault Diagnosis,"*

International Journal of Electrical Power & Energy Systems, Vol. 33, Dcc 2012, ISSN: 0142-0615, pp 1196- 1203.

[3] W. H. Tang, Z. Lu, Q. H. Wu. *"A Bayesian network approach to power system asset management for transformer dissolved gas analysis"*, 2008 Third International Conference on Electric Utility Deregulation and Restructuring and Power Technologies, 2008.

[4] Merita, Amit Kumar, R. N. Sharma, and Sushil Chauhan. *"Study and diagnosis of key gases to assess condition of power transformers"*, 2011 3rd International Conference on Electronics Computer Technology, 2011.

[5] Duval M, (2002) *"A review of fault detectable by gas-in-oil analysis in transformers,"* IEEE Electrical Insulation Magazine, 18(3), pp. 8-17.

[6] IEEE Guide for the Interpretation of Gases Generated in Oil-Immersed Transformers, *ANSI/IEEE Standard C57.104-1991.*

[7] IEC *"Mineral Oil Impregnated Electrical Equipment in service. Guide to the interpretation of Dissolved Free Gas Analysis"* CEI, *IEC 60599*, second edition 1999-03.

[8] Duvai M, and dePablo A, (2001), *"Interpretation of Oil in Gas Analysis using New IEC Publication 60599 and IEC TC 10 Databases,"* IEEE Electrical Insulation Magazine, 17(2), pp.31-41.

[9] Kingshuk Chatterjee, Vinay Kumar Jadoun, R k Jarial, Subham Dawn, "A Novel Prediction-reliability Based Graphical DGA Technique Using Multi-layer Perceptron Network &Gas Ratio Combination Algorithm" IET Science, Measurement& Technology, vol. 13, Iss. 6, pp. 836-842, 2019. 10.1049/iet-smt.2018.5397.

[10] Kingshuk Chatterjee, Vinay Kumar Jadoun, R K Jarial, "Emerging Trends for Determining Incipient Faults by Dissolved Gas Analysis", has been published in 3rd International Conference on Condition Assessment Techniques in Electrical Systems (CATCON 2017) [IEEE Conference record No. 41724], organized in Indian Institute of Technology Ropar, Rupnagar, Punjab, India, during 16th to 18th, November, 2017, pp. 63-68. 10.1109/CATCON.2017.8280185

IOP Conf. Series: Materials Science and Engineering 1228 (2022) 012028 doi:10.1088/1757-899X/1228/1/012028

A Recapitulation of Electric Spring for Demand Side Management & Power Quality Mitigation

Sanjana Satish Solanke[1], Vinay Kumar Jadoun[2], Jayalakshmi N. S.[3], Neeraj Kanwar[4], Ashish Shrivastava[5]

[1,2,3]Department of Electrical & Electronics, Engineering, Manipal Institute of Technology, Manipal Academy of Higher Education, Karnataka, India

[4,5]Department of Electrical Engineering, Manipal University Jaipur, India

solankesanjana10@gmail.com, vjadounmnit@gmail.com, jayalakshmi.ns@manipal.edu, nk12.mnit@gmail.com, rewa.ashish@gmail.com

Abstract. The Mechanical Springs act has developed into electric spring (ES). The efficient law developed by Hooke's, has been upraised with the new scientific application which pays vital role in this modern society. The main function of this highly new grid technology provides distributed energy storage, voltage regulation, amplify the power quality and stability in a highly and reliable way. So the improvised control scheme has been assist to open new and different avenues to utilize the electric spring to greater extent. Electric Spring is new and innovative concept by which modern society is able to renew natural energy in easy way. Power match between the generation facet as well as the demand facet will be effectively relieved. This theory transforms many non- critical loads with smart loads in the latest century that have automatically following power generation. It is a novel control approach which is needed in the new future smart grid. The previous concept of output voltage control turns into new one that is input voltage control which is based on reactive power to emphasize this work. Because of the rise in the growth of smart grids, micro grids and RESs, it seems to be observed that ES has been growing in popularity for the past few years. It is a form of power system stability solution and it will be ultimate demand side management which is ready to provide stability solutions.

1. Introduction

The application of sources of renewable energy (RESs) can be seen as significant incense, incensing has occurred recently and is likely to expand in the coming years. Recently, the largest amount of energy contribution has been made from environmentally friendly power sources such as RESs, for example, sunlight based and wind. It is commonly used to fulfill our everyday requirements in a simple way. There is a distant future that has been taken up by the exponential growth in more and more sources of

Content from this work may be used under the terms of the Creative Commons Attribution 3.0 licence. Any further distribution of this work must maintain attribution to the author(s) and the title of the work, journal citation and DOI.
Published under licence by IOP Publishing Ltd

IOP Conf. Series: Materials Science and Engineering 1228 (2022) 012028 doi:10.1088/1757-899X/1228/1/012028

renewable energy. As the power generation is intermittent in nature, it would be difficult to maintain power generation with the variance in load demand. In order to comply with power generation, creative control criteria are required to ensure the demand on load side. This is unlike the generation of power that has historically followed the need for loads as conventional power systems do. In order to implement this, we need some other loads that can comply with RES fluctuations and disruptions without affecting their operating performance. There are many distinct charges in a household, and they are like refrigerators, electric heaters, lighting, etc. It is considered that these loads are non- critical loads. There are several types of other load groups that are critical loads that are only for a stable operation it cannot withstand any fluctuations [1].

This is an unmanageable method for predicting production variability for renewable energy sources. In Smart Grid, attention was drawn towards this novel phenomenon in which load demand should be in cooperate with power generation[2]. Besides this, the varying outputs of the systems will be leading to voltage. The various Fluctuations and the effects of some connected equipment are making the stability worsen of the smart Grid [3] and because of this demand for new method of control [4]. In a smart grid, power quality is an essential concern because it will negatively impacts on the equipment's reliability and produces power economic losses. The use of intermittent wind and solar power presents new challenges to the reliability of the power grid, use is growing day by day, and the need for an instantaneous equilibrium of power supply and load demand [5]. Inclusion and management of new technologies, for example, Electric vehicles, RESs, non-linear loads, electronic power devices as exulted in electric grid modernization [1].

It's a new electric grid mostly referred to as smart grids and it should satisfy specific requirements, Features that enable optimum operation, monitoring, control, improved quality of power. A new device which is called as the electric spring (ES) and recently, this ES was introduced in PQ enhancement and demand side power management in real-time administration. Generally speaking, its entire configuration includes a various power converter which is coupled to a series connected non-critical load (NL).

It improves power quality and stability in same way as the FACTS devices. But with the additional property that makes the load demand to Track the generation of power firmly. These characteristics make it the right one to meet the demands of future smart grids with a rise in elevated source penetration of distributed system generation, as like wind turbines as well as photovoltaic rooftop panels.

It is worth remembering two significant points about ES:

1) Over the last few years, various types of ES have been created. The basic ES type is capable of providing well-regulated standard ac mains voltage within the system voltage tolerance band of +5 percent / -5 percent and adjustable mains voltage which most probably have higher voltage tolerance band of +10 percent /-20 percent.

   The standard mains for critical loads (ZC) (which requires strict voltage regulation) will be used in this situation, while the adaptive mains for non-critical loads (ZNC) will be used (those are having tolerant band of relatively large mains voltage fluctuation) [6][7].

2) Recent ES research shows, however, that if the adjustable mains voltage was regulated, having voltage tolerance band of +5 percent/-5 percent, distributed electric spring will though provide considerable real power as well as reactive power compensation to reduce mains voltage along with frequency fluctuations at the level of the power system[6][7].

IOP Conf. Series: Materials Science and Engineering 1228 (2022) 012028 doi:10.1088/1757-899X/1228/1/012028

2. Theory Of Electrical Spring

2.1. Electric Spring principle

An electric spring, equivalent to the mechanical spring, is the electrical mechanism that can be used for:

i) It offers protection for electrical voltage

ii) The storage of electrical energy

iii) It dampens electrical oscillations.

The electric spring guiding equation is as,

$q = C\, v_a$ (Inductive mode) (1)

$q = -\, C\, v_a$ (Capacitive mode) (2)

C is the capacitance of the capacitor, whereas q denotes the charge contained in the capacitor(C), and V_a is the difference in electrical potential around the capacitor [8].
Three separate modes are controlled by the electric spring. These operating modes are:

1. Capacitive mode of operation.
2. Inductive mode of operation.
3. Neutral mode of operation.

If the non-critical voltage of load drops down the reference value, the capacitive mode of the electric spring (ES) operates.

If the non-critical voltage of load rises above the reference value, a negative voltage will be produced and the inductive mode will start working.

When the voltage of the supply device is same as the reference voltage, then the electric spring (ES) will act in a neutral mode of operation and does not produce any voltage [9].

It has been imagined that if electrical spring were installed massively, they will provide the future smart grid with highly dispersed and stable support inside the distribution network, similar to the future smart grid.

Table 1. Differentiation of ES from mechanical spring [10].

Mechanical Spring	Electric Spring
Mechanical Support	Electric voltage support
Mechanical energy storage	Electrical energy storage
Mechanical oscillation dampening	Electrical oscillation dampening

As a VAR compensator unit, an electric spring differentiates by itself from all the various compensating devices by changing the control technology from the control of output voltage to the control of input voltage. In Fig.1 the similar nature of the ES will be observed.

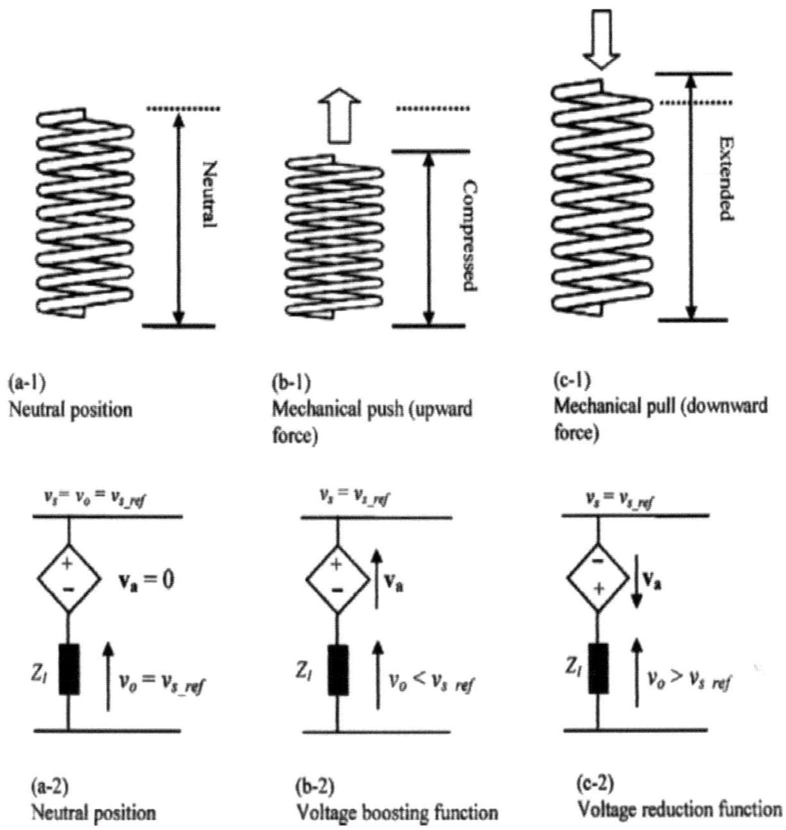

Fig.1. Electric Spring (ES) Scenario [11].

In series, the Electric Spring is related to the loads which will not emphasize with constant voltage. These are implemented as loads that are non-critical load (NCL). When we are assuming constant voltage to be preserved as Vs-ref in case of the non-critical load (VNC), then the VES (electric spring voltage) is to be regulated to meet with the necessity of three modes: neutral, boosting along with suppression mode, as it is shown in Fig. 1, b-1, b-2 and b-3 as well. The closeness between the mechanical spring and the ES is interpreted in Fig.1. In this the electric springs is used with a voltage source producing the VES which is the compensating voltage along with series with a (ZNC) non-critical load [12]-[13].

As the VS ac mains voltage differs from the VS reference value, ES will operate to restore VS in various operating modes. ES produces a support potential to raise the ac mains voltage in the voltage boosting mode of operation, while ES offers a suppressing voltage to moderate AC mains voltage in the voltage suppression mode of operation [14].Electric appliances with the electric spring envelope will adapt to the new generation by modern loads that follow their electricity generation profile. It has large-scale control technologies and broad domestic and industrial application.

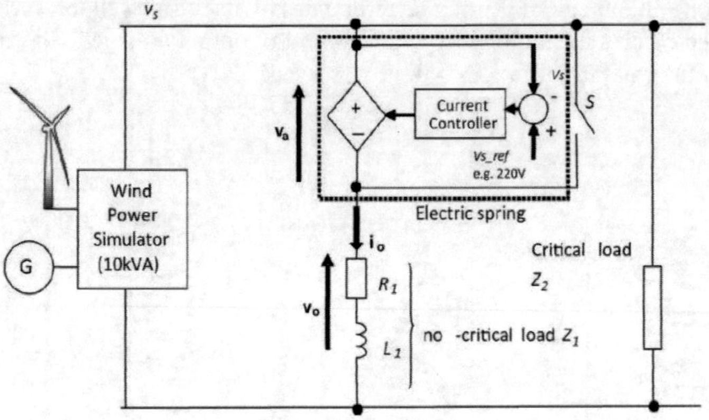

Fig.2. Descriptive visualization of ES operation [1].

The electrical device in Figure.2 is used to demonstrate the idea of an electric spring and its operating constraints. This system is made up of unreliable AC power supply provided by the manual power source which is ac in nature. An ES is to be mounted with series electric load (Z1). In this system, electric spring as well as Z1 together constructs as smart load. Load Z1 is called a non critical load (NCL) since it will be handled with any degree of voltage fluctuation for an AC voltage supply. The load Z1 is called a noncritical load. Electric water heater, refrigerator, as well as lighting systems provide examples of non-critical loads. The electrical load Z1 can normally be represented in series with the resistor R1 and inductor L1. Another electrical load Z2 which requires well controlled ac mains voltage called as a critical load (CL). The load can be split into 2 components, critical along with non-critical load. In that case, ES is linked with series non critical load, so we will ensure like that when line voltage supplying load will fluctuates, the power along with voltage at the CL remain constant. It is called "smart load" to arrange this sort of load. The load which is dissipative is considered as a NCL because it can be worked with some degree of the voltage fluctuation at an AC voltage supply [16]. The main objective of this ES is that to restore Vs back to actual value of the Vs-ref [18] AC mains voltage. The Va and Io vectors must be perpendicular to guarantee that the adjustable ac voltage source was lossless just same as ideal MS. The current Io can either bring the voltage Va lead to 90 degree (capacitive voltage boost) and lag to 90 (inductive voltage reduction mode)[18].

3. Methodology

Fig. 3 shows an Electric Spring schematic diagram linked to various multiple loads which are variable. ES was incorporated as an inverter controlled by voltage that generates a Ves voltage. The control was developed so that Vs would be less than reference voltage when the critical load voltage, Vs-ref; Ves lag Inc by 90 °, which improves Vs to be equal to Vs-ref. Similarly, Ves tends to lead Inc by 90 ° when Vs would be more than Vs-ref, which decreases Vs to Vs- ref [19].
Using Kirchhoff's Voltage Law we can say that,

$$Vs = Ves + Vnc \quad (3)$$

$$V_{pwm-ref} = \text{m}\sin(wt + \theta) \quad (4)$$

Whereas m will be the index modulation which is calculated by the compensation controller whereas θ is the Ves phase calculated by the network synchronization block. Ves has often been kept perpendicular to Inc, ensuring that ES' real power was zero, making this as a lossless device [21].

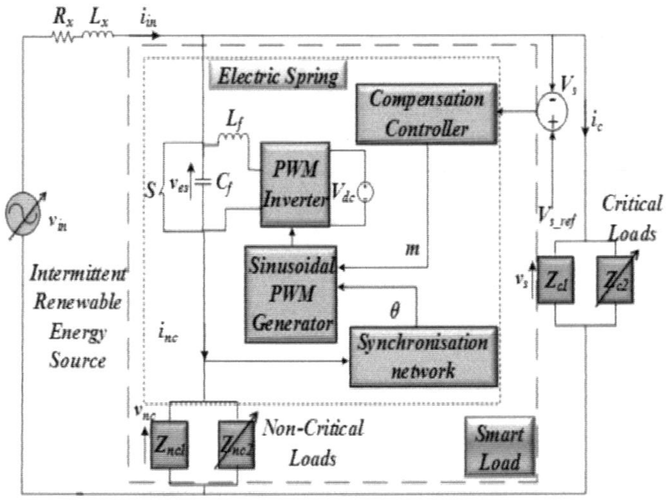

Fig.3 Electric Spring Schematic Illustration linked to IRESs [20].

The main observation is that the electric spring acts as an capacitor for the operation of boosting and a reduction mode of operation for inductor. ES's ability to regulate this voltage will be therefore depend on the circuit's collective reaction [22].

The block diagram for controlling ES is represented in Figure.3. There is a variation among the original mains voltage which is Vs and reference ac mains voltage that can be supplied to a compensation controller with such a PI controller rather a lead lag controller [23]. The control signal for the Pulse Width Modulated (PWM) generator was offered by the output of the compensation controller. A synchronization network is used for the PWM generator phase information that further produces the various switching signals for a power inverter's, as well as power semiconductor switches to produce a large sine wave PWM voltage waveform.

The power inverter's dc voltage is commonly produced from the capacitor that was charged with the help of the inverter switch's anti parallel diodes (like that of a diode rectifier with a dc capacitor output) [24]. A low pass LC filter filters the PWM voltage produced by power inverter so the additional ac voltage $V_a$will be developed. This analysis is to be done to discover various spring operating modes, including the requirements of:

i. Neutral mode of operation
ii. Capacitive mode of operation and
iii. Inductive mode of operation

4. Different Controlling Modes Of Electric Spring

The first electric spring paper was published in 2012[24]. Figures.4 (a) and 4(b), respectively, are based on simplified control schematics. This is very important to mark here that ES distinguishes itself to the adoption of a "input feedback and input voltage control" as seen in Figure 4(c) from previous RPC use. By controlling the Vs which is the input voltage and enabling the Vo which is the output voltage to fluctuate continuously (namely the new control of the input voltage), such an RPC would:

1. As an electric spring, it provides the large voltage support
2. At the same instance of time, the load power is formed to follow the available power produced from the source of renewable energy. Such a subtle shift in an RPC's control strategy from the output side control to an input side control provides new power and voltage control features and functionality [25].

There are many other types of shunt - connected and series - connected reactive power controller applications in power systems [27]-[28]. In the power industry, reactive power controllers (RPC) has been established over the last two decades to regulate flow of power on high voltage transmission lines[29] as well as these are equipped for dimming lighting systems.

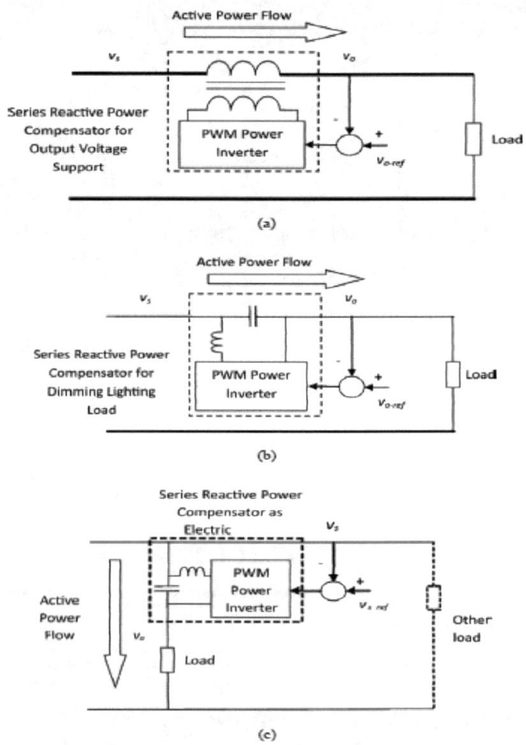

Fig.4 (a) Series RPC for output voltage support [26]. (b) Series RPC as central system of dimming [26]. (c) Series RPC as an ES [26].

Output side voltage of power controller is normally controlled when typical uses of reactive power controllers which are in series for voltage regulation are addressed. Their simpler control schematics, as in Figure 4(a) and 4(b), are highlighted. As well as comprehensive applications used for domestic and industrial purposes too [30]. In micro grids, several models of Electric Spring and related control methods are therefore being suggested and verified practically [31].

5. Comparison of Electric Spring (Es) With Facts

In general, FACTS and custom power devices include many factors same, but converters are the same, working for RPC and voltage stability. ES fundamentally varies in three ways from FACTS forms:

1. Compared to traditional FACTS devices that can able to only handle the reactive control, an Electric Spring provides a viable alternative to active and reactive management [32].
2. FACTS devices such as STATCOM as well as DVR provide a compensation of single point at the level of transmission and must therefore be mounted with far higher ratings, where ES works at several locations such as an inherent part of the partial load, which is also at the level of distribution and therefore less costly due to lower ratings [34].
3. Electric Spring is nothing but source of voltage controlled by current, while FACTS is a source of voltage controlled equipment [35].
4. ES can operate and install for voltage control on the distribution side, while FACTS devices offer centralized transmission level compensation, ES can also be installed in electrical equipment where the individual arrangement is needed for FACTS devices [36].
5. ES do provide voltage control for the non-critical load at input and at the critical load for output, while FACTS devices only provide voltage control at the output [37].

6. Application of ES

Electric Spring began to refer to issues such as momentary power imbalance and voltage control in the smart grid, but later expanded its authority to solve harmonic challenges, neutral current stabilization, correction of power factor, etc.[39]. Figure.5 depicts a brief overview of the applications of electric springs.

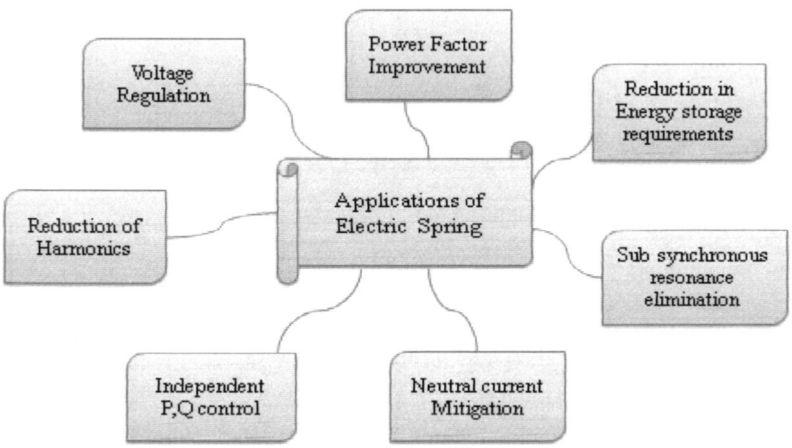

Fig.5 Applications of Electric Spring.

A) Voltage Regulation

The fundamental problem with which ES was established in the 1st place was voltage management. The goal behind the original version of ES was to mitigate disturbances in the mains voltage so that it could be regulated at reference voltage.[41]-[42]. This is accomplished by manipulating the reactive power stored in reactive energy storage component such as a capacitor. When the voltage is low, the electric spring gives reactive power; when the voltage is high, the spring absorbs reactive power. The capacitive & inductive modes of an electric spring (ES) will alter depending on the main voltage.

B) Moderation of Harmonics

Voltage variations can be tolerated by NCL connected in series with ES. Harmonics are generated by the majority of loads at the PCC (point of common coupling). In Electric Spring, the control technique achieves voltage and current values that are within their safe limits [43]. The essential component of quantity is retrieved using a low pass filter in the preparatory step. In most cases, ES is used in conjunction with a resonant controller & a second-order integrator [44]. To minimize harmonics in compensated voltages, current source inverters for ES were considered.

C) Power Compensation

With the addition of power compensation and 2nd & 3rd types of ES, active power & reactive power may be controlled in eight different ways [45]. It became feasible to regulate actual & reactive powers with in system by adjusting the voltage magnitude & phase angles of the electric spring.

D) Neutral Current Mitigation

Neutral current as a result of imbalance in voltage is one of the most common phenomena in three phase circuits. The adoption of the renewable energy will exacerbate the problem. To compensate for the negative sequence current, traditional solutions used are two phase or three phase transformers or rotating induction machines (IM) to alleviate imbalance in voltage. It is thought as less efficient & costly. By transferring a portion of the imbalance in power to non critical load, the ES will mitigates neutral current and achieves voltage equilibrium [46]-[47].

7. Future Scope

Electric spring technology has developed into different types, as reviewed in this work. Several models of Electrical Spring are actually being tested by China's southern power grid [23]. Without Information Communication Technology (ICT), Electrical Spring operates independently. This function will be useful as it guarantees a wide number of ES (electric springs) to hold up future power grid in the event that the chain of communication system collapses. During these days, renewable energy research has become more and more popular. The share of Renewable Energy Sources (RESs) generation will be increased in the future. Rapid exploration is currently needful since distributed control of the huge set of electric springs (ESs) within distribution networks will be able to provide several new methodologies as well as functions as that have not been investigated. For the instance, it is possible to reduce the loss of conduction power in a microgrid [40] and prevent overvoltage in a distribution network [40].

8. Conclusion

This analysis highlights the various methodologies and different electric spring control modes, the comparison of ES and FACTS, and the applications that are being applied to Electric Spring because it is simple to apply. We are progressing quickly toward increase RESs incorporation. Smart grid technology will allow good growth in the future, because of this system would be embraced by our environmental concern world. As our primary source for generation, the planet will use renewable energy, but there are a couple of reactive power fluctuations inside it. Thus, these issues can be resolved easily with the aid of this electric spring. With Electric Spring approaching the platform of various practical assessments by the power generation industry, Electric Spring technology is expected to provide a realistic solution to speed up the use of RESs without creating grid instability to tackle climate change.

Nomenclature

ES	Electric Springs
VAR	Reactive Power Compensators
RESs	Renewable Energy Sources
CL	Critical Load
IRESs	Intermittent Renewable Energy Sources
NCL	Non Critical Load
Vs	Main Voltage
PWM	Pulse-width-Modulation
RPC	Reactive Power Compensators
LC Filter	Low-pass filter
Va	Auxiliary Sinusoidal Voltage
Vo	Output Voltage
FACTS	Flexible AC Transmission System
STATCOM	Static synchronous compensator
DVR	Dynamic Voltage Restoration
ICT	Information Communication Technology
PI	Proportional Integral Controller
C	Capacitor

References

[1] H. Shu Yuen, L. Chi Kwan, and F. F. Wu, "Electric springs-a new smart grid technology," Smart Grid, IEEE Transactions on, vol. 3, no. 3, pp. 1552–1561, 2012.

[2] D. Westermann and A. John, "Demand matching wind power generation with wide-area measurement and demand-side management," IEEE Trans. Energy Convers., vol. 22, no. 1, pp. 145–149, Mar. 2007.

[3] R S Ravi Sankar, S V Jayaram Kumar, K KDeepika, "Model Predictive Current Control of Grid Connected PV Systems," Indonesian Journal of Electrical Engineering and Computer Science Vol. 2, No. 2, May 2016, pp. 285- 296.

[4] Mr.Nikhil Sunil Janwe, Prof.P.R.Jadhav,"Grid voltage stabilization by Electric Springs," 2019 5th International Conference for Convergence in Technology (I2CT) Pune, India. Mar 29-31, 2019.

[5] "International energy outlook 2016," May 2016. [Online].Available:http://www.eia.gov/forecasts/ieo/world.cfm

[6] J. Guo, B. Chaudhuri and S.Y.R. Hui, "Flexible demand through Point-ofload voltage control in domestic sector", IEEE Trans. Smart Grid, vol.10, no.4, pp. 4662 – 4672, Jul. 2019.

[7] T. Chen, J. Guo, B. Chaudhuri and R. S. Y. Hui, "Virtual Inertia from Smart Loads," in IEEE Transactions on Smart Grid, (2020 early access).
[8] Mr.Nikhil Sunil Janwe,Prof.P.R.Jadhav,"Grid voltage stabilization by Electric Springs," 2019 5th International Conference for Convergence in Technology (I2CT) Pune, India. Mar 29-31, 2019.
[9] Ankit Kumar Srivastava, Himanshu Shukla, A.N. Tiwari, "Analysis of Electric Spring for Power Quality improvement", 2018 5th IEEE Uttar Pradesh Section International Conference on Electrical, Electronics and Computer Engineering (UPCON)
[10] C. Subramani , K. R. Ramanand, Journal of Circuits, Systems, and Computers Vol. 27, No. 5 (2018).
[11] K.K.Deepika, Dr. J. Vijaya Kumar, Dr. G.Kesava Rao, "A Comprehensive survey on Electric Springs and it's Applications in Smart Grids'' 3rd IEEE International Conference on Recent Trends in Electronics, Information & Communication Technology (RTEICT-2018), MAY 18th & 19th 2018.
[12] Gang Ma, Guchao Xu, Yixi Chen, RongJu,"Voltage stability control method of electric springs based on adaptive PI controller", Electrical Power and Energy Systems 95 (2018) 202–212.
[13] X. Chen, Y. Hou, S.-c.Tan, C.-K. Lee, and S. Y. R. Hui, "Mitigating voltage and frequency fluctuation in microgrids using electric springs," IEEE Transactions on Smart Grid'', vol. 6, pp. 508-515,2015.
[14] S. Y. R. Hui, C. K. Lee, and F. F. Wu, "Electric springs-a new smart grid technology," IEEE Trans. Smart Grid, vol. 3, no. 3, pp.1552–1561, Sept 2012.
[15] Pallavi G Bendre, PramodJadhav, "Development and Control of Electric Spring to Improve the Stability of Future Smart Grid with Intermittent Renewable Energy Sources" (IJERT) Vol. 4 Issue 03, March-2015.
[16] C. K. Lee, S. N. Li, and S. Y. R. Hui, "A design methodology for smart LED lighting systems powered by weakly regulated renewable power girds," IEEE Trans. Smart Grid, vol. 2, no. 3, pp. 548–554, Sep. 2011
[17] N. R. Chaudhuri, C. K. Lee, B. Chaudhuri and S. Y. R. Hui, "Dynamic Modeling of Electric Springs'', IEEE Trans. Smart Grid, vol. 5, no. 5, pp. 2450-2458, 2014.
[18] Y. Yang, S. Ho, S. Tan, and S. Y. R. Hui, "Small-Signal Model and Stability of Electric Springs in Power Grids," IEEE Trans. on Smart Grid, vol. 9, no. 2, pp. 857-865, 2018.
[19] B. Sen, J. Soni, and S. K. Panda, "Performance of electric springs with multiple variable loads," in 2016 IEEE 8th International Power Electronics and Motion Control Conference (IPEMC-ECCE Asia), May 2016, pp. 3290–3295
[20] Binita Sen Student Member IEEE, Ren Kailin Student Member IEEE, Romika Sharma. "Performance Evaluation of Electric Spring: Effect of Load Variation on Voltage Regulation", 2016 IEEE International Conference on Sustainable Energy Technologies (ICSET)
[21] T. Siew-Chong, L. Chi Kwan, and S. Y. Hui, "General steady-state analysis and control principle of electric springs with active and reactive power compensations," Power Electronics, IEEE Transactions on, vol. 28, no. 8, pp. 3958–3969, 2013.
[22] X. Chen and Y. Hou, "Distributed control of multiple electric springs in microgrids," in 2015 IEEE Power Energy Society General Meeting, 2015, pp. 1–5.
[23] E. Areed, M. Abido and A. Al-Awami, "Switching model analysis and implementation of electric spring for voltage regulation in smart grids", IET Gener. Transm. Dis.: Special Issue: Smart Grid Voltage Control, vol.17, no.15, pp.3703-3712, 2017.
[24] L. Chi Kwan, N. R. Chaudhuri, B. Chaudhuri, and S. Y. R. Hui, "Droop control of distributed electric springs for stabilizing future power grid," Smart Grid, IEEE Transactions on, vol. 4, no. 3, pp. 1558–1566, 2013.

[25] R. Fares, "Renewable energy intermittency explained: Challenges, Solutions and Opportunities", March 11, 2015, ScientificAmerican[online]

[26] Pallavi G Bendre, PramodJadhav, "Development and Control of Electric Spring to Improve the Stability of Future Smart Grid with Intermittent Renewable Energy Sources" (IJERT) Vol. 4 Issue 03, March-2015.

[27] T. Shimakage, "Supply and demand control of distributed generators in a microgrid",Proc. IEEE INTELEC, 2008, pp. 1– 5.

[28] R. D. Patidar and S. P. Singh, "Active and reactive power control and quality management in dc-grid interfaced systems", ARPN J. Eng.Appl. Sci., vol. 4, no. 3, pp. 81–89, May 2009.

[29] W. Yan, S. Y. R. Hui, and H. Chung,Energy saving of large-scale high-intensity-discharge lamp lighting networks using a central reactive power control system, IEEE Trans. Ind. Electron., vol. 56, no. 8, pp. 3069–3078, Aug. 2009.

[30] Gang Ma, Guchao Xu, Yixi Chen, RongJu,"Voltage stability control method of electric springs based on adaptive PI controller", Electrical Power and Energy Systems 95 (2018) 202–212.

[31] Chi-Kwan Lee, "Electric Spring and Smart Load",10.1109/JESTPE.2020.3004164,IEEE Journal

[32] H. Shu Yuen, L. Chi Kwan, and F. F. Wu, "Electric springs-a new smart grid technology," Smart Grid, IEEE Transactions on, vol. 3, no. 3, pp. 1552–1561, 2012.

[33] S. C. Tan, C. K. Lee, and S. Y. R. Hui, "General steadystate analysis and control principle of electric springs with active and reactive power compensations," IEEE Trans. Power Electronics, vol. 28, no. 8, pp. 3958-3969, 2013 N. Tang, K. Yang, H. Huang and C.K. Lee, "The application of one-cyclecontrol technology in electric spring system", IOP Conference Series: Earth and Environmental Science, Volume 170, Issue 4

[34] X. Che, T. Wei, Q. Huo, and D. Jia, "A General Comparative Analysis of Static Synchronous Compensator and Electric Spring", ITEC AsiaPacific 2014, pp. 1–5, 2014

[35] X. Che, T. Wei, Q. Huo, and D. Jia, "A General comparative analysis of static synchronous compensator and electric spring", ITEC Asia-Pacific 2014, pp. 1-5, 2014.

[36] J. Soni, K. R. Krishnanand, and S. K. Panda, "Load-side demand management in building using controlled electric springs." 40th Annu. Conf. IEEE Ind. Electron. Soc., pp. 5376-5381, Oct. 2014.

[37] K.K.Deepika, Dr. J. Vijaya Kumar, "A Comprehensive survey on Electric Springs and it's Applications in Smart Grids", 2018 3rd IEEE International Conference on Recent Trends in Electronics, Information & Communication Technology (RTEICT-2018), MAY 18th & 19th 2018.

[38] S. Y. R. Hui, C. K. Lee, and F. F. Wu, "Electric springs-a new smart grid technology," IEEE Trans. Smart Grid, vol. 3, no. 3, pp.1552–1561, Sept 2012

[39] Y. Yang, Y. Qin, S.C. Tan; S.Y. Hui, "Reducing distribution power loss of islanded AC microgrids using distributed electric springs with predictive control", IEEE Trans. Ind Electron, 2020 (early access).

[40] T. Chen, Y. Zheng, B. Chaudhuri, S.Y.R. Hui, "Distributed electricspring-based smart thermal loads for overvoltage prevention in LV distributed network using dynamic consensus approach", IEEE Trans. Sustainable Energy, 2020 (early access) Gang Ma, Guchao Xu, Yixi Chen, Rong Ju,"Voltage stability control method of electric springs based on adaptive PI controller", Electrical Power and Energy Systems 95 (2018) 202–212.

[41] Gang Ma, Guchao Xu, Yixi Chen, Rong Ju,"Voltage stability control method of electric springs based on adaptive PI controller", Electrical Power and Energy Systems 95 (2018) 202–212.

[42] P. Kanjiya, and V. Khadkikar, "Enhancing power quality and stability of future smart grid with intermittent renewable energy sources using electric springs," *Internationa Conference on*

IOP Conf. Series: Materials Science and Engineering 1228 (2022) 012028 doi:10.1088/1757-899X/1228/1/012028

Renewable Energy Research and Applications., pp. 918–922, 2013.

[43] Ming-Hao Wang, Shuo Yan, Siew-Chong Tan, Shu Yuen (Ron) Hui, Hybrid-DC Electric Springs for DC Voltage Regulation and Harmon Cancellation in DC Microgrids," IEEE Trans. Power Electronics.

[44] Q. Wang, M. Cheng, and Y. Jiang, "Harmonics suppression for critical loads using electric springs with current-source inverters," IEEE J. Emerging Sel. Topics Power Electron., vol. 4, no. 4, pp. 1362–1369, Dec.2016.

[45] S. C. Tan, C. K. Lee, and S. Y. R. Hui, "General steady-state analysis and control principle of electric springs with active and reactive power compensations," IEEE Trans. Power Electronics, vol. 28, no. 8, pp 3958–3969, 2013.

[46] Q. Wang, M. Cheng, and Y. Jiang, "Harmonics suppression for critical loads using electric springs with current-source inverters," IEEE J. Emerging Sel. Topics Power Electron., vol. 4, no. 4, pp. 1362–1369, Dec.2016.

[47] S. Yan, S. C. Tan, C. K. Lee, B. Chaudhuri, and R. Y. Ron Hui, " Electric springs for reducing power imbalance in three-phase power systems,"IEEE Trans. Power Electron., vol. 30, no. 7, pp. 3601–3609, Aug. 2015.

Multidisciplinary avenues on the integration of industry 4.0 in the degree of mechanical engineering

Mayank Deshpande, Mayank Khandelwal and Kunal Singh

Department of Mechanical Engineering, Shri Ramdeobaba College of Engineering and Management, Nagpur – 440013, INDIA

Abstract. The world is moving towards Industry 4.0. There has been rapid growth in a different type of technologies being used in industries. Modern industries have been exploiting the digitalization of machines to increase efficiency and analyzing the data for various other purposes. Even though mechanical engineers and manufacturing move hand in hand, in this data-driven production era, mechanical engineers will have to adapt themselves to various other technologies from automation to data analytics for their efficient implementation. This paper proposes a conceptual framework for Industry 4.0, and also discusses how these technological advancements will assist mechanical engineers in various tasks like designing, machining, etc. The suggested framework covers the entire lifecycle from manufacturing to product deployment in the sense of Industry 4.0.

1. Introduction

Since the first industrial revolution just about 200 years ago, there have been various advances in the way goods are manufactured. From electrically run machines (the second industrial revolution) to extensive digitization (the third industrial revolution), the world has now come to the era of the fourth industrial revolution so-called – "Industry 4.0". The definitive goal of Industry 4.0 is to achieve mass customization of products according to the needs of people and consecutively making the social and economic environment more human-friendly [1]. This is achieved by a system having integration of machines and computer algorithms called Cyber-Physical System (CPS) [2].

Industry 4.0 marks the era of smart manufacturing where the machine learns by itself using various IoT (Internet of Things) based devices and automation & robotics are used to make manufacturing more precise and effortless. Even though Industry 4.0 relies on technologies like IoT, Big Data, Augmented Reality (AR) and Virtual Reality (VR), etc., it is important to note that in its roots it is a manufacturing revolution and hence its development and working will highly hinge on expertise of mechanical engineers. Mechanical engineers will play a very crucial role in the sustainability and efficient working of the integrated CPSs and Industry 4.0. Manufacturing companies are facing lots of difficulties in the implementation of Industry 4.0 technologies and hence require a stable approach for the shift. In this paper, first, we develop a conceptual framework having three layers: Smart Manufacturing, Smart working, and Smart Product. We then analyze how these layers play their role in the integration of the entire Industry 4.0. As a key finding, we then propose multidisciplinary avenues that enhance the processes taking place in these layers and the role of Mechanical Engineering in maintaining and controlling them.

Content from this work may be used under the terms of the Creative Commons Attribution 3.0 licence. Any further distribution of this work must maintain attribution to the author(s) and the title of the work, journal citation and DOI.
Published under licence by IOP Publishing Ltd

IOP Conf. Series: Materials Science and Engineering 1228 (2022) 012029 doi:10.1088/1757-899X/1228/1/012029

As the capabilities of industries are increasing, so is their ability to satisfy the needs of the customers. The main focus of industries in the upcoming era will be the customization of products according to the needs of the customers, that too at affordable costs. The technologies like IoT, AR, VR will play a major role in achieving the goal of mass customization.

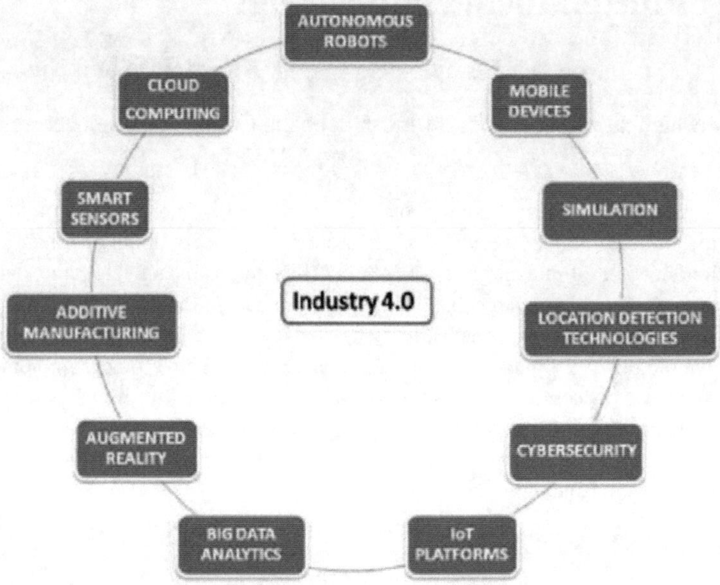

Figure 1. *Key technologies of Industry 4.0*

2. Smart Manufacturing

Manufacturing methods have evolved drastically since the first industrial revolution. Industries have been continuously adopting the new upcoming technologies to increase their chances of sustaining. One of the main goals of Industry 4.0 has always been enhancing the capabilities of the manufacturer to cope with the individual customer requirements, simultaneously maintaining very low production volume and

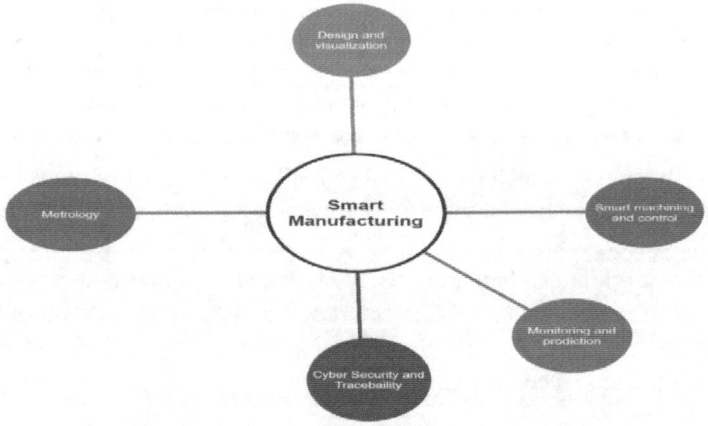

Figure 2. *Smart Manufacturing Topology*

still gain profit. Many concepts have emerged like factory-within-factory, which have resulted in a massive boost in productivity for many industries. In the longer term, smart manufacturing also aims to overcome bigger global challenges, like resource and energy efficiency, zero-waste manufacturing, allowing manufacturers to enjoy continuous productivity and efficiency gains [3]. The introduction of CPSs and Cyber-Physical Production Systems (CPPSs) play a major role in this smart manufacturing paradigm. Today engineers have many IT tools at their disposal like CAx, Big Data, Cloud computing, IoT which can help them achieve these goals. We have subdivided the smart manufacturing-related technologies into five main purposes:

2.1 Visualization and Design

Earlier the mechanical engineers used to make hand-drawn designs of objects, since then various technologies like Computer-Aided Design (CAD), Computer-Aided Manufacturing (CAM), Computer-Aided Engineering (CAE) have emerged which help to visualize the object more efficiently. The introduction of this CAx software's revolutionized the field of engineering design as more detailing can be achieved. This software plays a crucial role in the entire production cycle from product concept, designing, prototyping, and testing, for the programming of CNC machine tools. These CAD software's have the capabilities to interact with prototyping systems via 3D printing, integrated with CPS and AR [4]. Various other technologies like VR and AR have also been developed helping to visualize objects in real-time. AR and VR can provide new direction and opportunities for a revolutionary manufacturing training environment [5].

In Industry 4.0, with the extensive use of IoT technologies the design of the object can also be created from data obtained using a mix of sensors such as LIDAR, photogrammetry, and providing this directly to the machines for prototyping and testing. This data can later be imported to physical or cloud-based databases via an MQTT or HTTP data feed from sensor gateway or by directly connecting to another database via APIs. This will also help in creating digital twins of the object to get a real-time comprehensive view. Millions of data points, sensor data, maintenance, and simulation data can be used for reducing design flaws. Simulations will be done more rigorously in industries to mimic the actual physical environment of the object and assess its capabilities [6]. This will help in improving the user experience and quality of the product.

2.2 Smart machining and Control

In Industry 4.0, smart machining refers to the synchronized or collaborative working of the different types of machines that are used in factories. This can be achieved through smart robots and various sensor coordination. Various emerging manufacturing techniques follow the rules of smart machining like Additive Manufacturing (Layer by layer deposition of material for production) [7], Flexible Manufacturing System (FMS) [8], Reconfigurable Manufacturing System (RMS) [9]. Various Additive Manufacturing technologies like Selective Laser Sintering (SLS), Fused Deposition Modeling (FDM), 3D Printing (3DP) are used for the development of products presently, but one of the drawbacks of these techniques is that they cannot be used for mass production. FMS is a highly automated system that reacts to both predictable and unpredictable changes that take place in the system. Such a system can be used for the purpose of mass customization as per the individual need of a customer. These processes provide in-process quality control and hence render the need for post-process quality check.

Cloud-based manufacturing and control are some of the most efficient manufacturing methods. It uses Cloud computing for real-time production planning and scheduling and hence plays a very important role in reducing overall downtime. It is one of the most important parts of CPS. Cloud manufacturing thus requires extensive support from IoT and Cloud computing for its efficient working. Cloud computing also enables the capability of controlling and monitoring the machine remotely [10]. In the case of CNC, the operator can feed the program to the machine remotely via the cloud and adjust its runtime and scheduling. The machine will interpret and create an offline tool-path for production. The IoT-enabled sensor can be used in the machine to get the real-time data of production and this data can also be stored in the cloud or any kind of physical storage for future use. The data collected can also be used for creating a Digital Twin for a comprehensive representation of its attributes. CAM helps in determining the entire parameters like feed, depth of cut, tool-path, etc. during the design phase itself and the data is then provided to the machine for manufacturing of the product.

2.3 Monitoring and Prediction

Until now, if there is some kind of failure or damage in the system somewhere, the engineers could easily analyze the data to detect the problem and find its solution. But as the machining systems are getting more and more complex, the engineers might not be able to properly detect and solve the problem. A promising approach to this would be, intelligent condition monitoring systems (ICMS) [11]. Various IoT sensors are attached to the machine to get a comprehensive view of various factors like the direction of vibration, phase, voltage, amplitude, battery status and helps make real-time adjustments. This data is then used to predict the possible failures that might occur in the machinery. The predictive manufacturing methods are employed to reduce the downtime of the system and maintenance costs as much as possible. The area of prognostics and health management (PHM) has been a priority research area for various industries [12]. Digital Twins technology can also play a substantial role in PHM and predictive management as they are the exact digital prototype of the physical object.

Another major aim of Industry 4.0 is Energy management. There are various factors to look for when computing the energy efficiency of any machine. This can be easily taken care of by using IoT sensors to remotely monitor the factors that cause the efficiency of any system to increase or decrease. The data for these sensors can also be assessed using machine learning methods like Deep Neural Network (DNN) to determine the energy demand characteristics [13].

2.4 Cyber Security and traceability

As the overall framework of smart manufacturing is highly dependent on data, there will always be a risk of data theft, viruses, and malware, hacking which can cause heavy damage to the industry. The need for strong security protocols for avoiding these things becomes necessary. Many advanced internet protocols make use of artificial intelligence to track the behavioral pattern of these viruses and malwares to stop them from entering and corrupting the system [14]. Encryption of communication data along with stored data and data stream is necessary, to avoid the chances of data tampering [15]. Some methods presently used by the experts include creating a multiple-layer defensive firewall and isolating remote nodes in different networks but these methods are not going to be enough. In 2005 a virus named zotob worm attacked the servers of Daimler Chrysler, temporarily shutting down the entire production line of the plant, resulting in financial losses in thousands of dollars. At that time there was not much automation in plants but if such attacks happen in Industry 4.0 era, then it might result in losses in the range of millions of dollars.

Tracing technology can be used to monitor the security and condition of any smart product. Traceability of a product gives unique information about the product like its location information, temperature, pressure, etc. with the help of smart sensors attached to it [16]. Traceability also helps in enhancing vertical integration in the plant by providing information regarding materials used on the shop floor using various technologies like Radio Frequency Identification (RFID). RFID can also be used to give unique code to the product to easily access all the information regarding that product. Security and traceability are very important factors in increasing the efficiency of smart manufacturing. In the era of customer-specific marketing, tracking data if collected, can be used by the companies to decide their target audience and hence increase more sales. This will also help in the customization of the product according to the target audience beforehand.

2.5 Metrology

With such complex production systems in action, high accuracy and precision will be an obvious parameters in product manufacturing. The application of in-process metrology will be very crucial to obtain high-quality products at low production costs. In-process metrology employs the contactless measurement of products at different stages using laser triangulation or structural light scanners. These systems also help in real-time diagnostics and fault detection. The use of automation and robots increases the efficiency of the inspection process [17]. The robots are attached with structural light scanners in them which during production can assess the product quality in real-time, this also helps in monitoring the measurement of more intricate areas inaccessible present measuring instruments. The aim of automation in metrology is to achieve a high degree of accuracy with a very short measurement time. Industries often have a measurement cell where the robots have some kind of vision sensor attached to it, to identify different parts according to their particular assembly. These cells are usually controlled by PLC controllers [18].

3. Smart Working

Smart working is a different aspect of the future but at the end of the day, it is all about the stream of people working together in a very precise way, fully integrated with the best of technology to do amazing things. In a broad sense, smart working enumerates autonomy, flexibility to the workers to work where they want. This creates a new way of managing and regulating the relationships between company and worker with the help of technology, to provide a better working environment to the workers to enhance productivity. Overall, what's going on is the combination of the genius, good ideas, and common sense of every single worker working with different technologies, with one purpose in mind that is a minimum waste of effort and time.

Smart working also includes technology to support vertical integration within the organization. Everything from man, material, and machine are interconnected and thus each level of production has real-time information of the product. For smooth and proper working of the organization in industry 4.0, firstly, digitalization of documents is necessary, this can be achieved by "cloud management system", this will enable remote access of document and real-time data sharing within the workers and with the management [19]. Secondly, the development of system security is essential, to allow remote access without entailing the risk of information leaking. This altogether helps in improving the level of automation.

Even when there are technologies like robots, AR, VR, Big Data, etc. that can do various work very efficiently, but when it comes to developing things differently and more discretely expertise of an engineer is irreplaceable. The talk is about fine-tuning things, problem-solving, and the speed with which we can diagnose whether something is about to happen, and make interventions to keep the work going on. This will incredibly affect the overall efficiency of the factory. Above all this it is most important to nurture the human element i.e., to ensure that the workers are comfortable, safe, and well-supervised because only when all of these things come together it makes a happy working environment.

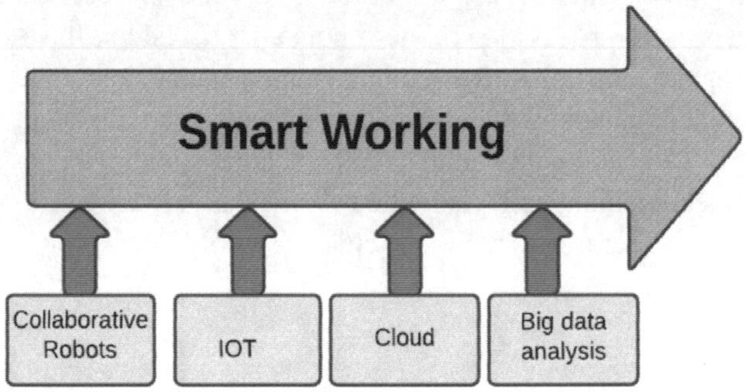

Figure 3. Technological input to smart working

3.1 Challenges Faced in smart Working

Smart working is a necessity for the upcoming industrial revolution. The above-mentioned technologies play a very crucial role in this paradigm and when used efficiently can improve the production efficiency of plants by more than 250%. The truth that smart working plants will be arranged as exceedingly complex, dynamic, and adaptable systems implies that they will need employees who are engaged to act as decision-makers and controllers [20]. The present workforce of engineers is not qualified enough to work efficiently enough in this environment. Therefore, there is a requirement for training and continuing professional development for Industry 4.0. In the future, the person having a foot in both Mechanical Engineering and a little bit of IoT, automation, and analytics is sure to excel in that paradigm.

4. Smart Products

Industry 4.0 lead to the emergence of many technological developments, that are combination of advanced digital technologies and optimized industrial manufacturing systems. Smart manufacturing to smart products depicts the beginning to end of industry 4.0 [21]. Demands for the smart product are increasing due to the rapid change in market requirements, which increases functional elements and becomes more complex due to the interaction of various technologies. Smart products can store, process, and communicate the data to industrial systems without any external means. IoT, CPS, and the processing of vast data (Big Data) are the modern information technologies that are used in industry 4.0. Some of the smart technologies that play a key role in the formation of Industry 4.0 are discussed.

CPS connects computational and physical aspects of the system and facilitates efficient communication between them [22]. CPS is the configuration of embedded systems, sensors, actuators including network access.

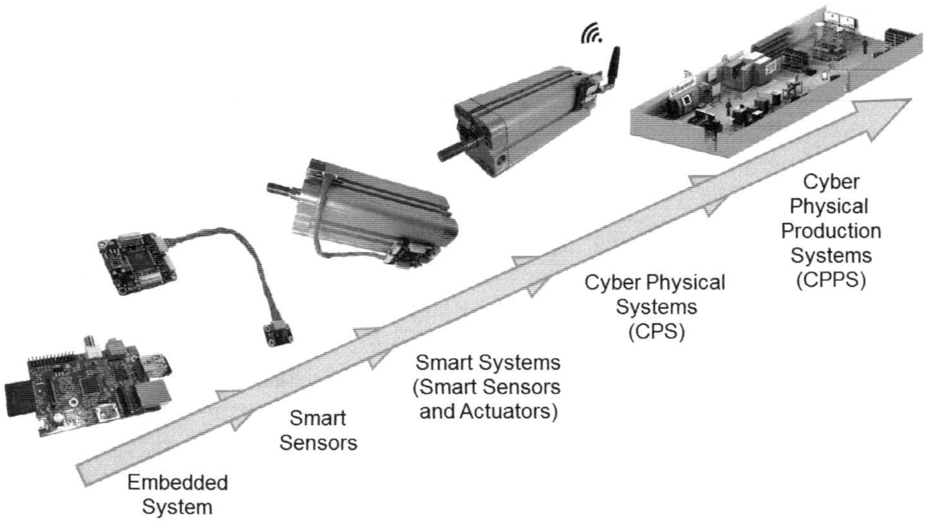

Figure 4. System ladder for CPPS

Internet technology is the future to upgrade and enhance the performance of CPS, which consists of three concepts: IoT, internet of services (IoS), internet of Data (IoD).

Internet technology offers a vast database for connecting and exchanging data with different systems over the internet. CPS systems have a similar pattern as IoT but Emphasize higher combination and coordination between physical and computational elements. Technologies like AR and VR played a major role in upgradations in prototyping and product development, which paved more opportunities for smart product development. This offered highly flexible and durable products at an affordable cost [23].

During the production processes, data generated by machines and sensors, that are also interlinked to each other are stored in a big field of data by Cloud computing. Analyzing enormous volume of data generated during production processes involving advanced, predictive, historical, and descriptive analysis or Big Data techniques. This makes it possible to examine the state and operation of machines.

This isn't about putting smart things into existing products but using smart technologies to create value for the organization, end-user, and the value chain. Necessary elements for manufacturing smart products are:

4.1 Smart Product Design

The design should be quick, easy & correct, should be designed to ask questions about how, what, when, which needs are to be acknowledged. The design should be of a type that allows automation of repeatable processes at an accelerated rate which would gather and process the data by computing system. This will in fact result in great savings and optimization of work.

4.2 Hybrid Prototyping

The modifications/prototyping in existing products in terms of smart technology should be assessed according to the client's requirements. Integration of prototyping and VR offers a solution to these problems. Both these technologies are called time compression technologies [24]. Rapid prototyping and VR produce a product directly & digitally, much quicker and highly similar to the real product without actually building a physical one, which allows to conduct tests and examine the products efficiently. Hybrid prototyping is a system by which traditional physical prototypes are connected with computer-generated visualizations allowing easier simulation processes. A major drawback of Hybrid Prototyping is the expenses in its working which hinders its manufacturing at a commercial level.

4.3 Smart production Control

The smart product should be designed using available resources and should achieve the aim of reducing the cost of production and increasing efficiency along with synchronization of material flow in the production system. The reactions that occur in internal and external assets in the production system affect its functioning. Proper controlled methods can help overcome this and permit smart production. [25-26]. Design and process knowledge is the main key of product to be manufactured and accordingly decisions are made. Clients' requirements are to be kept in mind while designing products. As a result, so many factors are taken to be cared of, which makes it important to follow certain well-defined production methods to undergo dynamic changes. Techniques like IoT helps in gathering informative data for production processes, getting the solution for the best material flow, conducting operations, and manufacturing devices. On one hand sensors and controllers work is to assess and analyze the production data, on the other hand, communication devices work is to provide a way of interaction between production resources at the same time. CPS system is designed for its autonomy in working, without any human intervention [27]. An idea of smart production when fully examined, realized, and implemented with all solutions, makes it possible to achieve the expected flexibility of a production system on an organizational level and not only on a technical level.

In 2016, Microsoft released a product called HoloLens. This is an AR device that is designed to be used in the business sector. This offers a hologram to be viewed with an image resolution, with 3D elements occurring naturally along with voice commands. These are enriched by capabilities based on the cloud and Artificial Intelligence. Another product was developed by Xiaomi in 2021 called "Mi Air Charge Technology". It produces a narrow wave beam which will start charging your mobile phone as soon as you go near it. Such futuristic technologies are developed keeping in mind the need of the customer.

5. Conclusion

As technological advances are taking place all around us, people are expecting things to be more suitable as per their personal needs. But mass customization of this quantity requires sturdy infrastructure that could support the minuscule changes in demands with out affecting the profits of the companies. In industry 4.0 the production techniques are more flexible for mass customization as well as there has also been the focus on energy and resource management. But with the introduction of systems like CPS, Cloud computing, Big Data, IoT, production methods are getting more and more complex, and the present workforce is not trained enough for handling such mechanisms. As stated earlier, Industry 4.0

in its truest sense is an industrial revolution and hence mechanical engineers will have a crucial role in its efficient functioning. The expertise of a mechanical engineer is required from designing a machine to getting the satisfactory output to analyzing and finding the solution in case of failure. In Industry 4.0 paradigm mechanical engineers will need to know apart from their specific fields, like data analysis, embedded system, automation, etc. to provide a significant contribution in the area. From the research done during the writing of this paper, it was found that there were more technologies for a single broader application, for example, technologies associated with efficient connectivity would be IoT/IoS, RFID, cyber security, machine learning, Interfaces, and these technologies assured proper interoperability, reliability, sustainability, decentralization, resilience, networkability of the system.

The success of the industries will mostly depend on their competence in adopting the new system and the flexibility of the dynamic market. It is also important for setting proper standards for communication and data transfer to ensure the interoperability and sustainability of the cyber-physical environment. These standards should be set keeping in mind the dynamics of production and should be more tilted towards precision and accuracy rather than supply and demand, this would make the route for innovation within the companies to meet the market demands satisfying both customers as well as the manufacturers. There are various challenges in the sustainable implementation of Industry 4.0. The population of unskilled and semi-skilled labor will be affected resulting in a huge global economic crisis, and hence it is necessary to ensure steady implementation of the system to ensure timely adaptability of the population. Other challenges include the cost of implementation, adaptability, regional problems, etc. Many factories have already started the transition towards Industry 4.0 and have contributed strongly to boosting the global economy. Huge data centers are being constructed all over the world to store the huge amount of data that is being generated.

There is a huge scope of research on this topic right from its implementation and effect on the global economy to energy management and eco-friendly energy production techniques. Researchers have already started talking about the 'Industry 5.0' and 'Society 5.0'. Industry 5.0 is mainly based on the efficient collaboration between humans and robots, but simultaneously increasing the level of automation using Big Data and Artificial Intelligence (AI). Many Industrialists have also been acknowledging the concept of "Dark Factory" (fully automated factories) and have been investing heavily in ongoing researches. Whatever the case may be, robotics and AI are going to be an eminent part of the future manufacturing era.

Acknowledgments
The authors would like to thank Mechanical Engineering Department, RCOEM for providing constant motivation to write the paper. The authors would also like to thank Prof. Vishal Shukla Sir for their valuable suggestions and for reviewing the final paper.

References

[1] Lasi, H., Fettke, P., Kemper and HG. et al. Industrie 4.0. Wirtschaftsinf **56**, 261–264 (2014). https://doi.org/10.1007/s11576-014-0424-4

[2] Dafflon, B., Moalla, N. and Ouzrout, Y. The challenges, approaches, and used techniques of CPS for manufacturing in Industry 4.0: a literature review. *Int J Adv Manuf Technol* **113**, 2395–2412 (2021). https://doi.org/10.1007/s00170-020-06572-4

[3] Adam Hamrol, Józef Gawlik and Jerzy Sładek. MECHANICAL ENGINEERING IN INDUSTRY 4.0. *mper*. DOI: 10.24425/mper.2019.129595

[4] Zheng, P., wang, H. and Sang, Z. et al. Smart manufacturing systems for Industry 4.0: Conceptual framework, scenarios, and future perspectives. Front. Mech. Eng. **13**, 137–150 (2018). https://doi.org/10.1007/s11465-018-0499-5

[5] Sudip Phuyal, Diwakar Bista and Rabindra Bista,. Challenges, Opportunities and Future Directions of Smart Manufacturing: A State of Art Review, Sustainable Futures, Volume **2**, 2020, 100023, ISSN 2666-1888, https://doi.org/10.1016/j.sftr.2020.100023

[6] Saurabh Vaidya, Prashant Ambad and Santosh Bhosle, Industry 4.0 – A Glimpse, *Procedia Manufacturing*, Volume **20**, 2018, Pages 233-238, ISSN 2351-9789, https://doi.org/10.1016/j.promfg.2018.02.034.

[7] Horst, Diogo & Duvoisin, Charles Vieira and Rogerio. (2018). Additive Manufacturing at Industry 4.0: a Review. *International Journal of Engineering and Technical Research*. 8. 3-8.

[8] Kashif Mahmood, Tatjana Karaulova, Tauno Otto and Eduard Shevtshenko, Performance Analysis of a Flexible Manufacturing System (FMS), *Procedia CIRP*, Volume **63**, 2017, Pages 424-429, ISSN 2212-8271, https://doi.org/10.1016/j.procir.2017.03.123.

[9] Jeff Morgan, Mark Halton, Yuansong Qiao and John G. Breslin, Industry 4.0 smart reconfigurable manufacturing machines, *Journal of Manufacturing Systems*, Volume **59**, 2021, Pages 481-506, ISSN 0278-6125, https://doi.org/10.1016/j.jmsy.2021.03.001.

[10] S. Suárez Fernández-Miranda, M. Marcos, M.E. Peralta and F. Aguayo, The challenge of integrating Industry 4.0 in the degree of Mechanical Engineering, *Procedia Manufacturing*, Volume **13**, 2017, Pages 1229-1236, ISSN 2351-9789, https://doi.org/10.1016/j.promfg.2017.09.039.

[11] Jay Lee, Behrad Bagheri and Hung-An Kao, A Cyber-Physical Systems architecture for Industry 4.0-based manufacturing systems, Manufacturing Letters, Volume 3, 2015, Pages 18-23, ISSN 2213-8463, https://doi.org/10.1016/j.mfglet.2014.12.001.

[12] Sameer Mittal, Muztoba Ahmad Khan, David Romero, and ThorstenWuest, Smart manufacturing: Characteristics, technologies and enabling factors, Journal of Engineering Manufacture, Volume **233(5)**, 2019, Pages 1342-1361, https://doi.org/10.1177/0954405417736547

[13] Baicun Wang, Fei Tao, Xudong Fang, Chao Liu, Yufei Liu and Theodor Freiheit, Smart Manufacturing and Intelligent Manufacturing: A Comparative Review, Engineering, Volume **7**, Issue 6, 2021, Pages 738-757, ISSN 2095-8099, https://doi.org/10.1016/j.eng.2020.07.017.

[14] Seeram Ramakrishna, Tham Chen Khong and Teo Kie Leong, Smart Manufacturing, Procedia Manufacturing, Volume **12**, 2017, Pages 128-131, ISSN 2351-9789, https://doi.org/10.1016/j.promfg.2017.08.017.

[15] V. Mullet, P. Sondi and E. Ramat, "A Review of Cybersecurity Guidelines for Manufacturing Factories in Industry 4.0," in IEEE Access, vol. **9**, pp. 23235-23263, 2021, doi: 10.1109/ACCESS.2021.3056650.

[16] Prinsloo J, Sinha S and von Solms B. A Review of Industry 4.0 Manufacturing Process Security Risks. Applied Sciences. 2019; 9(23):5105. https://doi.org/10.3390/app9235105

[17] Varshney, A., Garg, N. and Nagla, K.S. et al. Challenges in Sensors Technology for Industry 4.0 for Futuristic Metrological Applications. MAPAN 36, 215–226 (2021). https://doi.org/10.1007/s12647-021-00453-1

[18] Industry 4.0 implications in machine vision metrology: an overview, Procedia Manufacturing, Volume **41**, 2019, Pages 359-366, ISSN 2351-9789, https://doi.org/10.1016/j.promfg.2019.09.020

[19] Alejandro Germán Frank, Lucas Santos Dalenogare and Néstor Fabián Ayala, Industry 4.0 technologies: Implementation patterns in manufacturing companies, *International Journal of*

Production Economics, Volume **210**, 2019, Pages 15-26, ISSN 0925-5273, https://doi.org/10.1016/j.ijpe.2019.01.004.
[20] Prof. Dr. Henning Kagermann, Prof. Dr. Wolfgang Wahlster and Dr. Johannes Helbig, Recommendations for implementing the strategic initiative INDUSTRIE 4.0, 2013
[21] J.-G. Persson, *Procedia CIRP*, **50** (2016) 378–383.
[22] Baheti, R. and H. Gill. "Cyber-physical Systems." (2011).
[23] Elverum, C., Welo, T., and Tronvoll, S. (2016). Prototyping in New Product Development: Strategy Considerations. *Procedia CIRP*, **50**. doi: 10.1016/j.procir.2016.05.010
[24] Gorski F., Wichniarek R., Kuczko W., Zawadzki P. and Bun P., Strength of ABS parts produced by Fused Deposition Modelling technology – a critical orientation problem, *Adv. Sci. Technol. Res. J.*, 9/26, 12– 19, 2015
[25] Trojanowska J., Żywicki K., Varela M.L.R. and Macha- do J., Improving production flexibility in an industrial company by shortening change over time: a triple helix collaborative project, multiple helix ecosystems for sustainable competitiveness, *Springer International Publishing*, pp. 133–146, 2016.
[26] Trojanowska J., Żywicki K. and Pająk E., Influence of selected methods of production flow control on environment, *Information Technologies in Environmental Engineering*, **3**, 695–705, 2011.
[27] Seitza K.F. and Nyhuisa P., Cyber-physical production systems combined with logistic models – a learning factory concept for an improved production planning and control, The 5th Conference on Learning Factories 2015, *Procedia CIRP* **32**, pp. 92–97, 2015.

IOP Conf. Series: Materials Science and Engineering 1228 (2022) 012030 doi:10.1088/1757-899X/1228/1/012030

Smart Techniques for Fault Diagnosis in Power Transformer

Naveen Kumar Sharma[1], Deepika Bhalla[2], Manisha[3], Kulraj Kaur[4]
[1,2,3,4] I. K. Gujral Punjab Technical University, Kapurthala, Punjab

Abstract. In a power network installed transformers are among the most expensive equipment. Outage of a transformer causes substantial financial loss not only due to the cost involved in repair or replacement, but also due to loss of revenue until corrective action is taken. To avoid the financial and reliability loss in the past two decades the monitoring of in-service transformer health has become obligatory for the utilities. The tests that were earlier done were to assess the dame of permanent nature. However, it has been known for quite some time that system abnormalities, switching. Loading, ambient conditions contribute to accelerated aging that results in sudden failure or incipient faults. Monitoring the condition of in-service equipment can always reduce failure risk. For early detection of fault dissolved gas analysis, furan analysis and degree of polymerization are the recent techniques that are becoming prevalent. A real-time assessment of the transformer condition using these techniques is an important tool for enhancing system reliability.

Index Terms— Power Transformer, Sweep Frequency Response Analysis (SFRA), degree of polymerization (DP), Furan analysis.

1. Introduction

Transformer manufacturers have keen interest in the available best practices for the design and users are keen that the transformer is effectively managed throughout its life, giving good performance, there are no costly outrages, operating losses are minimum, so that there is a respectable return on investment and considerable revenue generated from the costly asset. A considerable amount of the initial cost of power system is spent on a transformer and it is critical component from the reliability point of view. The expected life of the transformer is totally dependent on the life of its insulation. The operating conditions and climatic factors have a considerable impact on the life of a transformer irrespective of how well it is designed for normal operation. As the load increases on an installed transformer its insulation degrades substantially. This degradation needs a regular assessment of the health of the transformer. In-order to increase the life expectancy of a transformer that has a compromised condition of its insulation dr-rating of the transformer is a general practice. The failure of power transformers could be due to the asymmetry caused due to tap-changer, imbalanced electromagnetic forces in winding, core heating, bushing failure. Chemical, electrical, mechanical, optical and thermal diagnostic methods have been developed and are available for assessment of the transformer condition. Large rating power transformers can be made available and reliable through indexing of the operating condition so as to take the additional risk [1].

2. Transformer Life Management

In the past two decades monitoring of transformer condition and the management of its life has gained acceptance due to technological advancements and economics associated with it. The basic objective of the life management of transformer life is to reduce the operation cost, along with it, to have the longest possible life. The longest possible life objective has resulted in extensive research in this area [1-3]. The

Content from this work may be used under the terms of the Creative Commons Attribution 3.0 licence. Any further distribution of this work must maintain attribution to the author(s) and the title of the work, journal citation and DOI.
Published under licence by IOP Publishing Ltd

IOP Conf. Series: Materials Science and Engineering 1228 (2022) 012030 doi:10.1088/1757-899X/1228/1/012030

life of the transformer is equal to the life of its insulation; this is in turn depends upon the electrical integrity and mechanical strength of the insulation. The degradation of the insulation is due to one or more of these three reasons: namely thermal, oxidative and hydrolytic degradation. The life of the paper used in the transformer insulation decides upon the age and life of the transformer [4]. Electromagnetic forces during operation, chemical reactions due to temperature increase, stresses due to voltage levels and contamination of oil or solid particle during manufacturing process are some of the prominent reasons that come under the ageing mechanisms.

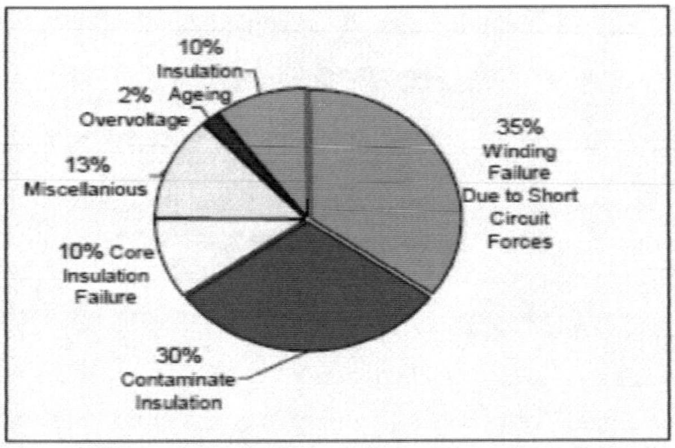

Figure 1: Failure statistics for large transformers in service between 15 and 25year [5].

Failure statistics for large transformers is shown in figure 1 shows that the large magnitude of electromagnetic forces during short-circuit faults cause 35% failures, insulation contamination cause 30% failure. 53% failures are due to mechanical reasons and 31% due to dielectric failure. Another survey conducted on the reasons for failure of large power transformers by CIGRE working group as shown in figure 2, concluded that 41% of the failures are caused due to on-load tap changers (OLTC), 19% are due to windings. For transformers that do not have OLTC; 33.3 % failure are at the terminals, 26.6 % failure at the windings, 11% due to accessories, 6.4 % due to the magnetic circuit, and the rest due to miscellaneous reasons.

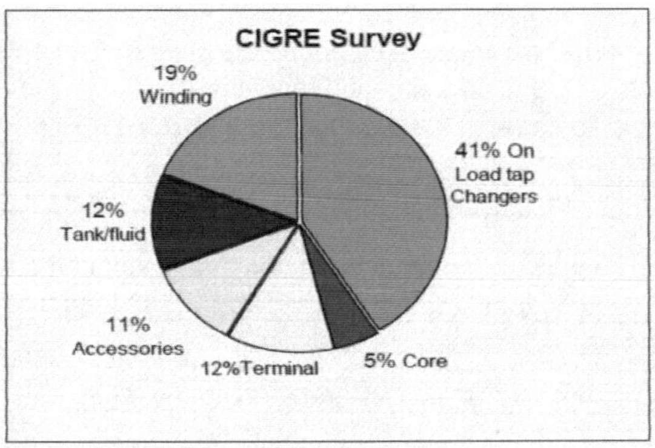

Figure 2: Percentage failure of power transformers (CIGRE survey) [6].

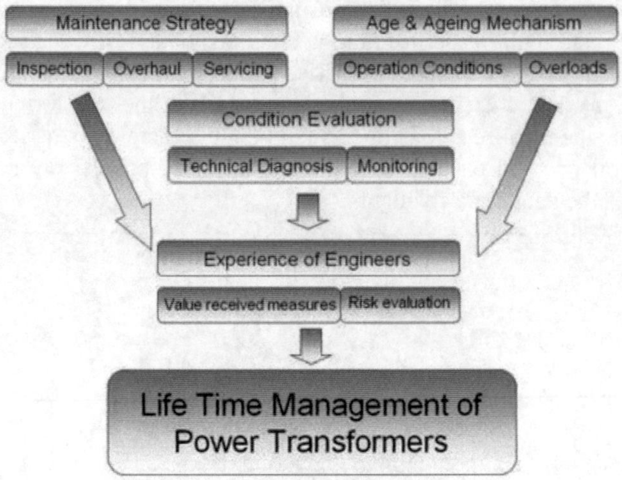

Figure 3: Life Time Management

3. Condition Monitoring of Transformers

The capital cost of installation of power equipment is considerable part of the power system, and power transformer being its critical component. The depreciation rate of power equipment is low and its is expected that they perform reasonably well of over tens of years. Power transformer was amongst the weak link of the electrical energy system and in case of any issue with its performance the thumb rule was replacement. The cost of the equipment was borne by the consumer hence the Utility was not financially affected by the replacement policy. With trading of power in an open market and a deregulated environment the approach has shifted towards from replacement, continuous monitoring the performance, diagnosis of incipient faults, and assessment of its life as shown in figure 3 [6].

For electrical utilities in developed economies, power equipment are now at the fag end of their design life and so are the installed transformers. The condition of the insulation of such transformers has aged with time due to the high operating temperature, oxidation and humidity. The sudden breakdown of insulation can result in power interruption, resulting in possible financial loss, and along with it the time-consuming physical inspection. To assess the condition of aged transformers and inherent design defects, now advanced methods are available such as frequency response analysis (FRA), frequency domain spectroscopy (FDS), advanced on-load tap changer (OLTC) and bushings diagnostics, and measurement of partial discharges.

The deregulation of energy sector has gravitated the need of reduction of costs, this affected the philosophies of management to shift from replacement towards maintenance and repair. There is concern of customer satisfaction in context with the power quality, and the unprecedented power flow patterns in the system need to be handled.

The advanced technologies along with the deregulation of the energy market are two major factors have resulted in change in the approach of the transformer users.

4. Sweep Frequency Response Analysis:

The transformer impedance changes when there is any movement of the windings, this could be due to electromagnetic forces due to internal or external faults, during transportation, or seismic activity. The Sweep Frequency Response Analysis (SFRA) is a diagnostic tool for detection of change in the impedance. After a through fault, the mechanical integrity of the transformer can be detected by this electrical test. The change in the response to a frequency indicated by the difference in signature may indicate damage to the transformer which can be confirmed by other tests or internal examination. Transformer testing and its maintenance can be done by SFRA [7].

IOP Conf. Series: Materials Science and Engineering 1228 (2022) 012030 doi:10.1088/1757-899X/1228/1/012030

4.1. Fundamentals of Frequency Response Analysis

The transformer is a complex mesh of RLC parameters; where the major contributors for the components are: copper in the windings for resistance, the coils of the windings for inductance, and the insulation layers for capacitance. There is insulation between the core and the windings, between the LV and HV windings, windings and the tanks, core and the tank, etc. These RLC components are distributed and for understanding and analysis point of view may be lumped. Figure 4, shows the equivalent circuit of the lumped RLC components, theses are used to develop the equivalent circuit that forms the basis of the principal of frequency response.

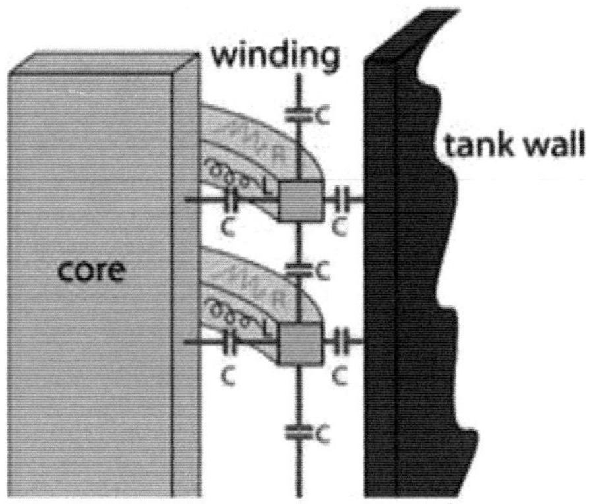

Figure 4: Network Behavior of a Transformer's Active Part

The value of inductance and capacitance if changed respond differently to a given frequency. Any change in position of the RLC network components due to electromagnetic forces, physical damage to transformer results in a change to the frequency response. The frequency response is used to identify the minute changes in the transformers RLC network [8-9]. The response to the frequency taken for the first time, or benchmark traces are used to detect any winding displacement in the future. Rather than using "impulse technique, for frequency domain measurements sweep frequency response analysis (SFRA) is used. The SFRA maintains the same energy level for each frequency, thus, provides accurate results that are consistent. This method covers a full dynamic range. The valid measurements are on account of a high signal-to-noise ratio across the entire 20 Hz to 2 MHz frequency range

The mechanical issues that are internal in nature in a transformer are movement of the core, winding displacement, winding deformation, partial winding collapse, turns shorted, open windings, buckling due to hoop, faulty core ground, faulty core ground. On comparing the baseline or benchmark or results of similar transformer unit with the transformer being assessed, the deviations are identified [10].

4.2. SFRA Analysis

For analysing the results of SFRA there is grading method needed. The ideal way of doing it is to compare the current results with those obtained previously, i.e. taking the previous result as a baseline. Practically, at times the field engineers do not have any baseline results, in case of such a situation there are three types of comparisons that can be done. The first being, comparing the results with another transformer of same design from the same manufacturer, second being, comparison of phase-to-phase short circuit test, and the third being, comparison of phase-to-phase results of open circuit test.

A prominent benefit of comparing the results with another transformer of same design from the same manufacturer, at times referred in literature to as sister transformer, is that a number of transformers can be compared simultaneously, as seen in figure 5. However, at times it is seen that, sister transformers show variation such a phase shift and form changes even when units of same manufacturer with successive serial number are compared [11].

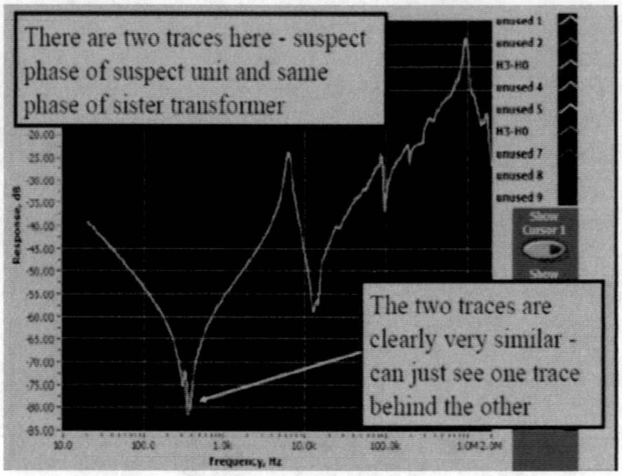

Figure 5: FRA comparison of identical sister unit transformers

The frequency band for probable fault for diagnosis of power transformer faults is give in table 1.

Table 1: Frequency bands of FRA for detection of transformer faults [12]

Frequency Band	Probable causes of variation
<2kHz	Residual magnetism, core deformation, open-circuits, and shorted turns
2kHz - 20kHz	Bulk winding movement relative to each other, clamping structure
20kHz - 400kHz	Deformation within the main winding and tap winding
400kHz - 2MHz	Movement of main and tap winding leads, axial shift

The SFRA is a major advancement in the field of condition monitoring of a transformer. It is not only an accurate method but also reliable on account of repeatable measurements. To carry out the condition assessment, first the initial record is obtained and stored, this is referred to as the initial signature, and is used for future reference when comparison is to be done. From the maintenance view point the test is carried out once in two years; the comparison with the initial signature is done for condition monitoring purpose. In event of a major external short-circuit, in the vicinity of the transformer the test is carried out immediately. The test is carried out even after transportation or re-location, and in event of an earthquake.

5. Furan analysis

The condition of paper insulation in a transformer that has been in operation for a while can be assessed by analysis of furanic compound. In the presence of transformer mineral oil, the chain scission of cellulosic insulating paper results in furan derivatives, which are degradation products. The ability of paper to effectively operate as an insulator can be assessed by its degradation. Paper insulation of a transformer in operation cannot be assessed; a convenient approach is to analyse the oil sample to assess the furanic compounds [13,14]. Figure 6 shows the degradation mechanism in a transformer.

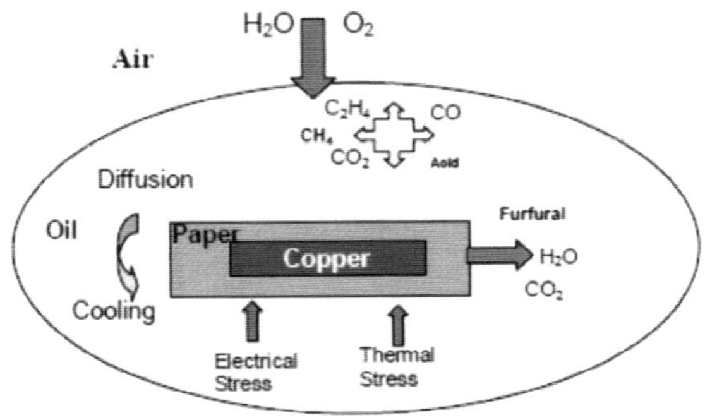

Figure 6: General schematic representation of degradation mechanism in transformer insulation [15]

During service the paper cellulose degrades slowly as the polymer chains break, and products are released into the oil. The degradation of paper is a result of heating, presence of moisture, oxidation. As the paper ages, cellulose chains break leading the change in the glucose molecules. This degradation reduces the mechanical strength of the paper, and the possibility of mechanical damage to the insulating paper increases, this results in putting the transformer to a risk of internal short-circuit faults. The electrical integrity is at risk as the paper in addition to insulation provides mechanical strength to the winding.

The condition of paper is assessed by measuring the concentration of furanic compounds or furans in the mineral oil [17]. The furans are partially soluble in mineral oil hence they are present not only in the paper but also in the oil. The degradation of insulation paper chemically changes the glucose molecules; the glucose monomer molecules are removed from the polymer chain and one of a series of related furanic compound [18]. There are five type of furanic compounds: Oxidation of the paper results in the formation of 5-Hydroxymethyl-2-furfural (5H2F), High moisture content results in formation of 2-Furfuryl alcohol (2FOL), Overheating and aging results in formation of 2-Furfural (2FAL), overheating results in formation of 5-Methyl-2-furfural (5M2F), and, electrical stresses result in formation of 2-Acetylfuran (2ACF). Degradation of paper insulation can lead to premature failure of the transformer. The rate of aging of the paper is indicated by the rate of change of furan concentration.

6. Degree of Polymerization

The degree of polymerization (DP) is the measurement of the tensile strength of insulating kraft paper, its required value is one of the key properties that the insulating paper need to possess. DP provides information of the average condition of the paper used in the transformer, this is used to assess the ageing in mechanical properties. For the purpose of interpreting the DP it is taken that the fresh/new kraft paper has DP value of 1200, when the value falls 250 and lower it is unfit for providing the required mechanical strength needed for electrical insulation. The values can be optimistic if the oil has been regenerated within the last two years [16,19] DP conveys that information need to evaluate the reliable life of kraft paper. For assessment 1200-700 DP value indicates healthy condition, 700-450 DP indicates moderate deterioration, 450- 250 DP indicates extensive deterioration, and any value less than 250 DP indicates end of life.

The average value of DP and the interaction between adjacent chains convey the physical strength of the paper insulation. The ageing reduces the burst strength of cellulose of the kraft paper. There occurs breakage of linkage by hydrolytic decomposition, the ring structure breaks and the end products are CO, CO2 and H2O [22]. CO2 is also produced when the transformer oil is heated in presence of oxygen. Thus, CO2 is produced both when oil is heated in presence of oxygen and cellulose degenerates. The ratio of CO2/CO, and the level of CO and CO2 are indicator for deciding further cellulose condition investigation [23]. Extensive research has been carried out based on it. A composite of cellulose, mineral oil and water nano particles were analysed for ac conductivity [24]. A relationship between DP and furan compounds for wet and dry insulation was developed [25]. The water content distribution along the windings of a transformer was investigated [26]. For improvement of the accuracy of the available methods, at a reference temperature of 20°C, the dependence of the degree of polymerization of Kraft paper to 2-FAL content in the oil, a method was developed [27]. The method converting experimental frequency dependence of the loss angle tangent measured by the FDS to the reference temperature is available [28-29]. The best method for determining the lifetime of cellulose is DP [4].

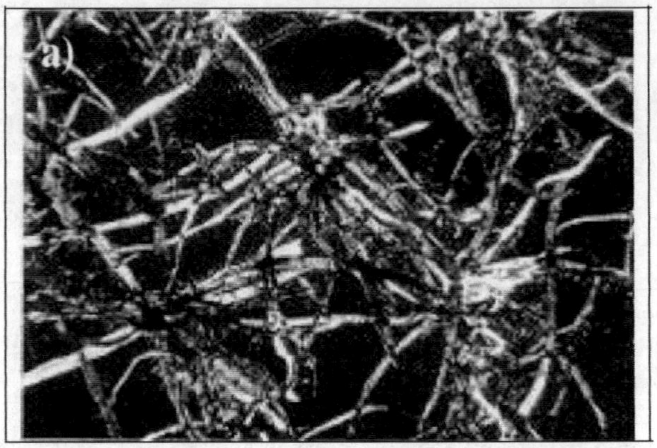

Figure 7: Microscopic view of new paper with high DP value

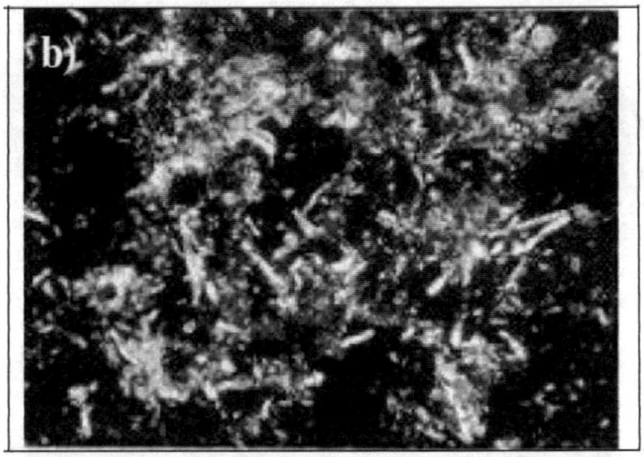

Figure 8: Microscopic view of aged paper with low DP value

The microscopic image of DP of new kraft paper is shown in figure 7. In service the paper will detoriated, the microscopic image of paper with low DP is shown in figure 8, such a condition of the paper indicates the end of the service life of the transformer [20]. Extensive study has been carried out for the DP based assessment of end of life and the value varies from 300 to 200 DP. Based on the accelerated ageing of paper immersed in oil, end of life is take as 50% of the original tensile strength. Different ranges of DP are given in table 2 [21]. The extent of cellulose degradation is a critical factor in estimating the condition and life expectancy of solid insulation. DP value indicates the mechanical ability to resist an electric short-circuit.

Table 2: Range of degree of polymerization and its interpretation

DP Range	Remark
<200	Test indicates extensive paper degradation exceeding the critical point. Strongly recommend that the transformer be taken out of service immediately and visually inspected.

200-250	The paper is near or at the critical condition. Recommend that the transformer be taken out of service as soon as possible and thoroughly inspected. Paper samples can be taken for direct DP testing.
260-350	The paper is approaching the critical condition. Suggest inspection be scheduled and/or re-sample within 1 year to reassess condition.
360-450	The paper is starting to approach the critical condition. Suggest a re-sample in 1-2 years time.
460-600	Significant paper deterioration but still well away from the critical point.
610-900	Mild to minimal paper ageing.
>900	No detectable paper degradation

7. Conclusion

The life expectancy of a transformer along with the reliability of power system can be met with if the condition of the transformer insulation health is assessed regularly. Natural aging occurs, however, accelerated aging due to unfavorable ambient conditions, ingress of moisture, oxidation and high operating temperatures can damage the transformer faster that it normally would outlive its life. This paper explains the various prevalent techniques used for condition assessment. Sweep frequency response analysis is an electrical method for condition assessment, by doing careful comparison with an benchmark, and degree of polymerization and furan analysis are powerful and dependable tools that can be applied to transformers of all size and construction type that are oil immersed. These techniques are reliable and can avoid the need of loss of assent and subsequent replacement.

References

[1] Sharma N.K., Banshwar A., Sharma B.B., Pathak M., Kumar S. (2022) "*DGA-Based Health Assessment of a 20 MVA Power Transformer*", Advances in Mechanical and Materials Technology. Lecture Notes in Mechanical Engineering, Springer, Singapore. https://doi.org/10.1007/978-981-16-2794-1_69

[2] P.N. Jarman, J.A. Lapworth, and A. Wolson, (1998), "Life assessment of 275 and 400 kV transmission transformers," presented at the International Council on Large Electric Systems (CIGRÉ), Paris, France.

[3] L. Pettersson, N.L. Fantana, and U. Sundermann, (1998) "Life assessment: Ranking of power transformers using condition based evaluation—A new approach," presented at the International Council on Large Electric Systems (CIGRÉ), Paris, France, 1998.

[4] L. Pettersson, (1990), "Estimation of the remaining service life of power transformers and their insulation," Electra, no. 133, pp. 65-71, 1990.

[5] O.N. Grechko and I. Kalacheva, (1996), "Current trends in the development of in-service monitoring and diagnostic systems for 110-750 kV power transformers," Applied Energy: Russian Journal of Fuel, Power, and Heat Systems, vol. 34, no. 5, pp. 84-97, 1996.

[6] Patel A., Sharma N. K., Banshwar A, Sharma B.B., Pathak M., (2020) "An evaluation of different health assessment methods on 50 MVA power transformer: a case study", IEEE students conference on engineering and systems (SCES), Prayagraj, India, pp 1–5.

[7] Simon A. Ryder, (2002), "Methods for comparing frequency response analysis measurement", IEEE International Symposium on Electrical Insulation, Boston, MA USA, April 7-10, 2002.

[8] M. Wang and A.J. Vandermaar K.D. Srivastava "Transformer Winding Movement Monitoring in Service—Key Factors Affecting FRA Measurements".

[9] C. Bengtsson, "Status and trends in transformer monitoring," IEEE Trans. on Power Delivery, vol. 11, no. 3, pp. 1379–1384, 1996.

[10] Ashok Kumar Yadav, Subash C. Taneja, "Transformer diagnostics testing by SFRA", The Journal of CPRI, vol.2, no.2, pp 177-185 September, 2005.

[11] Dick E.P., Erven C.C., (1978), "Transformer Diagnostic Testing by Frequency Response Analysis" IEEE Trans PAS-97, No. 6, pp 2144-2153.

[12] Suwarno, S.; Pasaribu, R.A. "Thermal Aging of Mineral Oil-Paper Composite Insulation for High Voltage Transformer" Int. J. Electr. Eng. Inform. 2016, 8, 820–835.
[13] G. C. M. A. W. T. John Scheirs, "Origin of furanic compounds in thermal degradation of cellulosic insulating paper," Journal of Applied Polymer Science, vol. 69, pp. 2541-2547, 1998.
[14] ASTM, "Standard Test Method for Furanic Compounds in Electrical Insulating Liquids by High-Performance Liquid Chromatography (HPLC)," D 5837 - 05, 2005.
[15] P. J. Baird, H. Herman, and G. C. Stevens, "Nondestructive condition assessment of insulating materials in power transformers," 2005, pp. 425-428 Vol. 2.
[16] T. K. Saha, "Review of Modern Diagnostic Techniques for Assessing Insulation Condition in Aged Transformers," IEEE Transaction on Dielectrics and Electrical insulation, Vol. 10, No. 5, pp. 903-916, 2003.
[17] L. R. Lewand, "Practical Experience Gained from Furanic Compound Analysis", Doble Engineering Company. - 73rd Annual International Doble Client Conference, 2006
[18] Griffin P. I. (1999), "Assessment of Paper and Other Cellulosic Materials in Service-Aged Transformers", Doble Engineering Company, Presentation at Southeastern Electric Exchange, 1999.
[19] Hohlein I, Kachler AJ, (2004), Impact of Moisture on DP of Solid Insulation and Furan Development in Oil at Transformer Service Temperatures," Cigre Session, D1- 309.
[20] B. Pahlavanpour and M. Eklund, "Thermal Ageing of Mineral Insulating Oil and Krafts Paper", Nynas Naphthenics AB, SE-14982 Nynashamn, Sweden.
[21] Mehanna N., Jaber A., Oweimreen G., Abulkibash A. , (2014), Assessment of dibenzyl disulfide and other oxidation inhibitors in transformer mineral oils. IEEE Trans. Dielectr. Electr. Insul. 21, 1095–1099.
[22] T. V. Oommen T. V. and T. A. Prevost, (2006), "*Cellulose insulation in oil-filled power transformers: part II-Maintaining insulation integrity and life*", IEEE Electrical Insulation Magazine, vol. 22, Issue 2, pp. 5 – 14, 2006.
[23] Gupta A., Jain K., Sood Y.R., Sharma N.K. (2019), "Comparative Study of Duval Triangle with the New DGA Interpretation Scheme", Advances in System Optimization and Control. Lecture Notes in Electrical Engineering, vol 509. Springer, Singapore. https://doi.org/10.1007/978-981-13-0665-5_24.
[24] P. Żukowski, K. Kierczyński, T. N. Kołtunowicz et al., (2019), "Application of elements of quantum mechanics in analysing AC conductivity and determining the dimensions of water nanodrops in the composite of cellulose and mineral oil," Cellulose, vol. 26, no. 5, pp. 2969–2985, 2019.
[25] Vasovic V., Lukic J., Lukic J. et al., (2019), "Aging of transformer insulation - experimental transformers and laboratory models with different moisture contents: part I—DP and furans aging profiles," IEEE Transactions on Dielectrics and Electrical Insulation, vol. 26, no. 6, pp. 1840–1846, 2019.
[26] Vasovic V., Lukic J., Lukic J., Mihajlovic D. et al., (2019), "Aging of transformer insulation of experimental transformers and laboratory models with different moisture contents: part II—moisture distribution and aging kinetics," IEEE Transactions on Dielectrics and Electrical Insulation, vol. 26, no. 6, pp. 1847–1852, 2019.
[27] Sharma, N.K., Tiwari, P.K., Sood, Y.R., (2011), Review of Artificial Intelligence Techniques Application to Dissolved Gas Analysis on Power Transformer. Int. J. Comput. Electr. Eng. 2011, 3, 577–582.
[28] P. Żukowski, T. N. Kołtunowicz, K. Kierczyński et al., (2016), "Dielectric losses in the composite cellulose–mineral oil–water nanoparticles: theoretical assumptions," Cellulose, vol. 23, no. 3, pp. 1609–1616, 2016.
[29] Banshwar A., Kumar Sharma N., Pathak M., Sharma B.B., Kumar S. (2022) "*An Investigation to Different Methods of Health Assessment in Power Transformers*", Advances in Mechanical and Materials Technology. Lecture Notes in Mechanical Engineering, Springer, Singapore. https://doi.org/10.1007/978-981-16-2794-1_66.

Equivalent models of solar photovoltaic cells and discuss the effects of resistance on maximum output power curve

Sunny Vaish[1], Naveen Kumar Sharma[2] and Gagandeep Kaur[3]

Department of Electrical Engineering, Main Campus, IKG PTU, Kapurthala-Jalandhar, Punjab[123].

Abstract. The usage of renewable energy for power generation plays a key role in power sector through reduces the noxious impacts of conventional fuels to generate power. Solar photovoltaic is a clean and emission-free renewable energy sources which is widely used for the power generation in the world. The solar photovoltaic array is made up of solar modules. These modules are act as an interfacing unit. The conversion process from sunlight to electricity is obtained through modules. But, the generation of power through involving the solar photovoltaic is dependent on variation of environments. The solar photovoltaic arrays have non-linear characteristics and depend on available solar insolation and operating temperature. A problem can be raised if these cells have not synchronized with each other. The variation of series resistance can cause the more absorption of current by the leakage current resistance. Due to this, a substantial reduction of current flows through the load circuit. So, the variation in shunt (parallel) resistance and series resistance can obtained a significant change in output characteristics. The data sheet provided by manufacturer of solar photovoltaic module contains current-voltage (I-V) and power-voltage (P-V) characteristics. The datasheet contains various parameters of solar module. But there are some parameters not provided in the datasheet by the manufacturer. This paper presents the effect of series resistance (Rs) and parallel resistance (Rp) on solar photovoltaic characteristics curves. The variation in characteristics curve (or maximum power output curve) well shown by the graphical representations.

Keywords. Solar Photovoltaic, Models, Series Resistance, Parallel Resistance, PV module

1. Introduction

The ever-increasing electricity demand is unable to accomplished by only conventional fuels. Solar and wind renewable energy sources are eco-friendly and able to overcome this problem [1]. To meet the energy requirements in the world and replacing the conventional energy sources to generate electricity in the power sector, a renewable energy sources is better option to produce electric energy and heat energy without affecting environment [2]. The hybrid energy system consisting different renewable energy sources and energy storage system technology reduces the dependency on the conventional energy source as well as handle the uncertainties of power [3]. According to report of International Renewable Energy Agency (IRENA) in 2021, the global renewable generation capacity is 27,99,094 MW in which total contribution of solar energy is near about 7,13,970 MW [4]. Solar energy is the clean and eco-friendly renewable energy source that has capable to accomplished the demand of power in the world. The solar energy source is intermittent in nature but extensively available in the day time at free of cost. The solar cells are the fundamental unit of solar power system. The solar cells in the series connection combine to form a module of solar photovoltaic. The solar modules are the basic and crucial unit which actually converts the solar radiation into electricity in the

Content from this work may be used under the terms of the Creative Commons Attribution 3.0 licence. Any further distribution of this work must maintain attribution to the author(s) and the title of the work, journal citation and DOI.
Published under licence by IOP Publishing Ltd

solar photovoltaic power system. The output power produced by the solar photovoltaic modules is directly proportional to the solar radiation coming from the sun. The Solar Photovoltaic (SPV) system generate power in the form of dc current directly when a suitable amount of insolation is incident on solar cells. They produce power without polluting environment.

For the studies of SPV system, a reliable and accurate PV modeling is necessary. The modeling of PV system is useful for understanding the influence of SPV generation penetration on the stability and ensuring the better power quality and reliability of the system [5]. The fundamental unit of SPV power system are Photovoltaic (PV) modules. The output characteristics of PV modules dependent on the solar irradiance, operating temperature and the output voltage of the photovoltaic module [6]. There are variety of studies comprises modeling of PV techniques by using various methods and findings parameters by involving different methods done by researchers and authors. By using 72-watt photovoltaic module, I-V and P-V characteristics studied by using a single diode model. The various steps are developed for modeling in MATLAB [7]. Three equivalent models (simplified, series and shunt models) of PV module are discussed in [8]. The first simplified PV models can be established by removing the series resistance (R_p)/parallel resistance (R_s) or both resistances. The second series model can be formed by adding resistance (called series resistance) in series with diode. The third shunt resistance can be formed by adding resistance (called parallel resistance) in parallel with diode. The parameters which are not provided by the manufacturer are calculate by using shunt model with the help of Newton Raphson method [8]. The complexity and high computational time of accurate PV models produced in the previous research. This paper explains the triple diode model, double diode model, ideal model, simplified model and practical model and introducing iterative method to obtain unknown parameters [9]. The output characteristics of one-diode, two-diode and single diode-based four parameter model comparison done by evaluating the unknown parameters through a developed algorithm [10]. A MATLAB/SIMULINK is used to done simulation and modeling of one and two diode model to find out the parameters of photovoltaic panel [11]. The implementation of MATLAB/SIMULINK in simulation of SPV cells/modules/arrays described by step by step to effectively understand a output characteristics graphs [12]. A simple model consists of one current source, a single diode and series resistance (R_s) is considered to determine the change in maximum power point due to solar irradiance and temperature [13]. The model is formed to get more precisely estimation of the performance of photovoltaic system at low level of irradiance. An improved two-diode solar photovoltaic model is used to evaluating the R_s and R_p [14]. A two-diode simplified novel model (consider only R_s, neglecting R_p proposed in which considering the current source and two ideal diodes and neglecting the parallel resistance and series resistance to decrementing the computational time [15].

There are several known and unknown parameters given in datasheet of solar photovoltaic modules which justifies the characteristics curves. The findings of unknown parameters are very important to study of output power characteristics. These unknown parameters are very crucial to determining the estimation of the behavior of the solar photovoltaic modules. This paper presents the effects of series resistance and parallel resistance. There are six sections in this paper. The first section presents the introduction of the papers. Literature review and equivalent models of solar photovoltaic cell presented in section second and third respectively. The fourth sections depict the parameters of solar photovoltaic cell/modules. The fifth and sixth section contains the characteristics of solar photovoltaic cell/modules and effect of series resistance on maximum output power.

2. Equivalent models of Solar Photovoltaic (SPV) cell

The combination of series and parallel SPV cells joined to form a SPV module. The modules are connected in series and parallel combination to form arrays [16]. The representation of equivalent ideal model for PV cell is given in Figure 1. A superlative solar cell consists of one current source that is connected parallel with the diode illustrated in Figure 1. The ideal solar cell described by an equation 1 can be written as

$$I = I_{pho} - I_d \tag{1}$$

IOP Conf. Series: Materials Science and Engineering 1228 (2022) 012031 doi:10.1088/1757-899X/1228/1/012031

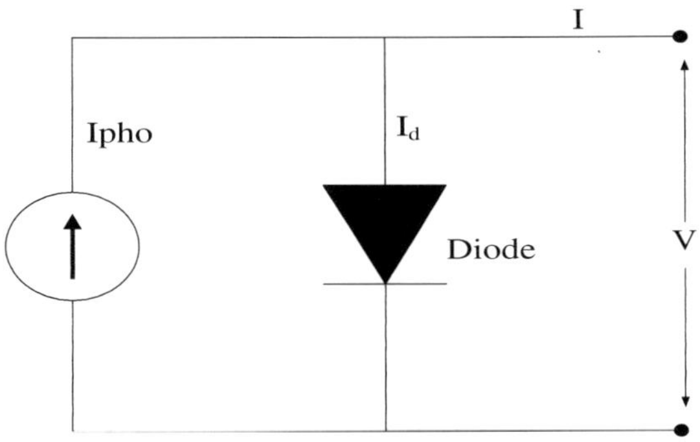

Figure 1. Equivalent ideal model for PV cell [16]

Where I_{pho} is photon current. The I_d is diode current and I_s is saturation current. Diode current is directly proportional with saturation current shown by equation (2). The basic mathematical equation which explains the current-voltage characteristics of SPV cell known as Shockley's diode current equation in semiconductor theory [18-19] described by equation (2) [20-21].

$$I_d = I_s \left[\exp\left(\frac{qV}{N_s KAT_o}\right) - 1\right] \tag{2}$$

Where I_s represents the reverse saturation current, q is charge of electron and K is Boltzmann constant which is equal to $1.602 \times 10^{-19}\ C$ and $1.38 \times 10^{-23}\ J/K$ respectively. V is applied voltage, T_o is the actual temperature of cell in Kelvin (K), A is ideality factor and N_s is number of PV cells that is connected in series. The I is output current of solar cell in simplified model that can be evaluated by putting the value of I_d in equation (1) from equation (2).

$$I = I_{pho} - I_s \left[\exp\left(\frac{qV}{N_s KAT_o}\right) - 1\right] \tag{3}$$

A current generated by photon is directly dependent on intensity of irradiance and its temperature. Solar cell provides good approximation of photon current in ideal case or equivalent ideal SPV circuit. There are some parameters which are neglects in the ideal case model. The output of SPV system in practical cases effected by these neglected parameters [18-19]. If R_s considered in simplified model for PV cell, the equation will be [20-21] I_p I_d I_p

$$I = I_{pho} - I_s \left[\exp\left(\frac{q\,(V + I.R_s)}{N_s KAT_o}\right) - 1\right] \tag{4}$$

The Figure 2 depicts the simplified model for PV cell. This simplified model consists of one current source with parallel diode and series resistance. The accuracy of such model is not more effective than practical model [8]. The practical model for PV cell is shown in Figure 3. Equation (5) represents the practical model of PV cell that is given below

$$I = I_{pho} - I_d - I_p \tag{5}$$

Where I_p represents the parallel resistor in practical model [13, 14].

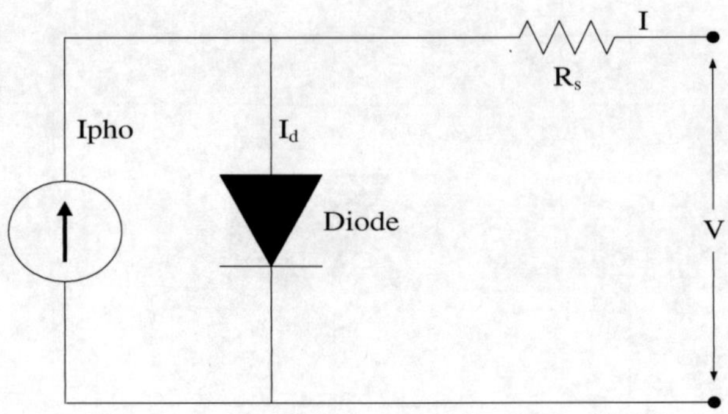

Figure 2. Simplified model for SPV cell [9]

By putting the value of I_p in the equation (5) from equation (6), the output current of practical model

$$I_p = {(V + I.R_s)}/{R_p} \tag{6}$$

can be calculated. The practical model for PV cell consists of a single current source with a parallel

$$I = I_{pho} - I_s \left[\exp\left(\frac{q\,(V+I.R_s)}{N_s K A T_o}\right) - 1\right] - {(V + I.R_s)}/{R_p} \tag{7}$$

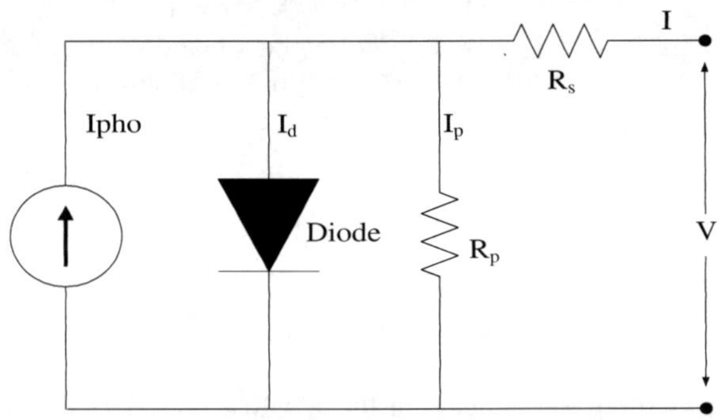

Figure 3. Practical model for PV modules [9]

diode a series resistance and shunt resistance. This model is widely used to determination of the accuracy and complexity of the model [22]. This is very important to implement an accurate model for optimum prediction of PV module or cell. The model which has both R_s and R_p (known as Practical model) produces optimum characteristics curves and offers the best match amongst the experiment data (evaluated) and the PV module data provided by the manufacturer of module [20].

3. Parameters of Solar Photovoltaic (SPV) cell/modules

The effective PV mathematical modeling is needed for valuable simulation and dynamic analysis. The performance of model can be estimated precisely for any photovoltaic system is carried out through optimum modeling. In the ideal case, a series resistance and shunt resistance can be neglected. The efficiency of the SPV cell is dependent on the series resistance and shunt resistance in practical case, due to this they cannot be neglected [22]. The datasheet of SPV module provides nominal power (P_m), maximum power current (I_{mp}), maximum power voltage (V_{mp}), short-circuit current (I_{sc}), open-circuit voltage (V_{oc}), number of cells connected in series (N_s), temperature coefficient of voltage (K_v), temperature coefficient of current (K_i) etc. at Standard Test Conditions (1000 W/m^2 at 25 °C) which are given in Table 1. There are some parameters which are not available in datasheet of SPV module.

IOP Conf. Series: Materials Science and Engineering 1228 (2022) 012031 doi:10.1088/1757-899X/1228/1/012031

The equation (7) describes the practical model. The shunt resistance, series resistance, ideality factor, reverse saturation current and photon current are not given in the datasheet of SPV modules. The unknown parameters are required to shown the good and effectiveness match between the modeled value and experimental value. The Table 1 depicts the datasheet and unknown parameters. The script file is formed in MATLAB software. The parameters value for script file is taken from table 1 and characteristics curves (I-V & P-V curves) are so obtained. There are three type of PV module which are taken into consideration. The monocrystalline, polycrystalline and thin film SPV module are used for simulation. The table 1 depicts the parameters of different SPV modules parameters and its unknown parameters.

Table 1. Different SPV modules parameters and its Unknown parameters

Parameters given in the datasheet by SPV module manufacturer			
Parameters/PV Module	**Monocrystalline** JAM5(L)-72-155 [24]	**Polycrystalline** NDQ2E3E [25]	**Thin Film** GE-CdTe78 [30]
I_{mp} (Maximum power current)	4.92	7.11	1.07
V_{mp} (Maximum power voltage)	35.08	22.8	72.4
Pmax (Nominal power)	155	162	78
I_{sc} (Short circuit current)	4.94	7.92	1.23
V_{oc} (Open circuit voltage)	43.68	28.4	94.0
K_V (Temp. Coefficient of voltage)	-0.297 %/°C	-0.104 %/°C	-0.29 %/°C
K_I (Temp. Coefficient of current)	0.037 %/°C	0.053 %/°C	0.04 %/°C
N_s (Number of cells connected in series)	72	48	116
Unknown parameters not provided in the datasheet by SPV module manufacturer [9]			
R_p (Shunt resistance)	333.663 ohm	98.7726	3989.9
R_s (Series resistance)	0.0991 ohm	0.1166	0.9091
A (Ideality factor)	1.8	1.5	2.5

4. Characteristics Curves of PV cell/Modules

4.1. I-V & P-V Characteristics of Different Modules

The simulated scripted file is used to plots the I-V & P-V curves in MATLAB/SIMULINK. The equation (7) gives the current (*I*) of practical model. The Figure 4 presented the I-V characteristics of monocrystalline, polycrystalline and thin film PV module.

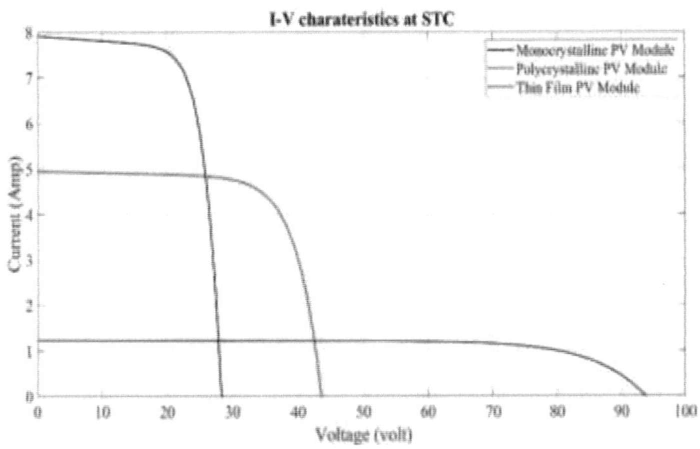

Figure 4 – I-V characteristics of monocrystalline, polycrystalline and thin film PV modules

The plotted I-V characteristics curves shows that the I-V curves pass exactly through the three main points of data provided by the manufacturer of SPV module that is (I_{sc}, 0), (0, V_{oc}) and ($I_{mp,}$ V_{mp}). The Figure 5 depicted P-V characteristics of monocrystalline, polycrystalline and thin film PV modules. Also, the plotted P-V characteristics shows that the P-V curve has passed accurately through the maximum power point at its corresponding voltage (V_{mp}).

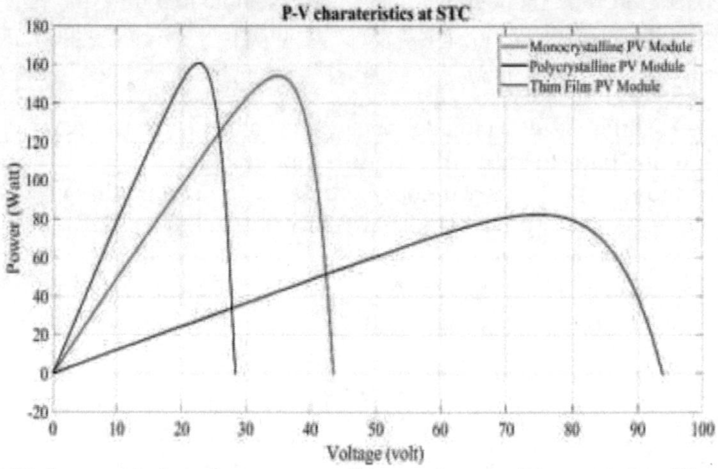

Figure 5. P-V characteristics of monocrystalline, polycrystalline and thin film PV modules

4.2. Different Modeled I-V & P-V Characteristics
There are various types of models explained in the paper. The different modeled such as Parallel/Shunt model, Series model and Practical model are described in this section. A current source and a single diode join to form a ideal model of photovoltaic cell where as series model consist of current source, a diode and series resistance and shunt model consist of current source, a diode and parallel resistance. The practical model is the combination of series model and shunt model having one current source and a diode. This type of model increases the accuracy that reveals the effectiveness of the model. The different model (Parallel, Series and Practical model) has different output curves. Figure 6 depicts the comparison of different modeled I-V and P-V characteristics of PV module.

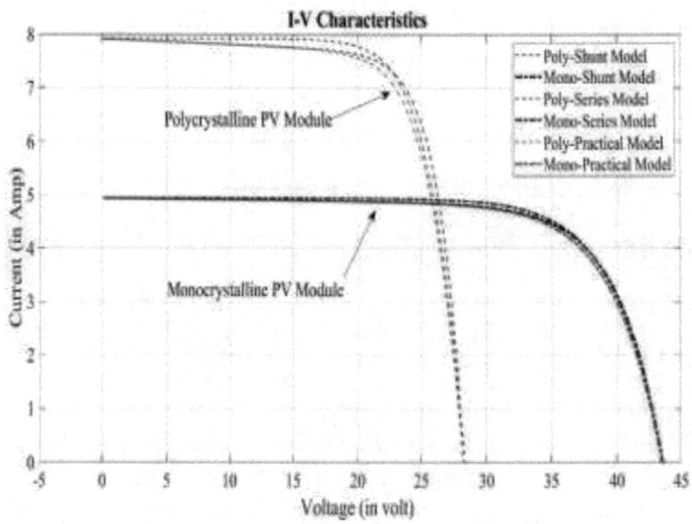

IOP Conf. Series: Materials Science and Engineering 1228 (2022) 012031 doi:10.1088/1757-899X/1228/1/012031

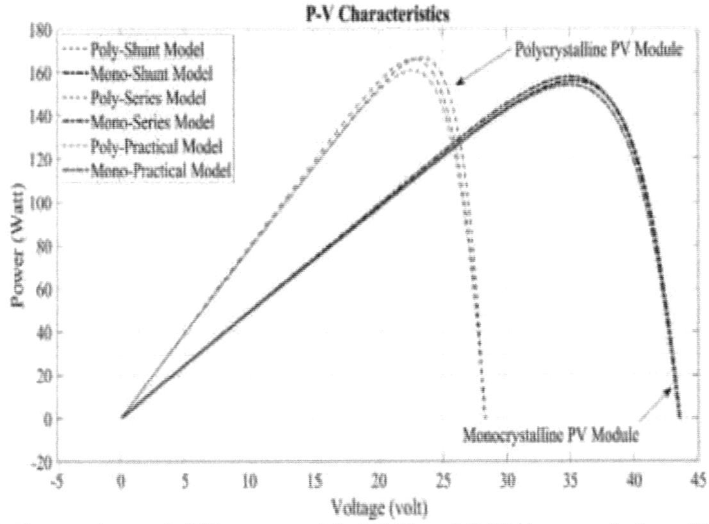

Figure 6. Comparison of different modeled I-V and P-V characteristics of PV module

5. Effect of Series Resistance and Parallel Resistance on Maximum Output Power

The current (I) in simplified model obtained through equation (4). The series resistance (R_s) is very crucial to determine for better match between the modeled and experimental values. The R_{sh} and R_s are the unknown parameters in the data given by manufacturer for SPV module. The merits and demerits, description of features and concepts of single diode series resistance, single diode shunt resistance as well as two diode models of SPV cells are discussed in [26]. The combination of SPV cells make to form an array. A problem can be raised if these cells have not synchronized with each other. The variation of series resistance can cause the more absorption of current by the leakage current resistance. Due to this, a substantial reduction of current flows through in load circuit. The variation in I-V curve can depicted by the change in shunt resistance of photovoltaic cell in the MATLAB program [27]. A reduction in open circuit voltage and short-circuit current if shunt resistance is very low and series resistance is very high respectively [28]. So, the variation in shunt resistance and series resistance can obtained a significant change in output characteristics. The quality of junction of cells plays a important role in the performance of cell [29].

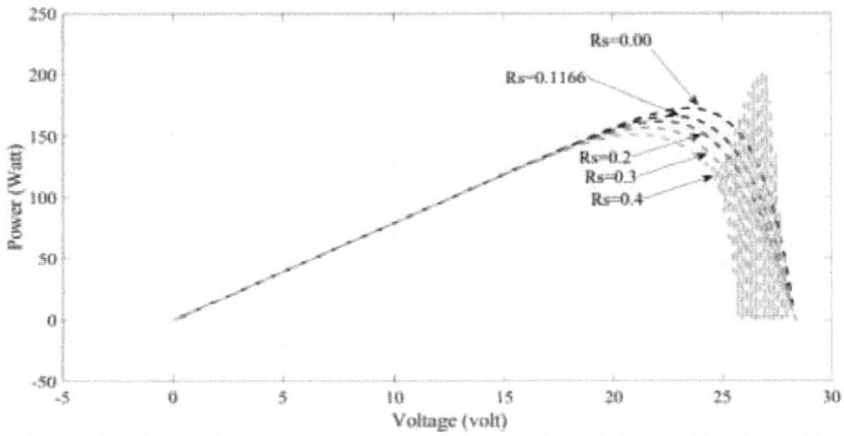

Figure 7. Change in maximum output power with variation in R_s (Series Resistance)

The Figure 7 depicts the change in maximum output power with variation in R_s (Series Resistance). Many authors used iterative method to varying a R_s. The R_s is increasing until the I-V

IOP Conf. Series: Materials Science and Engineering 1228 (2022) 012031 doi:10.1088/1757-899X/1228/1/012031

characteristics curves are visually fitting the experimental data. Once the R_s is decided then change R_p in the same manner. It is very poor and erroneous method to find out the R_s and R_p discretely. If a virtuous I-V model is required, then R_s and R_p cannot be judge separately [31]. Table 2 shows that different value taken for Series Resistance (R_s) to shows that the change in Maximum Power Curve.

Table 2. Different value taken for Series Resistance (R_s) to represent the change in Maximum Power Curve.

S.no.	Series Resistance (R_s)	Type of value	Maximum Power Curve with respect to Maximum Power Point (MPP)
1	0.00	Randomly	Move right side
2	0.1166	Calculated	Good match
3	0.20	Randomly	Move left side
4	0.30	Randomly	Move left side
5	0.40	Randomly	Move left side

In the Figure 7, a various value of series resistance (R_s) is taken to demonstrate the variation in maximum output power. When R_s is equal to 0.00, maximum power curve moves towards the right side of Maximum Power Point (MPP). If value of R_s increased slightly (0.00 ohms to 0.2, 0.3, 0.4), the maximum power curve moves towards the left side from MPP.

$$Pmax(m) = Pmax(e) = V_{mp}I_{mp} \text{ (At } V_{mp}I_{mp} \text{ point of I-V curve)} \tag{8}$$

Where $Pmax(e)$ is maximum experimental power from datasheet at the MPP and $Pmax(m)$ is maximum power calculated by I-V model. When R_s is taken which is equal to 0.1166 ohms, the maximum output power (nominal power) given in datasheet of solar photovoltaic module is generated. This value of R_s is suitable for good match between the experimental value and modeled value. On the basis of abovementioned variation in R_s, the value of R_s should follow the given condition in the equation (8). The change in maximum output power with variation

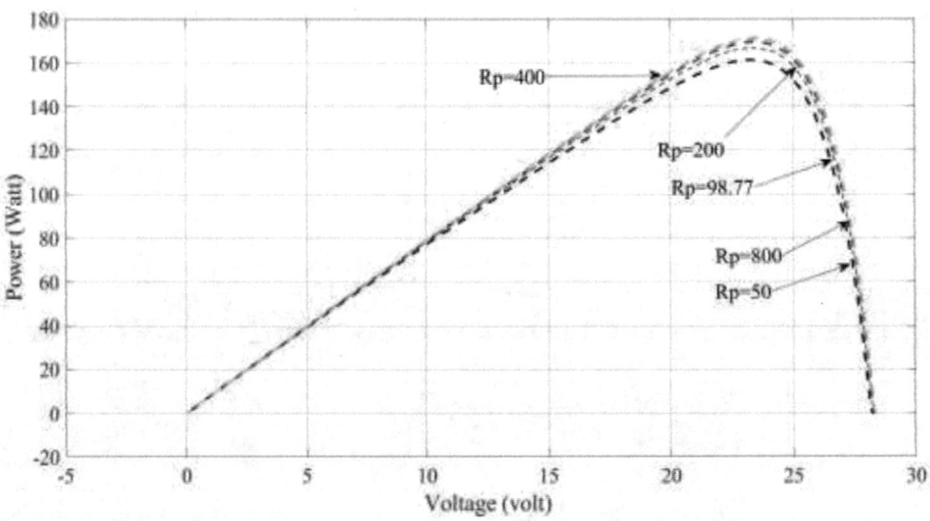

Figure 8. Change in maximum output power with variation in R_p (Parallel Resistance)

in R_p (parallel resistance) depicted in Figure 8. The unknown parameters of equation (7), can be determined by maintaining $Pmax(m) = Pmax(e)$ and solving the resulting equation for R_s. The relationship of R_s and R_p can be found by:

$$Pmax(m) = Vmp\left(I_{pv} - I_s\left[\exp\left(\frac{q\,(V_{mp}+I_{mp}.R_s)}{N_sAKT_o}\right) - 1\right] - {(V_{mp} + I_{mp}.R_s)}/{R_p}\right) = Pmax(e) \quad (9)$$

$$R_p = V_{mp}\left(V_{mp} + I_{mp}.R_s\right)/\left(V_{mp}.I_{mp} - V_{mp}.I_s\left[\exp\left(\frac{q\,(V_{mp} + I_{mp}.R_s)}{N_sAKT_o}\right) - 1\right] + V_{mp}.I_s - Pmax(e)\right. \quad (10)$$

If there is any value of series resistance (R_s), there will be a value of shunt resistance (R_p) which makes the mathematical characteristics curves across the experimental ($V_{mp}.I_{mp}$) point [31]. Table 3 shows that different value taken for Parallel Resistance (R_p) to represent the change in Maximum Power Curve.

Table 3. Different value taken for Parallel Resistance (R_p) to represent the change in Maximum Power Curve.

S.no.	Parallel Resistance (R_p)	Type of value	Maximum Power Curve with respect to Maximum Power Point (MPP)
1	50.00	Randomly	Move downwards
2	98.77	Calculated	Good match
3	200.00	Randomly	Move upwards
4	400.00	Randomly	Move upwards
5	800.00	Randomly	Move upwards

In the figure 8, a various value of parallel resistance (R_p) is taken to demonstrate to shown the change in maximum output power. When R_p is equal to 50.00, maximum power curve moves downwards from Maximum Power Point (MPP). If value of R_p increased (50.00 to 200.00, 400.00, 800.00), the maximum power curve moves upwards in the direction of MPP. When R_p is taken which is equal to 98.77, the maximum output power (nominal power) given in datasheet of solar photovoltaic module is generated. The value of R_p should be manipulated in accordance with R_s which follows the condition of equation (9) and (10). Due this, there will be a good match can be obtained between experimental values and modeled values.

6. Conclusion

The parameters provided by manufacturer of solar photovoltaic module is not sufficient for the good match between the experimental values and modeled value. The unknown parameters have great influenced in good match for practical model. The effects of series resistance and parallel resistance on maximum output power in practical model as well as different modules such as monocrystalline, polycrystalline and thin film also discussed in this paper. The study of ideal model (neglected both R_s and R_p), simplified model (consider R_s, neglected R_p) and practical model (consider both R_s and R_p) reveals the importance of unknown parameters. This paper helps to study the maximum output power with respect to variation in resistances (series & parallel resistance) and different modeled can be used to plot the characteristics curves.

References

[1] V. Khare, S. Nema and P. Baredar, "Solar-wind hybrid renewable energy system – A review", Renewable Sustainable Energy Reviews, vol. 58, pp. 23-33, 2016.
[2] Z. Gareiou, E. Drimili and E. Zervas, "12 - Public acceptance of renewable energy sources", Low carbon energy technologies in sustainable energy systems, pp. 309-327, 2021.

[3] R. Hemmati and H. Saboori, "Emergence of hybrid energy storage systems in renewable energy and transport applications-A review", Renewable Sustainable Energy Reviews, vol. 65, pp. 11-23, 2016.

[4] Renewable capacity statistics 2021, International Renewable Energy Agency (IRENA). Available at - https://www.irena.org/-/media/Files/IRENA/Agency/Publication/2021/Apr/IRENA_RE_Capacity_Statistics_2021.pdf (Accessed on 11 Oct 2021).

[5] H. Kanchev, D. Lu, F. Colas, V.D. Lazarova and B. Francois, "Energy management and operational planning of a microgrid with a PV-based active generator for smart grid applications", IEEE transactions on industrial electronics, vol. 58 (10), pp. 4583-4592, 2011.

[6] C. Keles, B.B. Alogoz, M. Akcin, A. Kaygusuz and A. Karabiber, "A photovoltaic system model for MATLAB/Simulink simulations", International Conference on Energy and Electrical Drives, pp.1643-1647, 2013.

[7] P.K. Dewangan, A. Goswami and S.K. Dewangan, "Mathematical model of 72 W cosmic module with MATLAB/Simulink", International journal of electrical and electronics engineering, vol. 2 (5), pp. 15-21, 2015.

[8] Y. Mahmoud and W. Xiao, "Evaluation of shunt model for simulating photovoltaic modules", IEEE journal of photovoltaics, vol. 8 (6), pp. 1818-1823, 2018.

[9] Y. Mahmoud and E.F. El-Saadany, "A photovoltaic model with reduced computational time", IEEE transaction on industrial electronics, vol. 62 (6), pp. 3534-3544, 2015.

[10] H.A.B. Siddique, P. Xu and R.W. De Doncker, "Parameter extraction algorithm for one-diode model of PV panels based on datasheet values", International conference of clean electrical power, Italy, 2013.

[11] A. Bouraiou, M. Hamouda, A. Chaker, M. Sadok, M. Mostefaoui and S. Lachtar, "Modeling and simulation of photovoltaic module and array based on one and two diode model using Matlab/Simulink", Energy procedia, vol. 74, pp. 864-877, 2015.

[12] X.H. Nguyen and M.P. Nguyen, "Mathematical modelling of photovoltaic cell/module/arrays with tags in Matlab/Simulink", Environmental system research, vol. 4, 2015.

[13] G. Walker, "Evaluating MPPT converter topologies using a Matlab PV model", Journal of electrical and electronics engineering, Australia, vol. 21, 2014.

[14] Z. Salam, K. Ishaque and H. Taheri, "An improved two - diode photovoltaic model for PV system", Joint international conference on power electronics, drives and energy systems & 2010 power India, New Delhi, India, 2010.

[15] B.C. Babu and S. Gurjar, "A novel simplified two-diode model of photovoltaic module", IEEE Journal of photovoltaics, vol. 4 (4), pp. 1156-1161, 2014.

[16] H. Tian, F.M. David, K. Ellis, E. Muljadi and P. Jenkins, "A cell to module to array detailed model for photovoltaic panels", Solar energy, vol. 86 (9), pp. 2695-2706, 2012.

[17] Vinod, R. Kumar and S.K. Singh, "Solar photovoltaic modelling and simulation – As a renewable energy solution", Energy reports, vol. 4, pp. 701-712, 2018.

[18] S. Silvestre, "Chapter 7 – Strategies for fault detection and diagnosis of PV systems", Advances in renewable energies and power technologies, vol. 1, pp. 231-255, 2018.

[19] A. Bellini, S. Bifaretti, V. Iacovone and C. Cornaro, "Simplified model of a photovoltaic module", Applied electronics IEEE, Pilsen, 2009.

[20] H. Bellia, R. Youcef and M. Fatima, "A detailed modelling of photovoltaic module

using MATLAB", NRIAG journal of astronomy and geophysics, vol. 3 (1), pp. 53-61, 2014.

[21] A. Zekry, A. Shaker and M. Salem, "Chapter 1 – Solar cells and arrays: principles, analysis and design", Advances in renewable energies and power technologies, vol. 1, pp. 3-56, 2018.

[22] X. Weidong, M.G.J. Lind, W.G. Dunford, and A. Capel, "Real-time identification of optimal operating points in photovoltaic power systems", IEEE Transaction on industrial electronics, vol. 53, pp. 1017-1026, 2006.

[23] S. Lyden, M.E. Haque, A. Gargoom, M. Nagnevitsky and P.I. Muoka, "Modelling and Parameter estimation of photovoiltaic cell", Australasian universities power engineering conference, Indonesia, 2012.

[24] Monocrystalline, JAM5(L)-72-155 photovoltaic module. Available at - https://www.enfsolar.com/pv/panel-datasheet/crystalline/14583. (Accessed on 11 Oct 2021)

[25] Polycrystalline, NDQ2E3E photovoltaic module. Available at - https://www.datasheets.com/en/part-details/nd162e1-sharp-52673060#datasheet. (Accessed on 11 Oct 2021)

[26] V.J. Chin, Z. Salam and K. Ishque, "Cell modelling and model parameters estimation techniques for photovoltaic simulator application", Applied Energy, vol. 154, pp. 500-5019, 2015.

[27] A.D. Dhass, E. Natarajan and L. Ponnusamy, "Influence of shunt resistance on the performance of solar photovoltaic cell", international Conference on Emerging Trends in Electrical Engineering and Energy Management, India, 2013.

[28] P. Singh and N.M. Ravindra, "Analysis of series and shunt resistance in silicon solar cells using single and double exponential models", Emerging Materials Research, vol. 1, pp. 33-38, 2011.

[29] J. Zhang, S.T. Lee and B. Sun, "Effect of series and shunt resistance on organic-inorganic hybrid solar cells performance", Electrochimica Acta, vol. 146, pp. 845-849, 2014.

[30] Thin film, GE-CdTe78 photovoltaic module. Available at - https://www.enfsolar.com/pv/panel-datasheet/Thin-film/38. (Accessed on 11 Oct 2021)

[31] M.G. Villalva, J.R. Gazoli and E.R. Filho, "Comprehensive approach to modelling and simulation of photovoltaic arrays", IEEE transactions on power electronics, vol. 24 (5), pp. 1198-1208, 2009.

Effect of Nanoparticle on the Performance of Water Lubricated Journal Bearing

Nathi Ram Chauhan[1], Arhantika Khera[2], Shubhi Agarwal[3], Mehak Bagri[4], Aditi Aggarwal[5]

[1-5]Department of Mechanical and Automation Engineering, IGDTUW

Mehakbagri1999@Gmail.com, akhera27@gmail.com

Abstract. Journal bearings have widespread use in machinery with rotating shafts. Hydrodynamic journal bearings have lubricating fluid present between shaft and bearing to reduce friction and allow smooth operation. The lubricating fluid develops pressure when bearing is in operation and this pressure is used to carry load on the shaft. The commonly used lubricants in the industry are petroleum based synthetic oils. To do away with the disposal issues and toxicity regarding synthetic oils, water is used as the lubricating fluid for this study. The effect of adding nanoparticles to water is studied and a program in MATLAB is developed to determine the characteristics of bearing, namely load carrying capacity, coefficients of stiffness and coefficients of damping for shaft speeds 500 rpm, 1000 rpm and 1500 rpm. It is observed that addition of nanoparticles results in increased magnitude of load carrying capacity and the bearing exhibits high value of damping and stiffness coefficients indicating stability of shaft for every speed.
Keywords- Journal bearing, nanolubricant, load carrying capacity, stiffness coefficient, damping coefficient

1. INTRODUCTION

Journal bearings are used in machinery with rotating components. They assist the rotor and transfer loads from rotating parts to the stationary frame. The machinery where they are most commonly present includes gearboxes, compressors, pumps etc. [1]. They are low clearance assemblies with a lubricating fluid present between the journal and the bearing.

Lubrication is used to minimise wear, friction between surfaces, and heating in components of machine that shift in relation to one another. Since the hydrodynamic film forms as the shaft rotates, a suitable fluid must be always available when the shaft is rotating to hold the bearing surface and shaft surface apart and prevent contact. This fluid supports the load acting on bearing. The viscosity of the lubricating fluid is an important characteristic that affects the capability of the bearing to withstand load. [2] The fluid used to lubricate journal bearing develops pressure due to the wedge shape the film forms and this pressure can carry load on the bearing. Hence, choice of fluid is critical for a bearing.

Water is one of the most easily sourced resources and has good cooling performance. It is also cheaper than synthetic lubricants and easy to incorporate in the design and dispose. [3] However, water as a journal bearing lubricant can carry only low-level loads compared to synthetic lubricants. To improve the characteristics of journal bearing, nanoparticles are added to water at distinct volume fractions and the load carrying capacity, damping coefficients and stiffness coefficients of bearing is evaluated.

Nanomaterial additives have gained popularity recently due to the enhancement they provide to fluids. Their distinct properties are high surface to weight ratio and high porosity. They are used as additives in coolants and lubricants and are reported to have increased thermal properties. [4]

A critical factor of bearing performance is the surface roughness of journal and bearing. Owing to the extremely small size of nano particles which are in the range 1-100nm, they can access the fine spaces on the surface of the metal and ameliorate the lubricating properties of the fluid and

Content from this work may be used under the terms of the Creative Commons Attribution 3.0 licence. Any further distribution of this work must maintain attribution to the author(s) and the title of the work, journal citation and DOI.
Published under licence by IOP Publishing Ltd

reduce friction between the contact surfaces. [5] In addition to this, any additives in bearing lubricant must be smaller than the minimum film thickness which is in micrometres. This makes particles of size in nanometres suitable additives.

Numerous authors have worked on hydrodynamic journal bearings and their performance characteristics. To name a few, Jamalabadi et al. [6] used Computational Fluid Dynamics (CFD) and Thermohydrodynamic (THD) to simulate and observe how nanoparticles affected the behaviour characteristics of plain journal bearings, Sadabadi and Nezhad [7] used Discrete Phase Modelling (DPM) of nano material in lubricating fluid to study the effect of inorganic fullerene, WS_2 on the load withstanding capability of the bearing designed for high loads. Kulkarni et al. [8] theoretically analysed the effect of axial magnetic field and a radial electric field on the stability of a fixed rotor held on MHD journal bearings by considering Routhe's criteria of stability. Kalakada et al. [9] analysed the dynamic and static behaviour of THD bearing that uses lubricating fluid with nanoparticles. Tang et al. [10] used deep reinforcement learning (DRL) on the active flow control of a CFD model over an array of Reynolds numbers with values 100, 200, 300 and 400. Concli [11] estimated the trajectory and equilibrium position in journal bearing using the combination of CFD and Analytical techniques. Gertzos et al [12] conducted the CFD simulation to study a hydrodynamically lubricated journal bearing with Bingham lubricant. In the study by S. Hother-Lushington [13] different significant factors of a lubricant like viscosity and flash point are compared for water with that of turbine oil. It in concluded that as water has lighter viscosity, its load carrying capacity is reduced which means water as lubricant is best when the speed is high, and load is low in a hydrodynamically lubricated bearing. One major advantage of water as lubricant instead of oil is that water has no flash point and is hence is fire resistant. So, in applications like high pressure where the components have higher temperature, water can be used without the risk of fire. To improve the tribological properties of water-based lubricants, there are various studies done to test the improvement of the addition of nanoparticles like Titanium dioxide, Silicon dioxide, Zinc oxideetc. Yuxiao Cui [14]et al concluded in their paper how adding SiO2 as an additive would reduce the Coefficient of friction and would act as an excellent anti wear additive. Various gaps were recognized upon extensive review of literature and therefore, supported the research.

Upon extensive literature review,it has been identified that the lubricating oils used in the industry are majorly petroleum based and according to a study, even the oils that are categorized biodegradable under the European Union Legislation may be made up of 50% of petroleum oil as base fluid [15]. The petroleum is a finite resource and fast depleting, in addition to which, the composition of such lubricating oils has harmful and toxic effects when released in the environment. Further, water as a lubricant has good cooling effects on bearing but has low load carrying capacity, thus limited use. Therefore, based on the literature, the performance of journal bearing has been evaluated using water as a lubricant with nanoparticles as additive.

2. ANALYSIS

2.1 KREIGER-DOUGHERTY MODEL OF VISCOSITY

To calculate the viscosity of nanofluid which is the resultant fluid after adding nanoparticles to water, the Kreiger-Dougherty Model of Viscosity is used. The effective viscosity of a nanofluid varies with respect to volume fraction of added nanoparticles. The Krieger-Dougherty viscosity model with monodispersed suspensions of hard spheres is given by:

$$\frac{\mu_{nf}}{\mu_{bf}} = \left(1 - \frac{\Phi_a}{\Phi_m}\right)^{-2.5\Phi_m} \quad (1)$$

where, μ_{nf} denotes the effective viscosity of desired nanofluid, μ_{bf}represents the viscosity of base fluid, water in our case, Φ_m denotes the maximum particle packing fraction, which varies from 0.495 to 0.54 and $\Phi_a$represents the effective volume fraction and is given by equation (2),

$$\Phi_a = \Phi' \left(\frac{a}{a_a}\right)^{1.3} \quad (2)$$

Here a_a is the radius of aggregate nanoparticles in nm, a is the radius of primary nanoparticles also in nm. Equation (1) can be rewritten as:

$$\frac{\mu_{nf}}{\mu_{bf}} = \left(1 - \frac{\Phi'}{\Phi_m}\left(\frac{a}{a_a}\right)^{1.3}\right)^{-1.25} \quad (3)$$

For this study, Φ_m is taken as 0.5 and the ratio $\frac{a}{a_a}$ is kept 0.2042 after referring to study on nanoparticles in journal bearing. [16]

2.2 REYNOLDS EQUATION FOR DISTRIBUTION OF PRESSURE AND LOAD CARRYING CAPACITY

Reynolds' equation is derived by considering the equilibrium of a fluid element and the continuity of flow in a column. Fluid velocities are derived from the equations of equilibrium of fluid element. When it comes to the iso-viscous case, which is the most common form of Reynolds equation, it can be approximated and the steady state Reynolds Equation for journal bearing [4], to find pressure P is as follows:

$$\frac{\partial}{\partial x}\left(h^3\frac{\partial P}{\partial x}\right)+\frac{\partial}{\partial z}\left(h^3\frac{\partial P}{\partial z}\right) = 6\mu U\frac{\partial h}{\partial x}) \qquad (4)$$

Here, P denotes pressure developed in the fluid, h denotes the height or thickness of fluid film, μ represents the viscosity of fluid and U denotes the speed of journal.
The formula for film thickness h is as follows at a given point in the bearing can be found by:

$$h = c(1 + e.cos\theta) \qquad (5)$$

Here, c represents the clearance between the shaft and bearing while e denotes the eccentricity between centres of journal and bearing. Thickness of the film varies with angle θ.
The load carrying capacity of film made of lubricating fluid is found by integrating the distribution of pressure generated in the fluid over the surface area of given bearing. Equation 6 is used to obtain the load that a bearing will support for a defined film size

$$W = \int_0^L \int_0^B pdxdy \qquad (6)$$

The load carrying capacity is thus dependent on the size of bearing, the viscosity of fluid used as lubricant viscosity and the velocity at which shaft rotates, hence these parameters can be changed suitably to improve the performance and characteristics of bearing.

3. METHODOLOGY

The bearing parameters are defined and the viscosity of nanofluid is found using Kreiger-Dougherty Model. This viscosity along with bearing parameters are fed as inputs to the MATLAB program. The program solves the partial differential Reynold's equation to calculate the pressure distribution in bearing using Finite Element Method. The fluid film of bearing is unwrapped and considered a cuboid with height equal to film thickness. The rectangular section is discretized into 49 nodes in x direction and 30 nodes in Z direction. Pressure is found at every element by running iterations using FEM till convergence is achieved and then integrated over the area of bearing. The program checks the stability according to Routh Hurwitz criterion and gives the load carrying capacity of the bearing.

All values are non-dimensionalized for the sake of standardisation of results.

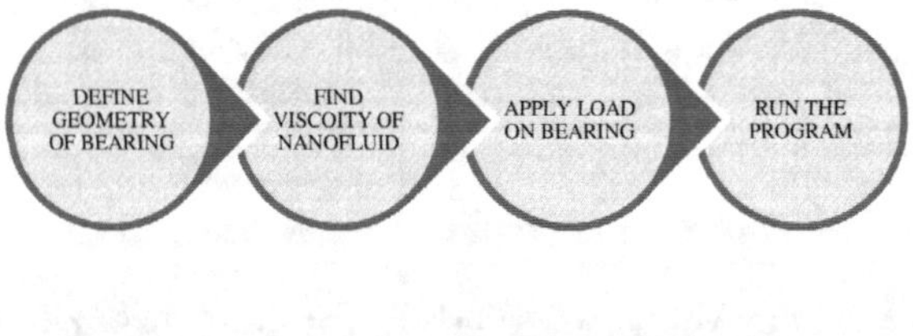

Figure 1: Methodology employed

3.1 DEFINE GEOMETRY OF BEARING

The $\frac{L}{D}$ ratio of bearing is kept 1 and with diameter and length of bearing, both equal to $50mm$. The eccentricity ratio is kept0.7. These values are based on commonly used bearing parameters according to literature. The values are non-dimensionalized before using them as inputs to the MATLAB program. The bearing parameters are specified in Table 1.

Table 1: Bearing parameters

PARAMETERS	DIMENSIONAL VALUES	NON-DIMENSIONAL VALUES
LENGTH OF BEARING, L	$50\ mm$	1
RADIUS OF BEARING, R	$25\ mm$	0.5
ECCENTRICITY RATIO, ε	0.7	0.7

3.2 FINDING EFFECTIVE VISCOSITY OF NANOFLUID

The value of μ_{bf} is kept $0.6531\ mPa.s$ in equation (3), which denotes the dynamic viscosity of base fluid, water, at $40°C$. Corresponding viscosities of nanofluid at different volume fraction of nanoparticles are given in Table 2.

Table 2: Viscosity of nanofluid at varying volume fraction of nanoparticles according to Kreiger-Dougherty Model

VOLUME FRACTION OF NANOPARTICLE ADDITIVE, Φ'	DYNAMIC VISCOSITY OF FLUID, $\mu_{nf}[mPa.s]$
0	$0.6531\ (Viscosity\ of\ plain\ water)$
0.1	0.6743
0.2	0.6969
0.3	0.7208

3.3 APPLICATION OF LOAD

The load is applied in $-Z$ direction of bearing. The load is varied to find the maximum load the bearing can withstand before becoming unstable. The load carrying capacity is determined for 500, 1000 and 1500 rpm respectively.

3.4 RUN THE PROGRAM

The MATLAB program solves the partial differential Reynolds equation to calculate the pressure distribution in bearing using Finite Element Method.

As the lubricant considered is a nanofluid, the program calculates the shear stress based on non-newtonian Power Law model of viscosity. Once the pressure is calculated, the program finds the damping and stiffness coefficients for the bearing.

4. RESULTS AND DISCUSSIONS

To validate the MATLAB program used, output data is compared to results from a study on water lubricated journal bearing conducted using CFD [17]. The parameters used in reference study are given in Table 3.

Table 3: Reference bearing parameters

	VALUES
LENGTH OF BEARING, L	$322\ mm$
RADIUS OF BEARING, R	$161\ mm$
RADIAL CLEARANCE, c	$1.3\ mm$
ECCENTRICITY RATIO, ε	0.85

The MATLAB program is run to find the maximum load the bearing can carry under the same conditions as reference study and the results are compared in Figure 2.

IOP Conf. Series: Materials Science and Engineering 1228 (2022) 012032 doi:10.1088/1757-899X/1228/1/012032

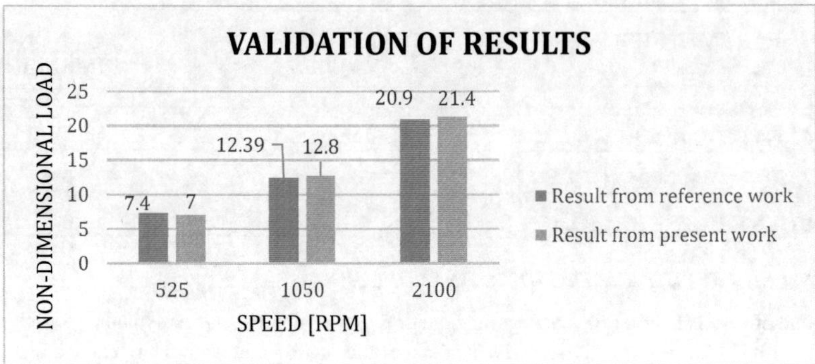

Figure 2: Comparison with reference study

It is observed that the obtained results using MATLAB program agree with reference values with an error of less than 4%. The program is hence validated.

The performance characteristics of journal bearing at different speeds are noted in Table 4 for each case of nanofluid. It is observed that each characteristic steadily increases in magnitude with increase in volume fraction of nanoparticles from 0 to 0.3.

Table 4: Performance characteristics for different lubrication cases at speeds 500, 1000, 1500 rpm

VOLUME FRACTION OF NANOPARTICLES	VISCOSITY OF FLUID [mPa.s]	SPEED OF SHAFT [rpm]	LOAD CARRYING CAPACITY [N]	STIFFNESS COEFFIICENTS [10^8 N/m]				DAMPING COEFFIICENTS [10^6 Ns/m]			
				Kxx	(-)Kzx	(-)Kxz	Kzz	Cxx	(-)Czx	(-)Cxz	Czz
		500	429.34	1.52	0.32	4.6	7.44	4.38	9.73	9.73	33.88
		1000	1664.8	11.1	2.17	33.05	52.95	5.31	11.54	11.54	40.43
0 [PLAIN WATER]	0.6531	1500	3691.01	35.25	6.57	103.79	165.06	11.3	24.05	24.05	84.36
		500	443.26	1.6	0.33	4.85	7.86	4.61	10.28	10.28	35.83
		1000	1706.92	11.38	2.22	33.88	54.29	8.17	17.76	17.76	62.18
0.1	0.6743	1500	3810.24	36.39	6.78	107.14	170.4	11.67	24.83	24.83	87.09
		500	458.13	1.65	0.34	5.01	8.13	4.61	10.28	10.28	35.83
		1000	1763.43	11.76	2.29	35	56.08	8.17	17.76	17.76	62.18
0.2	0.6969	1500	3936.96	37.6	7.01	110.7	176.06	11.67	24.83	24.83	87.09
		500	473.84	1.6	0.33	4.85	7.86	4.93	10.99	10.99	38.3
		1000	1824.06	11.38	2.22	33.88	54.29	8.73	18.97	18.97	66.45
0.3	0.7208	1500	4072.32	36.39	6.78	107.14	170.4	12.47	26.54	26.54	93.08

It is evident from Table 4 that the characteristic values increase with increase in shaft speed. The maximum load that bearing can carry is found to be at volume fraction of 0.3 nanoparticles at 1500 rpm journal speed and is 4072.32 $Newton$.

On observing Table 4, it is found that the stiffness coefficients are of high magnitude for each case. In case of plain water, it is slightly less than damping coefficients of nanofluids, however, overall, very little variation is observed. An average of only 6.2% increase in value of stiffness coefficient is observed when nanofluid with 0.3 volume fraction additive is compared to plain water at 1500 rpm. The magnitude of stiffness coefficient increases with speed of shaft.

It is further observed that the damping coefficient is highest for the case when water contains 0.3 volume fraction, it is equivalent for cases when volume fraction of nanoparticles is 0.1 and 0.2, and least for plain water lubricated bearing. The average maximum increase in damping coefficients from Table 4, is noted to be 10.49%. Additionally, it is observed that the magnitude of damping coefficient increases with speed of shaft. Symmetry is present in the magnitudes of Czx and Cxz as values are exact same for every case.

5. CONCLUSIONS

The performance characteristics of bearing with different nanofluid are compared to bearing lubricated by plain water and the following conclusions are drawn:

1. Properties of plain water can be enhanced by addition of small amounts of nanoparticles.

2. An average increase of 9.98% is observed in capacity of bearing to withstand load of at different speeds of shaft when 0.3 volume fraction nanoparticles are added to water.

3. An average increase of 6.2% is observed in magnitude of stiffness coefficients when 0.3 volume fraction nanoparticles are added to water.

4. An average increase of 10.49% is observed increase magnitude of damping coefficients when 0.3 volume fraction nanoparticles are added to water.

5. Water is cheaper, renewable, and easy to dispose, as compared to synthetic oils which are more expensive, finite in nature and may add toxicity to the environment. Nanofluids made of water can be used instead of the petroleum-based lubricants used till now in hydrodynamic journal bearings as they have enhanced properties.

References

[1] Sean DeCamillo. Basics of Hydrodynamic Bearings in Industrial Applications. 201

[2] Zhang, X., Yin, Z., Gao, G., & Li, Z. (2015). Determination of stiffness coefficients of hydrodynamic water-lubricated plain journal bearings. *Tribology International*, *85*, 37–47.

[3] Manshoor, B., Jaat, M., Izzuddin, Z., & Amir, K. (2013). CFD Analysis of Thin Film Lubricated Journal Bearing. *Procedia Engineering*, *68*, 56–62.

[4] Stachowiak, G. W., & Batchelor, A. W. (2016). *Engineering tribology*. Butterworth-Heinemann.

[5] N.N.M. Zawawi, W.H. Azmi, A.A.M. Redhwana, M.Z. Sharif. Thermo-Physical Properties of Metal Oxides Composite Nanolubricants. Journal of Mechanical Engineering Vol SI 5(1), 28-38, 2018 ISSN 1823-5514. Published: 2018-01-15

[6] Jamalabadi, M. Y. (2017). Effects of Nanoparticle Enhanced Lubricant Films in Dynamic Properties of Plain Journal Bearings at High Reynolds Numbers. *International Journal of Engineering and Technologies*, *13*, 1–23.

[7] Sadabadi, H., & Sanati Nezhad, A. (2020). Nanofluids for Performance Improvement of Heavy Machinery Journal Bearings: A Simulation Study. *Nanomaterials*, *10*(11), 2120.

[8] P. A. Kulkarni and B. V. A. Rao. Stability Behaviour of Finite MHD Journal Bearings. Mechanism and Machine Theory, 1977, Vol. 12, pp, 293-302.

[9] Sreedhar Babu Kalakada, Prabhakaran Nair Kumarapillai and Rajendrakumar Krishnan Perikinalil. Analysis of Static and Dynamic Performance Characteristics of THD Journal Bearing Operating Under Lubricants Containing Nanoparticles. International Journal of Precision Engineering and Manufacturing Vol. 13, No. 10, pp. 1869-1876, 2012.

[10] Hongwei Tang , Jean Rabault , Alexander Kuhnle , Yan Wang , and Tongguang Wang. Robust active flow control over a range of Reynolds numbers using an artificial neural network trained through deep reinforcement learning. Phys. Fluids 32, 053605 (2020)

[11] Concli, F. (2020). Journal Bearing: An Integrated CFD-Analytical Approach for the Estimation of the Trajectory and Equilibrium Position. *Applied Sciences*, *10*(23), 8573.

[12] K.P. Gertzos, P.G. Nikolakopoulos, C.A. Papadopoulos. CFD analysis of journal bearing hydrodynamic lubrication by Bingham lubricant. Tribology International 41 (2008) 1190– 1204.

[13] Hother-Lushington, S. (1976). Water lubricated bearings. Tribology International, 9(6), 257–260. doi:10.1016/0301-679x(76)90014-1

[14] Cui, Y., Ding, M., Sui, T., Zheng, W., Qiao, G., Yan, S., & Liu, X. (2020). Role of nanoparticle materials as water-based lubricant additives for ceramics. Tribology International, 142, 105978.

[15] Nowak, P., Kucharska, K., & Kamiński, M. (2019). Ecological and health effects of lubricant oils emitted into the environment. International Journal of Environmental Research and Public Health, 16(16), 3002.

[16] Suryawanshi, S.; Pattiwar, J. Effect of TiO2 Nanoparticles Blended with Lubricating Oil on the Tribological Performance of the Journal Bearing. Industrial Lubrication and Tribology, Vol. 40, pp. 370–391, 1980.

[17] Gengyuan, G., Zhongwei, Y., Dan, J., &Xiuli, Z. (2015). CFD analysis of load-carrying capacity of Hydrodynamic lubrication on a water-lubricated journal bearing. Industrial Lubrication and Tribology, 67(1), 30–37.

Java script animation of generator rotors under different modes of oscillation in two area four machine system

Aaqib Manzoor, Mairaj ud Din Mufti and Mohammad Younis Nabi

Department of Electrical Engineering National Institute of Technology Srinagar, India

Abstract: Computer technology advancements aid in the creation of instructional plans that allow students to study in a variety of ways. Animations and digital maps make it easier to understand abstract concepts and ensure that learning is permanent. Animation is a technique of creating continuous motion and shape change illusion. Interconnected power system often seems to be an intractably intricate morass of equations and ad hoc rules of thumb. Lack in visibility impedes learning about the power system phenomena. Animation can help us to convey information that would otherwise require a thousand or more words. This work is all about the development of a visualization tool in order to understand oscillatory dynamics of generator rotor. The power vs phase angle curve gradient interacting with generator rotor inertia forms an equivalent mass spring system. So, by this very reason different modes of oscillation in mass spring system are animated. A two area four machine system is modelled in MATLAB/Simulink by using Kron reduction technique and different modes of oscillation are analyzed. Later the animation of generator rotors of two area system are created with the help of HTML, Cascading Style Sheets (CSS) and Java Script. Each animation is coupled to the results obtained in MATLAB/Simulink under different modes of oscillation.

1. Introduction

Technology is rapidly evolving in the twenty-first century and these advancements are bringing many educational advances with them. Learning and teaching settings are also affected by the usage of technology in education as mentioned by *Bai* in [1]. It has been argued by *Lai et al.* [2] that the employment of animation in education results in a considerable favourable change in student's attitude and academic achievements. Animation, as technological tool used in education has shown to benefit students in terms of security, speeding and slowing time, examining very rare events, simplifying complicated systems etc as mentioned by *Betrancourt et al.* [3]. Visuality has been more prevalent over time as a result of new technology and needs [4]. Animation is at the forefront of visualization systems as a multimedia tool.

Animation is a technical method that, in general, creates the appearance of motion in the observer by sequencing still images created in an analog or digital environment. Popularity of animation has increased progressively as a teaching tool over the last five years. A research was conducted in [5] which reveals that 75% of engineering students are visual learners but 90% of instructions given to them is oral and written verbal communication. University of Arizona in [5] provided a tool to students which attracted the inductive and visual learners and helped them to understand the mathematics and physics of electric machine theory by these animations.

Swarthmore University has developed various kinds of animations of different mechanical and

Content from this work may be used under the terms of the Creative Commons Attribution 3.0 licence. Any further distribution of this work must maintain attribution to the author(s) and the title of the work, journal citation and DOI.
Published under licence by IOP Publishing Ltd

electromechanical systems using Java Script, HTML and CSS in order to understand the dynamics of these physical systems. Interconnected power system often seems to be an intractably intricate morass of equations and ad hoc rules of thumb. Lack in visibility impedes learning about the power system phenomena. In this work, animation of behavior of rotor under different modes of oscillation is presented. These animations are developed with the help of HTML, CSS and JAVA SCRIPT. These animations seem to offer the better representation of rotor dynamics under small signal stability. The goal of this work is to use animations to convey information that would otherwise require a thousand or more words or equations to describe. The detailed work can be summarized as:

- To do the modal analysis of mass spring system and two area four machine system.
- To develop animation of mass spring system for different modes of oscillations and in analogy animation of generator rotors for low frequency oscillations is developed.
- To develop the classical MATLAB model for two area four system and analyze different modes of LF oscillations.

2. Modal analysis and animation of mass spring system

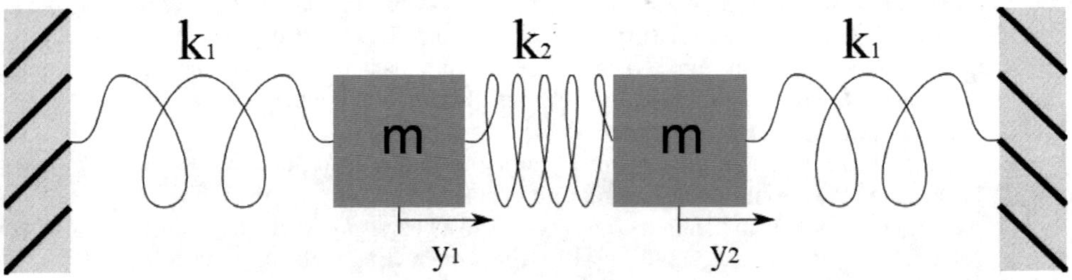

Figure 1. Mass spring system

In order to have a better understanding of normal modes of oscillation, we can take an example of two mass spring coupled system from [6] containing two distinct normal modes. Here y_1 and y_2 are position variables or state variables which are measured at equilibrium position of the masses.
Writing the equilibrium equation of two masses

$$m\ddot{y_1} = -k_1y_1 - k_2(y_1 - y_2) \tag{1}$$

$$m\ddot{y_2} = -k_1y_2 - k_2(y_2 - y_1) \tag{2}$$

These equations can be rewritten as:

$$m\ddot{y_1} = -(k_1 + k_2)y_1 + k_2y_2$$

$$m\ddot{y_2} = k_2y_1 - (k_1 + k_2)y_2$$

We can write the equations from a system of two homogeneous linear second order differential equations in matrix form

$$m\frac{d^2}{dt^2}\begin{bmatrix} y_1 \\ y_2 \end{bmatrix} = \begin{bmatrix} -(k_1 + k_2) & k_2 \\ k_2 & -(k_1 + k_2) \end{bmatrix}\begin{bmatrix} y_1 \\ y_2 \end{bmatrix}$$

Let us try a solution $y = Ve^{rt}$ and obtain eigenvalue problem $AV = \lambda V$ with $\lambda = ms^2$. The eigenvalues can be determined by

$$\det\,[A - \lambda I]=0$$

$$= (\lambda + k_1 + k_2)^2 - {k_2}^2 = 0$$

The two solutions for λ are:

$$\lambda_1 = -k_1 \ , \quad \lambda_2 = -(k_1 + 2k_2) \tag{3}$$

Corresponding values of s in $y = Ve^{st}$ with $s = \pm\left(\frac{\lambda}{m}\right)^{1/2}$ are:

$$s_1 = i\left(\frac{k_1}{m}\right)^{1/2} \quad \overline{s}_1$$

$$s_2 = i\left(\frac{k_1 + 2k_2}{m}\right)^{1/2} \quad \overline{s}_2$$

Since values of s are pure imaginary, so $y_1(t)$ and $y_2(t)$ will be oscillating in nature with angular frequency

$$\omega_1 = \left(\frac{k_1}{m}\right)^{1/2}$$

$$\omega_2 = \left(\frac{k_1 + 2k_2}{m}\right)^{1/2}$$

ω_1 and ω_2 are frequencies of two normal modes, the eigenvector corresponding to eigenvalue λ_1 is

$$-k_2 v_{11} + k_2 v_{12} = 0$$

$$v_{11} = v_{12} \tag{4}$$

Which means both the state variables are in phase with each other. This mode follows a motion where $y_1 = y_2$ and motion is independent of k_1 and k_2. The eigenvector corresponding to eigenvalue λ_2 is

$$k_2 v_{21} + k_2 v_{22} = 0$$

$$v_{21} = -v_{22} \tag{5}$$

Which means both the state variables are in phase opposition with each other. This mode follows a motion where $y_1 = -y_2$.

2.1. Animation of different modes in mass spring system

The animation of three different modes corresponding to mass spring damping are created with the help of MATLAB GUI. The snapshots of animation corresponding to different modes at different instants are shown in figure 2 to 4.

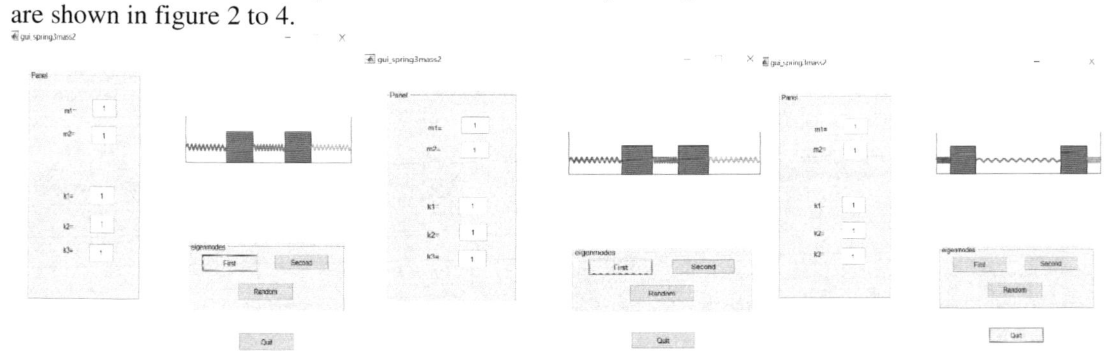

Figure 2. Animation of mode $1(y_1 = -y_2)$

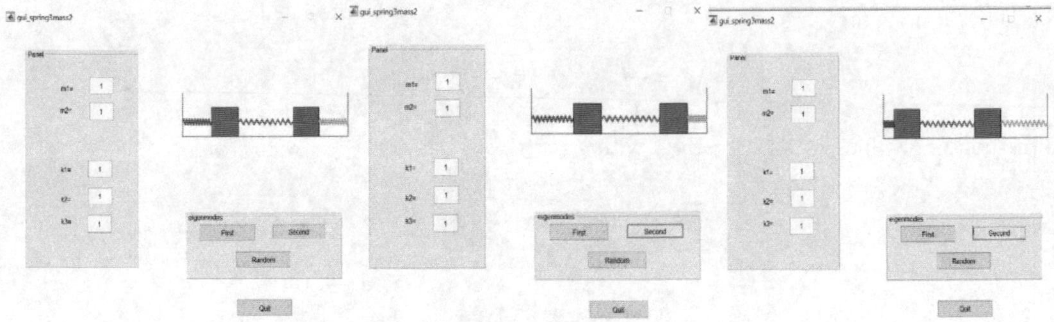

Figure 3. Animation of mode $2(y_1 = y_2)$

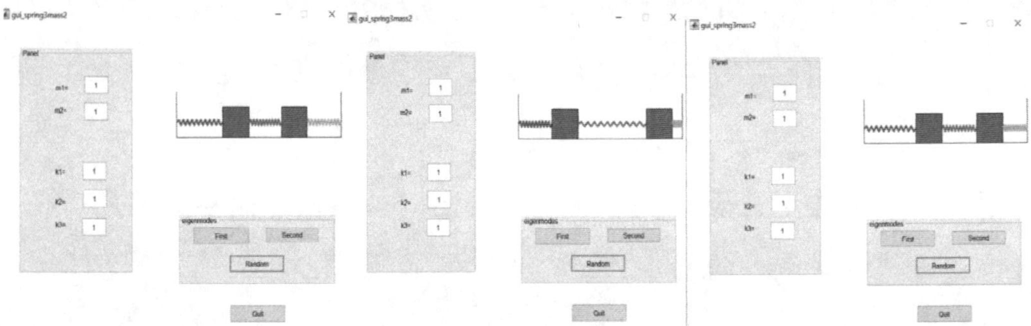

Figure 4. Animation of random mode

3. Modal analysis of two area system

The power systems are highly non-linear, so they are first linearized about the steady state equilibrium point. Once we have a linear model, the modal analysis can be done in order to characterize the oscillations. The interconnected system for study is taken from the work performed by *Rogers* [7]. The data of the two area four machine system is taken from power system toolbox which inturn is used to find the steady state operating point of the system by using load flow analysis. The mathematical modelling is done and simulated in MATLAB for four machines and the steady state operating parameters are obtained from load flow results. The state matrix is obtained from MATLAB by linear analysis of the model. This is done by changing each state by small amount and corresponding rate of change of each state is obtained. Then rate of change of the states divided by the perturbation will give us the column of the state matrix corresponding to the interrupted state. The states of the classical model of two area system are:

$$x' = [\Delta\delta_1 \;\; \Delta\omega_1 \;\; \Delta\delta_2 \;\; \Delta\omega_2 \;\; \Delta\delta_3 \;\; \Delta\omega_3 \;\; \Delta\delta_4 \;\; \Delta\omega_4]$$

The eigenvalues of the state matrix calculated by using **eig** function in MATLAB is below:

Table 1. Eigenvalues of two area four machine system

Eigenvalues
0
0
±4.75i
±14.42
±11.76

There are eight eigenvalues as the system has eight states. The three oscillatory or eigen modes are identified by complex eigenvalues. The frequencies of these oscillations can be found from the imaginary parts of eigenvalues.

$$f = \frac{im(\lambda)}{2\pi} = \frac{\omega}{2\pi}$$

From the above eigenvalues of our system the frequencies of oscillatory modes are 0.76Hz, 2.31Hz And 1.88Hz respectively. The eigenvectors corresponding to different eigenvalues are calculated and given in table 2.

Table 2. Eigenvectors for different modes of oscillation

Eigenvector for Local area1 mode	Eigenvector for Local area2 mode	Eigenvector for Inter area mode
0.7680	-0.0333	-0.4905
-0.0532i	0.0028i	0.1032i
-0.6359	-0.0044	-0.4477
0.0440i	0.0003i	0.0941i
-0.0020	-0.7196	0.5426
-0.0001i	0.0612i	-0.1141i
0.0303	0.6882	0.4711
-0.0021i	-0.0585i	-0.0991i

The right eigenvectors of the system reveal that:

- In inter-area mode which is the lowest frequency complex mode, the generators of area 1 are having speed deviations that are 180 degrees out of phase with that of generators in area 2. The eigenvector components corresponding to speed deviation are 90 degrees out of phase with the components corresponding to angle.
- In local area1 mode, the speed deviations of generators in area 1 are 180 degree out of phase with each other.
- In local area 2 mode, the speed deviations of generators in area 2 are in phase opposition with each other and speed deviations of generators in area1 are in phase with each other.

In order to excite a particular mode of oscillation from the given system, small perturbation to the mechanical input power of different generators is given. For local area 1 mode the input power to generator 1 is increased by small amount and input power to generator 2 is decreased by small amount. Similarly for local area 2 mode the input power to generator 3 is increased by small amount and input power to generator 4 is decreased by small amount. For inter-area mode the input power of generator 2 of area 1 is increased by small amount and input power to generator 4 of area 2 is decreased by small amount. The speed deviations in pu for different modes of oscillations are shown in figure 5 to 7.

IOP Conf. Series: Materials Science and Engineering 1228 (2022) 012033 doi:10.1088/1757-899X/1228/1/012033

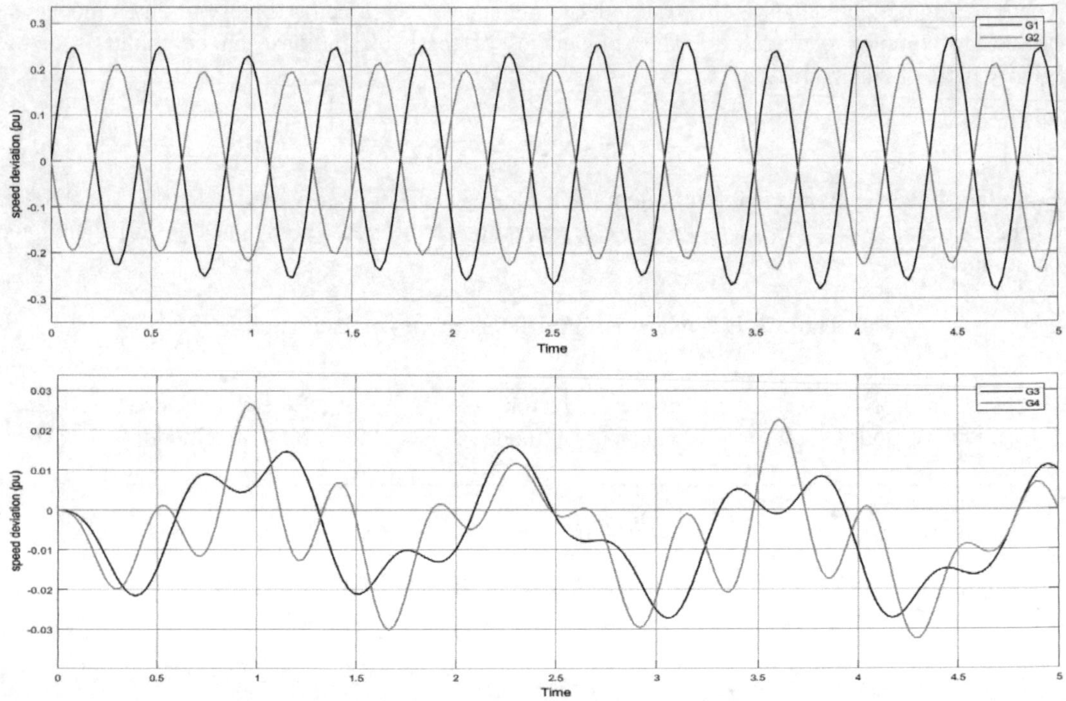

Figure 5. Speed deviations of generators in local area1 mode

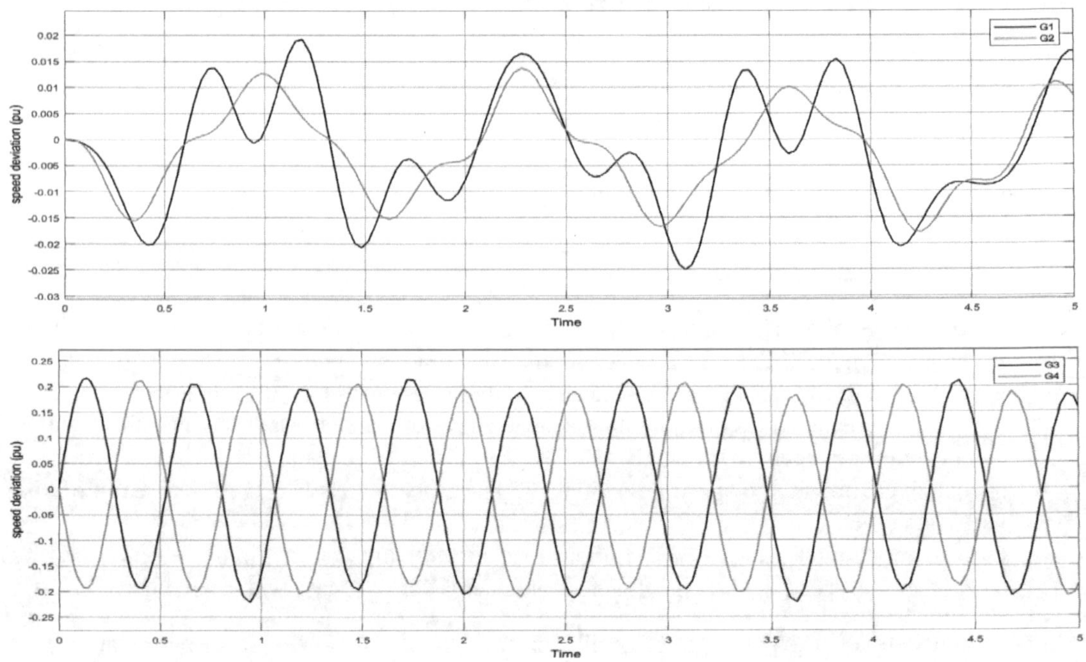

Figure 6. Speed deviations of generators in local area2 mode

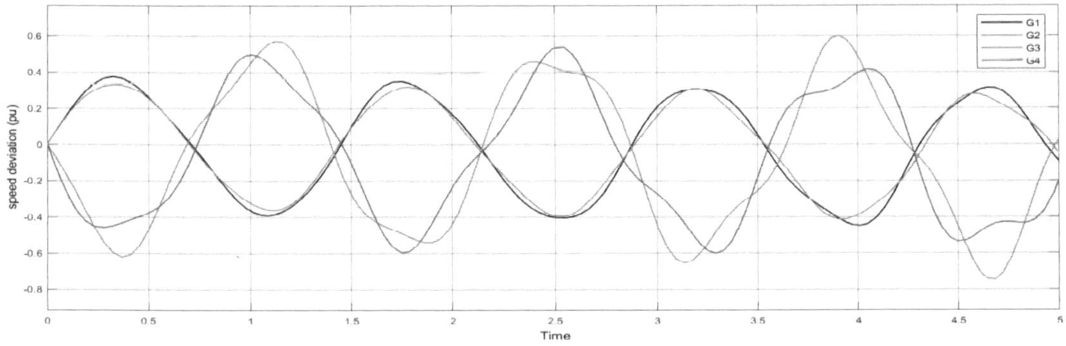

Figure 7. Speed deviations of generators inter area mode

3.1 Animation of rotor under different modes of oscillations

After the initiation of HTML5, the JavaScript nowadays becomes most popular scripting language for creating dynamic and interactive content. Several JavaScript libraries make the work of developers easy to create interactive animations. The JavaScript library we used in this work is P5.js. The main purpose of this is to make programming easy to use. Abundant availability and all over support are the big advantages of JavaScript. So, we can run p5.js programs on any web browser.

Visual Studio Code Editor is used in this work for writing the code. It supports developmental operations like task running, version control and debugging. The snapshots of animation of rotors for different modes are shown in figure 8. From these figures we can get an idea about different modes of oscillations from the position of windings of different generators. The results for different modes of oscillations from MATLAB Simulink are converted into .csv file and transferred into Visual Studio Code.

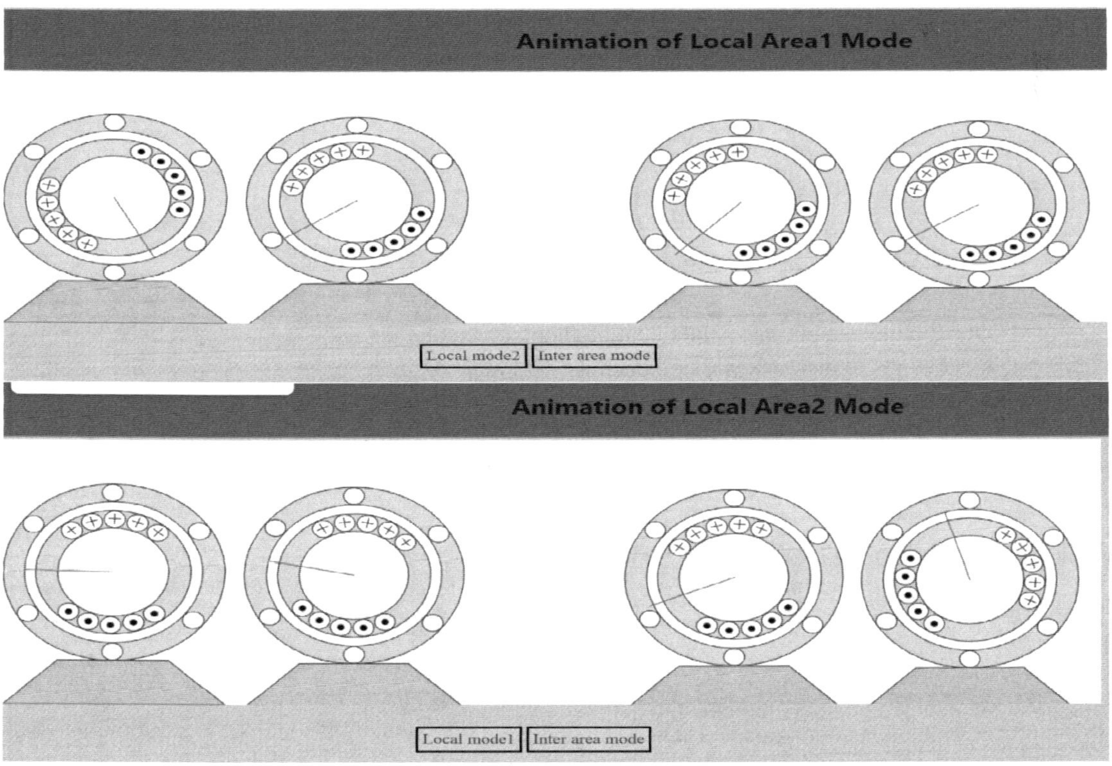

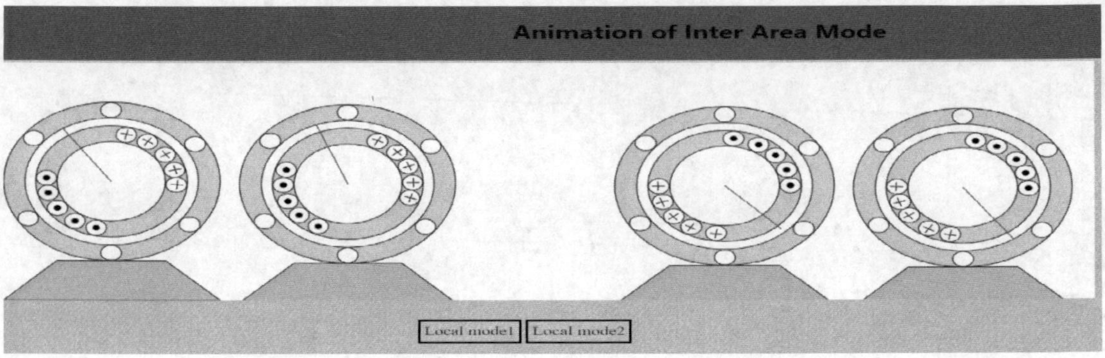

Figure 8. Animation of different modes

4. Conclusion

The animations developed in this work are very much helpful in order to understand the rotor dynamics under small signal stability. By using HTML, CSS and JavaScript, animation of different modes of oscillations in an interconnected power system are created, the dynamics of which is controlled by MATLAB/Simulink results of different modes. These animations give us a visualization effect of different modes of oscillations in interconnected power system, where we can see how generator rotors will behave when a particular mode is excited.

References

[1] Bai H 2018 Preparing teacher education students to use instructional technology in an asynchronous blended course *Innovative Practices in Teacher Preparation and Graduate-Level Teacher Education Programs* pp 603-619

[2] Lai Y L, Kuan K K, Hui K L & Liu N 2009 The effects of moving animation on recall, hedonic and utilitarian perceptions, and attitude *IEEE Transactions on Engineering Management*, **56** 3 pp 468-477

[3] Betrancourt M & Tversky B 2000 Effect of computer animation on users' performance A review *Le Travail Humain* **63** 4 pp 311-329

[4] Craig R and Amernic J H 2006 PowerPoint presentation technology and the dynamics of teaching *Innovative Higher Education* **31** 3 pp 147-160

[5] Karady G G and Tylavsky D *Use of Animation for Improvement of Student Understanding of Energy Conversion* (Arizona State University Tempe) AZ 85287

[6] Chasnov J R 2019 *Differential Equations for Engineers* (The Hong Kong University of Science and Technology Department of Mathematics)

[7] Rogers G 2000 *Power System Oscillations* (Norwell MA Kluwer)

[8] Chandravanshi M L and Mukhopadhyay A K 2013 Modal Analysis of Structural Vibration Proceedings of the *International Mechanical Engineering Congress & Exposition*

[9] Lengvarsky P and Bocko J 2013 Theoretical Basis of Modal Analysis *American Journal of Mechanical Engineering* **7** pp 173-179

[10] Andrew S E and Mark H R 1998 Virtual Experimental Modal Analysis proceedings IMAC XVI, *Mechanical Dynamics inc. Mesa Arizona Vibrant Technology Inc.* (Jamestown, California)

[11] Hermen V A 2001 Structural Dynamics Modeling using Modal Analysis Applications, Trends and Challenges Proceedings *IEEE Instrumentation and Measurement Technology Conference* (Budapest, Hungary)

[12] Lowe R K 2003 Animation and learning Selective processing of information in dynamic graphics *Learning and Instruction 13* 2 pp 157-176

[13] Rosen Y 2009 The effects of an animation-based on-line learning environment on transfer of knowledge and on motivation for science and technology learning. *Journal of Educational Computing Research* **40** 4 pp 451-467

[14] Xiao L 2013 Animation trends in education *International Journal of Information and Education Technology* **3** 3 pp 286-289

[15] Vegh L and Udavoras J 2020 Possibilities of Creating Interactive 2D Animations for Education Using HTML5 Canvas JavaScript Libraries *The 16th International Scientific Conference eLearning and Software for Education*

[16] Frisbie M 2019 *Professional JavaScript for Web Developers* (Wrox)

IOP Conf. Series: Materials Science and Engineering 1228 (2022) 012034 doi:10.1088/1757-899X/1228/1/012034

Research on concrete compressive strength in existing reinforced concrete elements with Schmidt hammer, ultrasonic pulse velocity method and destructive testing of cores

I Ivanchev[1]

[1] University of Architecture, Civil Engineering and Geodesy (UACEG), Faculty of Structural Engineering, Department Reinforced Concrete Structures, Sofia, Bulgaria, ivanchev_fce@uacg.bg

* Correspondence: ivanchev_fce@uacg.bg; Tel.: +359 988 364 706

Abstract. Concrete is a widely used material in construction. Quality control, maintenance and extending its service life are major issues after the construction stage and during the service of reinforced concrete buildings and facilities. This article presents the author's experimental results for determining the compressive strength of concrete for tests on 12 reinforced concrete beams at the age of concrete 1926th day. In previous studies, some of the beams were loaded to a stage corresponding to the yield strength of the longitudinal tensile reinforcement, and others to a loading stage before yield, reaching a maximum crack width of 0.3 mm. The reinforced concrete elements were kept indoors for two years, and for the next more than three years they were left outdoors, exposed to external atmospheric influences. In the research was used combination of non-destructive and partially destructive techniques. The compressive strength of the concrete was determined using the method of elastic rebound (Schmidt hammer), ultrasonic pulse velocity method and partially destructive techniques - core sampling. The relative errors in the values of compressive strength obtained by the different methods were calculated. A comparative analysis of the experimentally obtained results by different methods was made. The tests were performed according to European standards for non-destructive and destructive testing of reinforced concrete structures.

1. Introduction

Concrete is one of the oldest, most widely used and reliable materials in the construction of buildings and facilities due to its long service life, relatively low cost, fire resistance, the possibility buildings to be formed in the desired architectural form [1-4]. The analysis of the reinforced concrete structures in serviceability state shows that for almost each structure, especially for old buildings there is a decline of its existing technical condition as a result of thermal influences, freezing or thawing, loading or dynamic overloads, earthquakes, fire impacts and other factors. The assessment, inspection and diagnostics of the condition and quality assurance of reinforced concrete buildings and facilities after the stage of their construction and during their service are associated with many difficulties and measurement of many parameters. Planning for monitoring, maintenance and repair of damages can extend their service life. For this purpose it is necessary to be evaluated the strength, modulus of elasticity, homogeneity, carbonation of concrete, to be detected and measured the cracks in it and other

Content from this work may be used under the terms of the Creative Commons Attribution 3.0 licence. Any further distribution of this work must maintain attribution to the author(s) and the title of the work, journal citation and DOI.
Published under licence by IOP Publishing Ltd

parameters. Their determination always encounters some difficulties. In practice, two types of methods are used to study building structures: destructive and non-destructive, or a combination of both methods [1,3,5,6]. Destructive methods provide direct, accurate and reliable information on the quality of concrete, but they are more expensive and more labor-intensive. Based on development of digital electronics and computer innovations for reinforced concrete structures since 1980, especially for in-situ tests, which are increasingly important for reliable control, fully automated non-destructive methods (NDM) are used. They are simple, fast, reliable and relatively inexpensive and are important for monitoring and preventing irreversible damage [2]. The disadvantage is that they are indirect methods and the parameters of the concrete are determined by measuring other parameters and subsequent processing of the results. The effective use of non-destructive techniques supplements the results of the destructive tests of drilled cores [5]. Due to the fact that concrete is not a homogeneous material, a unique evaluation model for all concretes doesn't exist. The assessment of the current state of reinforced concrete structures [4,5,7] most often involves the complex application of various methods to obtain the most reliable results.

This article illustrates the results of the author's experimental tests for determining one of the most important properties of concrete: compressive strength [3]. The compressive strength of concrete has no fixed value during its service life and this is due to the above factors. The aim of the present study is to establish the current concrete compressive strength in 12 reinforced concrete beams at the age 1926th day.

2. The essence of the destructive and the non-destructive methods in the experiment

In the experiment were used visual, partially destructive and non-destructive methods [1,4,6,7,8]. All results contain a sum of random, methodological, instrumental and subjective errors.

The methods used in the experiment are:

2.1. Visual inspection

The initial and one of the most important non-destructive assessments of quality and the structural integrity of the concrete was done by visual inspection [1,2,4,5]. It provides valuable information of the condition of reinforced concrete structures, their safety, the presence of damages, the need for rehabilitation and/or retrofitting. With visual inspections it can be established the degree of damage and the decline of the quality of the materials, the presence of mechanical damages. Visual inspection was used to assess the visible concrete. Areas with damages were excluded from further research.

2.2. Non-destructive testing of concrete

Non-destructive techniques are widely used as indirect methods for evaluating the properties of concrete. There are many non-destructive methods for evaluating reinforced concrete structures. They use different effects such as resonance, vibrations, propagation velocity of ultrasonic waves, electromagnetic waves, infrared thermography [1,2,4,5] and others. In the present study, the combination of the following was used:

2.2.1. Method of elastic rebound (Schmidt hammer). It is a fast, cheap, convenient and very common diagnostic non-destructive method for determining the surface hardness of concrete, and hence the probabilistic compressive strength of concrete in new and existing reinforced concrete structures [4,5,6,9-12].

The Schmidt hammer measures the magnitude of the elastic rebound of a steel heavy body with spring action from a concrete surface, i.e. the shock pulse occurring in the concrete when force is applied is measured. The value of the rebound characterizes the degree of surface hardness of the material, and it is directly proportional to the probabilistic compressive strength of the concrete. The test is sensitive and the results can be affected by the carbonation of the concrete, the presence of moisture, air pockets, chemical impact. Therefore, it is recommended multiple measurements to be made in each test area.

The probabilistic compressive strength $f_{c,cube,Schmidt}$ was calculated by the formula [13]:

$$f_{c,cube,Schmidt} = kk_1R_s[MPa] \quad (1)$$

where: R_s- is the compressive strength reported depending on the magnitude of the rebound; k_1 – is a coefficient depending on the age of concrete; k – is an experimentally determined coefficient of concordance.

2.2.2. Non-Destructive Ultrasonic Pulse Velocity Method (UPVM). UPVM uses the propagation of ultrasonic waves introduced into concrete, where they propagate, dissipate and are reflected from the boundary between two environments [2,5,9,14-17].

The measuring instruments consist of an ultrasonic oscillation generator, transmitter, receiver, amplifier and reading device [1]. The generator generates high frequency signals with a frequency of 25-150 kHz. The transmitter is a piezoelectric crystal that converts these signals into ultrasonic mechanical pulses that propagate through the concrete, undergoing multiple reflections of the boundaries of the various concrete components within the reinforced concrete member. The source of ultrasonic pulses causes longitudinal, transverse and surface waves. The receiving piezoelectric transducer registers the longitudinal ultrasonic waves, which are the fastest and converts them into electrical signals [4,14,18]. The time or velocity of propagation of the ultrasonic signal in the mass of the samples is measured directly. The velocity of mechanical waves is very high and depends on the elastic properties and density of the medium in which they propagate.

Depending on the access to the reinforced concrete member, there are three ways of location of the two piezoelectric transducers: direct transition, semi-direct transition, indirect transition [1,2,15,19].

From the ultrasonic signal velocity it can be determined the probabilistic compressive strength, modulus of elasticity, homogeneity and quality of concrete, the presence of cracks, to be monitored the changes occurring over time in the properties of concrete. The accuracy of these tests is not very high. The results are influenced by: type and size of aggregates, class of cement, water-cement ratio, type, characteristics, parameters and age of the concrete, inclusions of air, presence of cracks, stratification of the concrete, technology of preparation and laying of the concrete mix, corrosion of the reinforcement, presence of reinforcing bars, temperature, etc. factors. The dependence of ultrasonic propagation on these factors is very complex and has been studied by many researchers [1,4,15], they clearly indicate the need for establishing correlations between ultrasonic measurements and standard tests by destruction of test specimens (cores), which were prepared of the same materials and composition as the concrete in the reinforced concrete members.

In this article, UPVM was used for determining probabilistic concrete compressive strength.

The concrete compressive strength based on the measured ultrasonic pulse velocity is most often determined by [1,4,15,19-22]:

$$\mathrm{f_{c,UPVM}} = \mathrm{cV_{UP}^{3,75}}, \quad (2)$$

where: $\mathrm{V_{UP}}$ is the ultrasonic pulse velocity in km/s; $\mathrm{f_{c,UPVM}}$ is in MPa; c is factor in the range of 0,158 to 0,231 [4].

2.2.3. Destructive method - test of cores drilled from the reinforced concrete structure. In the absence of reference values for the probabilistic compressive strength of concrete for existing reinforced concrete buildings and facilities, the test results of the so-called drilled cores are used [3,6,13]. The core drilling [5,23-26] is useful, reliable, expensive, time-consuming operation, but is currently considered the most reliable method for evaluating the concrete in situ. Widely used standards are: EN 12504-1: 2019 [31], EN 13791: 2007 [13] in Europe and ACI 214.4R: 2003, ASTM C42: 2020, ASTM C42 (136): 2018 and ACI 318 in the USA. The test specimens were taken from the structural members, if possible from slightly stressed, non-critical regions, so as not to weaken the structure and not to affect its safety, the areas with reinforcement were avoided. The test was performed in accordance with the standards EN 12390-3:2019 [32] and the compressive strength of the cores were obtained by dividing the maximum load by the cross-sectional area of the core based on the average diameter of the core. The result obtained must be interpreted very carefully [27]. The compressive strength obtained in the test differs from the corresponding in-situ strength in the structure in the place where the core was drilled. This difference is due to the influence of several factors such as the presence of moisture in the core, poor compaction of the concrete and the presence of caverns in the specimen, different diameter and length to diameter ratio (l/d) of cores, direction of drilling relative to the direction of laying the concrete mix, presence of reinforcement or other inclusions.

3. Experimental setup

Specimens used for the experimental study are:

3.1. Reinforced concrete beams

The concrete compressive strength tests were made on four specimens of reinforced concrete beams, which were made more than 5 years ago (1926 days).

The specimens differ in provided longitudinal reinforcement, concrete cover and reinforcement ratio: 3 pcs. beams with a cross section of 27/15cm, provided longitudinal bottom reinforcement of 2N12, stirrups ф8/10(15)cm, 3 pcs. beams with a cross section of 27/15cm, provided longitudinal bottom reinforcement of 2N18, stirrups N8/10(15)cm, 3 pcs. beams with a cross section of 30/15cm, provided longitudinal bottom reinforcement of 2N12, stirrups ф8/10(15)cm, 3 pcs. beams with a cross section of 30/15cm, provided longitudinal bottom reinforcement of 2N18, stirrups N8/10(15)cm. The spans are the same for all beams - 3 m. Their structural parameters were selected so as to correspond to the characteristic parameters of beams designed in practice (Figure 1). They were prepared of concrete class C25/30, fine fraction of coarse aggregate (d_{max} = 12 mm) and consistency S3. In previous studies, some of the beams were loaded to a stage corresponding to the yield strength of the longitudinal tensile reinforcement, and others to a loading stage before yield, reaching a maximum crack width of 0.3 mm.

For two years the reinforced concrete elements were kept indoors, and for the next more than three years they were left outdoors, exposed to external atmospheric influences.

Visually, the surface concrete was assessed to identify and determine the visible defects (pores, cracks, etc.). These regions were excluded from further studies of the concrete. For each of the beams, at least 3 surfaces have been determined where the non-destructive tests were initially carried out (with the method of elastic rebound and UPVM) and then test specimens (cores) were drilled for laboratory tests. To be avoided the presence of reinforcement in the cores, its position is predetermined.

Figure 1. Reinforced concrete beams - 12 pieces (personal archive).

3.2. Test specimens (cores)

For the study of the in-situ concrete compressive strength at the age 1926th day, 18 cores with a diameter (D) Ø100 mm and depth ranging of 151 mm to 156 mm were drilled from the reinforced concrete beams (Figure 2 - a, b) from places without visible on the surfaces cracks, pores, cavities, defects and reinforcement [28-30]. The measurements were with accuracy up to 1 mm. The choice of core diameter was regulated by the D/A ratio, where A is the maximum size of the coarse aggregate. According to EN 12504-1:2019 [31] this ratio must be equal to or greater than 3.0. Before taking the cores, the position of the reinforcement in the reinforced concrete elements was determined using the Profometer 5+ Scanlong of the company "Proceq". This was done in order to be avoided the presence of reinforcement in the cores.

(a) (b) (c)

Figure 2. (a) 18 pieces of drilled cores, (b) drilled concrete cores from reinforced concrete beams, (c) machine for drilling the cores (personal archive).

The taking of cores was performed according EN 12504-1: 2019 [31]. The cores were taken from the reinforced concrete beams with the help of cutting tool, fixed to the members (Figure 2 - c). In real conditions, the opening obtained when removing the core must be filled with concrete, epoxy or other suitable fillers [3]. Strength measurements were made after 24 hours, during which the cores were left to air dry [27]. EN 12504-1: 2019 requires the drilled cores used for determining the concrete compressive strength, not to contain reinforcement if possible. After visual inspection, 7 cores were removed, in which parts of reinforcement, air pores, defects, cracks, stratification of concrete and other defects were noticed (Figure 3).

Figure 3. Drilled cores with defects (personal archive).

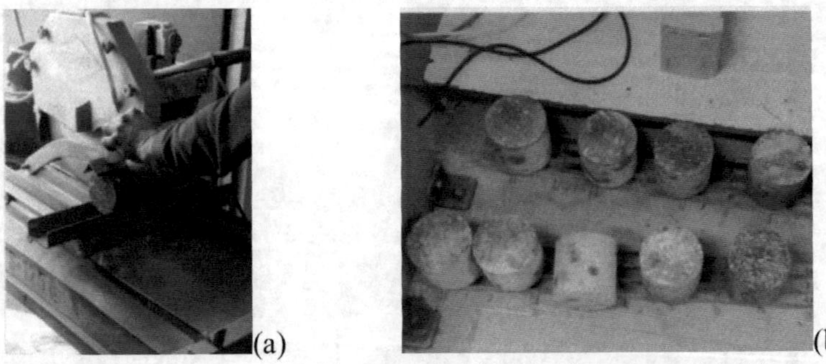

(a) (b)

Figure 4. Preparation of drilled cores: (a) trimming the ends; (b) cores, ready for experimental study (personal archive).

Before performing the experiments, the selected 11 concrete test specimens (cores) were prepared by trimming off their ends on both sides and they had become cylinders with dimensions of 100 mm depth and 100 mm diameter (Figure 4), i.e. the l/d ratio is equal to 1. Only the internal concrete of the reinforced concrete members was assessed for the obtained cores.

IOP Conf. Series: Materials Science and Engineering 1228 (2022) 012034 doi:10.1088/1757-899X/1228/1/012034

4. Experimental results

The experimental studies were conducted in an accredited mobile construction testing laboratory at Novostroy Control Ltd., Sofia. In this point the results of the experiment are presented, using different methods. The visual inspection was done directly due to the fact that all reinforced concrete members were fully accessible. The external visible surfaces were evaluated for smoothness, flatness, for identifying the visible defects - the presence of cracks, pores, insufficient concrete cover. Such regions were excluded from further studies of the concrete compressive strength. During this preliminary inspection, the places for testing were identified.

4.1. Determination of probabilistic concrete compressive strength by its surface hardness

The probabilistic compressive strength of concrete was obtained by determining its surface hardness by the elastic rebound method with a Schmidt OS8000 hammer (Figure 5) of the company Proceq, according to EN 12504-2:2012 [12].

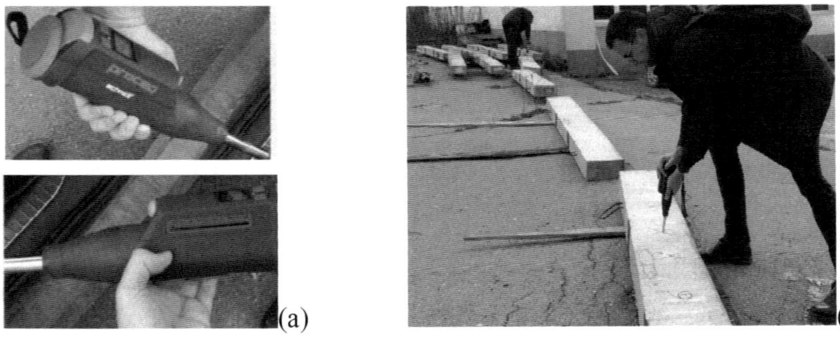

(a) (b)

Figure 5. Determination of the probabilistic compressive strength of concrete: (a) Schmidt 0S8000 hammer; (b) measurement (personal archive).

In the experimental researches was observed the requirement that the surface layer of concrete of the experimental points should be dry, with a smooth surface, without defects and as much as possible preserved. The probabilistic compressive strength was obtained after adjusting the compressive strength reported depending on the magnitude of the rebound *R*, taking into account the position of the instrument, with factor for the age of the concrete and adjustment factor for concordance according to standard dependence of the instrument.

Table 1. Probabilistic concrete compressive strength, determined by its surface hardness at age 1926th day. Maximum, minimum and mean value of R and $f_{c,cube,Schmidt}$.

№	1	2	3	4	5	6	7	8	9	10	11
median R	39.5	39.5	41	45	40.5	40.5	39	40	43	41.5	43
$f_{c,cube,Schmidt}$[MPa]	45.5	45.5	49.0	59.1	47.8	47.8	44.3	46.7	54.0	50.2	54.0
min R	39.0			max R		45.0		mean R		41.1	
min $f_{c,cube,Schmidt}$ [MPa]		44.3		max $f_{c,cube,Schmidt}$ [MPa]		59.1		mean $f_{c,cube,Schmidt}$ [MPa]			49.4

For each member on the two opposite sides of the regions where the cores were drilled, 10 measurements were made by the method of elastic rebound. In each set of tests, the direction was vertical. The median value of the rebound number R of the 10 tests on each of the beam surfaces was calculated. It was used for determining the probabilistic compressive strength of concrete $f_{c,cube,Schmidt}$. The minimum, maximum and mean values of $f_{c,cube,Schmidt}$ were obtained. The results are given in Table 1. The minimum value of the probabilistic concrete compressive strength, determined with a Schmidt hammer on age 1926th day is 44.3 MPa, the maximum is 59.1 MPa, and the mean value is 49.4 MPa.

4.2. Determination of the probabilistic concrete compressive strength with UPVM.

The measurements were performed on the drilled and prepared cores with a portable ultrasonic testing device Proceq Tico (Figure 6). The operating frequency of the transmitter and receiver is 54 kHz, the resolution is 0.1 μs. The distance between the transducers for the different test specimens was between 94 mm and 103 mm. These measurements were made immediately before the destructive tests. For optimal contact between the piezoelectric transducers and the surface of the cores, the surfaces of the transmitter, the receiver and the concrete were covered with a special coupling paste.

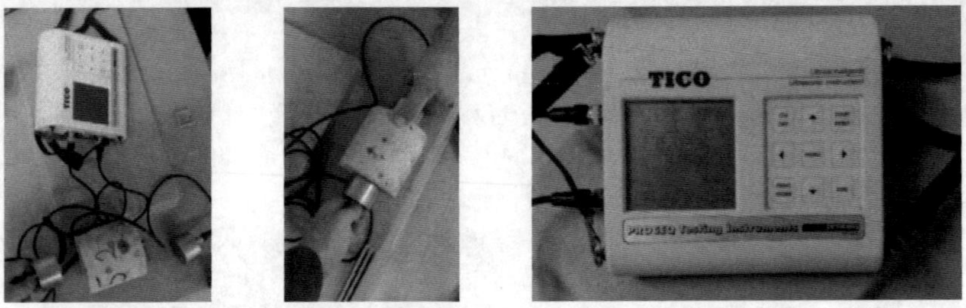

Figure 6. Measurement with device for ultrasonic control Proceq Tico (personal archive).

Table 2. Ultrasonic pulse velocity for age of concrete 1926th day.

№ of measurement	**1**	**2**	**3**	**4**	**5**	**6**	**7**	**8**	**9**	**10**	$\mathbf{V_{UP,mean}}$ **[km/s]**
№ of core					$\mathbf{V_{UP}}$ **[km/s]**						
1	4.625	4.605	4.610	4.590	4.600	4.602	4.628	4.595	4.595	4.603	4.605
2	4.617	4.605	4.607	4.590	4.600	4.587	4.620	4.615	4.623	4.615	4.608
3	4.742	4.736	4.739	4.754	4.741	4.728	4.710	4.761	4.742	4.736	4.739
4	4.896	4.880	4.893	4.876	4.880	4.865	4.850	4.864	4.876	4.880	4.876
5	4.711	4.667	4.667	4.715	4.686	4.689	4.695	4.686	4.690	4.708	4.692
6	4.680	4.667	4.715	4.667	4.640	4.640	4.630	4.625	4.667	4.667	4.660
7	4.557	4.592	4.592	4.590	4.572	4.574	4.569	4.580	4.585	4.569	4.578
8	4.640	4.650	4.640	4.672	4.691	4.689	4.618	4.620	4.618	4.642	4.648
9	4.771	4.821	4.808	4.743	4.739	4.720	4.810	4.741	4.783	4.785	4.772
10	4.766	4.776	4.762	4.730	4.778	4.760	4.757	4.766	4.760	4.762	4.762
11	4.855	4.863	4.850	4.874	4.852	4.828	4.872	4.739	4.879	4.917	4.853

Table 3. Values of probabilistic concrete compressive strength $f_{c,UPVM}$.

№	**1**	**2**	**3**	**4**	**5**	**6**	**7**	**8**	**9**	**10**	**11**
$\mathbf{f_{c,UPVM}}[MPa]$	48.5	48.6	54	60.1	52	50.7	47.4	50.2	55.4	55	59
$\mathbf{f_{c,UPVM,min}}[MPa]$	47.4			$f_{c,UPVM,max}$[MPa		60.1	$f_{c,UPVM,mean}$[MPa]			52.8	

For each of the drilled cores, 10 measurements of the ultrasonic pulse velocity were made according to EN 12504-4:2004 [20]. The transmitter and the receiver of the ultrasonic testing device were placed symmetrically against each other (Figure 6). Prior to each measurement, the equipment was calibrated with a reference cylindrical body with a known velocity. The results for the determined ultrasonic pulse velocity - 10 measurements of each specimen are given in Table 2.

The determined with UPVM probabilistic concrete compressive strength $f_{c,UPVM}$, minimum, maximum and mean value of $f_{c,UPVM}$ for the 11 measurements are shown in Table 3.

Analyzing the results obtained, it can be concluded that the minimum value of probabilistic concrete compressive strength determined by UPVM on 1926th day is 47.4 MPa, maximum value is 60.1 MPa, and mean value is 52.8 MPa.

4.3. Determination of concrete compressive strength by destructive testing of specimens (cores)

The compressive strength was determined according to EN 13791:2007 [13], EN 12504-1:2019 [20] and EN 12390-3:2019 [32]. According to EN 12504-1:2019, if the L/D ratio of the cylindrical test specimen (drilled core) is equal to 1, the resulting strength corresponds to the in-situ cube compressive strength. Before testing the cores, their mass and geometric dimensions were determined (Figure 7). Geometric dimensions, weight of test specimens, weight per unit volume of the 11 cores at age 1926th day are given in Table 4.

Figure 7. Determination of geometric dimensions and weight of specimens (personal archive).

Table 4. Geometric dimensions, weight of test specimens and weight per unit volume of concrete for 11 cores at the age 1926th day.

№ of core	Age	Primary Data						
		weight	diameter	length	weight per unit volume	weight	length	weight per unit volume
	days	G'	ϕ	L'	ρ'	G	L	ρ
		g	mm	mm	kg/m³	g	mm	kg/m³
1	1926	2728	99	154	2301	1792	99	2351
2	1926	2707	99	151	2329	1796	99	1357
3	1926	2805	99	152	2397	1746	94	2413
4	1926	2795	99	152	2389	1867	100	2425
5	1926	2876	99	156	2395	1890	100	2455
6	1926	2858	99	155	2395	1818	98	2410
7	1926	2777	99	152	2373	1835	99	2408
8	1926	2728	99	151	2347	1764	96	2387
9	1926	2786	99	151	2397	1838	100	2388
10	1926	2816	99	154	2375	1890	103	2384
11	1926	2795	99	153	2373	1822	99	2391

where: G'- weight of the test specimen after drilling it from the RC element; G- weight of the test specimen after the preparation for testing; ϕ- diameter of the test specimen; L'- depth of the test specimen after drilling it from the RC element; L- depth of the test specimen after the preparation for testing; ρ'- weight per unit volume of concrete determined after drilling the core from the RC element; ρ- weight per unit volume of concrete determined after preparation of the core for testing.

The drilled cores were tested for compression till failure on a calibrated testing machine CONTROLS Automax 5, type: 50-C46V2 (0 ÷ 2000) kN (Figure 8), meeting the requirements of EN 12390-4:2019 [33]. A constant loading rate of 0.5 MPa/s was selected according to the requirements of standard EN 12390-3:2019 [5,32]. The load was applied to the test specimen without hit and increased evenly until its failure, the load at destruction was reported. Maximum force F [kN] at failure, in-situ cube compressive strength of concrete $f_{c,cube,core}$ [N/mm²], minimum, maximum and mean value of $f_{c,cube,core}$ are shown in Table 5. The compressive strength is expressed with accuracy 0,1 MPa.

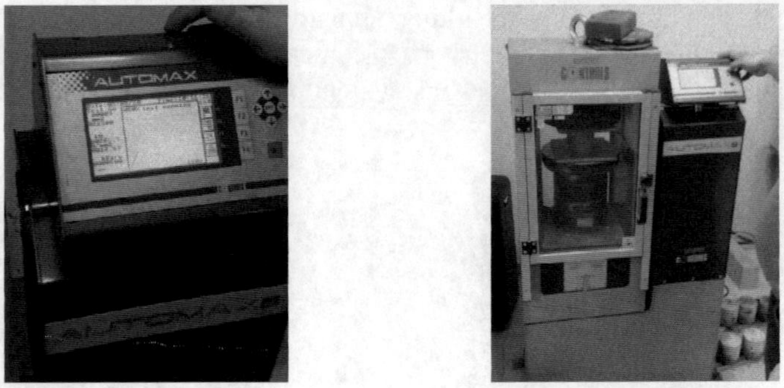

Figure 8. Destructive testing of cores (personal archive).

Table 5. Maximum force at failure, in-situ cube compressive strength of concrete $f_{c,cube,core}$, minimum, maximum and mean value of $f_{c,cube,core}$.

№ of core	Age	Maximum force at failure	Factor for the shape and size of specimen	Compressive strength
	days	F	K	$f_{c,cube,core}$
		[kN]	[-]	[MPa]
1	1926	298.9	1.15	44.7
2	1926	303.3	1.15	45.3
3	1926	331.9	1.15	49.6
4	1926	351.5	1.15	52.5
5	1926	322.3	1.15	48.2
6	1926	319.9	1.15	47.8
7	1926	283.9	1.15	42.4
8	1926	318.4	1.15	47.6
9	1926	336.4	1.15	50.3
10	1926	332.3	1.15	49.6
11	1926	339.7	1.15	50.7
$f_{c,cube,core}$	**max**	52.5	**min**	42.4
$f_{c,cube,core}$	**mean**	48.1		

K is factor, depending on the shape and size of test specimen

From the experimental studies it can be seen that the mean value of in-situ concrete compressive strength at the age 1926th day is 48.1 MPa, the minimum value is 42.4 MPa and the maximum value is 52.5 MPa.

4.4. Comparison of the experimentally obtained results for in-situ concrete compressive strength by different methods

The dependences of the in-situ concrete compressive strength, determined by the method of elastic rebound, by ultrasonic pulse velocity method (UPVM) and by the destructive testing of cores for the 11 measurement points are given in Figure 9. The lowest values of compressive strength are those determined by the destructive testing of the drilled cores, higher are those determined by Schmidt hammer and the highest values are obtained by UPVM.

In Figure 10 are given the values obtained for relative error of in-situ concrete compressive strength determined for the 11 measurement points by three methods: elastic rebound method, UPVM and by the destructive testing of cores. The values obtained by destructive testing of cores are taken as reference values. Smaller relative errors to the reference values are obtained by the method of elastic rebound - from 0.0% to 12.57% (average 3.5%). Bigger relative errors to the reference values are obtained by the UPVM - from 5.46% to 16.37% (average 9.8%).

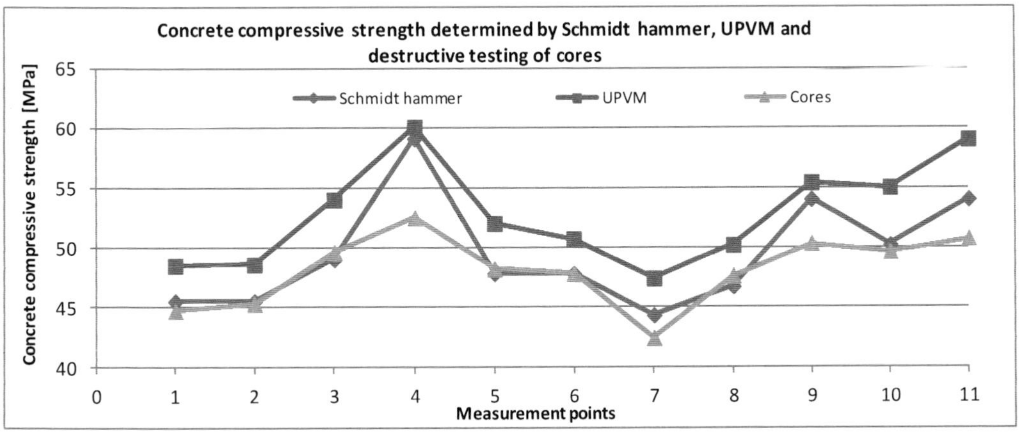

Figure 9. Dependence of in-situ concrete compressive strength for the 11 measurement points, obtained by destructive and non-destructive methods.

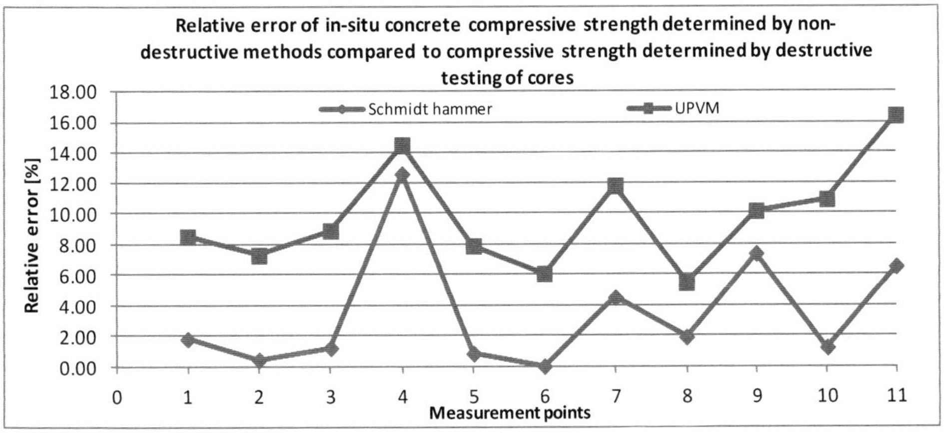

Figure 10. Relative error of in-situ concrete compressive strength determined by non-destructive methods compared to strength determined by testing of cores for the 11 measurement points.

5. Conclusions

Assessing the in-situ concrete compressive strength in existing structures is and will be a challenge for engineers. Tests of concrete at in-situ conditions on the 12 reinforced concrete beams, included in the present study were with goal determining the compressive strength of concrete using different methods. The use of a combination of visual, partially destructive and non-destructive methods for assessing the concrete compressive strength in reinforced concrete members is a common practice and provides effective solutions from both an economic and a technical point of view aiming increasing the accuracy of the assessment. The number of measurements required and the number of methods used must be as small as possible, aiming achieving a lower budget.

After analysis of the results obtained by different methods, it was found that the relative error between the probabilistic concrete compressive strength, determined by its surface hardness and reference values, determined by destruction of drilled cores, is from 0.0% to 12.57%, and the relative error between probabilistic compressive strength determined by UPVM and reference values is from 5.46% to 16.37%. Using only one method would not be sufficient for assessing the compressive strength. For example, increasing the moisture in concrete will increase the ultrasonic pulse velocity, but will reduce the rebound number of Schmidt hammer [11]. Based on results obtained from conducted research, the following conclusions could be done:

- A non-destructive test using a Schmidt hammer gives a good indication of the strength of the concrete. For age 1926 days, the determined concrete compressive strength with the destruction of cores is less than strength determined by the method of elastic rebound. The results obtained with Schmidt hammer were influenced by the presence of carbonation of concrete, moisture, air pockets and the differences in separate regions might be due to the different current state of the concrete over time. This method can be used for long-term measurements to determine the decline of the quality of concrete under the influence of the environment;
- UPVM provides reliable results for concrete compressive strength based on fast measurements with relatively inexpensive equipment. For age of concrete 1926th day, the determined compressive strength of concrete with partially destructive method is less than strength determined with UPVM. The results were influenced by a large number of factors that affect the velocity of the ultrasonic pulse, which limits its use [1,4,15].
- Testing cores takes a lot of time due to the necessity for drilling, preparation processes, testing. The obtained core strength differs from the real value of the in-situ strength in the region of structure from where the core was taken. Influence on the final results according to [5] may have the following factors: core diameter, the direction of drilling, and the length/diameter ratio of core. Testing more cores leads to a more accurate estimate, but the cost and destructive effect are the main limitations against excessive use of cores.

The use of combined methods for assessing in-situ compressive strength of concrete increases the accuracy and its determination becomes more rational and reliable. A combination of method of elastic rebound, UPVM and destructive testing of cores for assessing the strength of concrete is a good practice.

Acknowledgments
This research was supported by the research grant No KP-06-OPR01/3-2018 for the project "Creation of a flexible methodology for development and management of innovative project in scientific organizations" (2018/2021) funded by National Science Fund, "Competition for Financial Support for Basic Research Projects on Societal Challenges 2018", Ministry of Education and Science, Bulgaria.

References

[1] Khan A 2002 *Guidebook on Non-Destructive Testing of Concrete Structures IAEA*, (Vienna: Training Course Series **17**, IAEA) [CrossRef]
[2] Karaiskos G, Deraemaeker A, Aggelis D and Hemelrijck V 2015 Monitoring of concrete structures using the UPVM *Smart Mat. and Str.* **24**, 11 [CrossRef]
[3] Alwash M 2017 Assessment of concrete strength in existing structures using nondestructive tests and cores: analysis of current methodology and recommendations for more reliable assessment Mechanics [physics] Université de Bordeaux NNT : 2017BORD0587 [CrossRef]
[4] Dimov D 2016 *Investigation and Testing of Building Structures and Bridges*, 3rd ed.; Publisher: DL&M, Sofia, Bulgaria
[5] Masi A and Vona M 2009 Estimation of the in-situ concrete strength: Provisions of the European and Italian seismic codes and possible improvements ed E. Cosenza (Napoli: EC8 Perspectives from the Italian Standpoint Workshop) 67-77 [CrossRef]
[6] Nucera F and Pucinotti R 2009 Destructive and non-destructive testing on reinforced concrete structure: the case study of the museum of Magna Graecia in Reggio Calabria, *39th Int. Conf. and Exhibition*, Prague, Czech Republic [CrossRef]
[7] Jedidi M 2018 Evaluation of the concrete quality using destructive and non–destructive tests. *MOJ Civil Eng.* **4***(4),* 219-223 [CrossRef]
[8] Bohdan S and Tomasz K 2013 Determination of the influence of cylindrical samples dimensions on the evaluation of concrete and wall mortar strength using ultrasound method. *Proc. Engineering* **57**,1078 – 1085 [CrossRef]

[9] Wang X and Jiang K 2015 Quality control and evaluation methods of concrete engineering and its reliability analysis, *Open Civil Eng. J.* **9**, 639-643 [CrossRef]
[10] Macharia K 2016 Non-destructive testing of concrete structures using Schmidt hammer and Profometer 5+ Inst. of Nuclear Sc. and Tech., Un. of Nairobi [CrossRef]
[11] Mahmood A 2008 Structural health monitoring using non destructive testing of concrete Department of Civil Eng., National Institute of Technology, India [CrossRef]
[12] EN 12504-2:2012 Testing concrete in structures - Part 2: Non-destructive testing - Determination of rebound number
[13] BDS EN 13791:2007/NA:2011 Assessment of in-situ compressive strength in structures and precast concrete components. National Annex
[14] Lorenzi A, Tisbierek F and Pinto L 2007 Ultrasonic pulse velocity analysis in concrete specimens, *IV Conf. Panamericana de END Buenos Aires* [CrossRef]
[15] Naik T, Malhorta V and Popovics J 2004 The UPVM in *Handbook on Nondestructive Testing of Concrete*, ed. V. Malhorta and N Carino (USA: CRC Press) [CrossRef]
[16] Chakraborty J, Wang X and Stolinski M 2021 Damage detection in multiple RCS based on embedded ultrasonic sensors and wavelet transform *Buildings* **11**, 56 [CrossRef]
[17] Bogas J, Gomes M G and Gomes A 2013 Compressive strength evaluation of structural lightweight concrete by non-destructive ultrasonic pulse velocity method. *Elsevier, Ultrasonics* **53,** pp. 962-972 [CrossRef]
[18] Zatar W 2014 *Guidebook on non-destructive testing of concrete structures*, (West Virginia: Marshall University)
[19] Savaliya K, Thaker K and Dave U 2014 Comparison between different methods of ultrasonic pulse velocity tests on concrete, *I. J. of Eng. Res. and App.*, 41-44 [CrossRef]
[20] EN 12504-4:2004 Testing concrete - Part 4: Determination of ultrasonic pulse velocity.
[21] Hadianfard M and Jafari S 2016 Prediction of lightweight aggregate concrete compressive strength using ultrasonic pulse velocity test through gene expression programming, *Scientia Iranica* **23***(6),* 2506-2513 [CrossRef]
[22] Concu G and Trulli N 2018 Concrete defects sizing by means of ultrasonic velocity maps, *Buildings* **8,** *12, 176* [CrossRef]
[23] Ju M, Park K and Oh H 2017 Estimation of compressive strength of high strength concrete using ND technique and concrete core strength, *Appl. Sci.* **7,** 1249 [CrossRef]
[24] Brunarski L and Dohojda M 2016 An approach to in-situ compressive strength of concrete. *Bulletin of the Polish Ac. of Sc.,Ttechnical Sc.* **64,** *4,* pp. 687-695 [CrossRef]
[25] Pucinotti R and Lorenzo R 2005 In situ non-destructive testing: the steel and concrete resistance assessment of "ancient" r/c structures, *WIT Transactions on The Built Env. 83* [CrossRef]
[26] Tuncan M, Arioz O, Ramyar K and Karasu B 2008 Assessing concrete strength by means of small diameter cores, *Constr. and Building Mat.* **22,** 981–988 [CrossRef]
[27] Neville A 2001 Core tests: Easy to perform, not easy to interpret. *Concrete International* **23,** 11, pp. 59-68 [CrossRef]
[28] Celik Ah, Kilinc K, Tuncan M and Tuncan A 2012 Distributions of Compressive strength obtained from various diameter cores, *ACI Materials J.* **109***, 6* [CrossRef]
[29] Minutolo V, Ronza S, Eramo C and Zona R 2019 The use of destructive and ND testing in concrete strength assessment for a school building *Preprints* **1** [CrossRef]
[30] Dauji S, Bhalerao S, Srivastava P and Bhargava K 2019 Conservative characteristic strength of concrete from nondestructive and partially destructive testing, *J. of Asian Concrete Federation* **5,** *1*, pp. 25-39 [CrossRef]
[31] EN 12504-1:2019 Testing concrete in structures - Part 1: Cored specimens - Taking, examining and testing in compression
[32] EN 12390-3:2019 Testing hardened concrete - Part 3: Compressive strength of test specimens
[33] EN 12390-4:2019 Testing hardened concrete - Part 4: Compressive strength - Specification for testing machines

AUTHOR INDEX